NATURAL HAZARDS

Natural hazards afflict all corners of the Earth; often unexpected, seemingly unavoidable and frequently catastrophic in their impact.

This revised edition is a comprehensive, inter-disciplinary treatment of the full range of natural hazards. Accessible, readable and well supported by over 150 maps, diagrams and photographs, it is a standard text for students and an invaluable guide for professionals in the field.

Clearly and concisely, the author describes and explains how hazards occur, examines prediction methods, considers recent and historical hazard events and explores the social impact of such disasters. This revised edition makes good use of the wealth of recent research into climate change and its effects.

Edward Bryant is Associate Dean of Science at Wollongong University in Australia. Among his other publications is *Tsunami: The Underrated Hazard* (Cambridge University Press, 2001). He has particular interest in climatic change and coastal evolution.

Praise for the First Edition:
'Professor Bryant's heroic compilation is an excellent guide.'
Scientific American

NATURAL HAZARDS

SECOND EDITION

EDWARD BRYANT

PUBLISHED BY THE PRESS SYNDICATE OF THE UNIVERSITY OF CAMBRIDGE
The Pitt Building, Trumpington Street, Cambridge, United Kingdom

CAMBRIDGE UNIVERSITY PRESS
The Edinburgh Building, Cambridge CB2 2RU, UK
40 West 20th Street, New York, NY 10011–4211, USA
477 Williamstown Road, Port Melbourne, VIC 3207, Australia
Ruiz de Alarcón 13, 28014 Madrid, Spain
Dock House, The Waterfront, Cape Town 8001, South Africa

http://www.cambridge.org

First published by Cambridge University Press 1991
Second edition 2005

Printed in Australia by BPA Print Group

Typeface New Caledonia (*Adobe*) 10.5/14 pt *System* QuarkXPress® [MT]

A catalogue record for this book is available from the British Library

National Library of Australia Cataloguing in Publication data

Bryant, Edward, 1948– .
Natural hazards.

2nd ed.
Includes index.
ISBN 0 521 53743 6.

1. Hazardous geographic environments. 2. Natural disasters.
I. Title.

551

ISBN 0 521 53743 6 paperback

To Dianne, Kate and Mark

Contents

Illustrations

TABLES

Preface

I have reread the preface to the first edition many times: extreme events, dire warnings about Greenhouse warming, El Niño–Southern Oscillation prediction . . . Little has changed in the fifteen years since I wrote about them. I am still perplexed because extreme events continue to happen and global warming is no closer to occurring. As Sydney in February 2004 experienced a heat wave of a magnitude unprecedented since at least 1939, I was chasing my favourite research topic – cosmogenically induced mega-tsunami – on Stewart Island, New Zealand some two thousand kilometres away where an unprecedented cold snap was occurring. One event witnessed by four million people got all the publicity; the other played out in a remote cabin in front of half of dozen trekkers got none. Yet both climatic extremes were produced by the same pattern of atmospheric circulation controlled by the same sequence of mobile polar highs. Sydney lay on the equatorial 'greenhouse' side of the highs and Stewart Island lay on the poleward 'Ice Age' side. This book covers two of the phenomena I experienced in my February of extremes – mobile polar highs and tsunami. As with the first edition, the book does not cover the third phenomena, Greenhouse warming. This book is about everyday climatic and geological hazards that can be explained, predicted, and alleviated. Global warming is mentioned and is covered by the concept of changing hazard regimes. However, heat waves – and cold snaps – are about everyday hazards that we have lived with, will continue to experience, and hopefully can survive. These concepts are what this second edition is about.

In order to convey this point of view clearly in the book, adherence to academic referencing has been kept to a minimum. Usually each section begins by listing the major papers or books on a topic that have influenced my thinking and writing. Full reference to these publications can be found at the end of each chapter. I apologize to anyone who feels that their crucial work has been ignored; but the breadth of coverage in this textbook precluded a complete review of the literature on many topics including some of my own.

Manuscript preparation is quite different now from what it was in the late 1980s when the first edition was being published. For one thing, the software programs for scanning, image enhancement, graphics, and word processing are far more comprehensive and efficient at doing tasks. The diagrams in the first edition were hand-drawn, a technique that is rarely used today. Readers will find that many of those diagrams remain in this version. However, many have been revamped using graphics software. New computer-prepared diagrams have also been added. Word processing packages now allow spelling and grammar to be checked uniformly without the assistance of a copy-editor. Minor changes have been made to text retained from the first edition using this capability. The Internet was in it infancy when the first edition was prepared. It is now very easy to capture all the arguments or theories related to a hazard topic via this medium. Where the Internet was used to prepare the current edition, the web sites have been referenced in the text and their full addresses appended to individual chapters. The reader should be aware that some of these addresses may not be available to them because they have changed, or because of the lack of an archival tradition on the Internet.

Ted Bryant
9 August 2004

Acknowledgements

Several people generously provided data from their own research as follows: Jane Cook, formerly of the Faculty of Education, University of Wollongong for her Kiama landslip data in figure 2.11; Professor Eric Bird, formerly of the Department of Geography, University of Melbourne for updating table 8.2; Dr. Geoff Boughton, TimberED Services Pty Ltd., Glengarry, Western Australia, for information on the lower limits of seismically induced liquefaction; Col Johnson, New South Wales South Coast State Emergency Services for pointing out the importance of Critical Incident Stress Syndrome amongst rescue workers; Rob Fleming, Australian Counter Disaster College, Mt Macedon, Victoria for access to photographic files on hazards in Australia; and Bert Roberts, School of Earth and Environmental Sciences, University of Wollongong for pointing out the relative literature on uniformitarianism.

The following people or publishers permitted diagrams or tables to be reproduced: University of Wisconsin Press, for figure 2.4; Methuen, London for figures 3.12 and 3.14; Harcourt Brace Jovanovich Group, Academic Press, Sydney for figures 9.9 and 11.7; Butterworths, Sydney for figures 12.2 and 12.13; and Dr. John Whittow, former at the Department of Geography, University of Reading for figures 12.12 and 12.16. Acknowledgment is made to the following organizations who did not require permission on copyrighted material: NOAA for figures 3.7, 3.8, 4.6a & b, 4.9, 4.10, and 11.5; United States Jet Propulsion Laboratory for figure 8.4; the United States Geological Survey for figures 10.5, 10.7, 10.10, 10.14, 10.15, 11.2, 11.3, and 12.11; the United States Marine Fisheries Service, Seattle for Table 2.1; and the American Geophysical Union for Table 2.2. All of the above diagrams and tables are acknowledged fully in the text.

The following people or organizations provided photographic material: John Fairfax and Sons Limited, for figures 3.6, 7.9 and 13.1; Dr. Geoff Boughton, TimberED Services Pty Ltd., Glengarry, Western Australia, for figures 3.9a & b; the Canadian Atmospheric Environment Service for figures 3.22 and 3.24; the Australian Bureau of Meteorology for figure 3.29; John Telford, Fairymeadow, New South Wales for figure 8.5; Dr. Ann Young, for figure 8.13; Bob Webb, Mal Bilaniwskyj and Charles Owen, New South Wales Department of Main Roads, Wollongong and Bellambi Offices for figure 2.9a and b; the State Library of Victoria who kindly permitted the reproduction of William Strutt's painting entitled *Black Thursday* appearing as figure 7.1; N.P. Cheney, Director, and A.G. Edward, National Bushfire Research Unit, CSIRO, Canberra for figures 7.4a and 7.5 respectively; Jim Bryant, for figure 7.4b; Director Naval Visual News Service, US Navy Office of Information for figure 7.6; J.L. Ruhle and Associates, Fullerton, California for figure 11.12; A.G. Black, Senior Soil Conservator, Hawke's Bay Catchment Board (Hawke's Bay

Regional Council), Napier, New Zealand for figure 12.10; and Dr. Peter Lowenstein for figure 12.14, courtesy C.O. McKee, Principal Government Volcanologist, Rabaul Volcanological Observatory, Department of Minerals and Energy, Papua New Guinea Geological Survey.

I am indebted to the staff and associates of Cambridge University Press, Melbourne: Peter Debus who encouraged a second edition of the hazards text, Susan Keogh who oversaw the publishing process, Monique Kelly who processed the figures, and Malory Weston who copyedited the manuscript. The majority of this text has been presented as lecture material. I would like to thank many students in my first and second year Hazards and Climatology classes who challenged my ideas or rose enthusiastically to support them. Unfortunately, current students have no access to these subjects. I hope that this textbook makes up for the deficiency. Finally, I would like to thank my family who kindly tolerated my 'next textbook' and deflected many of the requests upon my time by friends who did not realize the magnitude of the task of writing a second edition.

Introduction to Natural Hazards

CHAPTER 1

RATIONALE

The field of environmental studies is usually introduced to students as one of two themes. The first examines human effects upon the Earth's environment, and is concerned ultimately with the question of whether or not people can irreversibly alter that environment. Such studies include the effect of human impact on climate, of land-use practices on the landscape in prehistoric and recent times, and of nuclear war upon the Earth's environment. The second theme totally disregards this question of human impact on the environment. It assumes that people are specks of dust moving through time subject to the whims of nature. In this sense, calamities are 'acts of God', events that make the headlines on the evening news, events you might wish on your worst enemy but would never want to witness yourself.

University and college courses dealing with this latter theme usually treat people as living within a hostile environment over which they have little control. Such courses go by the name of '"Natural" Hazards'. The difference between the two themes is aptly summarized by Sidle et al. (2003). Both themes describe hazards. The first theme can be categorized as chronic while the second is episodic or periodic. Chronic hazards would include desertification, soil *degradation*, and melting of permafrost. The causes could be due to humans or global warming. Periodic hazards are large magnitude events that appear over a short time period. They include phenomena such as earthquakes, tsunami, volcanic eruptions and flash *floods*. This book deals mainly with episodic hazards, events that wreak havoc upon humans. However, the message is not one of total doom and gloom. Humans choose hazardous areas because they often offer benefits. For example, ash from volcanoes produces rich soils that can grow three crops per year in the tropics, and *floodplains* provide easily cultivated agricultural land close to a water supply. Humans in these environments are forced to predict and avoid natural calamities such as landslides, cyclones, earthquakes, and drought. Or as Middleton (1995) states, 'a hazard should be seen as an occasionally disadvantageous aspect of a phenomenon, which is often beneficial to human activity over a different timescale.' For those hazards that cannot be avoided by humans, this book also examines how people can minimize their effects and rectify their negative consequences.

The book is organized around climatic and geological hazards. Intense storms and winds, oceanographic factors such as waves, ice and sea level changes, together with extreme precipitation phenomena, are all investigated under the heading of climatic hazards. Earthquakes, volcanoes, and land instability are examined under the topic of geological hazards. This book is not concerned overly with biological hazards such as plague, disease, and insect infestations; but one should realize that death and property loss from these latter perils can be just as severe as, if not worse than, those generated by geological and climatic hazards. A final section examines the social impact of hazards.

HISTORICAL BACKGROUND

The world of myths and legends

(Holmes, 1965; Day, 1984; Myles, 1985; Milne, 1986; Bryant, 2001)

Myths are traditional stories focusing on the deeds of gods or heroes, often to explain some natural phenomenon, while legends refer to some historical event handed down – usually by word of mouth in traditional societies. Both incorporate natural hazard events. Prehistoric peoples viewed many natural disasters with shock and as a threat to existence. Hazards had to be explained and were often set within a world of animistic gods. Such myths remain a feature of belief systems for many peoples across the world – from Aborigines in Australia and Melanesians in the South Pacific, to the indigenous inhabitants of the Americas and the Arctic Circle. With the development of written language, many of these oral myths and legends were incorporated into the writings of the great religions of the world. Nowhere is this more evident than with stories about the great flood. Along with stories of Creation, the Deluge is an almost ubiquitous theme. The biblical account has remarkable similarities to the Babylonian Epic of Gilgamesh. In this epic, Utnapishtim, who actually existed as the tenth king of Babylon, replaces Noah, who was the tenth descendant of Adam. The biblical account parallels the epic, which tells of the building of an ark, the loading of it with different animal species, and the landing of the ark upon a remote mountaintop. Babylonian civilization was founded in what was called Mesopotamia, situated in the basin of the Euphrates and Tigris Rivers, which historically have been subject to cataclysmic flooding. Clay deposits in excess of 2 m have been found at Ur, suggesting a torrential rain event in the headwaters of the valley that caused rivers to carry enormous *suspension loads* at high concentrations.

Such an event was not unique. The epic and biblical accounts are undoubtedly contemporaneous and, in fact, a similar major flood appears in the written accounts of all civilized societies throughout Middle Eastern countries. For example, broken tablets found in the library of Ashurbanipal I (668–626 BC) of Assyria, at Nippur in the lower Euphrates Valley, also recount the Babylonian legend. The Deluge also appears worldwide in verbal and written myths. The Spanish conquistadores were startled by flood legends in Mexico that bore a remarkable similarity to the biblical account. Because these staunchly Catholic invaders saw the legends as blasphemous, they ordered the destruction of all written native references to the Deluge. The similarities were most remarkable with the Zapotecs of Mexico. The hero of their legend was Tezpi, who built a raft and loaded it with his family and animals. When the god Tezxatlipoca ordered the floodwaters to subside, Tezpi released firstly a vulture and then other fowl, which failed to return. When a hummingbird brought back a leafy twig, Tezpi abandoned his raft on Mt Colhuacan. A Hawaiian legend even has a rainbow appearing, which signals a request for forgiveness by their god Kane for his destructiveness. Almost all legends ascribe the flood to the wrath of a god. The Australian Aborigines have a story about a frog named Tiddalik drinking all the water in the world. To get the water back, all the animals try unsuccessfully to get the frog to laugh. Finally, a cavorting eel manages the task. However, the frog disgorges more water than expected and the whole Earth is flooded. In some myths, floods are attributed to animals urinating after being picked up. Many myths are concerned with human taboo violations, women's menstruation, or human carelessness. Almost all stories have a forewarning of the disaster, the survival of a chosen few, and a sudden onset of rain. However, the Bible has water issuing from the ground, as do Chinese, Egyptian, and Malaysian accounts. The latter aspect may refer to earthquakes breaking *reservoirs* or impounded lakes. Many Pacific island and Chilean legends recount a swelling of the ocean, in obvious reference to *tsunami*. Despite the similarities of legends globally amongst unconnected cultures, there is no geological evidence for contemporaneous flooding of the entire world, or over continents. It would appear that most of the flood myths recount localized deluges, and are an attempt by human beings to rationalize the flood hazard.

Almost as ubiquitous as the legends about flood are ones describing volcanism or the disappearance of continents. North American natives refer to sunken land to the east; the Aztecs of Mexico believed they were descended from a people called Az, from the lost land of Aztlán. The Mayas of Central America referred to a lost island of Mu in the Pacific, which broke apart and sank with the loss of millions of inhabitants. Chinese and Hindu legends also recount lost continents. Perhaps the most famous lost continent legend is the story of Atlantis, written by Plato in his *Critias*. He recounts an Egyptian story that has similarities with Carthaginian and Phoenician legends. Indeed, if the Aztecs did migrate from North Africa, the Aztec legend of a lost island to the east may simply be the Mediterranean legend transposed and reoriented to Central America. The disappearance of Atlantis probably had its origins in the catastrophic eruption of Santorini around 1470 BC. Greek flood myths refer to this or similar events that generated tsunami in the Aegean. It is quite possible that this event entered other tales in Greek mythology involving fire from the sky, floating islands and darkening of the sky by Zeus. The Krakatau eruption of 1883 also spawned many legends on surrounding islands. Blong (1982) describes modern legends in Papua New Guinea that can be related to actual volcanic eruptions in the seventeenth century. The Pacific region, because of its volcanic activity, abounds in eruption mythology. For example, Pele, the fire goddess of Hawaii, is chased by her sister from island to island eastward across the Pacific. Each time she takes refuge in a volcano, she is found by her sister, killed, and then resurrected. Finally, Pele implants herself triumphantly in the easternmost island of Hawaii. Remarkably, this legend parallels the temporal sequence of volcanism in the Hawaiian Islands.

Earthquakes also get recounted in many myths. They are attributed by animistic societies in India to animals trying to get out of the Earth. Sometimes the whole Earth is viewed as an animal. Ancient Mexicans viewed the Earth as a gigantic frog that twitched its skin once in a while. Timorese thought that a giant supported the Earth on his shoulder and shifted it to the other side whenever it got too heavy. Greeks did not ascribe earthquakes to Atlas, but to Poseidon disturbing the waters of the sea and setting off earth tremors. The oscillating water levels around the Greek islands during earthquakes were evidence of this god's actions. In the Bible, earthquakes were used by God to punish humans for their sins. Sodom and Gomorrah were destroyed because they refused to give up sins of the flesh. The fire and brimstone from heaven, mentioned with these cities' destruction, were most likely associated with the lightning that is often generated above earthquakes by the upward movement of dust. Alternatively, it may have come from meteorite debris. The Bible also viewed earthquakes as divine visitation. An earthquake preceded the arrival of the angel sent to roll back Christ's tombstone (Matthew 28: 2).

Myths, legends, and the Bible pose a dilemma because they do not always stand up to scientific scrutiny as established in the last three centuries. Today an incident is not credible unless it can be measured, verified by numerous witnesses, viewed as repeatable, and certainly published in peer-reviewed literature. Media accounts, which may be cobbled together and subject to sensationalism because of the underlying aim to sell a product, are regarded with suspicion. There is distrust of events that go unpublished, that occur in isolated regions, or that are witnessed by societies with only an oral tradition occurring in isolated regions. Legends, however, can become credible. Hence, the Kwenaitchechat legend of a tsunami on the west coast of the United States suddenly took on scientific acceptability when it received front-page coverage in *Nature*. It was shown – using sophisticated computer modelling – that the source of a tsunami that struck the east coast of Japan on 26 January 1700 could have originated only from the Cascadian subduction zone off the west coast of the United States. The fact that the legend mentions water overrunning hilltops has still not been explained. Sometime in the future, we will find it just as hard to believe that published statements were accepted as articles of faith when they were scrutinized by only two referees who often came from an entrenched inner circle of associates. Certainly, the rise of information on the Internet is challenging twentieth century perceptions of scientific scrutiny.

Catastrophism *vs.* uniformitarianism

(Huggett, 1997)

Until the middle of the eighteenth century, western civilization regarded hazards as 'acts of God' in the strict biblical sense, as punishment for people's sins. On 1 November 1755, an earthquake, with a possible surface magnitude of 9.0 on the Richter scale, destroyed Lisbon, then a major center of European civilization. Shortly after the earthquake, a tsunami swept into the city and, over the next few days, fire consumed what was left. The event sent shock waves through the salons of Europe at the beginning of The Enlightenment. The earthquake struck on All Saints' Day, when many Christian believers were praying in church. John Wesley viewed the earthquake as God's punishment for the licentious behavior of believers in Lisbon, and retribution for the severity of the Portuguese Inquisition. Immanuel Kant and Jean-Jacques Rousseau viewed the disaster as a natural event, and emphasized the need to avoid building in hazardous places.

The Lisbon earthquake also initiated scientific study of geological events. In 1760, John Mitchell, Geology Professor at Cambridge University, documented the spatial effects of the earthquake on lake levels throughout Europe. He found that no seiching was reported closer to the city than 700 km, nor further away than 2500 km. The seiching affected the open coastline of the North Sea and shorelines in Norwegian fjords, Scottish lochs, Swiss alpine lakes, and rivers and canals in western Germany and the Netherlands. He deduced that there must have been a progressive, wave-like tilting movement of the Earth outward from the center of the earthquake, different to the type of wave produced by a volcanic explosion.

Mitchell's 1760 work on the Lisbon earthquake effectively represents the separation of two completely different philosophies for viewing the physical behavior of the natural world. Beforehand, the *Catastrophists* dominated geological methodology. These were the people who believed that the shape of the Earth's surface, the stratigraphic breaks appearing in rock columns, and the large events associated with observable processes were cataclysmic. Catastrophists believed events had to be cataclysmic to allow the geological record to fit with the date for the Earth's creation of 4004 BC, as determined by Biblical genealogy. Charles Lyell, one of the fathers of geology, sought to replace this 'catastrophe theory' with gradualism – the idea that geological and geomorphic features were the result of cumulative slow change by natural processes operating at relatively constant rates. This idea implied that processes shaping the Earth's surface followed laws of nature as defined by physicists and mathematicians as far back as Bacon. William Whewell, in a review of Lyell's work, coined the term *uniformitarianism*, and subsequently a protracted debate broke out about whether or not the slow processes we observe at present apply to past unobservable events. The phrase 'The present is the key to the past' also arose at this time and added to the debate.

In fact, the idea of uniformitarianism involves two concepts. The first implies that geological processes follow natural laws applicable to science. There are no 'acts of God'. This type of uniformitarianism was established to counter the arguments raised by the Catastrophists. The second concept implies constancy of rates of change or material condition through time. This concept is nothing more than inductive reasoning. The type and rate of processes operating today characterize those that have operated over geological time. For example, waves break upon a beach today in the same manner that they would have one hundred million years ago, and prehistoric tsunami behave the same as modern ones described in our written records. If one wants to understand the sedimentary deposits of an ancient tidal estuary, one has to do no more than go to a modern estuary and study the processes at work. Included in this concept is the belief that physical landscapes such as modern floodplains and coastlines evolve slowly.

Few geomorphologists or geologists who study earth surface processes and the evolution of modern landscapes would initially object to the above concept of constancy. However, it does not withstand scrutiny. For example, there is no modern analogy to the nappe mountain building processes that formed the European Alps or to the *mass extinctions* and sudden discontinuities that have dominated the geological record. Additionally, no one who has witnessed a *fault* line being upthrusted during an earthquake, or Mt St Helens wrenching itself apart in a cataclysmic eruption, would agree that all landscapes develop slowly. As Thomas Huxley so aptly worded it, gradualists had saddled themselves with the tenet of *Natura non facit saltum* – Nature does not make sudden jumps. J. Harlen Bretz from the University of Chicago challenged this tenet in the 1920s. Bretz attributed the formation of the scablands of eastern Washington to catastrophic floods. For the next forty years, he bore the ridicule and invectiveness of the geological establishment for proposing this radical idea. It was not until the 1960s that Bretz was proved correct when Vic Baker of the University of Arizona interpreted space probe images of enormous channels on Mars as features similar to the Washington scablands. At the age of 83, Bretz finally received the recognition of his peers for his seminal work.

Convulsive events are important climatic and geological processes, but are they likely to occur today? This is a major question cropping up throughout this book. The climatic and geological processes responsible for most hazards can be described succinctly. However, do these processes necessarily account for events occurring outside the historical record (in some places this is a very short record indeed)? In some cases, the evidence for catastrophic events cannot be explained. For example, there is ongoing debate whether or not mega-tsunami or super storms can move boulders to the top of cliffs 30 or more meters high. The mega-tsunami theory has been attacked – mainly because no historical tsunami generating similar deposits has been witnessed. However, the same argument can be applied to the alternative hypothesis of mega-storms. The dilemma of changing hazard regimes will be discussed at the end of this book.

The relationship between humans and natural hazards (Susman et al., 1983; Watts, 1983)

The above concepts emphasize the natural aspect of natural hazards. Disasters can also be viewed from a sociological or humanistic viewpoint. An extensive chapter at the end of this book deals with the human response to natural hazards at the personal or group level. While there are great similarities in natural hazard response amongst various cultures, societies and political systems, it will be shown that fundamental differences also occur. This is particularly evident in the ways that countries of contrasting political ideology, or of differing levels of economic development, cope with drought. Both Australia and Ethiopia were afflicted by droughts of similar severity at the beginning of the 1980s. In Ethiopia, drought resulted in starvation followed by massive international appeals for relief. However, in Australia no one starved and drought relief was managed internally.

In dealing historically with the scientific and mathematical description of natural hazards, it is apt to discuss a sociological viewpoint formulated in the twentieth century. This sociological viewpoint states that the severity of a natural hazard depends upon who you are, and to what society you belong at the time of the disaster. It is exemplified and expressed most forcibly by Marxist theory. Droughts, earthquakes and other disasters do not kill or strike people in the same way. The poor and oppressed suffer the most, experience worst the long-term effects, with higher casualties more likely as a result. In Marxist philosophy, it is meaningless to separate nature from society: people rely upon nature for fulfillment of their basic needs. Throughout history, humans have met their needs by utilizing their natural environment through labor input. Humans do not enter into a set contract with nature. In order to produce, they interact with each other, and the intrinsic qualities of these interactions determine how individuals or groups relate to nature. Labor thus becomes the active and effective link between society and nature. If workers or peasants are able to control their own labor, they are better adapted to contend with the vagaries of the natural environment. The separation of workers from the means of production implies that others, namely those people (capitalists or bosses) who control the means of production, govern their relationship with nature. This state of affairs is self-perpetuating. Because capitalists control labor and production, they are better equipped to survive and recover from a natural disaster than are the workers.

The main point about the Marxist view of natural hazards is that hazard response is contingent upon the position people occupy in the production process. This social differentiation of reactions to natural hazards is a widespread phenomenon in the Third World, where it is typical for people at the margins of society to live in the most dangerous and unhealthy locations. A major slum in San Juan, Puerto Rico, is frequently inundated at high tide; the poor of Rio de Janeiro live on the precipitous slopes of Sugarloaf Mountain subject to *landslides*; the slum dwellers of Guatemala live on the steeper slopes affected by earthquakes; the Bangladeshi farmers of the Chars (low-lying islands at the mouth of the Ganges–Brahmaputra Delta in the Bay of Bengal) live on coastal land subject to storm surge flooding.

The contrast in hazard response between the Third World and westernized society can be illustrated by comparing the effects of Cyclone Tracy – which struck Darwin, Australia, in 1974 – to the effects of hurricane Fifi, which struck Honduras in September of the same year. Both areas lie in major tropical storm zones, and both possess adequate warning facilities. The tropical storms were of the same intensity, and destroyed totally about 80 per cent of the buildings in the main impact zones. Over 8000 people died in Honduras, while only 64 died in Darwin. A Marxist would reason that the destitute conditions of under-development in Honduras accounted for the difference in the loss of life between these two events. Society is differentiated into groups with different levels of vulnerability to hazards.

One might argue, on the other hand, that many people in Third World countries, presently affected by major disasters, simply are unfortunate to be living in areas experiencing large-scale climatic change or *tectonic* activity. After all, European populations were badly decimated by the climatic cooling that occurred after the middle of the fourteenth century – and everyone knows that the Los Angeles area is well overdue for a major earthquake. One might also argue that Third World countries bring much of the disaster upon themselves. In many countries afflicted by large death tolls due to drought or storms, the political systems are inefficient, corrupt, or in a state of civil war. If governments in Third World countries took national measures to change their social structure, they would not be so vulnerable to natural disasters. For example, Japan, which occupies a very hazardous earthquake zone, has performed this feat through industrialization in the twentieth century. One might further argue that Australians or Americans cannot be held responsible for the misery of the Third World. Apart from foreign aid, there is little we can do to alleviate their plight.

The Marxist view disagrees with all these views. Marxism stipulates that western, developed nations are partially responsible for, and perpetuate the effect of disasters upon, people in the Third World – because we exploit their resources and cash-cropping modes of production. The fact that we give foreign aid is irrelevant. If our foreign aid is being siphoned off by corrupt officials, or used to build highly technical projects that do not directly alleviate the impoverished, then it may only be keeping underprivileged people in a state of poverty or, it is suggested, further marginalizing the already oppressed and thus exacerbating their risk to natural hazards. Even disaster relief maintains the status quo and generally prohibits a change in the status of peasants regarding their mode of production or impoverishment.

The basis of Marxist theory on hazards can be summarized as follows:

- the forms of exploitation in Third World countries increase the frequency of natural disasters as socio-economic conditions and the physical environment deteriorate;
- the poorest classes suffer the most;
- disaster relief maintains the status quo, and works against the poor even if it is intentionally directed to them; and
- measures to prevent or minimize the effects of disasters, which rely upon high technology, reinforce the conditions of underdevelopment, exploitation and poverty.

Marxist philosophy does not form the basis of this book. It has been presented simply to make the reader aware that there are alternative viewpoints about the effects of natural hazards. The Marxist view is based upon the structure of societies or cultures, and how those societies or cultures are able to respond to changes in the natural environment. The framework of this book, on the other hand, is centered on the description and explanation of natural hazards that can be viewed mainly as uncontrollable events happening continually over time. The frequency and magnitude of these events may vary with time, and particular types of events may be restricted in their worldwide occurrence. However, natural hazards cannot be singled out in time and space and considered only when they affect a vulnerable society. An understanding of how, where and when hazards take place can be achieved only by studying all occurrences of that disaster. The arrival of a tropical cyclone at Darwin, with the loss of 64 lives, is just as important as the arrival of a similar-sized cyclone in Honduras, with the loss of 8000 lives. Studies of both events contribute to our knowledge of the effects of such disasters upon society and permit us to evaluate how societies respond subsequently.

Hazard statistics (Changnon & Hewings, 2001; CRED, 2002; Balling & Cerveny, 2003)

Figure 1.1 plots the number of hazards by region over the period 1975–2001. The map excludes epidemics or famine. Despite its large population, Africa is relatively devoid of natural hazards. This is due to the tectonic stability of the continent and the fact that tropical storms do not develop here, except in the extreme south-east. However the African population is affected disproportionately by drought. The frequency of events over the rest of the world reflects the dominance of climate as a control on the occurrence of natural hazards. This aspect is illustrated further for the twentieth century in Table 1.1. The most frequent hazards are *tornadoes*, the majority of which occur in the United States. Over the latter half of the twentieth century, their frequency – more than 250 events per year – exceeds any other natural hazard. Climatically induced floods and tropical cyclones follow this phenomenon in frequency. Tsunami, ranked fourth, is the most frequent geological hazard, followed by damaging earthquakes with nine significant events per year. All totalled, climatic hazards account for 86.2 per cent of the significant hazard events of the twentieth century.

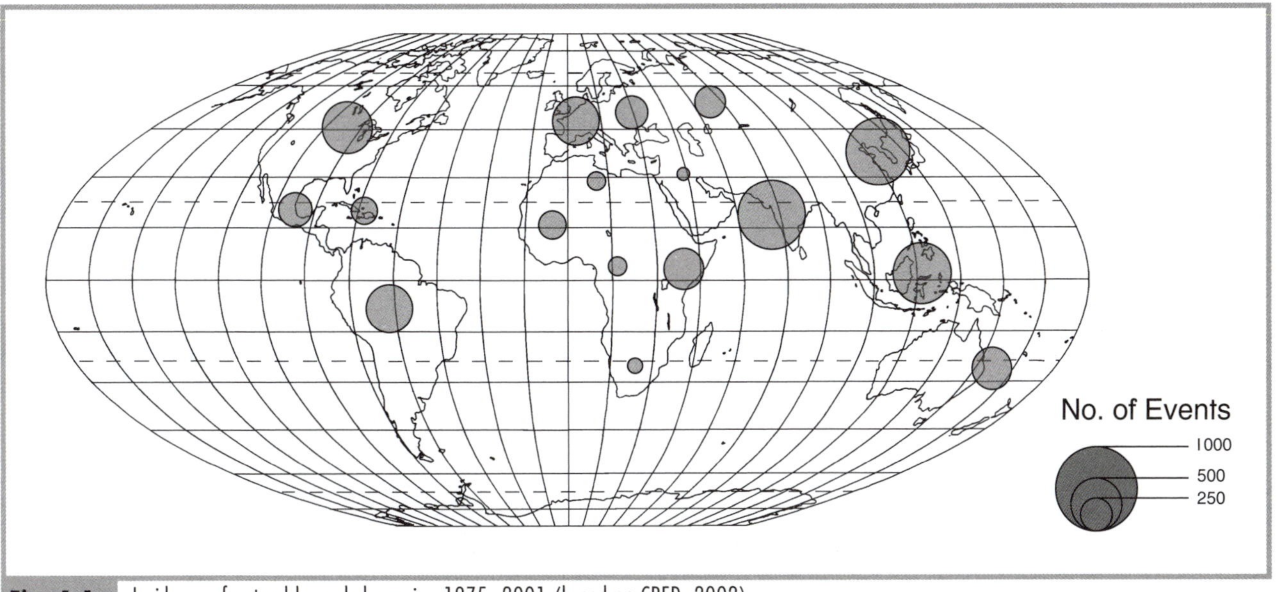

Fig. 1.1 Incidence of natural hazards by region 1975–2001 (based on CRED, 2002).

Table 1.1 Frequency of natural hazards during the twentieth century. Tsunami data comes from the Intergovernmental Oceanographic Commission (2003). Tornado statistics for the USA are for the 1990–1995 period only and derived from the High Plains Regional Climate Center (2003). All other data based on WHO (2002).

Type of Hazard	No of Events
Tornadoes (US)[1]	9476
Flood	2389
Tropical Cyclone	1337
Tsunami	986
Earthquake	899
Wind (other)	793
Drought	782
Landslide	448
Wild fire	269
Extreme temperature	259
Temperate winter storm	240
Volcano	168
Tornadoes (non-US)	84
Famine	77
Storm surge	18

[1] for F2–F5 tornadoes 1950–1995

The economic relevance of the biggest hundred of these hazard events is summarized over the same period in Table 1.2. Values are reported in US dollars and do not include inflation. Earthquakes, while ranking only fifth in occurrence, have been the costliest hazard ($US249 billion), followed by floods ($US207 billion), tropical storms ($US80 billion), and windstorms ($US44 billion). In total, the hundred most expensive natural disasters of the twentieth century caused $US631 billion damage. The single largest event was the Kobe earthquake of 20 January 1995, which cost $US131.5 billion. While this event is familiar, the second most expensive disaster of the twentieth century – floods in the European part of the former Soviet Union on 27 April 1991 costing $US60 billion – is virtually unknown.

Table 1.2 Cost of natural hazards, summarized by type of hazard for the 100 biggest events, 1900–2001 (based on WHO, 2002).

Type	Cost
Earthquake	$248,624,900,000
Flood	$206,639,800,000
Tropical storm	$80,077,700,000
Wind Storm	$43,890,000,000
Wild Fire	$20,212,800,000
Drought	$16,800,000,000
Cold wave	$9,555,000,000
Heat wave	$5,450,000,000
Total	$631,250,200,000

Figure 1.2 presents the number of hazards reported each year over the twentieth century. Apparently, natural hazards have increased in frequency over this

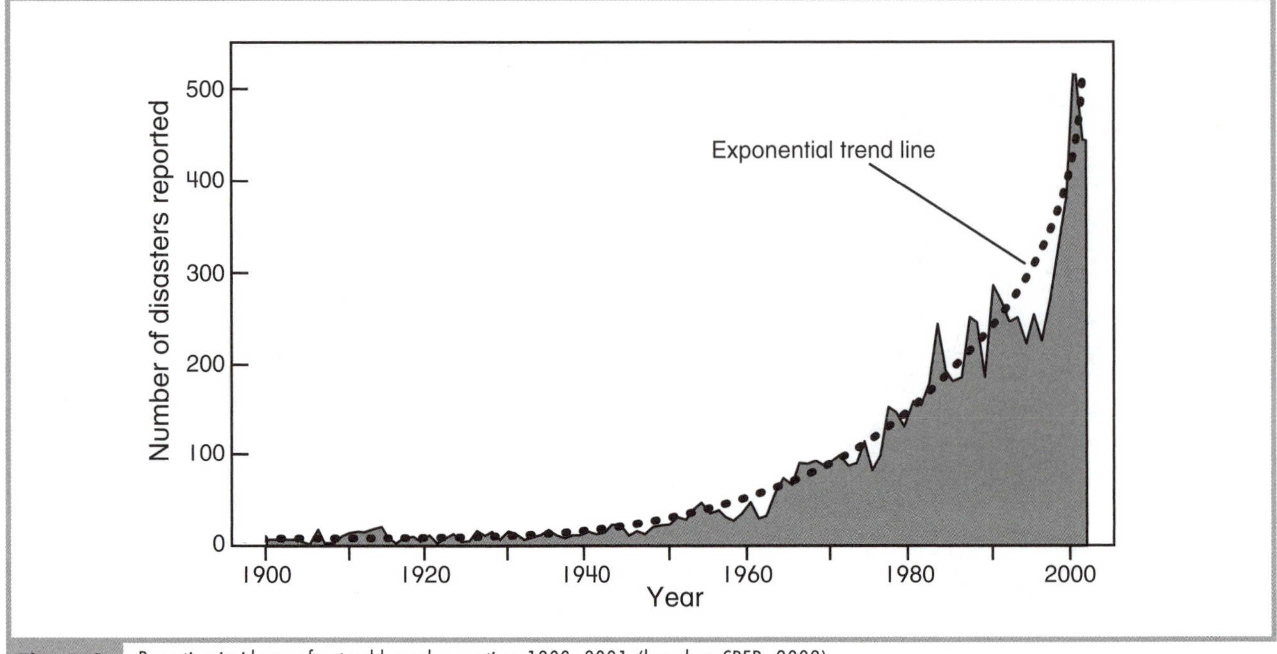

Fig. 1.2 Reporting incidence of natural hazards over time 1900–2001 (based on CRED, 2002).

time span from around ten events per year at the beginning of the century to over 450 events at the end. While increased vigilance and perception of natural hazards accounts for part of this increase, other data suggest that natural hazards are on the increase. The fact that these figures include both geological and climatic hazards rules out such a simple explanation as global warming as a major cause of this increased incidence. On the other hand, these figures should be treated with some caution because they do not always hold under scrutiny. For example, in the United States, the magnitude and frequency of thunderstorms, hailstorms, intense tornadoes, hurricanes and winter storms have not increased, but may have decreased. Over the same period, there is no increase in the cost of climatic hazards after the figures are adjusted for inflation. Instead, damaging events appear to recur at twenty-year intervals around 1950–1954, 1970–1974, and 1990–1994. This timing coincides with the peak of the 18.6 M_N *lunar cycle*, which will be described in Chapter 2.

Figure 1.3 presents the number of deaths due to natural hazards over the twentieth century. The time series does not include biological hazards such as epidemics or insect predation; however, it does include drought-induced famines. The number of deaths is plotted on a logarithmic scale because there have been isolated events where the death toll has exceeded the long-term trend by several million. On average 275 000 people have died each year because of natural hazards. If famine is removed from the data set, this number declines to 140 200 deaths per year. This reduction suggests that famine kills 134 800 people per year, roughly 50 per cent of the total number. Although the greatest number of deaths occurred in the 1930s – a figure that can be attributed to political upheaval and civil war – there is no significant variation over the twentieth century. This fact suggests that improved warning and prevention of natural hazards has balanced population increases resulting in a constant death toll. Many world organizations would consider several hundred thousand deaths per year due to natural hazards as unacceptable.

Table 1.3 presents the accumulated number of deaths, injuries and homeless for each type of hazard for the twentieth century. Also presented is the largest event in terms of death for each category. The greatest hazard during the twentieth century was flooding; however much of this was due to civil unrest. Half of the 6.9 million death toll occurred in China in the 1930s where neglect and deliberate sabotage augmented deaths. Earthquakes and tropical cyclones account for the other significant death tolls of the twentieth century. Interestingly, during the first three years of the twenty-first century 4242 deaths were caused by cold waves. This is 60 per cent of the total for the whole of the twentieth century despite the perception of global warming. In contrast, the death toll from a heat wave in France in 2003 resulted in 15 000

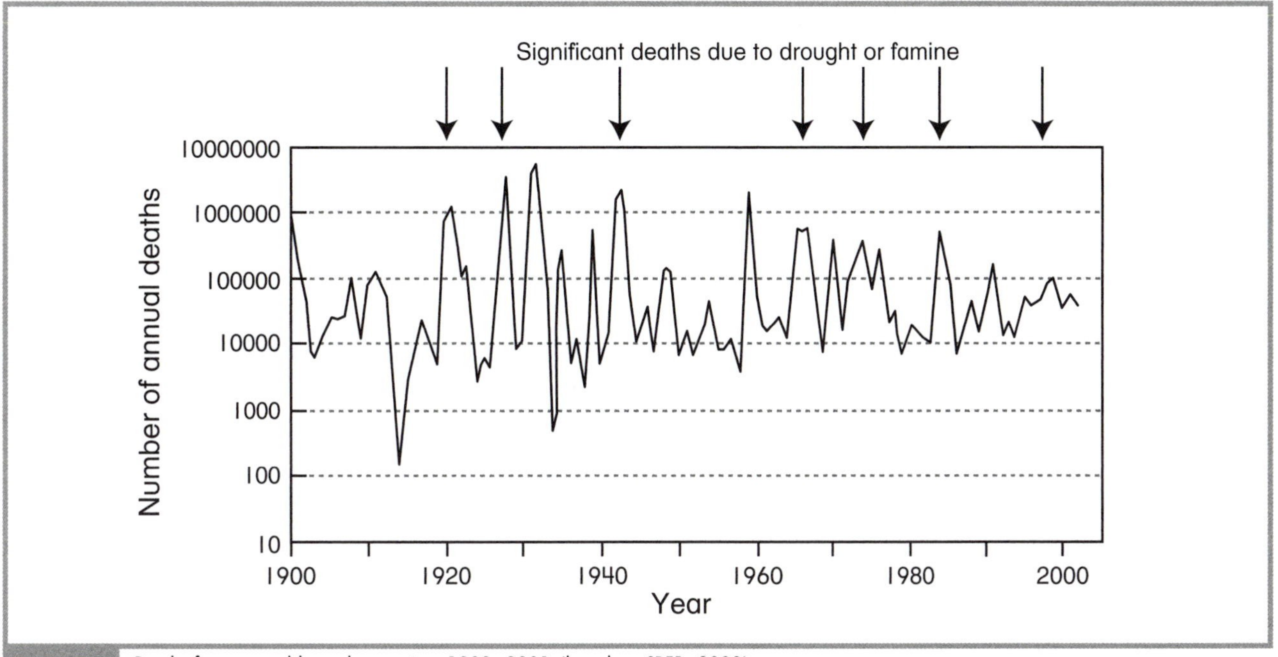

Fig. 1.3 Deaths from natural hazards over time 1900–2001 (based on CRED, 2002).

deaths, more than could be attributed to this cause during the whole of the twentieth century. Obviously, the statistics presented in Table 1.3 underestimate actual deaths, mainly because data were simply not collected for some hazards until recent times.

Table 1.3 also presents the number of injured and displaced people due to natural catastrophes in the twentieth century. Often hazard statistics concentrate upon death, not taking into account the walking wounded and the homeless who put a greater, more lasting burden, upon society. Eighteen times more people were made homeless by floods in the twentieth century than were killed. This ratio rises to 30 and 344 times respectively for tropical cyclones and extra-tropical storms. For example, the January 1998 ice storm that paralyzed Montreal killed twenty-five people. However up to 100 000 people were made homeless for up to a month afterwards because of the failure of electricity supplies as temperatures dipped to –40°C.

Table 1.3 Number of people killed, injured or displaced due to natural hazards during the twentieth century (based on WHO, 2002).

Type of Hazard	Deaths	Injuries	Homeless	Largest death toll event and date	Death toll
Floods	6 851 740	1 033 572	123 009 662	China, July 1931	3 700 000
Earthquakes	1 816 119	1 147 676	8 953 296	Tangshan, China, July 1976	242 000
Tropical cyclones	1 147 877	906 311	34 272 470	Bangladesh, Nov 1970	300 000
Volcano	96 770	11 154	197 790	Martinique, May 1902	30 000
Landslides, avalanches, mud flows	60 501	8 071	3 759 329	Soviet Union, 1949	12 000
Extra-tropical storms	36 681	117 925	12 606 891	Northern Europe, Feb 1953	4 000
Heat wave	14 732	1 364	0	India, May 1998	2 541
Tsunami	10 754	789	–	Sanriku Japan, Mar 1933	3 000
Cold wave	6 807	1 307	17 340	India, Dec 1982	400
Tornado	7 917	27 887	575 511	Bangladesh, Apr 1989	800
Fires	2 503	1 658	140 776	USA, Oct 1918	1 000
Total	10 052 401	3 257 714	183 533 065		

CHAPTER OUTLINES

While this book has been written assuming little prior knowledge in Earth science at the university level, there are occasions where basic terminology (jargon) has had to be used in explanations. More explicit definitions of some of these terms can be found in the glossary at the end of the book. You may have noticed already that some words in this book have been italicized. These terms are defined or explained further in the glossary. Note that book titles, names of ships and botanical names are also italicized, but they do not appear in the glossary.

Individual chapters are arranged around a specific concept or type of hazard. The mechanisms controlling and predicting this particular hazard's occurrence are outlined and some of the more disastrous occurrences worldwide are summarized. Where appropriate, some of humankind's responses to, or attempts at mitigating, the hazard are outlined.

The hazards covered in this book are summarized in Table 1.4, with a chapter reference for each hazard shown in the last column. The hazards, assessed subjectively, are listed in the order that reflects the emphasis given to each in the text. Table 1.4 also grades the hazards, on a scale of 1 to 5, as to degree of severity, time span of the event, spatial extent, total death toll, economic consequences, social disruption, long-term impact, lack of prior warning or suddenness of onset, and number of associated hazards. The overall socio-economic–physiological impact of the hazard is ranked using these criteria. The most important global hazard is drought, followed by tropical cyclones, regional floods, and earthquakes. Three of these four top hazards rank highest for cost (Table 1.2) and human impact (Table 1.3). These latter two tables do not evaluate drought because it develops so slowly and insidiously that it is often ignored when event statistics are collected. Of the top ten hazards, six are climatically induced. Table 1.4 has been constructed for world hazards. The ranking differs for individual countries and latitudes. For example, the ranking of hazards in the United States in terms of economic loss is tropical storms, floods, severe thunderstorms, tornadoes, extra-tropical storms, hail, and wind.

The hazards in the book are organized under two main parts: climatic hazards and geological hazards. Climatic hazards are introduced in Chapter 2, which outlines the mechanisms responsible for climatic variability. This chapter covers the basic processes of air movement across the surface of the globe, the concept of mobile polar highs, air–ocean temperature interactions resulting in the Southern, North Atlantic and North Pacific oscillations, and the effect of astronomical *cycles* (such as sunspots and the 18.6-year lunar tide) on the timing of climate hazard events.

A description of large-scale storms is then covered in Chapter 3. This chapter discusses the formation of large-scale tropical vortices known as tropical cyclones. Tropical cyclones are the second most important hazard, generating the greatest range of associated hazard phenomena. Familiar cyclone disaster events are described together with their development, magnitude and frequency, geological significance, and impact. The response to cyclone warnings in Australia is then compared to that in the United States and Bangladesh. Extra-tropical cyclones are subsequently described with reference to major storms in the northern hemisphere. These storms encompass rain, snow, and freezing rain as major hazards. Storm surges produced by the above phenomena are then described. Here, the concepts of probability of occurrence and exceedence are introduced. The chapter concludes with a description of dust storms as a significant factor in long-term land degradation.

Chapter 4 summarizes smaller hazards generated by wind. Attention is given to the description of thunderstorms and their associated hazards. These include tornadoes, described in detail together with some of the more significant events. The chapter concludes with a description of the measures used to avert tornado disasters in the United States.

These chapters set the scene for two chapters dealing with longer lasting climatic disasters, namely drought and floods. Chapter 5 deals with impact of human activity in exacerbating drought and people's subsequent responses to this calamitous hazard. Emphasis is placed on pre- and post-colonial influences in the Sahel region of Africa, followed by a discussion of modern impacts for countries representative of a range of technological development. The second half of the chapter deals with human response by a variety of societies, from those who expect drought as a natural part of life, to those who are surprised by its occurrence and make little effort to minimize its impact. These responses cover Third World African countries as well as developed westernized countries such as the United States, England, and Australia. The chapter concludes by describing the way

Table 1.4 Ranking of hazard characteristics and impact.

1 – largest or most significant   5 – smallest or least significant

Rank	Event	Degree of severity	Length of event	Total areal extent	Total loss of life	Total economic loss	Social effect	Long term impact	Suddenness	Number of associated hazards	Chapter location
1	Drought	1	1	1	1	1	1	1	4	3	5, 6
2	Tropical cyclone	1	2	2	2	2	2	1	5	1	2
3	Regional flood	2	2	2	1	1	1	2	4	3	5, 8
4	Earthquake	1	5	1	2	1	1	2	3	3	10, 11
5	Volcano	1	4	4	2	2	2	1	3	1	10, 12
6	Extra-tropical storm	1	3	2	2	2	2	2	5	3	2
7	Tsunami	2	4	1	2	2	2	3	4	5	11
8	Bushfire	3	3	3	3	3	3	3	2	5	9
9	Expansive soils	5	1	1	5	4	5	3	1	5	13
10	Sea-level rise	5	1	1	5	3	5	1	5	4	4
11	Icebergs	4	1	1	4	4	5	5	2	5	4
12	Dust storm	3	3	2	5	4	5	4	1	5	4
13	Landslides	4	2	2	4	4	4	5	2	5	13
14	Beach erosion	5	2	2	5	4	4	4	2	5	4
15	Debris avalanches	2	5	5	3	4	3	5	1	5	13
16	Creep & solifluction	5	1	2	5	4	5	4	2	5	13
17	Tornado	2	5	3	4	4	4	5	2	5	3
18	Snowstorm	4	3	3	5	4	4	5	2	4	7
19	Ice at shore	5	4	1	5	4	5	4	1	5	4
20	Flash flood	3	5	4	4	4	4	5	1	5	8
21	Thunderstorm	4	5	2	4	4	5	5	2	4	7
22	Lightning strike	4	5	2	4	4	5	5	1	5	7
23	Blizzard	4	3	4	4	4	5	5	1	5	7
24	Ocean waves	4	4	2	4	4	5	5	3	5	4
25	Hail storm	4	5	4	5	3	5	5	1	5	7
26	Freezing rain	4	4	5	5	4	4	5	1	5	7
27	Localized strong wind	5	4	3	5	5	5	5	1	5	3
28	Subsidence	4	3	5	5	4	4	5	3	5	13
29	Mud & debris flows	4	4	5	4	4	5	5	4	5	13
30	Air supported flows	4	5	5	4	5	5	5	2	5	13
31	Rockfalls	5	5	5	5	5	5	5	1	5	13

the international community alleviates drought. It is only apt that this section emphasizes the inspirational and enthusiastic role played by Bob Geldof during the Ethiopian drought of the mid-1980s. The reaction to his initiatives by individuals, regardless of their economic or political allegiance, gives hope that the effects of even the worst natural calamities affecting humanity can be minimized.

Chapter 6 examines flash flood and regional flooding events that are usually severe enough to invoke states of emergency. In addition, regional flooding accounts for the largest death toll registered for any type of hazard. This chapter begins by introducing the climatic processes responsible for flash flooding. This is followed by examples of flash flooding events in the United States and Australia. Regional flooding is discussed in detail for the Mississippi River in the United States, the Hwang Ho River in China, and Australia where floods are a general feature of the country.

Drought-induced bushfires or forest fires are then treated as a separate entity in Chapter 7. The conditions favoring intense bushfires and the causes of such disasters are described. Major natural fire disasters, including in the United States and Australia, are then presented. Two of the main issues discussed in this chapter are the continuing debate over *prescribed burning* as a strategy to mitigate the threat of fires (especially in urban areas), and the regularity with which fire history repeats itself. Both the causes of fires and the responses to them appear to have changed little between the ends of the nineteenth and twentieth centuries.

These climatic chapters are followed by Chapter 8's discussion of aquatic hazards – encompassing a variety of marine and lacustrine phenomena. Most of these hazards affect few people, but they have long-term consequences, especially if global warming becomes significant over the next 50 years. Because of their spatial extent and long period of operation, most of the phenomena in this chapter can be ranked as middle-order hazards. Waves are described first with the theory for their generation. The worldwide occurrence of large waves is discussed, based upon ship and satellite observations. This is followed by a brief description of sea-ice phenomena in the ocean and at shore. One of the greatest concerns today is the impending rise in sea level, supposedly occurring because of melting of icecaps or *thermal expansion* of oceans within the context of global warming. Worldwide sea level data are presented to show that sea levels may not be rising globally as generally believed, but may be influenced in behavior by regional climatic change. Finally, the chapter discusses the various environmental mechanisms of sandy beach erosion (or accretion) caused mainly by changes in sea level, rainfall, or storminess. This evaluation utilizes an extensive data set of shoreline position and environmental variables collected for Stanwell Park Beach, Australia. Much of the discussion in this chapter draws upon the author's own research expertise.

Geological hazards – forming the second part of this book – are covered under the following chapter headings: causes and prediction of earthquakes and volcanoes; earthquakes and tsunami; volcanoes; and land instability. Chapter 9 presents the worldwide distribution of earthquakes and volcanoes and a presentation of the scales for measuring earthquake magnitude. This is followed by an examination of causes of earthquakes and volcanoes under the headings of plate boundaries, hot spots, regional faulting, and the presence of reservoirs or dams. Next, the presence of clustering in the occurrence of earthquakes and volcanoes is examined, followed by a discussion of the long-term prediction of volcanic and seismic activity. The chapter concludes with a presentation of the geophysical, geochemical, and geomagnetic techniques for forecasting volcanic and seismic activity over the short term.

Two chapters that are concerned separately with earthquakes and volcanoes then follow this introductory chapter. Both hazards, because of their suddenness and high-energy release, have the potential to afflict human beings physically, economically, and socially. Both rank as the most severe geological hazards. Chapter 10 first describes types of seismic waves and the global seismic risk. Earthquake disasters and the seismic risk for Alaska, California and Japan are then described. The associated phenomenon of *liquefaction* or *thixotropy*, and its importance in earthquake damage, is subsequently presented. One of the major phenomena generated by earthquakes is tsunami. This phenomenon is described in detail, together with a presentation of major, worldwide tsunami disasters. The prediction of tsunami in the Pacific region is then discussed. Chapter 11, on volcanoes, emphasizes types of volcanoes and associated hazardous phenomena such as lava flows, tephra clouds, pyroclastic flows and base surges, gases and acid rains, lahars (mud flows), and glacier bursts. The major disasters of Santorini (~1470 BC), Vesuvius (79 AD), Krakatau (1883), Mt Pelée (1902), and Mt St Helens (1980) are then described in detail.

The section on geological hazards concludes, in Chapter 12, with a comprehensive treatment of land instability. This chapter opens with a description of basic soil mechanics including the concepts of stress and strain, friction and cohesion, shear strength of soils, pore-water pressure, and rigid, elastic, and plastic solids. Land instability is then classified and described under the headings of expansive soils, creep and *solifluction*, mud and debris flows, landslides and slumps, rockfalls, debris avalanches, and air-supported flows. Many of these hazards rank in severity and impact as middle-order events. For example, expansive soils, while not causing loss of life, rank as one of the costliest long-term hazards because of their ubiquitousness. For each of these categories, the type of land instability is presented in detail, together with a description of major disasters. The chapter concludes with a cursory presentation on the natural causes of land subsidence.

Because the emphasis of the previous chapters has been on the physical mechanisms of hazards, the book closes in Chapter 13 with a discussion of the social impact of many of these phenomena. Here is emphasized the way hazards are viewed by different societies and cultures. This section is followed by detailed descriptions of responses before, during, and after the event. As much as possible, the characterization of responses under each of these headings is clarified using anecdotes. It is shown that individuals or families have different ways of coping with, or reacting to, disaster. The chapter concludes by emphasizing the psychological ramifications of a disaster on the victims, the rescuers, and society.

REFERENCES AND FURTHER READING

Balling, R.C. and Cerveny, R.S. 2003. Compilation and discussion of trends in severe storms in the United States: Popular perception *v.* climate reality. *Natural Hazards* 29: 103–112.

Blong, R.J. 1982. *The Time of Darkness, Local Legends and Volcanic Reality in Papua New Guinea.* Australian National University Press, Canberra.

Bryant, E.A. 2001. *Tsunami: The Underrated Hazard.* Cambridge University Press, Cambridge.

Changnon, S.A. and Hewings, G.J.D. 2001. Losses from weather extremes in the United States. *Natural Hazards Review* 2: 113–123.

CRED 2002. *EM-DAT: The OFDA/CRED International Disasters Data Base.* Center for Research on the Epidemiology of Disasters, Université Catholique de Louvain, Brussels, <http://www.em-dat.net/>

Day, M.S. 1984. *The Many Meanings of Myth.* Lanham, NY.

High Plains Regional Climate Center 2003. *Annual average number of strong–violent (F2–F5) tornadoes 1950–1995.* University of Nebraska, Lincoln, http://www.hprcc.unl.edu/nebraska/USTORNADOMAPS.html

Holmes, A. 1965. *Principles of Physical Geology.* Nelson, London.

Huggett, R. 1997. *Catastrophism: Asteroids, Comets and Other Dynamic Events in Earth History.* Verso, London.

Intergovernmental Oceanographic Commission 2003. *On-line Pacific…, Atlantic…, and Mediterranean Tsunami Catalog.* Tsunami Laboratory, Institute of Computational Mathematics and Mathematical Geophysics, Siberian Division, Russian Academy of Sciences, Novosibirsk, Russia, <http://omzg.sscc.ru/tsulab/>

Middleton, N. 1995. *The Global Casino.* Edward Arnold, London.

Milne, A. 1986. *Floodshock: the Drowning of Planet Earth.* Sutton, Gloucester.

Myles, D. 1985. *The Great Waves.* McGraw-Hill, NY.

Sidle, R.C., Taylor, D., Lu, X.X., Adger, W.N., Lowe, D.J., de Lange, W.P., Newnham, R.M. and Dodson, J.R. 2003. Interaction of natural hazards and society in Austral-Asia: evidence in past and recent records. *Quaternary International* 118–119: 181–203.

Susman, P., O'Keefe, P. and Wisner, B. 1983. Global disasters, a radical interpretation. In Hewitt, K. (ed.) *Interpretations of Calamity.* Allen and Unwin, Sydney, pp. 263–283.

Watts, M. 1983. On the poverty of theory: natural hazards research in context. In Hewitt, K. (ed.) *Interpretations of Calamity.* Allen and Unwin, Sydney, pp. 231–261.

CLIMATIC HAZARDS

PART 1

Mechanisms of Climate Variability

CHAPTER 2

INTRODUCTION

Climatic hazards originate with the processes that move air across the Earth's surface due to differential heating and cooling. Surprisingly, examination of these processes has focused upon heating at the tropics and downplayed the role of cold air masses moving out of polar regions due to deficits in the radiation balance in these latter regions. Fluctuations – in the intensity of pulses of cold air moving out of polar regions or of heating at the equator – and the location of the interaction between these cold and warm air masses, are crucial factors in determining the magnitude, frequency, and location of mid-latitude storm systems. While most of these factors are dictated by internal factors in the Earth–atmosphere system, modulation by 11-year geomagnetic cycles linked to solar activity and by the 18.6 year M_N *lunar tide* also occurs. This chapter examines these processes and mechanisms. The responses in terms of centers of storm activity will be examined in the following chapter.

MODELS OF ATMOSPHERIC CIRCULATION AND CHANGE

(Bryson & Murray, 1977; Lamb, 1982)

How air moves

Barometric pressure represents the weight of air above a location on the Earth's surface. When the weight of air over an area is greater than over adjacent areas, it is termed 'high pressure'. When the weight of air is lower, it is termed 'low pressure'. Points of equal pressure across the Earth's surface can be contoured using *isobars*, and on weather maps contouring is generally performed at intervals of 4 *hectopascals* (hPa) or millibars (mb). Mean pressure for the Earth is 1013.6 hPa. Wind is generated by air moving from high to low pressure simply because a pressure gradient exists owing to the difference in air density (Figure 2.1): the stronger the pressure gradient, the stronger the wind. This is represented graphically on weather maps by relatively closely spaced isobars. In reality, wind does not flow down pressure gradients but blows almost parallel to isobars at the surface of the Earth because of *Coriolis force* (Figure 2.1). This force exists because of the rotation of the Earth and is thus illusionary. For example, a person standing perfectly still at the pole would appear to an observer viewing the Earth from the Moon to turn around in a complete circle every

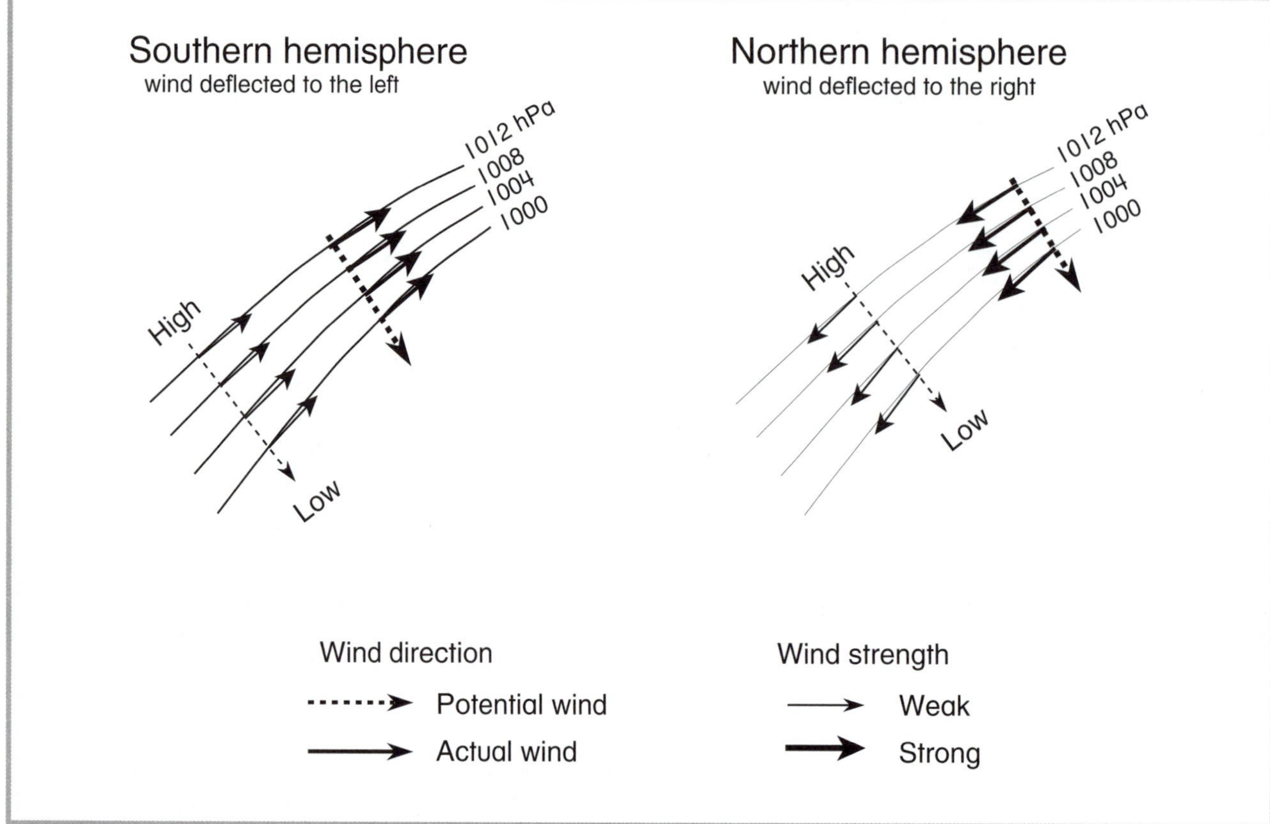

Fig. 2.1 Wind movements relative to isobars, with and without Coriolis force (i.e. potential and actual).

24 hours as the Earth rotates. However, a person standing perfectly still at the equator would always appear to be facing the same direction and not turning around. Because of Coriolis force, wind tends to be deflected to the left in the southern hemisphere and to the right in the northern hemisphere. Coriolis force can be expressed mathematically by the following equation:

$$\text{Coriolis force} = 2\,\omega\,\sin\phi v \qquad 2.1$$

where ω = rate of spin of the Earth
$\sin\phi$ = latitude
v = wind speed

Clearly, the stronger the pressure gradient, the stronger the wind and the more wind will tend to be deflected. This deflection forms a *vortex*. Also, the stronger the wind, the smaller and more intense is the resulting vortex. Very strong vortices are known as hurricanes, typhoons, cyclones, and tornadoes. Coriolis force varies across the surface of the globe because of latitude and wind velocity. Coriolis force is zero at the equator (sin 0° = 0), and maximum at the pole (sin 90° = 1). The equator is a barrier to inter-hemispheric movement of storms because vortices rotate in opposite directions in each hemisphere. For example, tropical cyclones cannot cross the equator because they rotate clockwise in the southern hemisphere and anti-clockwise in the northern hemisphere.

Palmén–Newton model of global circulation

Upward movement of air is a prerequisite for vortex development, with about 10 per cent of all air movement taking place vertically. As air rises, it spirals because of Coriolis force, and draws in adjacent air at the surface. Rising air is unstable if it is warmer than adjacent air, and instability is favored if *latent heat* can be released through condensation. The faster air rises, the greater the velocity of surface winds spiraling into the center of the vortex. Of course, in the opposite manner, descending air spirals outward at the Earth's surface. These concepts can be combined to account for air movement in the *troposphere*. The Palmén–Newton *general air circulation* model is one of the more thorough models in this regard (Figure 2.2). Intense heating by the sun, at the equator, causes air to rise and spread out poleward in the upper troposphere. As this air moves towards the poles, it cools through long wave

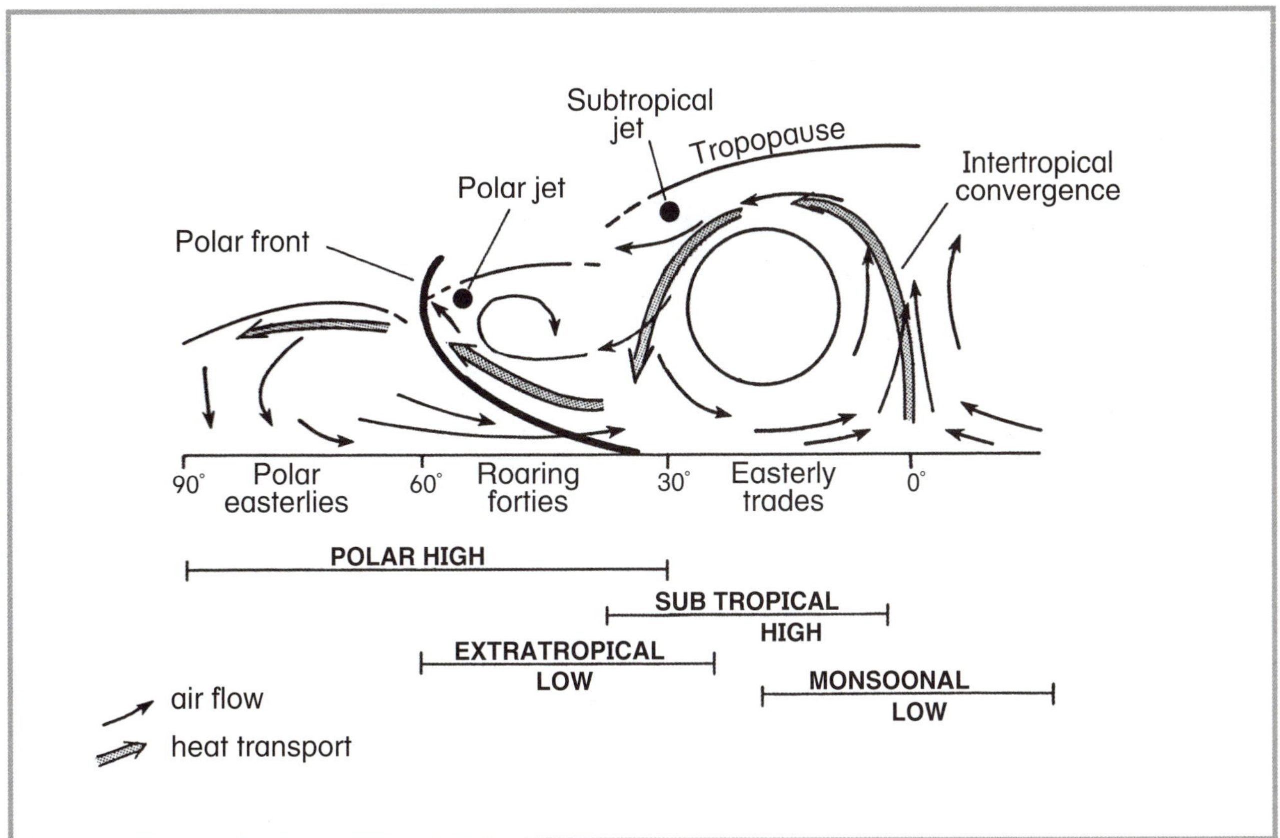

Fig. 2.2 Palmén–Newton schematic model of general air circulation between the pole and the equator.

emission and begins to descend back to the Earth's surface around 20–30° north and south of the solar equator, forming high pressure at the ground. Upon reaching the Earth's surface, this air either moves poleward or returns to the equator to form a closed circulation cell, termed a *Hadley cell*. Because of Coriolis force, equatorial moving air forms two belts of easterly *trade winds* astride the equator. Air tends to converge towards the solar equator, and the uplift zone here is termed the intertropical convergence. Over the western sides of oceans, the tropical easterlies pile up warm water, resulting in *convection* of air that forms low pressure and intense instability. In these regions, intense vortices known as *tropical cyclones* develop preferentially. These will be described in Chapter 3.

At the poles, air cools and spreads along the Earth's surface towards the equator. Where cold polar air meets relatively warmer, subtropical air at mid-latitudes, a cold *polar front* develops with strong uplift and instability. Tornadoes and strong westerlies can be generated near the polar front over land, while intense extra-tropical storms develop near the polar front, especially over water bodies. Tornadoes will be described in Chapter 4, while extra-tropical storms are described in the next chapter. A belt of strong wind and storms – dominated by westerlies poleward of the polar front – forms around 40° latitude in each hemisphere. These winds, known as the roaring forties, are especially prominent in the southern hemisphere where winds blow unobstructed by land or, more importantly, by significant mountain ranges. Maps of global winds support these aspects of the Palmén–Newton model (Figure 2.3).

The Palmén–Newton model has two limitations. First, the location of pressure cells within the model is based upon averages over time. Second, because the model averages conditions, it tends to be static, whereas the Earth's atmosphere is very dynamic. Additionally, the concept of Hadley circulation is overly simplistic. In fact, rising air in the tropics is not uniform, but confined almost exclusively to narrow updrafts within thunderstorms. At higher latitudes, large-scale circulation is distorted by relatively small eddies. Other factors also control winds. For example, over the Greenland or Antarctic icecaps, enhanced radiative cooling forms large pools of cold air, which can accelerate downslope under gravity because of the low *frictional* coefficient of ice. Alternative models that overcome some of these limitations will be presented.

A

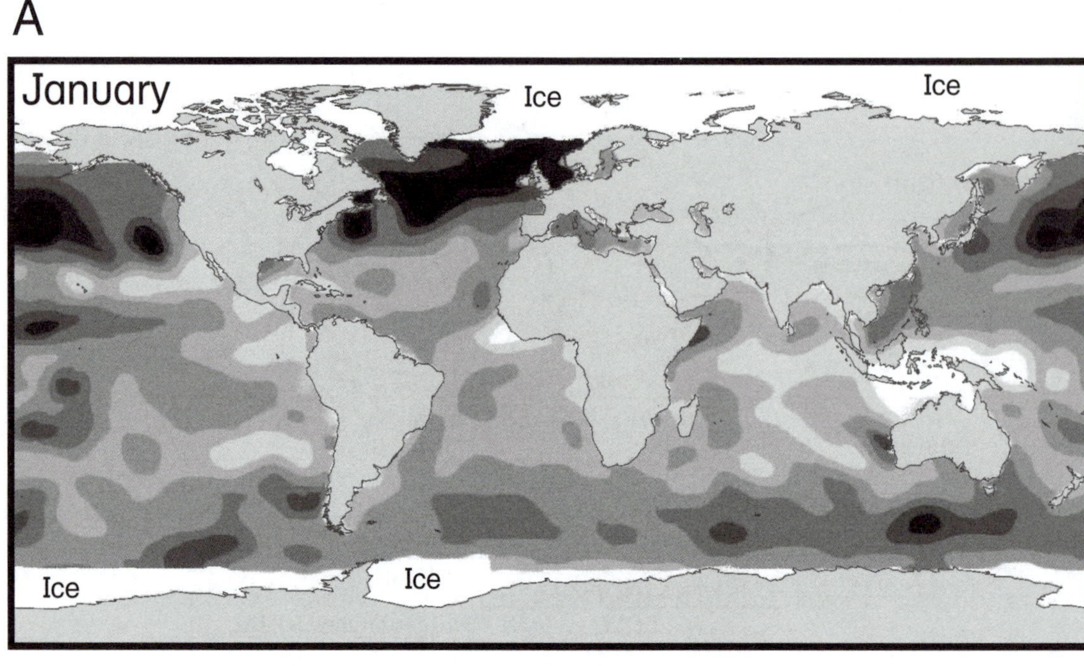

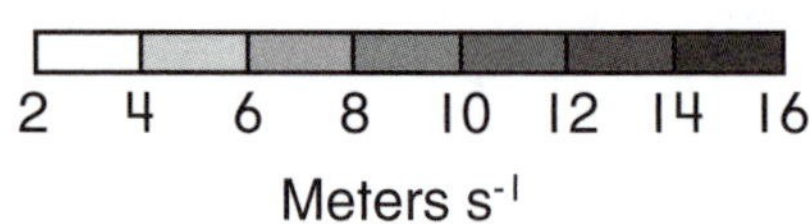

B

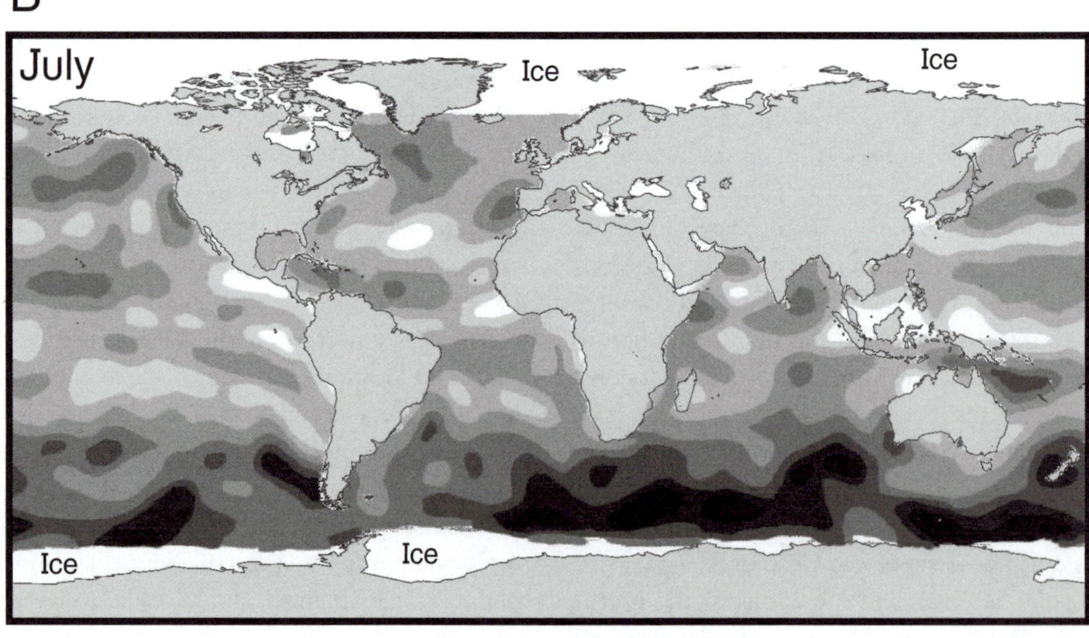

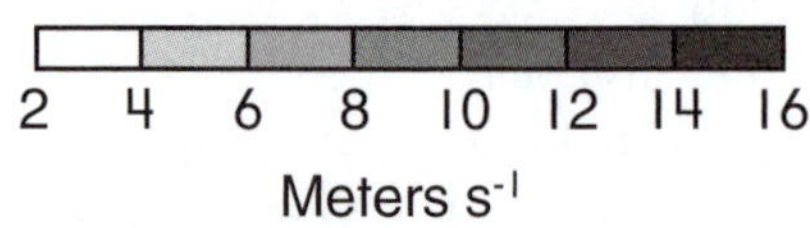

Fig. 2.3 Location of mid-latitude westerly and easterly trade wind belts based on TOPEX/POSEIDON satellite observations A) January 1995, B) July 1995 (Jet Propulsion Laboratory, 1995a, b).

Changes in jet stream paths

Strong winds exist in the upper atmosphere adjacent to the *tropopause* boundary (Figure 2.2). The more significant of these is the polar *jet stream* on the equatorial side of the polar front. This jet stream consists of a zone of strong winds no more than 1 km deep and 100 km wide, flowing downwind over a distance of 1000 km or more. Wind speeds can reach in excess of 250 km hr^{-1}. The polar jet stream is most prominent and continuous in the northern hemisphere. Here, it forms over the Tibetan Plateau, where its position is linked with the seasonal onset and demise of the Indian *monsoon*. The jet loops northward over Japan and is deflected north by the Rocky Mountains. It then swings south across the Great Plains of the United States, north-eastward (parallel to the Appalachian Mountains), and exits North America off Newfoundland, dissipating over Iceland. Both the Tibetan Plateau and the Rocky Mountains produce a resonance effect in wave patterns in the northern hemisphere, locking high pressure cells over Siberia and North America with an intervening low pressure cell over the north Pacific Ocean. Barometric pressure, measured along the 60°N parallel of latitude through these cells, reveals a quasi-stable planetary or *Rossby wave*.

In the northern hemisphere, the jet stream tends to form three to four Rossby waves extending through 5–10° of latitude around the globe (Figure 2.4). Rossby waves move a few hundred kilometres an hour faster than the Earth rotates and thus appear to propagate from west to east. Any disturbance in a Rossby wave thus promulgates downwind, such that a change in weather in North America appears over Europe several days afterwards. This aspect gives *coherence* to extreme events in weather across the northern hemisphere. Because there is a time lag in changes to Rossby waves downwind from North America, forecasters can predict extreme events over Europe days in advance. In some winters, the waves undergo amplification and loop further north and south than normal. Many researchers believe that changes in this looping are responsible, not only for short-term drought (or rainfall), but also for semi-permanent climatic change in a region extending from China to Europe. Sometimes the looping is so severe that winds in the jet stream simply take the path of least resistance, and proceed *zonally* (parallel to latitude), cutting off the loop. In this case, high and low pressure cells can be left stranded in location for days or weeks, unable to be shifted east by the prevailing westerly airflow. High-pressure cells are particularly vulnerable to this process and form *blocking highs*. Blocking highs deflect frontal lows to higher or lower latitudes than normally expected, producing extremes in weather. Blocking is more characteristic of winter, especially in the northern hemisphere. It also tends to occur over abnormally warm seas, on the western sides of oceans.

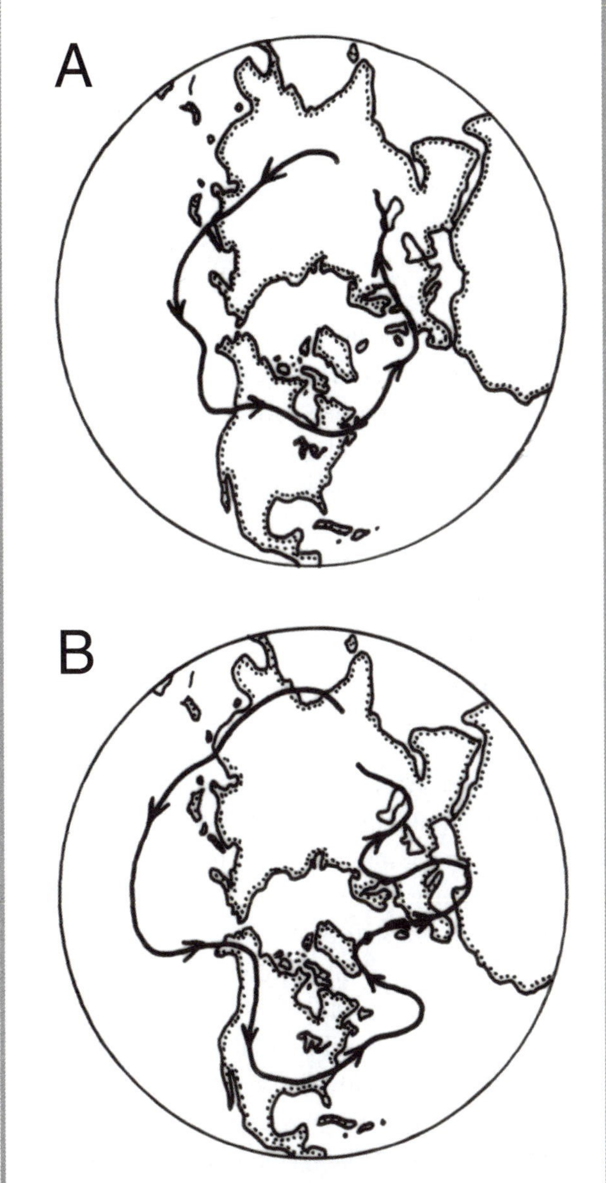

Fig. 2.4 Path of jet stream in the northern hemisphere showing likely Rossby wave patterns A) for summer, B) for extreme winter (Bryson & Murray, ©1977 with the permission of The University of Wisconsin Press).

Expansion of the westerlies because of enhanced looping of Rossby waves can be linked to recent failure of monsoon rains and resulting drought in the *Sahel* in the early 1970s. In the Sahara region, rain normally falls as the *intertropical convergence* moves seasonally northward. If westerlies expand, then such movement of the intertropical convergence is impeded and drought conditions prevail. Drought in Britain can also be linked to displacement northward of the jet stream and polar-front lows – shifting into the Arctic instead of crossing Scotland and Norway. This displacement leaves Britain under the influence of an extension of the Azores high-pressure system that is stable and rainfall-deficient. Blocking of this high leads to drought for several months. Blocking highs off the Californian coast were also responsible for drought in the mid-United States in 1997, while highs over eastern Australia have led to some of the worst droughts in that continent's recorded history. The social response to these types of droughts will be examined in Chapter 5.

Mobile polar highs (Leroux, 1993, 1998; Bryant, 1997)

The monsoonal circulation described above does not fit well within the Palmén–Newton general circulation model. In fact, the positioning of Hadley cells, and the semi-permanency of features such as the Icelandic or Aleutian Lows, are statistical artefacts. There is not an Icelandic Low, locked into position over Iceland. Nor are there consistent trades or polar easterlies. Air circulation across the surface of the Earth is dynamic, as is the formation and demise of pressure cells. In reality, the Icelandic Low may exist as an intense cell of low pressure for several days, and then move eastward towards Europe within the westerly air stream. Climate change in the Palmén–Newton model implies the movement, or change in magnitude, of these centers of activity. For example, weakening of the Icelandic Low conveys the view that winter circulation is less severe, while expansion of Hadley cells towards higher latitudes suggests that droughts should dominate mid-latitudes.

There are other problems with terminology in the Palmén–Newton model; these have an historical basis. The depression, extra-tropical cyclone, or polar low was initially explained as a thermal phenomenon, and then linked to frontal uplift along the polar front, to upper air disturbances, and recently to planetary waves or Rossby waves. The jet stream is supposedly tied to the polar front, and related to *cyclogenesis* at mid-latitudes. However, the connection between the jet stream and the polar front is approximate, and a clear relationship has not been established between the jet stream in the upper troposphere and the formation of low pressure at the Earth's surface. Indeed, no one theory adequately explains the initial formation of lows at mid-latitudes. The Palmén–Newton general circulation model is a good teaching model, but it is not an ideal model for explaining the causes of climate change.

Conceptually, the Palmén–Newton model in its simplest form has two areas of forcing: upward air movement at the equator from heating of the Earth's surface, and sinking of air at the poles because of intense cooling caused by long wave emission. At the equator, air moves from higher latitudes to replace the uplifted air, while at the poles subsiding air moves towards the equator as a cold dense mass hugging the Earth's surface. The lateral air movement at the equator is slow and weak, while that at the poles is strong and rapid. Polar surging can reach within 10° of the equator, such that some of the lifting of air in this region can be explained by the magnitude and location of polar outbursts. Hence, the dominant influence on global air circulation lies with outbursts of cold air within polar high-pressure cells. Each of these outbursts forms an event termed a *mobile polar high*. There is no separate belt of high pressure or Hadley cell in the subtropics. These cells, as they appear on synoptic maps, are simply statistical averages over time of the preferred pathways of polar air movement towards the equator.

Mobile polar highs developing in polar regions are initially maintained in position by surface cooling, air subsidence and advection of warm air at higher altitudes. When enough cold air accumulates, it suddenly moves away from the poles, forming a 1500 m thick lens of cold air. In both hemispheres, polar high outbreaks tend to move from west to east, thus conserving vorticity. Additionally, in the southern hemisphere, pathways and the rate of movement are aided by the formation of *katabatic winds* off the Antarctic icecap (Figure 2.5). In the northern hemisphere, outburst pathways are controlled by topography with mobile polar highs tending to occur over the Hudson Bay lowland, Scandinavia and the Bering Sea (Figure 2.5). The distribution of oceans and continents, with their attendant mountains, explains why the mean trajectories followed by these highs are always the same. For instance, over Australia, a polar outbreak always

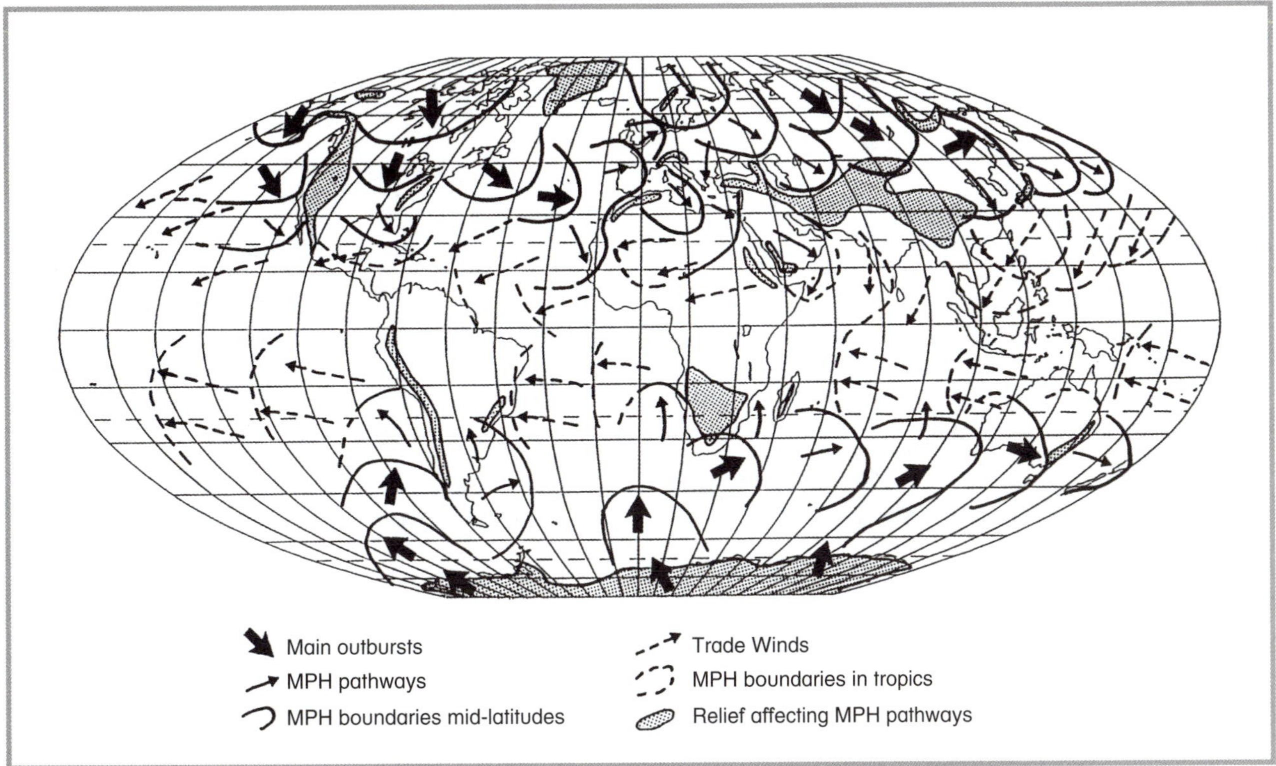

Fig. 2.5 Pathways for mobile polar highs (MPH) and resulting trade wind circulation in the tropics (based on Leroux, 1993).

approaches the continent from the southwest, and then loops across the continent towards the equator, and drifts in the Hadley belt out into the Tasman Sea. Highs rarely move directly northward across the continent, or sweep up from the Tasman Sea. In North America, mobile polar highs regularly surge southward across the Great Plains for the same reason.

Because mobile polar highs consist of dense air, they deflect less dense, warm air upward and to the side. The deflection is greatest in the direction they are moving. Hence, the polar high develops an extensive bulbous, high-pressure vortex, surrounded downdrift by a cyclonic branch or low-pressure cell (Figure 2.6). Typically, the high-pressure cell is bounded by an arching, polar cold front with a low pressure cell attached to the leading edge. However, in the northern hemisphere, individual outbreaks tend to overlap so that the low-pressure vortex becomes contained between two highs. This forms the classic V-shaped, wedged frontal system associated with *extratropical lows or depressions* (Figure 2.6). Thus lows, and upper westerly jet streams, are a product of the displacement and divergence of a mobile polar high. The intensity of the low-pressure cell becomes dependent upon the strength of the polar high, and upon its ability to displace the surrounding air. Strong polar highs produce deep lows; weak highs generate weak lows. In a conceptual sense, a deeper Icelandic or Aleutian Low must be associated with a stronger mobile polar high. If a mobile polar high is particularly cold, it can cause the air above to cool and settle. This creates low pressure above the center of the high-pressure cell, which can develop into a trough and then a cell as upper air flows inwards. If a mobile polar high has lost its momentum and stalled forming a 'blocking high', the low-pressure cell can intensify, propagate to the ground, and generate a surface storm. This process occurs most often on the eastern sides of continents adjacent to mountain ranges such as the Appalachians in the United States and the Great Dividing Range in Australia. These storms will be discussed further in the next chapter.

Mobile polar highs tend to lose their momentum and stack up or *agglutinate* at particular locations over the oceans. A blocking high is simply a stagnant mobile polar high. Air pressure averaged over time thus produces the illusion of two stable, tropical high-pressure belts – known as Hadley cells – on each side of the equator. Mobile polar highs can propagate into the tropics, especially in winter. Here, their arrival tends to intensify the easterly trade winds. More important, yet little realized, is the fact that strong

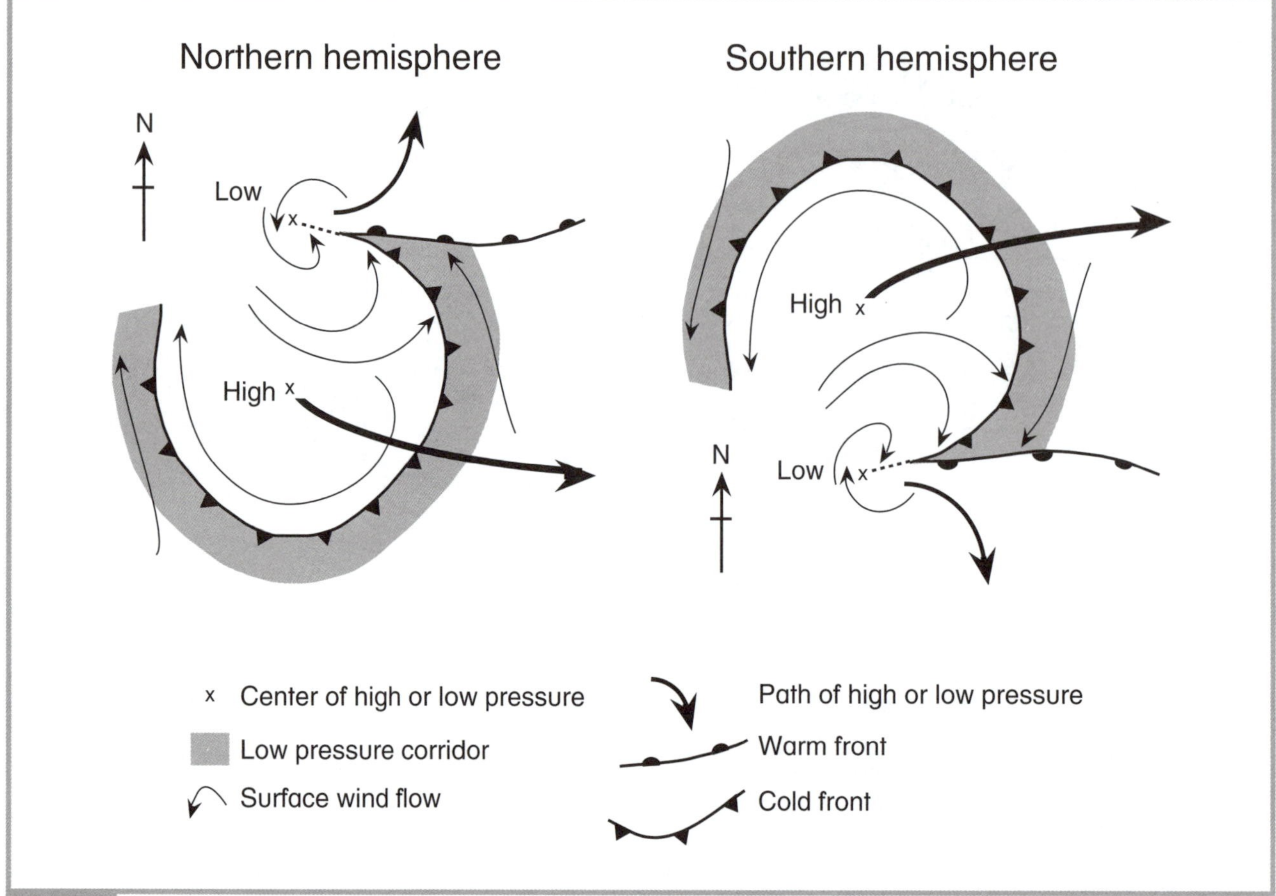

Fig. 2.6 Schematic view of the dynamic structure of a mobile polar high and its associated mid-latitude low (based on Leroux, 1993).

polar highs produce stronger monsoons, although in a more restricted tropical belt.

The Southern Oscillation

Introduction

The Earth's general atmospheric circulation in the tropics and subtropics, as shown above, can be simply described. However, the actual scene is slightly more complex. The intensity of mobile polar highs within a hemisphere varies annually with the apparent movement of the sun north and south of the equator. There is also a subtle, but important, shift in the intensity of heating near the equator. In the northern hemisphere summer, heating shifts from equatorial regions to the Indian mainland with the onset of the Indian monsoon. Air is sucked into the Indian subcontinent from adjacent oceans and landmasses, to return via upper air movement, to either southern Africa or the central Pacific. In the northern hemisphere winter, this intense heating area shifts to the Indonesian–northern Australian 'maritime' continent (Figure 2.7A), with air then moving in the upper troposphere, to the east Pacific. Convection is so intense that updrafts penetrate the tropopause into the *stratosphere*, mainly through *supercell thunderstorms*. These convective cells are labeled *stratospheric fountains*.

Throughout the year, mobile polar highs tend to stack up over the equatorial ocean west of South America. The center of average high pressure here shifts seasonally less than 5° in latitude. The highs are locked into position by positive *feedback*. Cold water in the eastern Pacific Ocean creates high pressure that induces easterly airflow; this process causes upwelling of cold water along the coast, which cools the air. On the western Pacific side, easterlies pile up warm water, thus enhancing *convective instability*, causing air to rise, and perpetuating low pressure. Air is thus continually flowing across the Pacific, as an easterly trade wind from high pressure to low pressure. This circulation is zonal in contrast to the *meridional airflow* inherent within the Palmén–Newton general circulation model. The persistent easterlies blow warm surface water across the Pacific Ocean, piling it up in

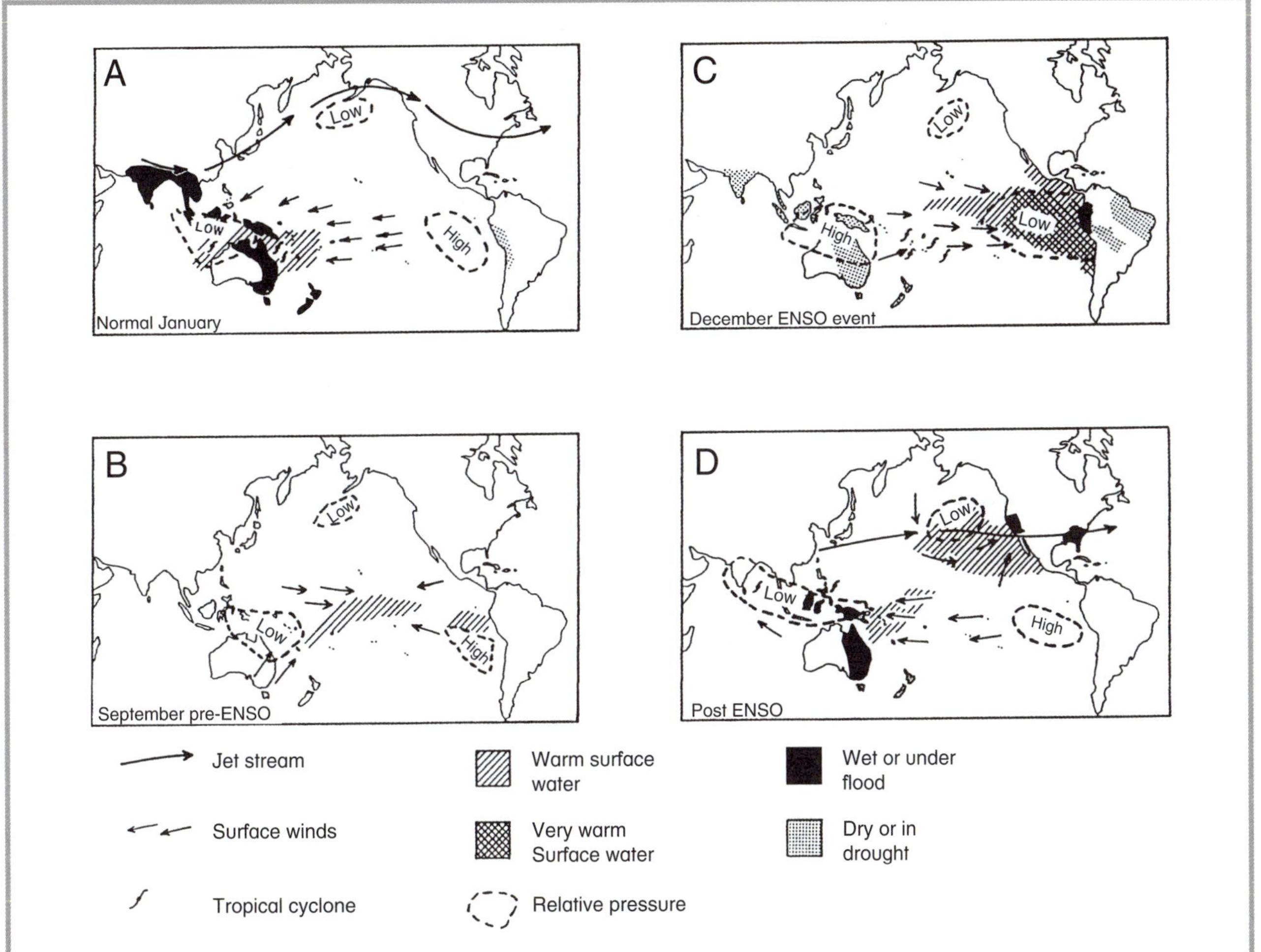

Fig. 2.7 Idealized representation of evolution and effects of an El Niño–Southern Oscillation event.

the west Pacific warm pool in the Philippine Sea. The supra-elevation of sea level in the west Pacific Ocean is approximately 1 cm for each 1°C difference in temperature between the west Pacific and the South American coast. Normally this supra-elevation amounts to 13–20 cm.

This atmospheric system is very stable, existing beyond the annual climatic cycle. The easterly trade wind flow is termed *Walker circulation*, after the Indian meteorologist, Gilbert Walker, who described the phenomenon in detail in the 1920s–1930s. For some inexplicable reason, this quasi-stationary heating process weakens in intensity, or breaks down completely, every 3–5 years – a frequency very suggestive of a chaotic system. High pressure can become established over the Indonesian–Australian area, while low pressure develops over warm water off the South American coast. In the tropics, the easterly winds abate and are replaced by westerly winds. The rainfall belt shifts to the central Pacific and drought replaces normal or heavy rainfall in Australia. Such conditions persisted in the Great World Drought of 1982–1983, which affected most of Australia, Indonesia, India, and South Africa.

Because it tends to fluctuate, this phenomenon is called the *Southern Oscillation* (SO). Simply subtracting the barometric pressure for Darwin from that of Tahiti produces an index of the intensity and fluctuations of the SO. This value is then *normalized* to a mean of 0.0 hPa and a standard deviation of 10. This normalized index, from 1851 to the present, is presented in Figure 2.8.

Warm water usually appears (albeit in reduced amounts) along the Peruvian coast around Christmas each year. This annual warming is termed the *El Niño*, which is Spanish for 'Christ Child'. However, when Walker circulation collapses, this annual warming becomes exaggerated, with sea surface temperatures increasing 4–6°C above normal and remaining that way for several months. This localized, above average,

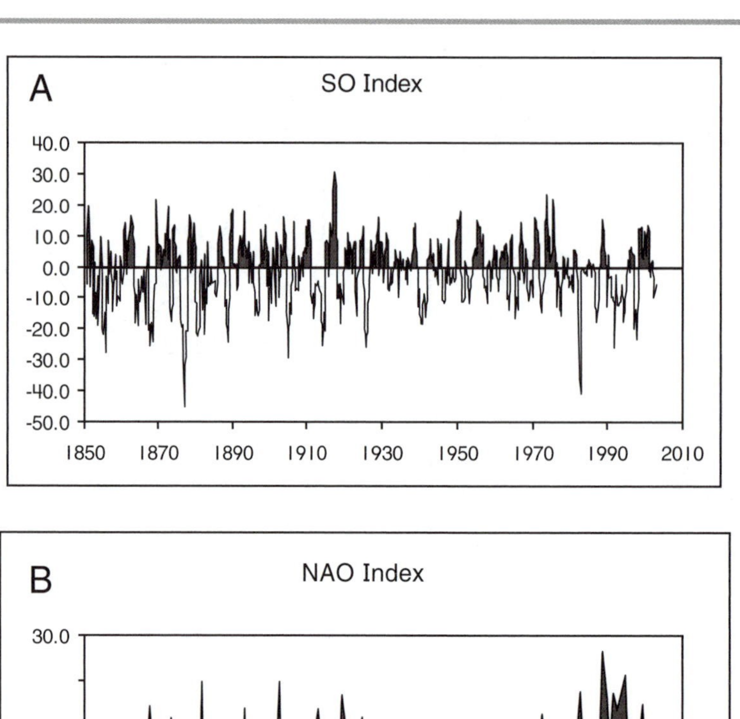

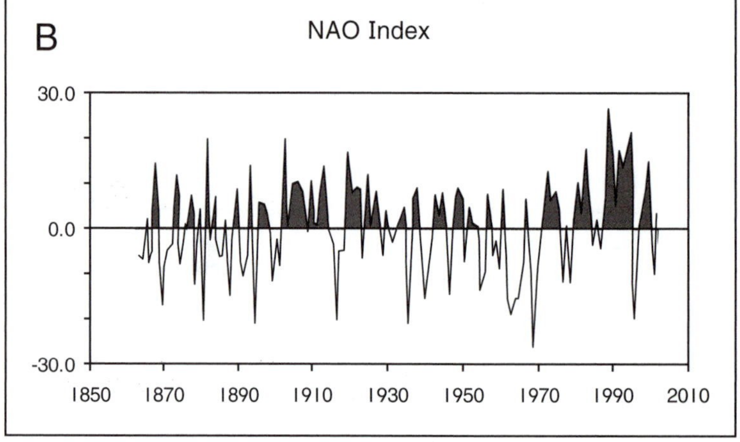

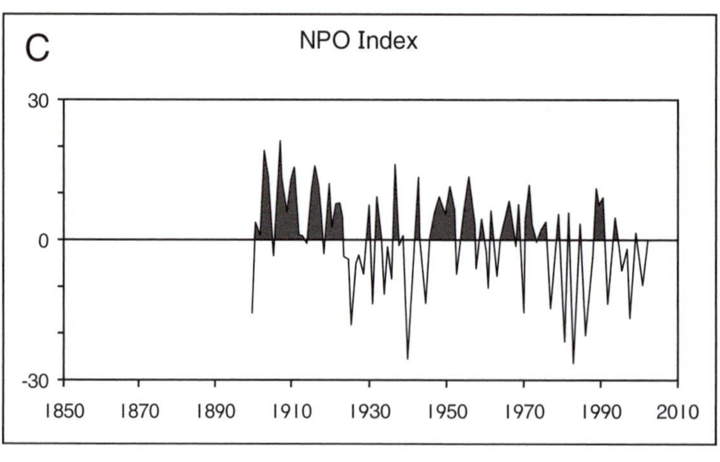

Fig. 2.8 Indices of atmospheric pressure oscillations over time A) the Southern Oscillation, B) the North Atlantic Oscillation (from Hurrell, 2002a), and C) the North Pacific Oscillation (from Hurrell, 2002b). The indices have been normalized to a mean of 0.0 hPa and a standard deviation of 10. See relevant sections in the text for further explanation.

warming is termed an El Niño event. In the early 1960s, Jacob Bjerknes recognized that warmer waters, during some El Niño events, were linked to the Southern Oscillation and the movement of warm water eastward from the west Pacific Ocean. These more widely occurring events are termed El Niño–Southern Oscillation, or for brevity, *ENSO* events. Greater amounts of warm water – along what is normally an arid coastline – lead to abnormally heavy rainfall, causing floods in the arid Andes rainfall region and drought in southern Peru, Bolivia and north-eastern Brazil. For example, in 1982–1983, in the coastal desert region of Ecuador and Peru, flash floods ripped out roads, bridges and oil pipelines. Towns were buried in mud, and shallow lakes appeared in the deserts. Irrigation-based agriculture was devastated, first by the flooding, and then by swarms of insects that proliferated under the wet conditions. The warm water also forces fish feeding on plankton to migrate to greater depths, and causes vast fish kills. Guano-producing birds feeding on anchovies die in their thousands, causing the collapse of the anchovy and fertilizer industries in Peru.

These climatic changes worldwide can be responsible for the increase of such isolated phenomena as snakebites in Montana, funnel-web spider bites in Sydney, and schizophrenia in Brazil (the latter caused by a lack of UV light due to increased cloudiness). More significantly, these changes can generate extreme economic and social repercussions, including damage to national economies, the fall of governments, and the deaths of thousands through starvation, storms and floods.

El Niño–Southern Oscillation (ENSO) events

(Philander, 1990; Glantz et al., 1991; Allan et al., 1996; Glantz, 1996; Bryant, 1997; Couper-Johnston, 2000)

Zonal, ocean–atmosphere feedback systems in the southern hemisphere are powered by three tropical heat-sources located over Africa, South America and the Indonesian–Australian maritime continent. The latter site, by far the strongest of the three systems, initiates Walker circulation. While it is relatively stable, Walker circulation is the least sedentary, because it is not anchored directly to a landmass. It is highly migratory during transition periods of the Indian monsoon, in either March–April, or August–September. Six *precursors* to the failure of Walker circulation stand out. First, sea-ice in the Antarctic and snow cover in central Asia tend to be more extensive beforehand. Second, the strength of the Indian monsoon relies heavily upon formation of an upper air, thermally driven anticyclone (between 100 and 300 hPa) over the Tibetan Plateau that leads to the formation of an easterly upper jet stream over southern India. Zonal easterly winds at 250 hPa over the Indian monsoon area also have been found to weaken up to two months before the onset of an ENSO event. Third, the Southern Oscillation appears to be a combination of the 2.2 year Quasi-Biennial Oscillation and a longer period cycle centered on a periodicity of five years. Fourth, changes in the behavior of southern hemisphere, mobile polar highs are linked to ENSO events. Strong Walker circulation is related to strong westerlies between 35° and 55°S. These westerlies appear to lock, over the Indonesian–Australian 'maritime' landmass, with heat lows developing in the austral spring. If the normally strong latitudinal flow in the mobile polar highs is replaced by *longitudinally* skewed circulation in the Australian region, then low pressure can be forced into the Pacific during the transition from northern to southern hemisphere summer. Increased southerly and south-westerly winds east of Australia preceded the 1982–1983 ENSO event, and flowed towards a South Pacific Convergence Zone that was shifted further eastwards. While highs were as intense as ever, they had a stronger meridional component, were displaced further south than normal, and were stalled over eastern Australia. Fifth, ENSO events are more likely to occur one year after the south polar vortex is skewed towards the Australian–New Zealand sector. A significant *correlation* has been found amongst this eccentricity, the Southern Oscillation, and the contemporaneous occurrence of rainfall over parts of southern hemisphere continents. Sixth, excess salinity north and south of the equator is correlated with warmer waters, presaging an ENSO event about twelve months later. The sinking of this cold salty water around the equator draws in warmer, less saline, surface water triggering the ENSO event. Finally, the causes may be interlinked. For example, more snow in Asia the winter before Walker circulation collapses may weaken the summer jet stream, leading to failure of the Indian monsoon. Alternatively, more sea-ice in the Antarctic may distort the shape and paths of mobile polar highs over the Australian continent, or displace the south polar vortex.

Whatever the cause, the movement eastward of low pressure, beyond the Australian continent, leads to westerly airflow at the western edge of the Pacific Ocean. Because there is no easterly wind holding supra-elevated water against the western boundary of the Pacific, warm water begins to move eastward along the equator (Figure 2.7b). Normally, the *thermocline* separating warm surface water from cooler water below is thicker (200 m) in the west Pacific than in the east (100 m). As warm water shifts eastward, the thermocline rises in the west Pacific. One of the first indications of an ENSO event is the appearance of colder water north of

Darwin, Australia. In the east Pacific, a layer of warm water extending 50–100 m below the ocean surface swamps cold water that is normally at the surface and maintaining the high-pressure cell off the South American coast (Figure 2.7C). As a result the thermocline deepens. As low pressure replaces high pressure over this warm water in the east Pacific, the easterlies fail altogether because there is no pressure difference across the Pacific to maintain them. Westerly winds may even begin to flow at the equator. So strong is the change that the Earth's rotation speeds up by 0.5–0.7 milliseconds as the braking effect of the easterly trades diminishes.

Cyclone development follows the location of the pool of warm water as it moves eastward towards Tahiti and Tonga. The 1982–1983 event caused widespread destruction in these areas, with Tonga receiving four tropical cyclones, each as severe as any single event in the twentieth century. The Tuamotu Archipelago, east of Tahiti, was devastated by five cyclones between January and April 1983. Not since 1906 had a cyclone occurred this far east in the tropics. In unusual circumstances, paired tropical cyclones may develop on each side of the equator in November. Because winds in tropical cyclones in each hemisphere rotate in opposite directions, coupled cyclones can combine (like an egg beater) to generate, along the equator, strong westerly wind that may accelerate water movement eastward.

At the peak of an ENSO event, a *Kelvin wave* is trapped along the east Pacific coast. Warm water and elevated sea levels spread north and south, reaching as far north as Canada. In the 1982–1983 event, temperatures were 10°C above normal and sea levels rose by 60 centimetres along the South America coastline, and by 25–30 cm along the United States west coast. The intense release of heat by moist air over the central Pacific Ocean causes the westerly jet stream in the upper atmosphere, and the position of the wintertime Aleutian Low, to shift towards the equator (Figure 2.7D). Westerly surface winds and the jet stream, rather than being deflected north around the Rocky Mountains, cross the continent at mid-latitudes. This can lead to dramatic changes in climate. For example, in the 1982–1983 ENSO event, coastal storms wreaked havoc on the luxurious homes built along the Malibu coastline of California while associated rainfall caused widespread flooding and landslides. The storms were made all the more destructive because sea levels were elevated by 25–30 cm along the entire United States west coast, from California to Washington State. Heavy snowfalls fell in the southern Rocky Mountains, record-breaking mild temperatures occurred along the American east coast and heavy rainfall fell in the southern United States.

Finally, with the spread of warm water in the northern Pacific, easterly air circulation begins to re-establish itself in the tropics. The depression of the thermocline in the east Pacific during an ENSO event causes a large wave, also called a Rossby wave, to propagate westward along the thermocline boundary. This wave is reflected off the western boundary of the Pacific and, as it returns across the Pacific, it slowly raises the thermocline to its pre-ENSO position. However, the return to Walker circulation can be anything but normal. The pool of warm water now moving back to the west Pacific drags with it exceptional atmospheric instability, and a sudden return to rainfall that breaks droughts in the western Pacific. In 1983, drought broke first in New Zealand in February, one month later in eastern Australia with rainfalls of 200–400 mm in one week, and then – over the next two to three months – in India and southern Africa.

The Southern Oscillation causes extreme, short-term climate change over 60 per cent of the globe, mainly across the southern hemisphere and certainly in North America. However, some events influence Europe. For example, the El Niño of 1997–1998 caused widespread flooding in central Europe concomitantly with flooding in central Africa, South-East Asia, and Peru. The climatic effects of an ENSO event are widespread, and its influence on both extreme climate hazards and short-term climate change is significant. Of South American countries, southern Peru, western Bolivia, Venezuela, and north-eastern Brazil are affected most by ENSO events. The 1997–1998 ENSO event destroyed $US1 billion in roads and bridges in Peru alone. The widest impact occurs over southern Africa, India, Indonesia, and Australia. For instance, in Indonesia, over 93 per cent of monsoon droughts are associated with ENSO events, and 78 per cent of ENSO events coincide with failure of the monsoon. In Australia, 68 per cent of strong or moderate ENSO events produce major droughts in the east of the continent. ENSO events have recently been shown to contribute to drought in the African Sahel region, especially in Ethiopia and Sudan.

Once an ENSO event is triggered, the cycle of climatic change operates over a minimum of two years. In historical records, the longest ENSO event lasted four years: from early 1911 to mid-1915. However, recent events have dispelled the belief that ENSO events are biennial phenomena. Following the 1982–1983 ENSO event, the warm water that traveled northward along the western coast of North America slowly crossed the north Pacific, deflecting the Kuroshio Current a decade later. As a result, northern Pacific sea surface temperatures increased abnormally, affecting general circulation across the North American continent. It is plausible that the drought on the Great Plains of the United States in the summer of 1988, and even the flooding of the Mississippi River Basin in the summer of 1993 (attributable by most to the 1990–1995 ENSO event) were prolonged North American climatic responses to the 1982–1983 El Niño. Warm water left over from the 1982–1983 ENSO event persisted in the north Pacific until the year 2000. The oceanographic effects of a major El Niño thus can have large decadal *persistence* outside the tropics.

The 1990–1995 ENSO event was even more anomalous. This event appeared to wane twice, but continued with warm central Pacific waters for five consecutive years, finally terminating in July 1995 – an unprecedented time span. Probability analysis indicates that an ENSO event of five years' duration should only recur once every 1500–3000 years. The changes in climate globally over these five years have been as dramatic as any observed in historical records. Two of the worst cyclones ever recorded in the United States, Hurricanes Andrew in Florida and Iniki in Hawaii, both in August 1992, occurred during this event. Hurricane Andrew was unusual because Atlantic hurricanes should be suppressed during ENSO events. The Mississippi River system recorded its greatest flood ever in 1993 (and then again in 1995), surpassing the flood of 1973. Record floods devastated western Europe in 1994–1995. Eastern Australia and Indonesia suffered prolonged droughts that became the longest on record. Cold temperatures afflicted eastern North America in the winter of 1993–1994, together with record snowfall, while the western half of the continent registered its highest winter temperatures ever. Record high temperatures and drought also occurred in Japan, Pakistan, and Europe in the summer of 1994.

Not all of the climatic extremes of 1990–1995 were consistent with that formulated for a composite ENSO event. For example, the Indian monsoon operated normally in 1994, while eastern Australia recorded its worst drought. Even locally within Australia, while most of the eastern half of the continent was in drought in 1992–1993, a 1000 km^2 region south of Sydney received its wettest summer on record. It is questionable whether or not all of these climatic responses can be attributed to the 1990–1995 ENSO, but they have occurred without doubt in regions where ENSO linkages or teleconnections to other climate phenomena operate. Additionally, the unprecedented nature of the 1982–1983 and 1990–1995 ENSO events may be linked to significant volcanic eruptions. While the Mexican El Chichon eruption of 1981 did not trigger the 1982–1983 El Niño, it may have exacerbated its intensity. Similarly, the 1990–1995 event corresponded well with the eruption and subsequent global cooling generated by significant volcanic eruptions in 1991 of the Philippines' Mt Pinatubo and Chile's Mt Hudson and, in 1992, of Mt Spurr in Alaska.

The events of the late twentieth century appear exceptional. They are not. Historical and proxy records are showing that mega-ENSO events occur every 400–500 years. For example, around 1100 AD, rivers in the Moche Valley of Peru reached flood levels of 18 m, destroying temples and irrigation canals built by the Chimu civilization. In the same location, the worst floods of the twentieth century (in 1926) reached depths of only 8 m. The latter event resulted in massive fires in the Rio Negro catchment of the Amazonian Basin. However, similar if not more extensive fires have occurred in 500, 1000, 1200, and 1500 AD. The 500 and 1100 AD events appear in palaeo-records from Veracruz and Mexico. In Veracruz, the latter events produced flooding and laid down sediments over a metre thick. Compared to this, the 1995 event, which has been viewed as extreme, deposited only 10–15 cm of sediment. Other mega-ENSO events occurred around 400, 1000, 1600, 2400, 4800, and 5600 BC in Veracruz. As with the prolonged 1990–1995 ENSO event, it is probable that more than one ENSO event was involved at these times. The recent prolonged 1990–1995 ENSO cycle may not be unusual, but on geological timescales represents the upper end of a shifting climate hazard regime.

La Niña events

Exceptionally 'turned on' Walker circulation is now being recognized as a phenomenon in its own right.

The cool water that develops off the South American coastline can drift northward, and flood a 1–2° band around the equator in the central Pacific with water that may be as cold as 20°C. This phenomenon is termed *La Niña* (meaning, in Spanish, 'the girl') and peaks between ENSO events. The 1988–1990 La Niña event appears to have been one of the strongest in 40 years and, because of enhanced easterly trades, brought record flooding in 1988 to the Sudan, Bangladesh and Thailand in the northern hemisphere summer. In the Sudan, the annual rainfall fell in fifteen hours, destroying 70 per cent of that country's houses. Regional flooding also occurred in Nigeria, Mozambique, South Africa, Indonesia, China, Central America, and Brazil. An unusual characteristic of this event was the late onset of flooding and heavy rain. Both the Bangladesh and the Thailand flooding occurred at the end of the monsoon season. Bangladesh was also struck by a very intense tropical cyclone in December 1988, outside the normal cyclone season.

In eastern Australia this same La Niña event produced the greatest rainfall in 25 years, with continuation of extreme rainfall events. Over 250 mm of rainfall on 30 April 1988 caused significant damage along a stretch of the coastal road between Wollongong and Sydney. Numerous landslides and washouts undermined the roadway and adjacent main railway line, leading to a total road repair cost, along a 2 kilometer stretch of coastline, of $A1.5 million (Figure 2.9). Slippage problems were still in evidence there 15 months later. The Australian summers of 1989–1990 not only witnessed record deluges but also flooding of eastern rivers on an unprecedented scale. Lake Eyre filled in both years, whereas it had filled only twice in the previous century. Towns such as Gympie, Charleville (Queensland), and Nyngan (New South Wales) were virtually erased from the map.

Fig. 2.9A The La Niña event of 1987–1988 was one of the strongest at the end of the twentieth century. Extreme rainfall fell throughout eastern Australia.
A) Fans deposited on the coastal road between Wollongong and Sydney following more than 250 mm of rainfall on 30 April 1988.

Fig. 2.9B B) The same rainfall event in A) caused numerous landslides and washouts that undermined the roadway and adjacent main railway. The total cost in road repairs along this 2 km stretch of coastline was $A1.5 million. Slippage problems as a result have continued for 15 years (photographs by Bob Webb, courtesy of Mal Bilaniwskyj and Charles Owen, New South Wales Department of Main Roads, Wollongong and Bellambi Offices).

Global long-term links to drought and floods

(Quinn et al., 1978; Pant & Parthasarathy, 1981; Bhalme et al., 1983; Pittock, 1984; Adamson et al., 1987)

Where the Southern Oscillation is effective in controlling climate, it is responsible for about 30 per cent of the variance in rainfall records and some of the largest death tolls in the resulting famines. Over 10 million people died in the state of Bengal during the 1769–1770 ENSO event, while the event of 1877–1878 – one of the strongest ENSO events recorded – was responsible for the deaths of 6 and 13 million people in India and China, respectively. In China people ate the edible parts of houses, while in the Sudan there were so many corpses that vultures and hyenas became fussy about the ones they would eat.

Severe droughts can occur simultaneously around the globe. In the twentieth century, the 1997–1998 ENSO event brought severe drought to South America from Colombia to north-eastern Brazil, the Caribbean, Central America, Hawaii, northern China and Korea, Vietnam, Indonesia, Papua New Guinea, and Bangladesh (East Pakistan). Southern Peru, western

Bolivia, Venezuela, and north-eastern Brazil are usually affected by drought during an ENSO event; however, greatest devastation occurs in the western part of the cell in southern Africa, India, Indonesia, and Australia. There is a strong coherence in the timing of both flood and drought on the Nile in Africa, the Krishna in central India and the Darling River in Australia. As well, rainfall over parts of Australia, Argentina and Chile, New Zealand and southern Africa coincides (Figure 2.10). Most notable are the years 1894, 1910, 1917, 1961 and 1970. There is an even better correspondence between droughts in India and 'turned off' Walker circulation, or ENSO events. The severe droughts in India of 1899, 1905, 1918, 1951, 1966, and 1971 all occurred before, or at times when, El Niño events were at their peak around Christmas or when the Southern Oscillation index was negative. All of these years, except for 1951, were also times when discharges on the Nile, Krishna, and Darling River systems were either very low, or nonexistent.

A similar relationship can be traced back even further in Indonesia. This is surprising because most climatic classifications label Indonesia as a tropical country with consistent yearly rainfall. Table 2.1 summarizes the relationship between monsoon droughts and ENSO events during the period 1844–1983. In Indonesia, over 93 per cent of monsoon droughts are associated with El Niño events and 78 per cent of ENSO events coincide with failure of the monsoon. The relationships are so strong that in Java the advent of an ENSO event can be confidently used as a prognostic indicator of subsequent drought. In recent years, both India and Indonesia have established distribution infrastructures that prevent major droughts from causing starvation on as large a scale as they once did; however, the 1982–1983 event caused the Indonesian economy to stall as the country shifted foreign reserves towards the purchase of rice and wheat.

Links to other hazards

(Gray, 1984; Marko et al., 1988; Couper-Johnston, 2000)

The effect of ENSO events and the Southern Oscillation goes far beyond just drought and flood. They are responsible for the timing of many other natural hazards that can influence the heart and fabric of society. Three culturally diverse examples illustrate this

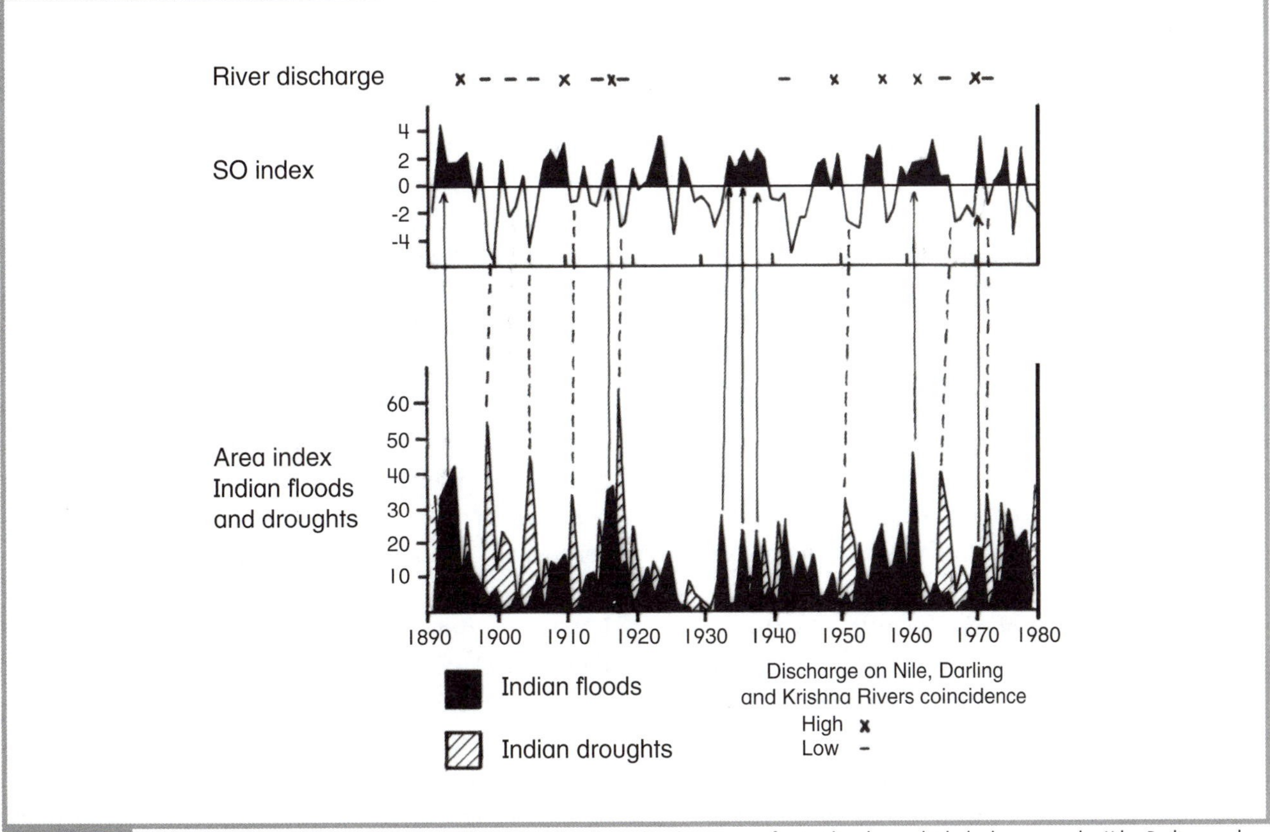

Fig. 2.10 Southern Oscillation index, drought and rainfall indices for India plus timing of coinciding low or high discharges on the Nile, Darling, and Krishna Rivers, 1890–1980 (based on Bhalme et al., 1983; Adamson et al., 1987).

Table 2.1 Comparison of drought in Indonesia with El Niño–Southern Oscillation events, 1844–1983 (Quinn et al., 1978).

1844–96		1902–83	
Drought	**El Niño event**	**Drought**	**El Niño event**
1844	1844	1902	1902
1845	1845–46	1905	1905
1850	1850	1913–14	1914
1853	none	1918–19	1918–19
1855	1855	1923	1923
1857	1857	1925–26	1925–26
1864	1864	1929	1929–30
1873	1873	1932	1932
1875	1875	1935	none
1877	1877–78	1940	1939–40
1881	1880	1941	1941
1883	none	1944	1943–44
1884–85	1884–85	1945–46	1946
1888	1887–89	1953	1953
1891	1891	no data	1954–55
1896	1896	1976	1976
		1982–83	1982

fact. First is the effect of ENSO events on civilization in Peru. The story is one of successive development and collapse of city-states. By the end of the eleventh century, most of Peru's desert coast had come under the influence of the Chimu civilization. They developed a vast network of irrigation canals stretching hundreds of kilometres between river valleys. Around 1100 AD, extraordinary flooding exceeding 18 m depths wiped out all of their engineering achievements in the Moche Valley. Rival city-states were also affected; however the Chimu undertook military action to expand their area of influence so that they could relocate. While the rival states sank into revolt and anarchy, the Chimu conquests succeeded to rival the Incas in power. This example was but one phase in a sequence of rapid cultural advancement, abandonment, and expansion of opportunistic city-states triggered by ENSO events along the west coast of South America. Over four centuries, this process has covered an area of 30 000 km^2 of coastal plain.

The second example is the historical development of Ethiopia. The ENSO events of 1888 and 1891 killed a third of this country's population at a time when the Italians hoped to establish a colonial empire in the Horn of Africa. Other European countries – mainly the British – had taken advantage of ENSO-related droughts to expand their empires in Africa. However, Emperor Menilek took advantage of his weakened enemies to consolidate his position as the country's absolute ruler. When the La Niña of 1892 brought rain and harvests to feed his army, Menilek was able to defeat the Italians at the Battle of Adowa and become the first African ruler to repel colonialism in Africa.

Third is the effect of the 1982–1983 ENSO event on Australia. This brought the worst drought Australia has experienced. Approximately $A2000 million was erased from the rural economic sector as sheep and cattle stocks were decimated by the extreme aridity. Wheat harvests in New South Wales and Victoria were, respectively, 29 per cent and 16 per cent of the average for the previous five years. The resulting lack of capital expenditure in the rural sector deepened what was already a severe recession, and drove farm machinery companies such as International Harvester and Massey Ferguson into near bankruptcy. By January 1983, dust storms in Victoria and New South Wales had blown away thousands of tons of topsoil. Finally, in the last stroke, the Ash Wednesday bushfires of 16 February 1983, now viewed as one of the greatest natural *conflagrations* observed by humans, destroyed another $A1000 million worth of property in South Australia and Victoria. At the peak of the event, Malcolm Fraser's Liberal–National Party coalition government called a national election. The timing could not have been worse, and the government of the day was swept convincingly from power.

Other natural hazards are associated with the Southern Oscillation. This is exemplified in Figure 2.11 for Australia between 1851 and 1980. (The index here is part of the one used in Figure 2.8.) Shaded along this index are the times when eastern Australia was either in drought or experiencing abnormal rainfall. Sixty-eight per cent of the strong or moderate ENSO events between 1851 and 1974 produced major droughts in eastern Australia. Sixty per cent of all recoveries after an ENSO occurrence led to abnormal rainfall and floods. Figure 2.11 also includes the number of tropical cyclones in the Australian region between 1910 and 1980; the discharge into the Murray River of Victoria's central north Campaspe River between 1887 and 1964; beach change at Stanwell Park, south of Sydney, between 1930 and 1980, described in Chapter 8; rainfall this century at Helensburgh, south of Sydney; and the

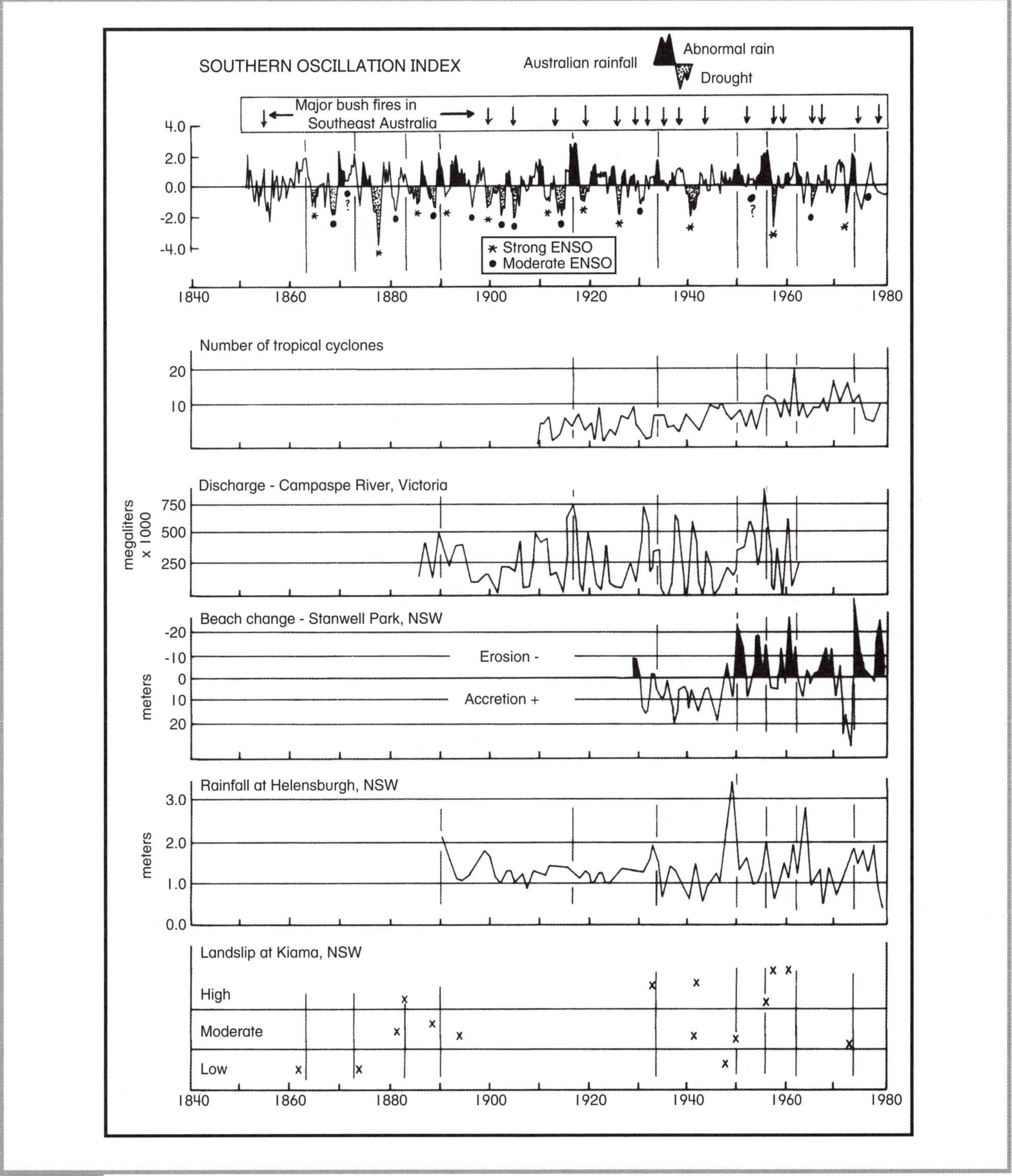

Fig. 2.11 Time series of the Southern Oscillation index, major bushfires (from Luke & McArthur, 1978), Australian cyclone frequency (from Lourensz, 1981), Campaspe River discharge (from Reichl, 1976), beach change at Stanwell Park (from Bryant, 1985), rainfall near Sydney, and landslip incidence at Kiama (courtesy Jane Cook, School of Education, University of Wollongong).

incidence of landslip in the Kiama area, south of Sydney, between 1864 and 1978. Drawn on the diagram are lines passing through all data at the times when the Southern Oscillation was strongly 'turned on'. At this time, tropical cyclones were more frequent, the Campaspe River discharge high, Sydney rainfall heavy, beach erosion at Stanwell Park more severe, and Kiama landslip more frequent. The diagram indicates that 'turned on' Walker circulation results in this predictable geomorphic response. The 1933, 1951 and 1961 periods of strong

Walker circulation show up in all five time series at both regional and localized levels. The reverse conditions apply during ENSO events, except for the frequency of tropical cyclones in the Australian region; historically this is not well correlated to ENSO events. This is owing to the fact that many cyclones are still counted within the Australian region even though they have shifted eastward during these events. More importantly, the Southern Oscillation appears to switch months in advance of the arrival of the meteorological conditions driving these responses. For instance, the heavy rainfall in eastern Australia, beginning in the autumn of 1987, was predicted 12 months in advance. Unfortunately, while the timing of rain can be forecast, its precise location or the amounts that may fall remain unpredictable because of the spatial variability of precipitation along the coast. Even during the worst of droughts, some areas of eastern Australia received normal rainfall.

Links to other hazards are also appearing globally as researchers rapidly investigate teleconnections between hazards and the Southern Oscillation. For example, droughts are often followed by extensive fires in Florida, eastern Russia, and Indonesia. There is also heavier precipitation, mainly falling as snow in north-eastern United States and south-eastern Canada. The 1997–1998 ENSO event produced five days of *freezing rain* in this region that downed 120 000 km of power and telephone lines, paralyzing the city of Montreal. The duration and incidence of tropical cyclones in the equatorial Atlantic is profoundly depressed during ENSO years (Figure 2.12). Here, upper tropospheric winds between 0° and 15°N must be easterly, while those between 20° and 30°N must be westerly, before easterly wave depressions and disturbances develop into tropical cyclones (hurricanes). During ENSO events, upper westerlies tend to dominate over the Caribbean and the western tropical Atlantic, giving conditions that suppress cyclone formation. Years leading up to an ENSO event have tended to produce the lowest number of tropical cyclone days in the Atlantic over the past century. Additionally, since 1955, the number of icebergs passing south of 48°N has been highly correlated to the occurrence of ENSO events. ENSO events occur concomitantly with a strengthened Icelandic Low that results in stronger winds along the North American east coast. This exacerbates the production and southward movement of icebergs. Both the 1972–1973 and 1982–1983 ENSO events produced over 1500 icebergs south of 48°N, a number that dramatically contrasts with non-ENSO years when fewer than one hundred icebergs per year were recorded this far south. Early detection of the onset of ENSO events can thus provide advance warning not only of drought and rainfall over a significantly large part of the globe, but also of the occurrence of numerous other associated hazards.

Other Oscillation phenomena

(Trenberth & Hurrell, 1994; Hurrell, 1995; Villwock, 1998; Stephenson, 1999; Boberg & Lundstedt, 2002; Hurrell et al., 2002)

North Atlantic Oscillation

There are two other regions: the north Atlantic and Pacific Oceans where pressure oscillates on the same

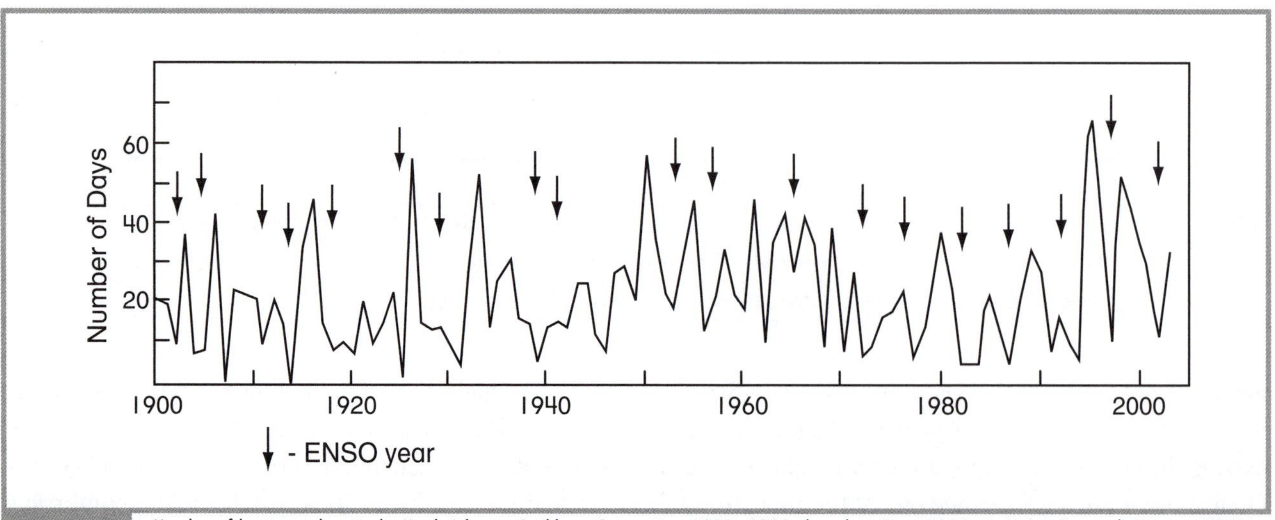

Fig. 2.12 Number of hurricane days in the North Atlantic–Caribbean Sea region, 1900–2003 (based on Gray, 1984; Hurricane Research Division, 2003).

scale as the Southern Oscillation. Variations in the intensity of atmospheric circulation at these locations are known as *North Atlantic Oscillation* (NAO) and the *North Pacific Oscillation* (NPO). Whereas the Southern Oscillation characterizes zonal pressure fluctuations in the tropics across the Pacific, the NAO and NPO are meridional air circulation phenomena reflecting inter-annual variability in the strength of Rossby waves in the northern hemisphere. The North Atlantic Oscillation also reflects major temperature differences in winter. An index of the NAO since 1864 has been constructed using the normalized atmospheric pressure between the Icelandic Low, measured at Stykkisholmur/Reykjavik, Iceland, and the Azores High, measured at Lisbon (Figure 2.8). Positive values of the index indicate stronger-than-average westerlies. Air temperature fluctuations are best characterized by the winter temperature between Jakobshavn, West Greenland, and Oslo, Norway. Significant environmental changes throughout the twentieth century have been associated with the NAO. These include variations in wind strength and direction, ocean circulation, sea surface temperatures, precipitation, sea-ice extent, and changes in marine and freshwater ecosystems. The North Atlantic Oscillation influences an area from Siberia to the eastern seaboard of the United States. Given the magnitude and extent of the NAO, it is surprising that little attention was given to it until the 1990s.

The NAO undergoes large variation from year to year and, more importantly, at decadal time scales. However, large changes can also occur during a winter season. Hence it is difficult to characterize northern hemisphere winters as completely severe or benign. This probably reflects the fact that climate processes are random over time. Large amplitude anomalies in wintertime stratospheric winds precede anomalous behavior of the NAO by one to two weeks. Changes to the strength and pathway of mobile polar highs and the amplitude of Rossby waves in the polar jet stream also correspond to winter anomalies. Similar to the Southern Oscillation, short-term alterations are also preceded by changes in sea surface temperature about nine months in advance. Together the NAO and SO account for 50 per cent (34 per cent and 16 per cent, respectively) of the variation in winter temperatures in the northern hemisphere outside the tropics. However, the Southern and North Atlantic Oscillations act independently of each other. The eastern United States and adjacent Atlantic region are the only areas where both phenomena always act together.

There are two distinct decadal phases associated with the NAO. When the index is positive, atmospheric pressure is low around Iceland and high in the Azores. Sea surface temperatures off the United States eastern seaboard and western Europe are warmer than normal, while those in the western north Atlantic are cooler than average (Figure 2.13a). These aspects enhance meridional air circulation in the atmosphere, strengthening westerly winds at temperate latitudes and forcing warm, moist air from the Atlantic Ocean over northern Europe. Hence land temperatures are warmer adjacent to these pools of warmer ocean waters, leading to milder winters. Temperatures are cool over North Africa, the Middle East, eastern Canada, and Greenland. The flux of heat across the north Atlantic produces more intense and frequent storms, a fact that has been more obvious since 1970 (Figure 2.14). As a result, wave heights increase in the north-east Atlantic and decrease south of 40°N latitude. This is paralleled by changes in precipitation – wetter over Scandinavia and drier in the Mediterranean. Over northern Africa, more dust moves from the Sahara into the Atlantic Ocean. Sea-ice also extends further south over the west Atlantic, while retreating east of Greenland. These effects are either wind- or temperature-driven. In the negative phase, the pressure gradient between Iceland and the Azores is weakened, the Mediterranean region is wetter and warmer, and northern Europe is cooler and drier (Figure 2.13b). While storms off the east coast of the United States may be more frequent, they are virtually absent off western Europe. These periods of anomalous circulation patterns have persisted over long periods. For example, the NAO index was positive from the turn of the twentieth century until about the 1930s. During the 1960s, the NAO index was negative with severe winters across northern Europe. Since the early 1970s, circulation has been locked into a positive phase with warmer temperatures than average over Europe and cooler ones over eastern Canada and Greenland.

North Pacific Oscillation

The North Pacific Oscillation (NPO) measures the strength and position of the Aleutian low-pressure system, mainly in winter. Because there are few recording stations in the north Pacific, an index

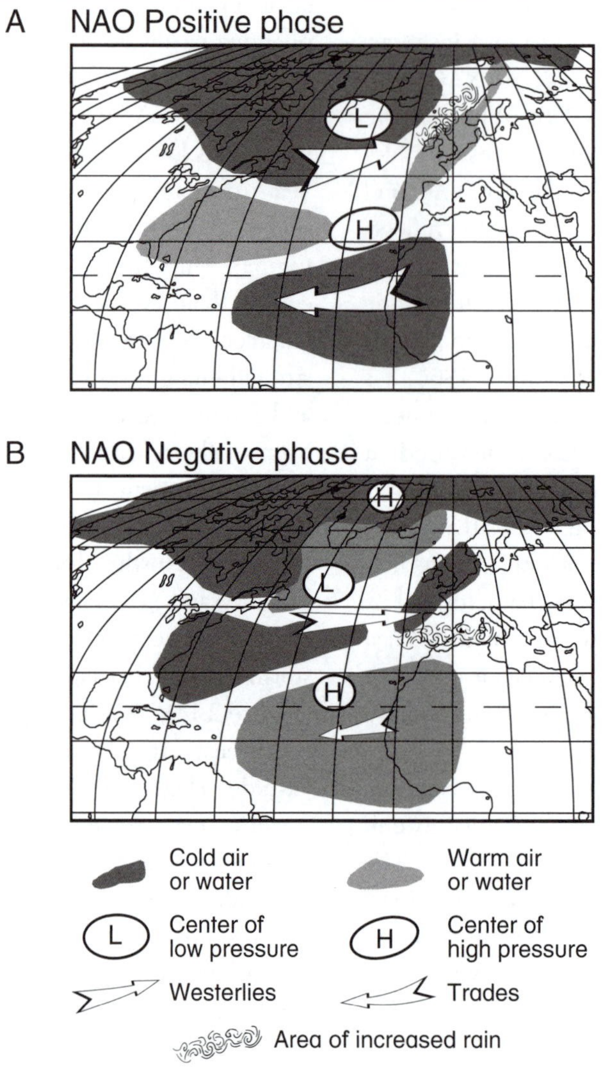

Fig. 2.13 Atmospheric circulation patterns and climatic impacts associated with A) positive and B) negative phases of the North Atlantic Oscillation (based on Stephenson, 1999).

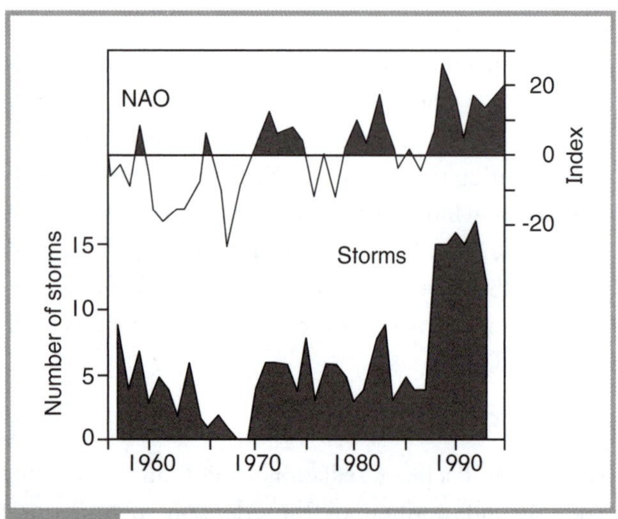

Fig. 2.14 The number of winter storms in the north Atlantic between 1956 and 1994 with pressure below 950 hPa (Villwock, 1998).

characterizing this oscillation has been built using average surface air pressure over the region 30°N–65°N, 160°E–140°W (Figure 2.8). Over the north Pacific Ocean, atmospheric pressure near the Aleutian Islands varies out-of-phase with that to the south, seesawing around the mean position of the Pacific subtropical jet stream. Over North America, variations in atmospheric pressure over western Canada and the north-western United States are negatively correlated with those over the south-eastern United States; but positively correlated with the subtropical Pacific. Changes in these centers occur concomitantly with changes in the amplitude of Rossby waves and the intensity of mobile polar highs. The NPO is linked to changes in tropical Pacific sea surface temperatures associated with the ENSO phenomenon.

In this respect, the NPO is probably the temperate north Pacific extension of the Southern Oscillation, with sea surface temperature changes in the north Pacific lagging those in the tropics by three months. However, the NPO does have its own independent features. Of all atmospheric indices, the North Pacific Oscillation shows the greatest well-defined trend over time (Figure 2.8). The index has been steadily declining, although there is significant year-to-year variation. Concomitantly with this trend, the Aleutian Low has intensified and shifted eastward in winter. As a result, storm tracks have shifted southward. This has resulted in warmer and moister air being transported northward along the west coast of North America into Alaska. Sea surface temperatures in the north Pacific have cooled, but sea-ice has decreased in the Bering Sea.

ASTRONOMICAL CYCLES

Solar cycles

(Shove, 1987; Hoyt & Schatten, 1997; Boberg & Lundstedt, 2002)

Sunspots are surface regions of intense disturbance of the sun's magnetic field: 100–1000 times the sun's average. The spots themselves appear dark, because the magnetic field is so strong that convection is inhibited and the spot cools by radiation emission. However, the total amount of radiation given out by the sun increases towards peaks in sunspot activity. The sun has a 22-year magnetic cycle, consisting of two 11-year sunspot cycles. In Figure 2.15, these are plotted back to 1500 AD. The 22-year periodicity known as the *Hale Cycle* is evident as an enhanced peak in sunspot numbers in this figure. There are also other periodicities, at 80–90 years and 180 years, in the number of sunspots. The interaction of these cycles has produced periods where there was little sunspot activity. Two of these, the Maunder and Dalton Minimums, in the late seventeenth and early nineteenth centuries, respectively, are evident in Figure 2.15. Both are correlated to periods of cooler climate, at least in the northern hemisphere. The *Maunder Minimum* is also known as the *Little Ice Age*. Periods of high activity correlate with warmer climate. This occurred in the thirteenth century during the Medieval Maximum and over the last 150 years concurrent with modern global warming.

Other phenomena characterize our sun. *Solar flares* – representing ejection of predominantly ionized hydrogen in the sun's atmosphere at speeds in excess of 1500 km s^{-1} – develop in sunspot regions. Flares enhance the solar wind that ordinarily consists of electrically neutral, ionized hydrogen. Accompanying any solar flare is a pulse of electro-magnetic radiation that takes eight minutes to reach the Earth. This radiation, in the form of soft X-rays (0.2–1.0 nm wavelengths), interacts with the Earth's magnetic field, increasing ionization in the lowest layer of the ionosphere at altitudes of 65 km. The enhanced solar wind arrives one to two days after this magnetic pulse and distorts the magnetosphere, resulting in large, irregular, rapid worldwide disturbances in the Earth's geomagnetic field. Enhanced magnetic and ionic currents during these periods heat and expand the upper atmosphere. The solar wind also modulates cosmic rays affecting the Earth. When the solar wind is strong, cosmic rays are weak – resulting in decreased cloud formation and warmer surface air temperatures. Cloud cover decreases around 3–4 per cent between troughs and peaks in *geomagnetic activity*. Solar flare activity and the strength of the solar wind are weakly correlated to the number of sunspots (Figure 2.15b). Thus, geomagnetic activity is a better indicator of solar influence on climate than the number of sunspots. While solar activity on a timescale of weeks can affect the temperature structure of the stratosphere, there is no proven mechanism linking it to climatic change over longer periods near the surface of the Earth. This fact has tended to weaken the scientific credibility of research into the climatic effects of solar cycles.

Despite this, the literature is replete with examples showing a correlation between climate phenomena and solar activity in the form of sunspots. Much of the emphasis has been upon a solar sunspot–global temperature association. High sunspot activity leads to warmer temperatures and more precipitation, although this relationship is consistent neither over time nor the surface of the globe. This discussion is beyond the scope of this text and the curious reader should see Hoyt and Schatten (1997) for further information. More substantial is the fact that thunderstorm, lightning and tropical cyclone activity worldwide increases during periods of sunspot activity. In addition, 15 per cent of the variation in the position of storm tracks in the north Atlantic and Baltic Sea can be

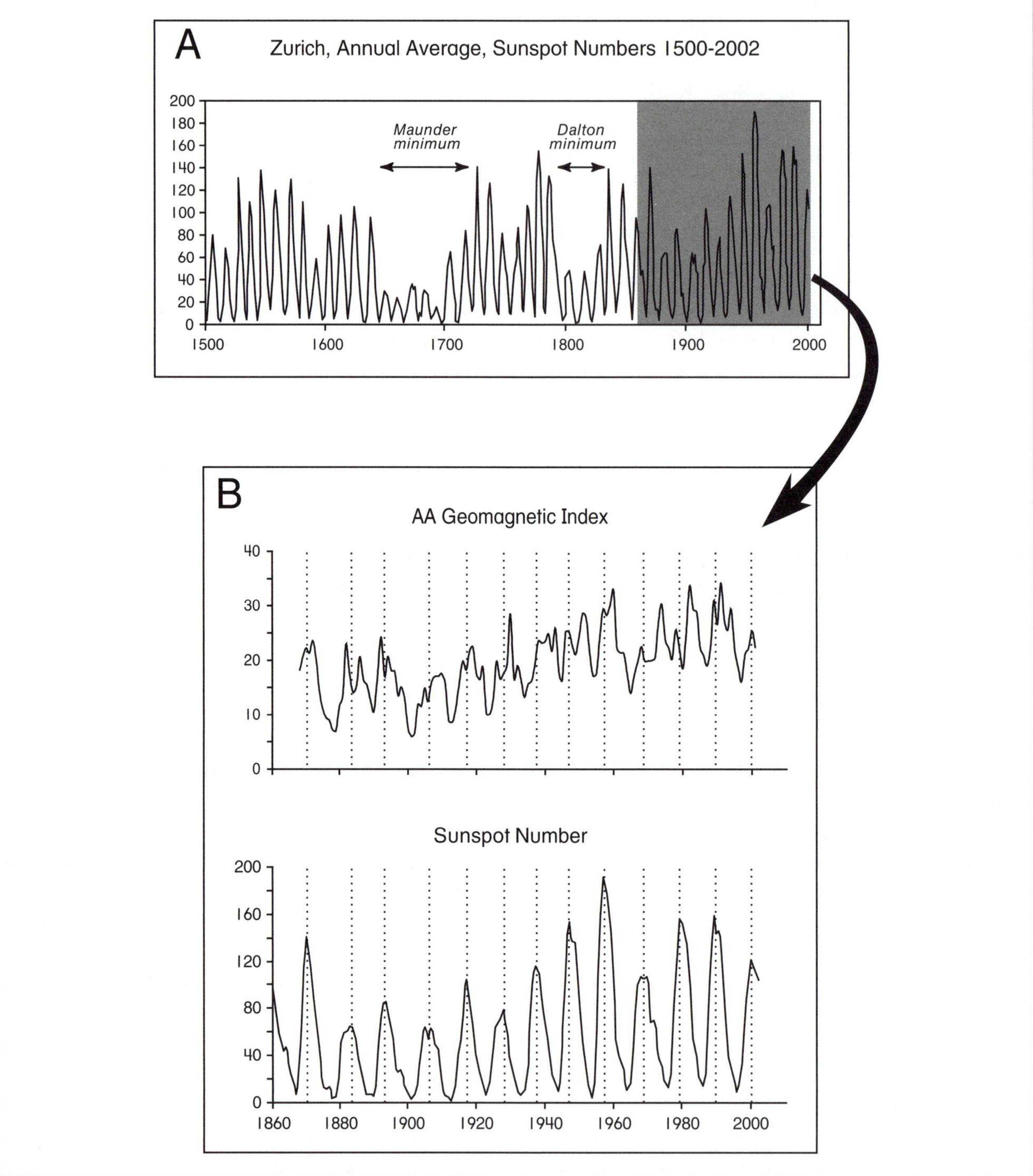

Fig. 2.15 A) Time series of the annual average, Zurich sunspot number between the years 1500 and 2002. Values between 1500 and 1749 are from Shove (1987), between 1750 and 2002 from the National Geophysical Data Centre (2003a). B) Detail of the Zurich sunspot record between 1860 and 2002 with the annual AA geomagnetic index superimposed. Latter from National Geophysical Data Centre (2003b).

accounted for by solar variations. Storm tracks shift southward by 3–4° latitude in both regions when sunspot activity is highest. Forest fires are more numerous in North America at peaks in the sunspot cycle. The decadal variance in the North Atlantic Oscillation is also positively correlated to enhanced geomagnetic activity. Finally, there are solar periodicities in the frequency of earthquakes and volcanic eruptions. The reason for the latter is beyond the scope of this text.

The 18.6-year M_N lunar cycle

(Tyson et al., 1975; Currie, 1981, 1984; Wang & Zhao, 1981)

The 18.6-year lunar tidal cycle (M_N) represents a fluctuation in the orbit of the moon. The moon's axis of orbit forms a 5° angle with the sun's equator as shown in Figure 2.16; however, with each orbit the moon does not return to the same location relative to the sun. Instead, it moves a bit further in its orbit. The process is analogous to a plate spinning on a table. The plate may always be spinning at the same angle relative to the table, but the high point of the plate does not occur at the same point. Instead, it moves around in the direction of the spin. The moon's orbit does the same thing, such that 9.3 years later the high point of the orbit is at the opposite end of the solar equator, and 9.3 years after that it returns to its original position. Thus, there is an 18.6-year perturbation in the orbit of the moon. This perturbation appears trivial until you consider the moon's orbit in relation to the Earth (Figure 2.17). Relative to the Earth, the solar equator moves seasonally, reaching a maximum of 23.46°N of the equator on 22 June and a minimum of 23.46°S of the Earth's equator on 22 December. If the moon–sun orbital configuration in Figure 2.16a is superimposed onto the Earth, then the moon is displaced in its orbit closest to the Earth's poles, 28.5° (23.5° + 5°) north and south of the equator. The moon's gravitational attraction on the Earth is slightly greater when the moon is displaced poleward. In 9.3 years' time, the moon will move to the opposite side of the sun's equator. Relative to the Earth, the moon will be closer to the equator, or in its minimum position (Figure 2.17b). Although it is only 3.7 per cent the daily effect, the gravitational effect of the M_N tide has two to three years to act on any standing, planetary wave phenomena.

Standing atmospheric waves, such as Rossby waves embedded in the jet stream (Figure 2.4), are amplified by the M_N lunar tide. An 18–20 year periodicity appears in sea level, air pressure, and temperature records at locations beneath the jet stream from Japan to Scandinavia. Detailed examination of the occurrence of drought on the United States Great Plains since 1800 shows a strong link between drought and the 18.6-year M_N lunar cycle. A 20-year cycle is also evident in rainfall over the Yangtze River Basin of China, coinciding with the American pattern and the Indian flood record. In the southern hemisphere, a 20-year periodicity appears in summer rainfall in

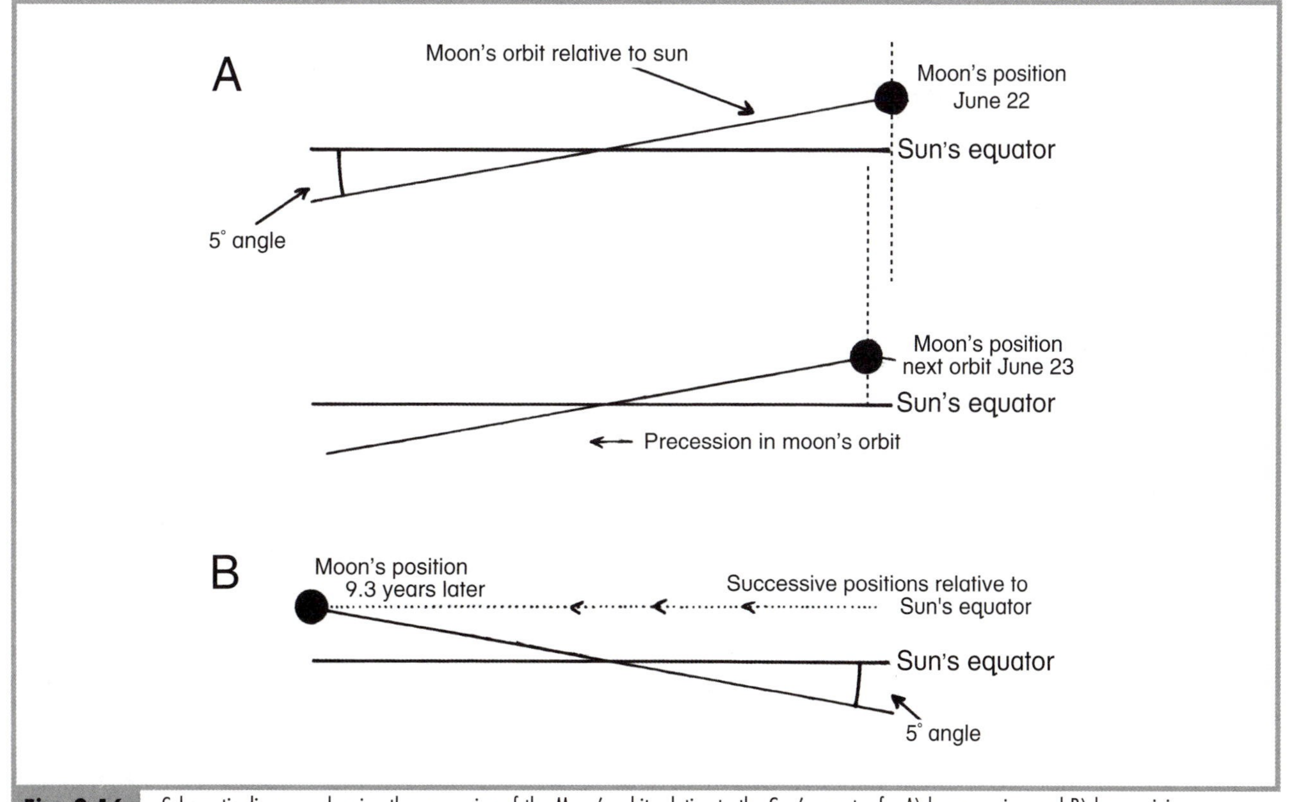

Fig. 2.16 Schematic diagrams showing the precession of the Moon's orbit relative to the Sun's equator for A) lunar maxima and B) lunar minima.

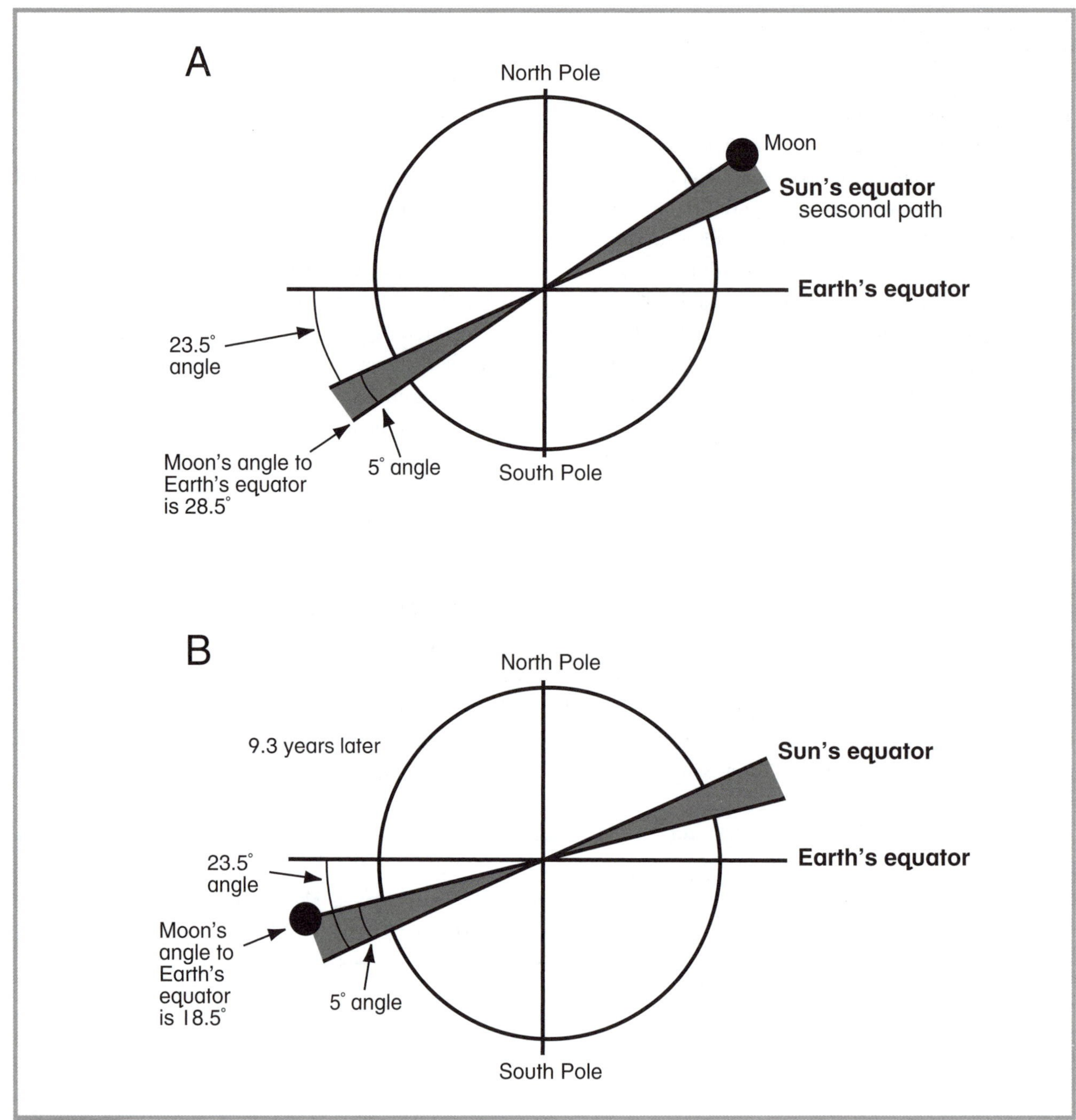

Fig. 2.17 Schematic presentations of the 18.6-year lunar orbit relative to the Earth for A) lunar maxima and B) lunar minima.

southern Africa and eastern Australia. A sampling of these associations worldwide is presented for the western United States and Canada, northern China, India, the Nile region of Africa, and the mid-latitudes of South America (Table 2.2). In all cases, except for India, the mean discrepancy between the occurrence of the hazard and peaks in the lunar tide is less than one year. Not only is the 18.6-year cycle dominant, but the data also show temporal and spatial *bistable phasing*. In this process, drought may coincide with maxima in the 18.6-year lunar cycle for one period, but then suddenly switch to a minimum. This 'flip-flop' happens every 100–300 years, with the most recent occurring at the turn of the twentieth century in South America, China, Africa, and India. On the Great Plains, no bistable 'flip-flop' has occurred since 1657; however, the last lunar cycle maximum in 1991 witnessed flooding instead of drought in this region. At present, disparate regions of the globe's major wheat growing areas undergo synchronous periods of drought that have major implications for the price of grain and the relief of famine-stricken regions.

Table 2.2 Timing of flood (F) and drought (D) in North America, northern China, Patagonia, the Nile valley, India and the coincidence with the 18.6 year lunar tide

Lunar Maxima	Canadian Prairies	United States Great Plains	N. China	Patagonian Andes	Nile in Africa	India
1583	1581D		1582D			
1601	1601D		1600D	1606F		
1620	1622D		1620D	1621F		
1638	1640D		1640D	1637F		
1657	1652F		1659D	1656F		
1676	1674F		1678D	1675F		
1694	1694F			1693F		
1713	1712F			1710F	1713D	
1731	1729F			1727F	1733D	
1750	1752F			1752D	1750D	
1768	1768F			1768D	1766D	
1787	1786F			1784D	1789D	
1806	1805F	1805D	1806F	1802D	1806D	
1824	1823F	1824D	1823F	1822D	1822D	
1843	1843F	1844D	1846D	1841D	1837D	
1861	1859F	1861D	1862D	1864D		
1880	1881F	1879D	1881D	1880D	1885D	
1899	1900F	1901D	1900D	1895D	1902D	1902D
1918	1916F	1919D	1918D	1918F	1917F	1913F
1936	1932F	1935D	1935F	1937F	1936F	1939F
1955	1953F	1955D	1954F	1956F	1953F	1958F
1973	1975F	1975D	1974F	1973F	1975F	1976F
1991	1995F	1993F	1990F	1992F	1996F	1990F

Sources: Currie (1984) and various internet searches. Since 1980, plotted rainfall anomalies for each region using http://climexp.knmi.nl/fieldplot.cgi?someone@somewhere+gpcp

CONCLUDING COMMENTS

The dominant factor controlling air movement across the surface of the Earth is the movement of cold polar air in the form of mobile polar highs from the poles towards the equator. This drives global air circulation and satisfies the requirement in the atmosphere to balance the inequality in heating and cooling between the equator and poles, respectively. The concept of mobile polar highs implies that the climate of high latitudes is the key to world climate and global climate change. Superimposed upon this meridional exchange are zonal processes controlled by ocean–atmosphere interactions. Chief amongst these is the Southern Oscillation. While droughts and floods have plagued people throughout recorded history, and are probably responsible for the greatest loss of life of any natural hazards, the realization is that they represent inseparable hazards that follow each other in time as night does day. Not only are the two events linked, but their occurrence is also remarkably coincident across the globe. They are intricately connected to the Southern Oscillation and other spatial cyclic phenomena such as the North Atlantic and North Pacific Oscillations. Other hazards – such as tropical and extra-tropical cyclones, wave erosion, and land instability – are also correlated to the Southern Oscillation, and thus subject to prediction. At mid-latitudes in the northern hemisphere, storminess is more important. Here, the North Atlantic Oscillation is the controlling factor. Similar influences may occur in western North America due to changes in sea surface temperature in the north Pacific.

There is now indisputable proof that droughts and subsequent heavy rainfall periods are cyclic, not only in semi-arid parts of the globe, but also in the temperate latitudes. This cyclicity correlates well with the 18.6-year lunar tide in such diverse regions as North America, Argentina, northern China, the Nile Valley, and India. There is also evidence that the 11-year sunspot cycle is present in rainfall time series in some countries. Long-term astronomical periodicities, especially in the northern hemisphere, permit scientists to pinpoint the most likely year that drought or abnormal rainfall will occur, while the onset of an ENSO event with its global consequences permits us to predict climatic sequences up to nine months in advance. These two prognostic indicators should give countries time to modify both long- and short-term economic strategies to negate the effects of drought or heavy rainfall.

Unfortunately, very few governments in the twentieth century have lasted longer than 11 years – the length of the shortest astronomical cycle. Realistically, positive political responses – even in the United States – to dire drought warnings are probably impossible. Societies must rely upon the efficiency and power of their permanent civil services to broadcast the warnings, to prepare the programs reducing the effects of the hazards, and to pressure governments of the day to make adequate preparations.

Finally, it should be realized that the astronomical cycles and oscillations such as the Southern or North Atlantic Oscillations account for only 15–30 per cent of the variance in rainfall in countries where their effects have been defined. This means that 70 per cent or more of the variance in rainfall records must be due to other climatic factors. For example, during the drought that affected eastern Australia in 1986–1987, Sydney was inundated with over 400 mm of rainfall, breaking the previous 48-hour rainfall record, and generating the largest storm waves to affect the coast in eight years. This randomness exemplifies the importance of understanding regional and local climatic processes giving rise to climatic hazards. These regional and local aspects will be described in the following chapters.

REFERENCES AND FURTHER READING

Adamson, D., Williams, M.A.J. and Baxter, J.T. 1987. Complex late Quaternary alluvial history in the Nile, Murray–Darling, and Ganges Basins: three river systems presently linked to the Southern Oscillation. In Gardiner, V. (ed.) *International Geomorphology* Pt II, Wiley, NY, pp. 875–887.

Allan, R., Lindesay, J. and Parker, D. 1996. *El Niño Southern Oscillation and Climatic Variability*. CSIRO Publishing, Melbourne.

Bhalme, H.N., Mooley, D.A. and Jadhav, S.K. 1983. Fluctuations in the drought/flood area over India and relationships with the Southern Oscillation. *Monthly Weather Review* 111: 86–94.

Boberg, F. and Lundstedt, H. 2002. Solar wind variations related to fluctuations of the North Atlantic oscillation. *Geophysical Research Letter* 29(15) 10.1029/2002GL014903.

Bryant, E.A. 1985. Rainfall and beach erosion relationships, Stanwell Park, Australia, 1895–1980: worldwide implications for coastal erosion. *Zeitschrift für Geomorphologie Supplementband* 57: 51–66.

Bryant, E.A. 1997. *Climate Process and Change*. Cambridge University Press, Cambridge.

Bryson, R. and Murray, T. 1977. *Climates of Hunger*. Australian National University Press, Canberra.

Couper-Johnston, R. 2000. *El Nino: The Weather Phenomenon That Changed the World*. Hodder and Stoughton, London.

Currie, R.G. 1981. Evidence of 18.6 year M_N signal in temperature and drought conditions in N. America since 1800 A.D. *Journal Geophysical Research* 86: 11055–11064.

Currie, R.G. 1984. Periodic (18.6-year) and cyclic (11-year) induced drought and flood in western North America. *Journal Geophysical Research* 89(D5): 7215–7230.

Glantz, M.H. 1996. *Currents of Change: El Niño's Impact on Climate and Society*. Cambridge University Press, Cambridge.

Glantz, M.H., Katz, R.W. and Nicholls, N. 1991. *Teleconnections Linking Worldwide Climate Anomalies: Scientific Basis and Societal Impact*. Cambridge University Press, Cambridge.

Gray, W.M. 1984. Atlantic seasonal hurricane frequency Part I: El Niño and 30 mb quasibiennial oscillation influences. *Monthly Weather Review* 112: 1649–1667.

Hoyt, D.V. and Schatten, K.H. 1997. *The Role of the Sun in Climate Change*. Oxford University Press, Oxford.

Hurrell, J.W. 1995. Decadal trends in the North Atlantic Oscillation: Regional temperatures and precipitation. *Science* 269: 676–679.

Hurrell, J.W. 2002a. *North Atlantic Oscillation (NAO) indices information: Winter (Dec–Mar)*. <http://www.cgd.ucar.edu/~jhurrell/nao.stat.winter.html>

Hurrell, J.W. 2002b. *North Pacific (NP) index information*. <http://www.cgd.ucar.edu/~jhurrell/np.html>

Hurrell, J.W., Kushnir, Y., Ottersen, G. and Visbeck, M. (eds) 2002. *The North Atlantic Oscillation: Climatic Significance and Environmental Impact*. American Geophysical Union, Washington.

Hurricane Research Division 2003. *What are the most and least tropical cyclones occurring in the Atlantic basin and striking the USA?*. United States Department of Commerce. <http://www.aoml.noaa.gov/hrd/tcfaq/tcfaqE.html>

Jet Propulsion Laboratory 1995a. *TOPEX/POSEDON: Wind Speed, January*. <http://education.gsfc.nasa.gov/experimental/all98invProject.Site/Pages/trl/inv4-7WIND_SPEED_JAN.html>

Jet Propulsion Laboratory 1995b. *TOPEX/POSEDON: Wind Speed, July*. <http://education.gsfc.nasa.gov/experimental/all98invProject.Site/Pages/trl/inv4-7WIND_SPEED_JUL.html>

Lamb, H.H. 1982. *Climate, History and the Modern World*. Methuen, London.

Lamb, H.H. 1986. The causes of drought with particular reference to the Sahel. *Progress in Physical Geography* 10(1): 111–119.

Leroux, M. 1993. The Mobile Polar High: a new concept explaining present mechanisms of meridional air-mass and *energy* exchanges and global propagation of palaeoclimatic changes. *Global and Planetary Change* 7: 69–93.

Leroux, M. 1998. *Dynamic Analysis of Weather and Climate: Atmospheric Circulation, Perturbations, Climatic Evolution*. Wiley-Praxis, Chichester.

Lourensz, R.S. 1981. *Tropical Cyclones in the Australian Region July 1909 to June 1980*. Australian Bureau of Meteorology, Australian Government Publishing Service, Canberra.

Luke, R.H. and McArthur, A.G. 1978. *Bushfires in Australia*. Australian Government Publishing Service, Canberra.

Marko, J.R., Fissel, D.B. and Miller, J.D. 1988. Iceberg movement prediction off the Canadian east coast. In El-Sabh, M.I. and Murty, T.S. (eds) *Natural and Man-made Hazards*. Reidel, Dordrecht, pp. 435–462.

National Geophysical Data Center 2003a. *Sunspot numbers*. <ftp://ftp.ngdc.noaa.gov/STP/SOLAR_DATA/SUNSPOT_NUMBERS>

National Geophysical Data Center 2003b. *AA Index*. <ftp://ftp.ngdc.noaa.gov/STP/SOLAR_DATA/RELATED_INDICES/AA_INDEX/>

Pant, G.B. and Parthasarathy, B. 1981. Some aspects of an association between the Southern Oscillation and Indian summer monsoon. *Archiv für Meteorologie, Geophysik und Bioklimatologie*. B29: 245–252.

Quinn, W.H., Zopf, D.O., Short, K.S. and Kuo Yang, R.T.W. 1978. Historical trends and statistics of the southern oscillation, El Niño, and Indonesian droughts. *U.S. Fisheries Bulletin* 76: 663–678.

Philander, S.G. 1990. *El Niño, La Niña and the Southern Oscillation*. Academic, San Diego.

Pittock, A.B. 1984. On the reality, stability and usefulness of southern hemisphere teleconnections. *Australian Meteorological Magazine* 32(2): 75–82.

Reichl, P. 1976. The Riverine Plains of northern Victoria. In Holmes, J.H. (ed.) *Man and the Environment: Regional Perspectives*. Longman, Hawthorn, pp. 69–95.

Shove, D.J. 1987. Sunspot cycles. In Oliver, J.E. and Fairbridge, R.W. (eds) *Encyclopedia of Climatology*. Van Nostrand Reinhold, New York, pp. 807–815.

Stephenson, D.B. 1999. *The North Atlantic Oscillation thematic web site*. <http://www.met.rdg.ac.uk/cag/NAO/>

Trenberth, K.E. and Hurrell, J.W. 1994. Decadal atmospheric–ocean variations in the Pacific. *Climate Dynamics* 9: 303–319.

Tyson, P.D., Dyer, T.G.S., and Mametse, M.N. 1975. Secular changes in South African rainfall 1880–1972. *Quarterly Journal of the Royal Meteorological Society* 101: 817–833.

Villwock, A. 1998. *CLIVAR initial implementation plan: Chapter 5, The North Atlantic Oscillation*. <http://www.clivar.org/publications/other_pubs/iplan/iip/pd1.htm>

Wang, S-W. and Zhao, A-C. 1981. Droughts and floods in China 1440–1979. In Wigley, T.M.L., Ingram, M.J. and Farmer, G. (eds) *Climate and History, Studies in Past Climates and Their Impact on Man*. Cambridge University Press, Cambridge, pp. 271–288.

Large-scale Storms as a Hazard

CHAPTER 3

INTRODUCTION

The previous chapter concentrated upon large-scale pressure patterns and their changes across the surface of the Earth. While these patterns regionally control precipitation, large-scale vortices amplify the amounts and add a wind component to produce devastating storms. In addition, moderate-to-strong winds on a regional scale have the capacity to suspend large quantities of dust. This material can be transported thousands of kilometers, a process that represents a significant mechanism for the transport of sediment across the surface of the globe. Dust storms are exacerbated by drought in low rainfall, sparsely vegetated regions of the world. For that reason, they pose a slowly developing but significant hazard in sparsely settled semi-arid regions. Unfortunately, many of these areas have been marginalized by the degrading agricultural practices of humans over thousands of years. However, in some regions where settlement has occurred only in the last 200 years, we are presently witnessing this marginalization process. Dust storms are one of the most prominent and long-lasting signatures of human impact on the landscape.

In this chapter, the process, magnitude, and frequency of *tropical cyclones* are described first, followed by a description of some of the more familiar cyclone disasters. This section concludes with a comparison of the human response to cyclones in Australia, the United States, and Bangladesh – formerly East Pakistan. Extra-tropical cyclones are then described with particular reference to major storms in the northern hemisphere and exceptional development of lows or *bombs* off Japan, the United States, and the east coast of Australia. Storm surges driven over bodies of water by these weather phenomena are then described. It is appropriate at this stage to introduce more formally the concepts of probability of occurrence and exceedence using storm surges as an example. These concepts are central to evaluating just how big and frequent is a particular hazard event. Finally, the chapter concludes with a description of dust storms and their impact.

TROPICAL CYCLONES

Introduction (Nalivkin, 1983)

Tropical cyclones are defined as intense cyclonic storms that originate over warm tropical seas. In North America, the term 'hurricane' is used because cyclone refers to an intense, counterclockwise-rotating, extra-tropical storm. In Japan and South-East Asia, tropical cyclones are called 'typhoons'. Figure 3.1 summarizes the hazards relating to tropical cyclones. These can be grouped under three headings: storm surge, wind, and rain effects. Storm surge is a phenomenon whereby water is physically piled up along a coastline by low pressure and strong winds. This leads to loss of life

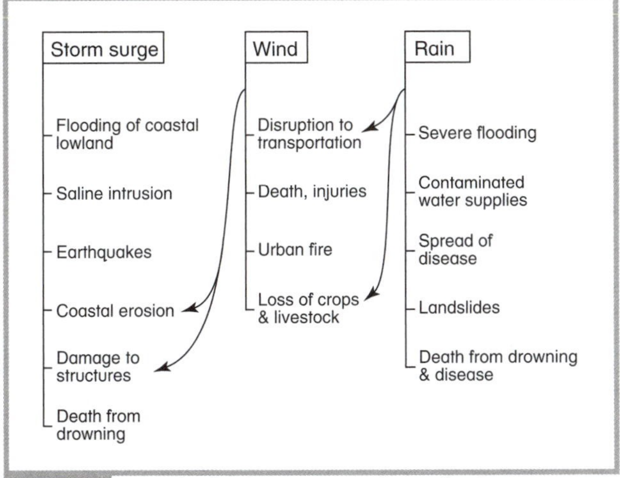

Fig. 3.1 Hazards related to occurrence of tropical cyclones.

through drowning, inundation of low-lying coastal areas, erosion of coastline, loss of soil fertility due to intrusion by ocean saltwater, and damage to buildings and transport networks. In the largest cyclones, winds can exceed 300 km hr^{-1}. High wind velocities can directly cause substantial property damage and loss of life, and constitute the main agent for crop destruction. Surprisingly, strong winds – simply because they are so strong – can also exacerbate the spread of fires in urban and forested areas, even under heavy rainfall.

On average, a tropical cyclone can dump 100 mm per day of rain within 200 km of the eye, and 30–40 mm per day at distances of 200–400 km. These rates can vary tremendously depending upon local topography, cyclone motion, and the availability of moisture. In 1952, a tropical cyclone at Reunion Island dropped 3240 mm of rain in three days. Rainfall is responsible for loss of life, property damage, and crop destruction from flooding – especially on densely populated floodplains. Contamination of water supplies can lead to serious disease outbreaks weeks after the cyclone. Heavy rain in hilly or mountainous areas is also responsible for landslides or mudflows as floodwaters in stream and river channels mix with excess sediment brought down slopes. The destruction of crops and saline intrusion can also result in famine that can kill more people than the actual cyclone event. This was especially true on the Indian subcontinent during the latter part of the nineteenth century.

Earthquakes are not an obvious consequence of cyclones; however, there is substantial evidence for their occurrence during cyclones. Pressure can vary dramatically in a matter of hours with the passage of a cyclone, bringing about a consequentially large decrease in the weight of air above the Earth's surface. The deloading can be as much as 2–3 million tonnes km^{-2} over a matter of hours. In addition, storm surges about 6–7 m in height can occur in shallow seas with a resulting increase in pressure on the Earth's surface of 7 million tonnes km^{-2}. In total, the passage of a cyclone along a coast can induce a change in load on the Earth's crust of 10 million tonnes km^{-2}. In areas where the Earth's *crust* is already under strain, this pressure change may be sufficient to trigger an earthquake. The classic example of a cyclone-induced earthquake occurred with the Tokyo earthquake of 1923 (see Figure 3.2 for the location of major placenames mentioned in this chapter). A typhoon swept through the Tokyo area on 1 September, and an earthquake followed that evening. The earthquake caused the rupture of gas lines, setting off fires that were fanned by cyclone-force winds through the city on 2 September. In all, 143 000 people lost their lives, mainly through incineration. The events of this tragedy will be described in more detail in Chapter 10. There is also evidence that tropical cyclones have triggered earthquakes in other places along the western margin of the Pacific *Plate* and along plate boundaries in the Caribbean Sea. In Central America, the coincidence of earthquakes and cyclones has a higher probability of occurrence than the joint probability of each event separately.

The Tokyo earthquake is also unusual for another reason. The typhoon winds blew the fires out of control. This dichotomous occurrence of torrential rain and uncontrolled fire is quite common in Japan. Typhoons moving north-east in the Sea of Japan produce very strong winds in the lee of mountains. In September 1954 and again in October 1955, cyclone-generated winds spread fires that destroyed over 3300 buildings in Hokkaido and 1100 buildings in Niigata, respectively. Nor is the phenomenon of cyclone–fire unique to Japan. The great hurricane of 1938 that devastated New England, in the United States, started fires in New London, Connecticut, that raged for six hours, and would have destroyed the city of 30 000 people if it were not for the fact that the fires were turned back on themselves as the winds reversed direction with the passage of the cyclone.

Mechanics of cyclone generation

(Anthes, 1982; Nalivkin, 1983; Gray, 1984; Gray et al., 1994; Chang et al., 2003; Simpson, 2003)

Tropical cyclones derive their energy from the evaporation of water over oceans, particularly the western

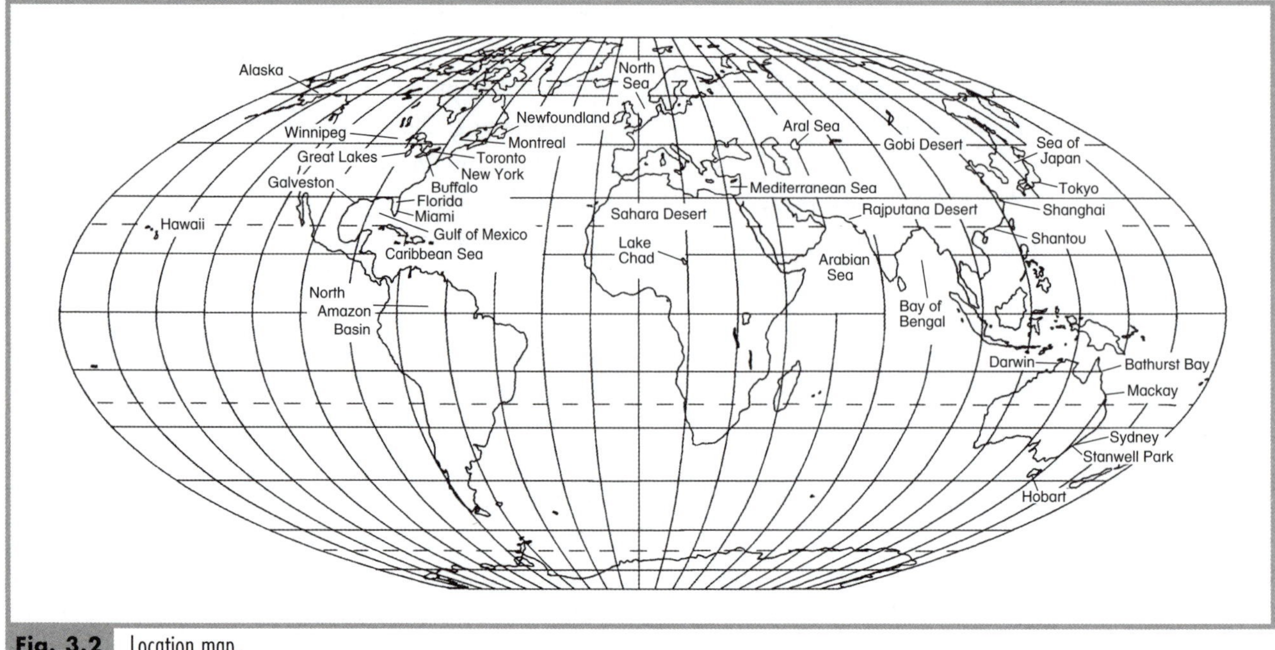

Fig. 3.2 Location map.

parts, where surface temperatures are warmest. As the sun's apparent motion reaches the summer *equinox* in each hemisphere, surface ocean waters begin to warm over a period of two to three months to temperatures in excess of 26°C. Tropical cyclones, thus, develop from December to May in the southern hemisphere and from June to October in the northern hemisphere. At the same time, easterly trades on the equatorial side of agglutinated mobile polar highs intensify and blow these warm surface waters to the westward sides of oceans. Hence, the warmest ocean waters accumulate to a great depth in the Coral Sea off the east coast of Australia, in the Caribbean, and in the west Pacific Ocean, south-east of China. A deep layer of warming reduces the upward mixing of cold sub-surface water that can truncate the cyclone before it fully develops. Warm water also develops in shallow seas because there is a smaller volume of water to heat up. Figure 3.3 presents the origin of tropical cyclones. The majority originate either at the western sides of oceans, or over shallow seas, where temperatures are enhanced by one of the above processes.

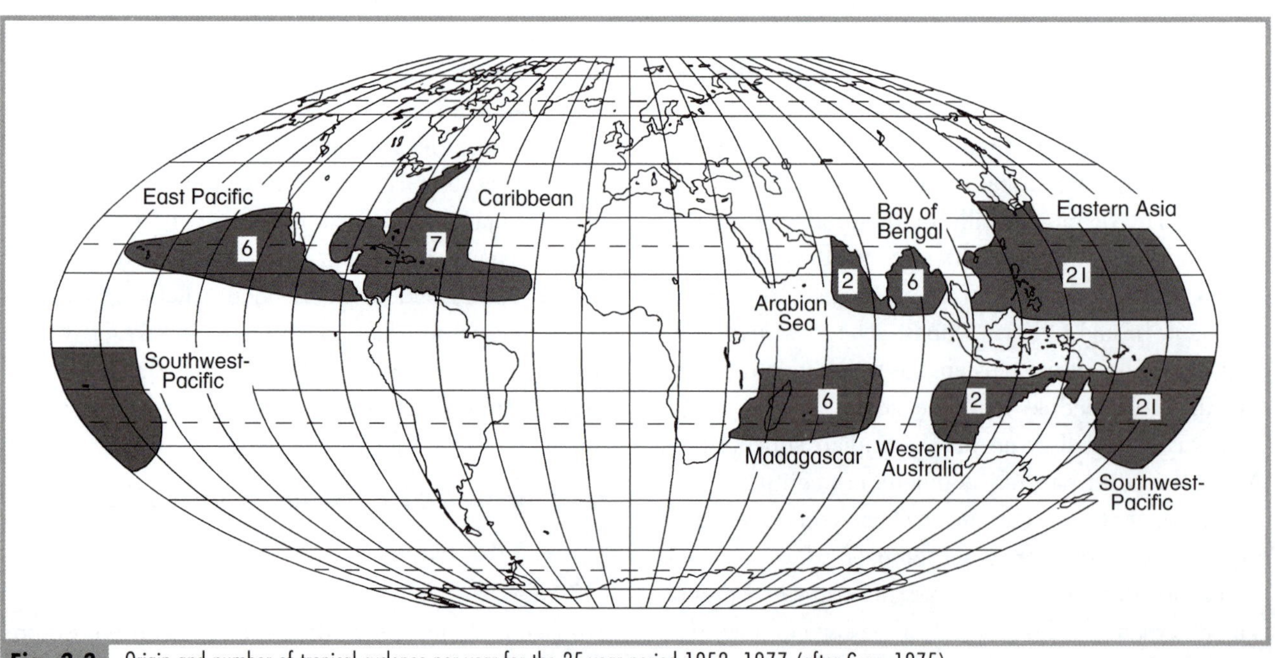

Fig. 3.3 Origin and number of tropical cyclones per year for the 25-year period 1952–1977 (after Gray, 1975).

At least eight conditions are required to develop a tropical cyclone. First, efficient convective instability is required causing near-vertical uplift of air throughout the troposphere. Cyclonic depressions begin to develop when ocean temperatures exceed 26–27°C because the lower atmosphere becomes uniformly unstable at these values. With further temperature increases, the magnitude of a tropical cyclone intensifies. Tropical cyclones tend not to form above 30°C because smaller-scale turbulence begins to dominate with the breakdown of near-vertical air movement. Once developed, cyclones rapidly decay below temperatures of 24°C. Cyclones rarely develop poleward of 20° latitude because ocean temperatures seldom reach these latter crucial values. These relationships have implications for global warming. While the area of ocean favorable for cyclone formation may increase, there is an upper temperature limit constraining growth. Tropical cyclones can also dramatically cool sea surface temperature by 3–4°C because of the enormous amount of heat involved in evaporating ocean water. Hurricane Gilbert cooled the ocean from 31°C to 26°C off the coast of Mexico in 1988.

Second, there must be convergence of surface air into areas of warm water. The formation of numerous thunderstorms over warm waters provides this mechanism. Convergence causes upward movement of air and creates instability in the upper atmosphere. However, in the case of tropical cyclones, this instability is enhanced by the release of latent heat of evaporation. Because of warm ocean temperatures and intense solar heating, evaporation at the sea surface is optimum. Each kilogram of water evaporated at 26°C requires 2425 kilojoules of heat energy. As air rises, it cools, and if water vapor condenses, then this energy is released as latent heat of evaporation. Vast amounts of heat energy are thus transferred to the upper atmosphere causing convective instability that forces air upwards. Figure 3.4 illustrates the type of temperature structure that can develop in a tropical cyclone because of this process. The temperatures shown schematically represent the maximum possible temperature, in degrees Celsius, if all evaporated moisture were condensed. The release of latent heat of condensation in the zone of uplift can raise air temperatures up to 20°C above those at equivalent levels. Convective instability will persist as long as this heating remains over a warm ocean.

Third, a trigger mechanism is required to begin rotation of surface air converging into the uplift zone. This rotation most likely occurs in an *easterly wave*

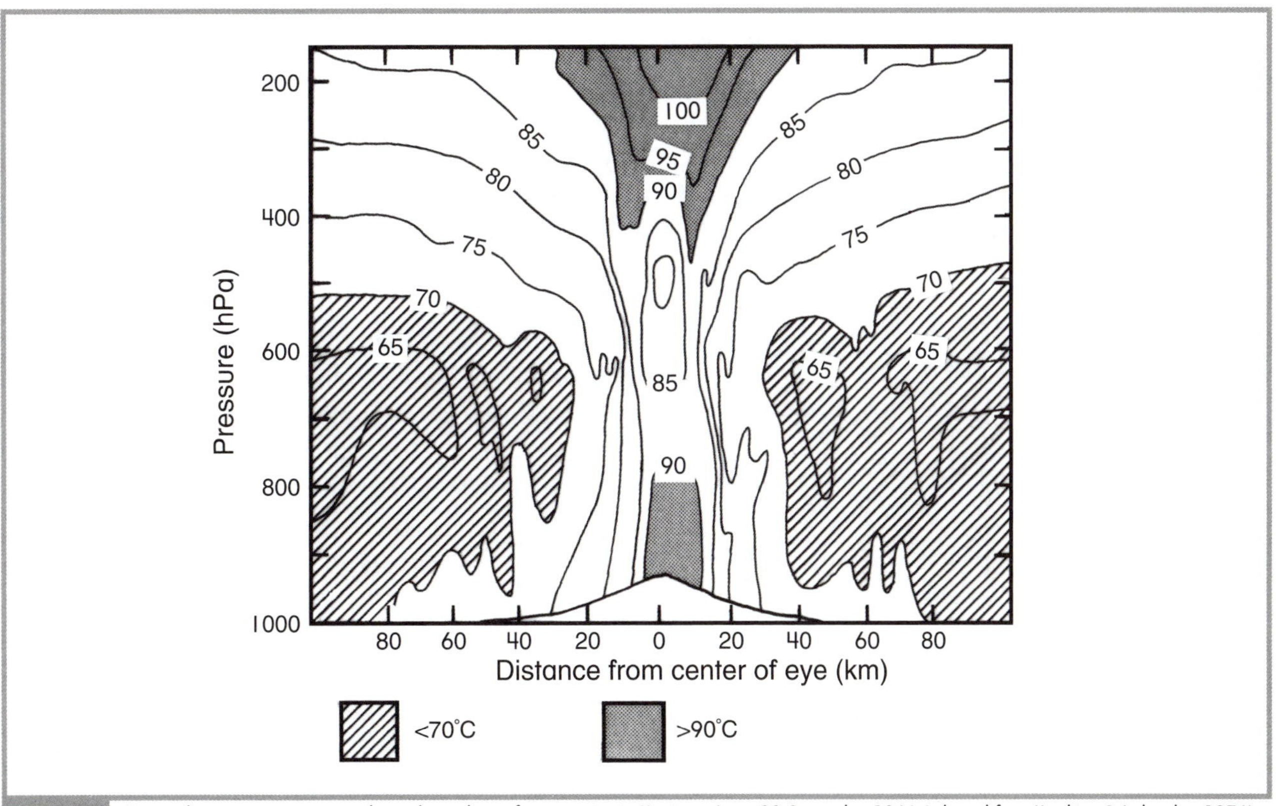

Fig. 3.4 Potential temperature structure due to latent heat of evaporation in Hurricane Inez, 28 September 1966 (adapted from Hawkins & Imbembo, 1976).

developed within the trade winds. On surface charts, this wave is evident as an undulation in isobars parallel to degrees of latitude.

Fourth, while easterly waves begin rotation, Coriolis force has to be sufficient to establish a vortex. Cyclones rarely form within 5° of the equator. Nor can cyclones cross the equator, because cyclones rotate in opposite directions in each hemisphere. Since observations began (in 1886 in the Atlantic and 1945 in the Pacific), only two tropical cyclones – Typhoon Sarah in 1956 and Vamei in 2001 – have traveled within 3° of the equator. Typhoon Vamei was unprecedented in that its center came within 1.5° of the equator east of Singapore and its zone of convection actually spilled across the equator into the southern hemisphere. This tropical cyclone's development was forced by a strong mobile polar high that swept out of China into the tropics.

Fifth, cyclones cannot be sustained – unless the land is already flooded with water – at temperatures above 24°C because they are dependent upon transfer of heat, from the surface to the upper atmosphere, through the process of evaporation and release of latent heat of condensation.

Sixth, tropical cyclones must form an *'eye'* structure (Figure 3.5). If the convective zone increases to around 10–100 km distance, then subsidence of air takes place in the center of the zone of convection, as well as to the sides. Subsidence at the center of uplift abruptly terminates convection, forming a 'wall'. Upwardly spiraling convection intensifies towards this wall, where surface winds become strongest. Once the convective wall forms with subsidence in the core, the cyclone develops its characteristic 'eye' structure. Subsidence causes stability in the eye, cloud evaporates, and calm winds result. The 'eye' is also the area of lowest pressure.

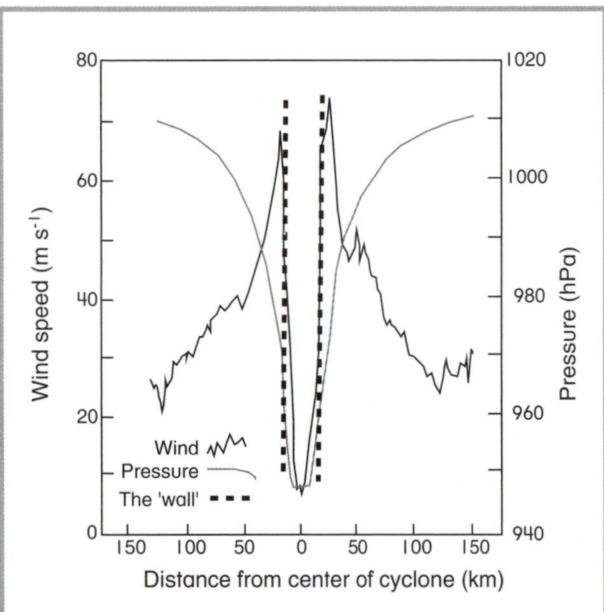

Fig. 3.5 Wind and pressure cross sections through Hurricane Anita, 2 September 1977 (based on Sheets, 1980).

Seventh, cyclones cannot develop if there are substantial winds in the upper part of the troposphere. If horizontal wind speed aloft increases by more than 10 m s^{-1} over surface values, then the top of the convective column is displaced laterally, and the structure of the 'eye' cannot be maintained. Cyclones tend to develop equatorward of, rather than under, the direct influence of strong westerly winds.

Finally, the central pressure of a tropical cyclone must be below 990 hPa. If pressures are higher than this, then uplift in the vortex is not sufficient to maintain the 'eye' structure for any length of time. Some of the lowest pressures recorded for tropical cyclones are presented in Table 3.1. The record low pressure measured in a tropical cyclone in the Pacific was 870 hPa in Typhoon Tip, north-east of the Philippines, in October 1979. In North America, Hurricane Gilbert in September 1988 produced the lowest pressure of 880 hPa just before making landfall over Jamaica. Note that cyclones with the lowest pressure are not always the most destructive in terms of the loss of human lives. This is mainly owing to the fact that not all cyclones occur over populated areas.

Magnitude and frequency

(Dvorak, 1975; Simpson & Riehl, 1981; Gray, 1984; Donnelly et al., 2001; Ananthaswamy, 2003; Australian Bureau of Meteorology, 2003)

Tropical cyclones occur frequently. Figure 3.3 also summarizes the average number of cyclones per year worldwide in each of seven main cyclone zones. While the numbers in any one zone appear low, when summed over several decades, cyclone occurrence becomes very probable for any locality in these zones. For example, consider Hurricane Hazel, which I witnessed as a child in 1954 when I was living at the western end of Lake Ontario, Canada – a non-cyclone area 600 km from the Atlantic Ocean. Hurricane Hazel was a rogue event and is still considered the worst storm witnessed in southern Ontario. This cyclone not only traveled inland, but it also crossed the Appalachian Mountains, which should have diminished its winds and rainfall. In fact, Hurricane

Table 3.1 Lowest central pressures recorded in tropical cyclones.

Event	Location	Date	Pressure (hPa)
1 Typhoon Tip	NE of Philippines	October 1979	870
2 Typhoon June	Guam	November 1975	876
3 Typhoon Nora	NE of Philippines	October 1973	877
4 Typhoon Ida	NE of Philippines	September 1958	877
5 Typhoon Rita	NE of Philippines	October 1978	878
6 unnamed	Philippines	August 1927	887
7 Hurricane Gilbert	Caribbean	September 1988	902
8 Typhoon Nancy	NW Pacific	September 1961	888
9 Labour Day storm	Florida	September 1935	892
10 Typhoon Marge	NE of Philippines	August 1951	895
11 Hurricane Allen	Caribbean	August 1980	899
12 Hurricane Linda	Baja Peninsula	September 1997	900
13 Hurricane Camille	Gulf of Mexico	August 1969	905
14 Hurricane Mitch	Caribbean	October 1998	905
15 Typhoon Babe	NE of Philippines	September 1977	906
16 unnamed	Philippines	September 1905	909
17 Cyclone Vance	Western Australia	March 1999	910
18 Typhoon Viola	NE of Philippines	November 1978	911
19 Cossack Cyclone	Australia	June 1881	914?
20 Hurricane Janet	Mexico	September 1955	914
21 Cyclone Mahina	Australia	March 1899	914?

Hazel's path over the Great Lakes was not that rare an event. Over a 60-year period, 14 cyclones have crossed the Appalachians into the Great Lakes area. Not all were of the magnitude of Hurricane Hazel, but all brought heavy rain and winds. In Australia, at least one cyclone every 20 years has managed to penetrate to Alice Springs in the center of the continent, and it is possible for west-coast cyclones to cross the whole continent before completely dissipating their winds and rain. In the New York region, major hurricanes, though rare, have struck twice in the last thousand years. The historical 1821 hurricane produced winds reaching 180–210 km hr^{-1}. Worse was a storm sometime between 1278 and 1438. Were it to recur, it would flood major cities along the eastern seaboard. Even southern California is not immune. It lies close to Mexico, the spawning center of some of the most intense tropical cyclones in the world. Fortunately, the majority track west into uninhabited parts of the Pacific Ocean. However, a tropical storm from this region moved into southern California in 1939, killing 45 people. Since 1902, the remnants of forty hurricanes have dropped rain on the state.

The actual number of cyclones within the zones in Figure 3.3 is highly variable. In Australia between 1909 and 1980, there have been on average ten cyclones per year. The number has been as low as one per year and as high as 19. In the Caribbean region, the presence of tropical cyclones has fluctuated just as much between 1910 and 2003 (Figure 2.12). In 1950, the aggregated number of days of hurricanes totaled 57, while no hurricanes were recorded at all in 1914. These patterns for both areas are not random, but linked to the occurrence of El Niño–Southern Oscillation (ENSO) events. Recent research indicates that hurricanes in the United States also follow decadal cycles. Periods with numerous and intense hurricanes were 1865–1900, 1930–1960, and 1992 onwards. Quiescent times were 1850–1865, 1900–1930, and 1960–1992.

The zones depicted in Figure 3.3 are also restrictive. Tropical cyclones have appeared in the 1970s and 1980s outside these limits, and outside the usually defined cyclone seasons for each hemisphere. The cyclone season around Australia usually lasts from 1 December to 1 May. Cyclone Namu devastated the Solomon Islands as late as 21 May 1986 and the 1987

Australian season ended with Cyclone Blanche on 26 May. The occurrence of these storms has effectively meant that the cyclone season for Australia extends to the end of May. Australian cyclones are also increasing in frequency. The worst season to date occurred in 1989, which witnessed more than 14 cyclones in the Coral Sea. Certainly, the number of cyclones affecting central Pacific islands such as Fiji and Tonga has increased significantly in the 1980s – coincidently with more frequent ENSO events.

Several scales measure the intensity of tropical cyclones. These are presented in Table 3.2. The most comprehensive is the Dvorak Current Intensity, which categorizes the magnitude of both tropical storms and cyclones. The scale is related to central pressure and differs between the Atlantic and the north-west Pacific oceans. Scaled storms are more intense in the latter. The schema utilizes satellite photography and is calculated using either cloud patterns discernible on visible images or the difference between the temperature of the warm eye and the surrounding cold cloud tops. Unfortunately, the Dvorak scale does not increment at regular intervals of pressure or wind speed. The most familiar scale is the Saffir–Simpson scale developed in the 1970s. The scale has five categories and applies only to hurricanes. The scale has the advantage of being related to property damage, so it can be used in a predictive capacity. A category 1 storm produces minimal damage, while a category 5 storm equates with the most destructive hurricanes witnessed in the United States. These will deroof most buildings, flatten or remove most vegetation, and destroy windows and doors. Note that Hurricane Andrew – which struck the Florida Peninsula in 1992 and is the second costliest hurricane in United States history – has recently been reclassified as a category 5 event. The Saffir–Simpson scale can also be related to storm surge potential depending upon the width of the *continental shelf* and how the storm tracks. For example, a category 5 hurricane can produce storm surge in excess of 5.5 m. This aspect is very convenient for laypeople in flood-prone areas. A variation of this scale is used in the southern hemisphere, mainly Australia. The Australian scale includes higher wind speeds than the Saffir–Simpson scale and is related to maximum wind gust. Unfortunately, wind gauges in the tropics of Australia are sparsely distributed so that it is difficult to verify the scale. The accuracy and viability of wind gauges is also questionable at wind speeds towards the upper end of the scale.

World cyclone disasters

(Cornell, 1976; Anthes, 1982; Whipple, 1982; Nalivkin, 1983; Holthouse, 1986)

Tropical cyclones are responsible for some of the largest natural disasters on record. In the United States, the deadliest hurricane struck Galveston Island,

Table 3.2 Scales for measuring cyclone intensity (Dvorak, 1975; Simpson and Riehl, 1981; Australian Bureau of Meteorology, 2003).

Mean sea level pressure (hPa) Atlantic	Mean sea level pressure (hPa) NW Pacific	Dvorak Current Intensity Scale	Mean wind speed (km hr-1)	Saffir-Simpson Scale	Wind speed (km hr-1)	Storm surge height (m)	Australia-Southern Hemisphere Scale	Strongest gust (km hr-1)
	1	46.3						
	1.5	46.3						
1009	1000	2	55.6					
1005	997	2.5	64.8					
1000	991	3	83.3				1	125
994	984	3.5	101.9					
987	976	4	111.1	1	119–153	1.2–1.5	2	125–170
979	966	4.5	142.6					
970	954	5	166.7	2	154–177	1.8–2.4	3	170–225
960	941	5.5	188.9	3	178–209	2.7–3.6		
948	927	6	213.0					
935	914	6.5	235.2	4	210–249	3.9–5.5	4	225–280
921	898	7	259.3	5	>249	>5.5		
906	879	7.5	287.0				5	>280
890	858	8	314.8					

Texas, on 8 September 1900, killing 6000–8000 people. A rapidly rising storm surge trapped inhabitants, who then either drowned as waves swept through the city or were killed by the 170 km hr^{-1} winds. Apathy and ignorance exacerbated the death toll. There have been at least four cyclones in recorded history that, as single events, killed over 300 000 people – all but one occurred over the Indian subcontinent. The most destructive cyclone in the twentieth century was the one that struck Bangladesh – formerly East Pakistan – in November 1970, killing over 500 000 people. It is memorable for two reasons. Firstly, most people were killed by the storm surge rather than strong winds or floodwaters. Secondly, although viewed by inhabitants as a rare event, the disaster was repeated in May 1985 and on 29 April 1991, resulting in another 100 000 and 140 000 deaths, respectively. The press also graphically documented the 1970 disaster. The ensuing ineffectual rescue and reconstruction effort by the Pakistan government led to political upheaval and independence for Bangladesh. Ironically, while Bangladesh obtained its independence because of a cyclone, ultimately it became the poorest country in the world, mainly because it never recovered economically from the disaster.

Historically, cyclones account for the greatest number of deaths resulting from any short-term natural hazard. Flooding over swathes of 400–500 km along the cyclone track causes most of this death toll. The most horrific cyclone-induced flooding occurred in China along the Chang River between 1851 and 1866. During this period, 40–50 million people drowned on the Chang River floodplain because of tropical cyclone-related rainfall. Another 1.5 million died along the Chang River in 1887 and, in 1931, 3–4 million Chinese drowned on the Hwang Ho River floodplain because of cyclone flooding.

Tropical cyclones have even changed the course of history in the Far East. In 1281 AD, Kublai Khan invaded Japan, and after seven weeks of fierce fighting was on the verge of breaking through Samurai defenses – until a typhoon swept through the battlefield. The storm destroyed most of the 1000 invading ships and trapped 100 000 attacking soldiers on the beaches, where they either drowned in the storm surge or were slaughtered by the Japanese. The Japanese subsequently believed their homeland was invincible, protected by the Kamikaze (or Divine Wind) that had saved them from this invasion.

Tropical cyclones can even affect non-tropical areas. As mentioned above, Hurricane Hazel swept through southern Ontario, Canada, in October 1954. This storm occurred during a period of remarkable superstorms that afflicted Japan and North America between 1953 and 1957. Hazel drifted northward along the eastern coast of the United States, where it killed 95 people, raised a storm surge of 5 m, and produced the strongest winds ever recorded at Washington and New York. It then turned inland over the Appalachians and headed across the Great Lakes to Toronto, 600 km from the nearest ocean. Here 78 people lost their lives in the worst storm in 200 years.

In Australia, on 4–5 March 1899, Cyclone Mahina (or the Bathurst Bay Cyclone) crossed the northern Great Barrier Reef and killed 407 people. The storm surge reached 4 m, which is high for Australia. Small cyclones, with eyes less than 20 km in diameter, are called 'howling terrors' or 'kooinar' by the Aborigines and appear the most destructive. For instance, the Queensland town of Mackay, with 1000 inhabitants, was destroyed in 1918 by one of these small cyclones. Cyclone Ada in 1970 was so small that it slipped unnoticed through the meteorological observation network, and then surprised and wrecked holiday resorts along the Queensland coast. The Yongala Cyclone of 1911, when it reached landfall south of Cape Bowling Green, cleanly mowed down mature trees 2 m above the ground in a 30-kilometre swathe. In recent times, Cyclone Tracy has been the most notable cyclone. Tracy, with an eye diameter of 12 km, obliterated Darwin on Christmas Day 1974 (Figure 3.6), killing over 60 people and reaching wind speeds over 217 km hr^{-1}. In world terms, its wind speeds were small and the loss of life meager. There was also no surge damage because the tidal range in Darwin is about 7 m and the cyclone occurred at low tide, with a storm surge of 4 m. Tracy is one of the best-documented Australian cyclones, with its graphic evidence of wind damage forming a benchmark for engineering research into the effects of strong winds. It also resulted in one of the largest non-military air evacuations in history, as the majority of the population of 25 000 was airlifted south to gain shelter and avoid the outbreak of disease.

Hurricane Andrew, 24 August 1992

(Gore, 1993; Santana, 2002)

In recent times, two events are noteworthy: Hurricane Andrew in the United States in 1992 and Hurricane

Fig. 3.6 Damage done to a typical suburb in Darwin by Cyclone Tracy, December 1974 (photograph courtesy John Fairfax and Sons Limited, Sydney, FFX Ref: Stevens 750106/27 #4).

Mitch in Central America in 1998. Hurricane Andrew struck after a 27-year lull in intense cyclones in the United States. The storm began on 13 August 1992 as a region of thunderstorms off western Africa and rapidly developed into an easterly wave. It became a weak tropical cyclone on 17 August; but by 23 August it had developed central winds of 240 km hr^{-1} and a central pressure of 922 hPa. Andrew made landfall in Dade County, 32 km south of Miami, just after midnight on Monday 24 August (Figure 3.7). While up to one million people had evacuated to shelters in the Miami area, most of the 370 000 residents of Dade County were still in their homes under the assumption that they were not in the path of the hurricane. Warnings were issued to them on television only minutes before winds began gusting to 280 km hr^{-1} – simultaneously with the arrival of a storm surge of 4.9 m. This made it a category 5 hurricane. Over 25 500 homes were destroyed and another 101 000 severely damaged. Many of the homes were of substandard construction for a hurricane-prone area. For example, plywood had been stapled to supports instead of being secured using screws or bolts. Much of the damage was caused by vortices that developed in the main wind stream, then tightened, and wound up to speeds of 120 km hr^{-1} in 10–20 seconds. Where the wind in these mini-cyclones flowed with the main air stream, wind velocities exceeded 200 km hr^{-1} – with buildings torn apart, seemingly sporadically, and debris flung at tornadic speeds (Figure 3.8). The official death toll was only 55, although there is evidence that it may have been higher because the region contained a high number of illegal immigrants. Over $US30 billion worth of property was destroyed. The hurricane then swept into the Gulf of Mexico and, two days later, came ashore in Louisiana, killing 15 people and causing another $US2 billion of destruction. This was the biggest insurance disaster in the country. What followed became one of the biggest federal government 'non-responses' in history. Looting was rampant and temperatures and humidity soared; but for a week state and federal authorities ignored the extent of the disaster. Over 250 000 people were homeless, there was no food or safe drinking water, and sanitation was failing due to food rotting in disconnected freezers and a lack of serviceable toilets. Usually, when response to a disaster requires resources beyond the capability of a state government, assistance is provided by the Federal Emergency Management Agency (FEMA). FEMA was established in 1979 as an independent federal agency reporting directly to the President. It has a clear mission to reduce loss of life and property and protect critical infrastructure from all types of hazards through a comprehensive, risk-based, emergency management program of mitigation, preparedness, response, and recovery. However, it had no capability to provide emergency accommodation, food, and medical assistance on the scale needed in southern Florida. Even private relief agencies such as the Red Cross did not know how to distribute donations they were receiving. Finally, Dade County's Emergency Director, Kate Hale, made a passionate public plea: 'Where the hell is the cavalry on this one?' Only then were military personnel increased to 16 000 to assist with clean-up and relief. So poor was FEMA's response that Congress considered eliminating the agency.

Fig. 3.7 NOAA composite satellite image of Hurricane Andrew's path, 22–25 August 1992 (NOAA, 2002).

Fig. 3.8 Sheet of plywood driven through the trunk of a Royal Palm by extreme winds during Hurricane Andrew in South Florida (NOAA, 2003).

Hurricane Mitch, October/November 1998

(Lott et al., 1999)

Hurricane Mitch was the second most destructive hurricane ever recorded in the Atlantic Ocean. It began as a tropical depression on 21 October 1998 and wreaked havoc across the Caribbean for two weeks before dissipating in the Atlantic on 4 November. During this time, it reached wind speeds of 290 km hr^{-1}, making it a rare Saffir–Simpson category 5 event. Mitch made landfall on 28 October over Honduras, slowed its westward progress and dumped, during the subsequent week, up to 1.9 m of torrential rain on the mountains of Central America. This led to catastrophic flooding and mudslides in Honduras, Nicaragua, Guatemala, Belize, and El Salvador. Hundreds of thousands of homes were destroyed and up to 18 000 people died, making the storm the worst in the Americas since 1780. Over three million people were made homeless. In Nicaragua, damage was worse than that resulting from the 1972 Managua earthquake. So calamitous was the disaster that appeals for international aid were made by all countries affected by the hurricane. In Honduras, at least 30 per cent of the population was homeless, 60 per cent had no access to safe water, 66 per cent of the crops were destroyed, and 100 per cent of the road network rendered impassable. The World Food Programme estimated that 12 per cent of the population required long-term food supplies, while UNICEF estimated that recovery would take 40 years. The disaster epitomizes the effects of severe natural hazards upon developing countries characterized by poverty, civil war, and economies dominated by a single commodity primary industry.

Impact and response

(Burton et al., 1978; Simpson & Riehl, 1981; The Disaster Center, 1998)

The human impact of cyclones can be assessed in terms of deaths, financial damage, reliance upon international aid, and the prospects of further impoverishment. The response to cyclones depends mainly on these four factors. Deaths due to cyclones in the United States decreased dramatically in the twentieth century to fewer than a 100 people per year, mainly because of a policy of total evacuation together with earlier warning systems. Costs, on the other hand, have increased considerably – as illustrated by Hurricane Andrew – because of the escalating development in cyclone-prone areas after the Second World War. Of course, the damage bill from a tropical cyclone also depends upon the type of building construction. Cyclone Tracy – the most destructive cyclone experienced in Australia – destroyed only 10 per cent of the value of infrastructure in Darwin. While most houses were damaged or destroyed, larger commercial and public buildings with a high capital investment were not badly damaged. Other non-residential buildings were mildly affected because they had to comply with strict building codes formulated to minimize the effect of cyclone winds.

The impact of cyclones goes far beyond just deaths and building damage. In developing countries, destruction of infrastructure and primary agriculture can lead to a decrease in exports and gross national product, while increasing the likelihood of forfeiture of international loan repayments. Figure 3.1 indicates that cyclones can cause increased soil salinity through storm-surge inundation, an effect that can have long-term economic consequences for agricultural production – with such consequences virtually immeasurable in monetary terms. Contamination of water supplies and destruction of crops can also lead to disease and starvation. In fact, some of the large death tolls caused by cyclones on the Indian subcontinent in the nineteenth century were due to starvation afterwards. In certain cases, the destruction of crops can affect international commodity markets. For example, Hurricane Flora's passage over Cuba in 1963 wiped out the sugar cane crop and sent world sugar prices soaring. Of all hazard events, the effects of tropical cyclones differ substantially from each other depending upon the development of a country's economy, transport network and communication infrastructure. This can be illustrated most effectively by comparing the response and impact of tropical cyclones for three countries: Australia, the United States, and Bangladesh.

Australia

(Australian Bureau of Meteorology, 1977; Western & Milne, 1979)

One of the effects of tropical cyclones not covered in Figure 3.1 is long-term human suffering caused by dislocation. Studies of hazards where people are dispossessed indicate that the incidence of psychosomatic illness (illness, such as by anxiety, induced by changes in mental well-being) increases afterwards. More importantly, people who have lost a relative recover more completely than people who have lost the family home. A death can be mourned during a funeral occurring within several days, but a family home has to be rebuilt over months or years. Australia has experienced this phenomenon on a large scale. In northern coastal Australia, it is accepted that cyclones will inevitably occur. Evacuation is normally not encouraged except for the most hazardous areas next to the ocean. Strict building codes have been established to minimize damage and loss of life. Such codes are policed and adhered to. A network of remote-controlled, satellite and staffed stations has been established to detect and track cyclones. These observations are continually reported over radio and television if a cyclone is imminent. Some criticism has been expressed regarding the Australian Bureau of Meteorology's inability to predict the landfall of all cyclones, and there is some fear that the staffed stations are being run down as a cost saving measure. However, it should be realized that tropical cyclones in Australia, as elsewhere, often travel unpredictably. Residents have been instructed on how to prepare and weather a cyclone. For example, most people upon hearing about the forecast arrival of a cyclone will remove loose objects in the house and yard that can become missiles in strong winds, will tape windows, and will store at least three days' supply of water and food.

Cyclone Tracy, which struck Darwin on 25 December 1974, demonstrated that apathy can negate all these preventative procedures. The cyclone also led to large-scale psychological damage that cannot be measured in economic terms. Despite three days of warnings, and because of the approaching Christmas holiday, only 8 per cent of the population was aware of the cyclone. About 30 per cent of residents took no precautions to prepare for the storm's arrival; fewer than 20 per cent believed that people took the cyclone warnings seriously; and 80 per cent of residents later complained that they had not been given enough warning. Fortunately, Australia had established a disaster coordination center that, depending on the severity of the disaster, could respond in varying degrees to any event. The director of this center, once requested to take charge, could adopt dictatorial powers and even override federal government or judicial decisions. This organization responded immediately to Cyclone Tracy. Within 24 hours, communications had been re-established with the city even though it was the height of the Christmas vacation season. Major-General Stretton, the head of the National Disasters Organisation, reached Darwin within 24 hours on a military aircraft with emergency supplies loaded at Sydney and Mt Isa. Stretton took total control of Darwin and made the decision to evacuate non-essential residents because of the serious threat to health. The cyclone destroyed 37 per cent of homes and severely damaged most of the remainder (Figure 3.9). There was no shelter, water, power, or sanitation. Within a few days, most of the food in deep-freezers began to decay, so that a stench of rotten meat hung over the city. Stray cats and dogs were shot and special army teams were organized to clean up rotten food. In addition, the decision was made to facilitate evacuation to permit orderly and manageable rebuilding. Commercial planes were commandeered for the evacuation and details taken of all vehicles leaving Darwin by road. Volunteer emergency relief organizations ensured that all evacuees had food, adequate transport, accommodation, and money. The response in the first week was efficiently organized and carried out.

In the medium term, the evacuation caused more problems than it solved. A major problem was the fact that, while residents were efficiently and safely evacuated, no register of their whereabouts was established. Over 25 000 people were flown out, mainly to southern cities of their choice 3000–4000 km away. Communication with, and among, evacuees became impossible. Hence, there was no way to contact people personally after the cyclone to assist them with adjustment to new circumstances. Residents who stayed in Darwin suffered the least, while those who did not return in the medium term suffered the most. Non-returned

Fig. 3.9 Effects of wind damage caused by Cyclone Tracy striking Darwin, Australia, 25 December 1974.
A) High winds in excess of 217 km hr^{-1} tended to detach outside walls from the floor, leaving only the inner framework surrounding the bathroom.

Fig. 3.9 B) In extreme cases, the high winds progressively peeled off the complete building, leaving only the iron floor beams and support pillars. (Photographs courtesy of Assoc. Prof. Geoff Boughton, TimberED Services Pty Ltd., Glengarry, Western Australia.)

evacuees reported greater negative changes in lifestyle. These people also tended to be disadvantaged by the storm. They were more likely not to have insured houses for cyclone damage and more likely to have suffered total property loss. They ended up with less income after the event and suffered the greatest stress. While the stress may have existed beforehand or been exacerbated by the disaster, the difference between this group and the others was striking. Both psychosomatic illness and family problems increased significantly for non-returnees. The evacuation reinforced and increased the levels of stress and anxiety residents were experiencing because of their confrontation with Cyclone Tracy. Some of the stress was a consequence of the dictatorial powers assumed after

Tracy by a few. Unintentionally, a *victim–helper relationship* became established to the point that little encouragement was given to victims to play an active role in their own recovery. Disaster victims in Australia are resourceful. To treat them otherwise impedes the social recovery from disaster.

The long-term response to Cyclone Tracy was also negative. The decision to rebuild and the method of allowing rebuilding did not take into consideration the long-term effects on human mental and physical health. Large sums of money were poured into rebuilding lavish government offices and facilities. The rebuilding effort was taken over by outsiders, and the growth of Darwin beyond 1974 population levels was encouraged. The rebuilding became a 'sacred cow', with the visible commitment to the rebuilding by the federal government leading to unlimited growth. Locals who still lived in Darwin objected to redevelopment plans, stating that they had no say in the range of plans put forward. Residents evacuated to the south had no say in the redevelopment at all. Only men were allowed to return to the city to rebuild. The men, isolated from their families for up to one year, grouped together socially. When their families finally returned home, many of these men could not break back into family groupings, most of which had completely changed (for example, children had physically and mentally grown in the absence of a male role figure). Even though within the first year, returnees suffered less physically and emotionally than non-returnees, returnees in the long term suffered the same consequences as non-returnees. By 1985, over 50 per cent of the families who experienced Tracy had broken up. The personal problems resulting from Cyclone Tracy are now being imparted to the second generation, the children, to be passed on in future years to another generation unless rectified. Today, Darwin is a transformed city. It is estimated that 80 per cent of the people now living there, never lived through Cyclone Tracy.

United States

(American Meteorological Society, 1986; Carter, 1987)

The United States government believes that early warning and monitoring is the best method to reduce loss of life due to tropical cyclones. Weather satellites were originally deployed to predict and track cyclones, thus giving enough warning to permit the total evacuation of the population, either horizontally or vertically. Horizontal evacuation involves moving large numbers of people inland from flood-prone coastal areas. In August 1985, millions of people were evacuated, some more than once, as Hurricane Danny wandered unpredictably around the Florida coast and then headed into the Gulf States. This type of evacuation can be especially difficult where traffic tends to jam on escape routes. The alternative is vertical evacuation into high-rise, reinforced buildings. This approach is taken in Miami Beach, where limited causeways to the mainland restrict the number of people who can flee, where the population is elderly and hence physically immovable, and where safe structures can easily accommodate the majority of the population. There is a concerted effort to build coastal protection works and even try to modify the intensity of hurricanes through cloud seeding. A high priority is also put on research activities that model, predict, or quantify hurricanes and related hazards. Some east coast states, such as North Carolina and Maine, have development plans that exclude development from low-lying coastal zones subject to storm surge and wave erosion due to hurricanes. These setback lines are equivalent to 30–100 years of average erosion.

The above measures have effectively reduced the loss of life due to tropical cyclones in the United States. However, there are serious flaws with the evacuation procedures. Most evacuations actually take up to 30 hours to carry out, whereas the general populace believes that it would require less than one day to flee inland to safety. The evacuation system also relies heavily upon warnings of hurricane movement put out by the National Hurricane Center, which attempts to give 12 hours' warning of hurricane movements. In fact, hurricanes move unpredictably and can shift dramatically within six hours. At one point in time, Hurricane Hazel moved at speeds of 50 km hr^{-1}. In the United States, the time required for evacuation currently falls well short of the monitoring and forecast lead times for hurricanes. There is also the problem that 80 per cent of the 50 million people living today in hurricane-affected areas have never experienced an evacuation. These same people, together with developers, have become complacent about the hurricane threat because the frequency and magnitude of hurricanes has decreased in recent decades. The 1970s, in fact, experienced the lowest frequency of hurricanes ever recorded. At present along the whole eastern seaboard and much of the Gulf Coast of the United States, record construction has taken place with dwellings being built either at the back of the beach, or just landward of ephemeral dunes. Studies have shown that there is a general lack of awareness of cyclone danger and evacuation procedures. This is despite wide publicity, as the hurricane season approaches, about the dangers of cyclones. Publicity includes television specials, distribution of free literature, and warnings on supermarket bags and parking tickets. These warnings are meeting with resistance because, in the absence of any storm, municipalities view the publicity as bad for the tourist trade and damaging to property investment.

The realities of evacuation, and the poor public perception of the ferocity of hurricanes in the United States, implies that – as far as the cyclone hazard is concerned – many people are sitting on a time bomb. When a large tropical cyclone strikes the United States east coast, evacuation procedures may fail and the resulting loss of life may reach unprecedented levels.

Hurricane Andrew also exposed flaws in the federal government's capability to respond to a large disaster. When the response to a disaster requires resources beyond the capability of a state government, the Federal Emergency Management Agency (FEMA) can assist. FEMA failed its mandate during Hurricane Andrew. While FEMA has been overhauled since 1992, the prospect still looms that two large hurricanes may occur so close together that they will overwhelm the capacity of any of the United States' resources to respond to the disaster in the short term.

Bangladesh (East Pakistan)

(Burton et al., 1978; Blong & Johnson, 1986; Milne, 1986)

Third World countries suffer exorbitantly from tropical cyclones because of depressed economic conditions and the lack of an effective domestic response capability. In 2000, cyclones Eline and Hudah devastated Mozambique, which has one of the lowest living standards in the world. In 2003, Cyclone Zoe swept the outer islands of Tikopia, Anuta, and Fatutaka in the Solomon Islands. As the government was bankrupt and wracked by lawlessness, the inhabitants were forced to fend for themselves. But nowhere has the plight of impoverished countries been exemplified more than in Bangladesh. The Bay of Bengal has a history of destructive storm surges. The 300 000 death toll from Calcutta's October 1737 cyclone was caused by a surge, 13 m high, racing up the Hooghly River. In 1876, a storm surge near the mouth of the Meghna River killed 100 000 people, while another 100 000 died in the subsequent famine. Small cyclone-related surges drowned 3000 people on the Ganges River in October 1960 and another 15 000 people in May 1965. The Chars, the deltaic islands in the mouth of the Ganges, were first viewed as potentially arable in the nineteenth century (Figure 3.10). Since 1959, 5500 km of dykes and embankments have been built to protect low-lying saltwater marshland from flooding, with the World Bank financing the scheme. The reclaimed land, however, was not protected against storm surge. Settlement, mainly by illiterate farmers, was permitted and encouraged to relieve overcrowding in the country. By the end of the twentieth century, over four million people lived in areas at high risk from storm surge. Many itinerant workers searching for temporary employment flock into the Chars at the time of harvesting and fish processing, swelling the population by 30 per cent.

The 13 November 1970 cyclone was detected by satellite three days in advance, but a warning was not issued by the Pakistani Meteorological Bureau until the evening that the cyclone struck. Even then, the message was given to the sole radio station after it had shut down for the day at 11:00 pm. If the warning had been broadcast, the majority of the population, being asleep, would not have heard it. Much of the Chars' reclaimed land had inadequate transport links across channels to higher ground, such that effective evacuation was impossible. Recent migrants had no knowledge of the storm surge hazard and were not prepared, or organized, for evacuation. In the middle of the night a storm surge, 15 m high, struck the southern Chars, obliterating 25 islands, swamping a further 2000 and drowning approximately 500 000 people. Over 400 000 hectares of rice paddies were inundated with salt water and 1 000 000 head of livestock killed. Over 50 million people were affected by the storm surge, flooding, and winds.

The pressure for land in Bangladesh is so great that the deltaic land was resettled rapidly after the 1970 event. Nothing had really changed by 1985 and, in May of that year, a similar cyclone caused the same sort of destruction. Again, while there was adequate satellite warning of the approaching cyclone, evacuation was impossible. Many illiterate, itinerant workers knew nothing of the hazard, the method of escape, or the

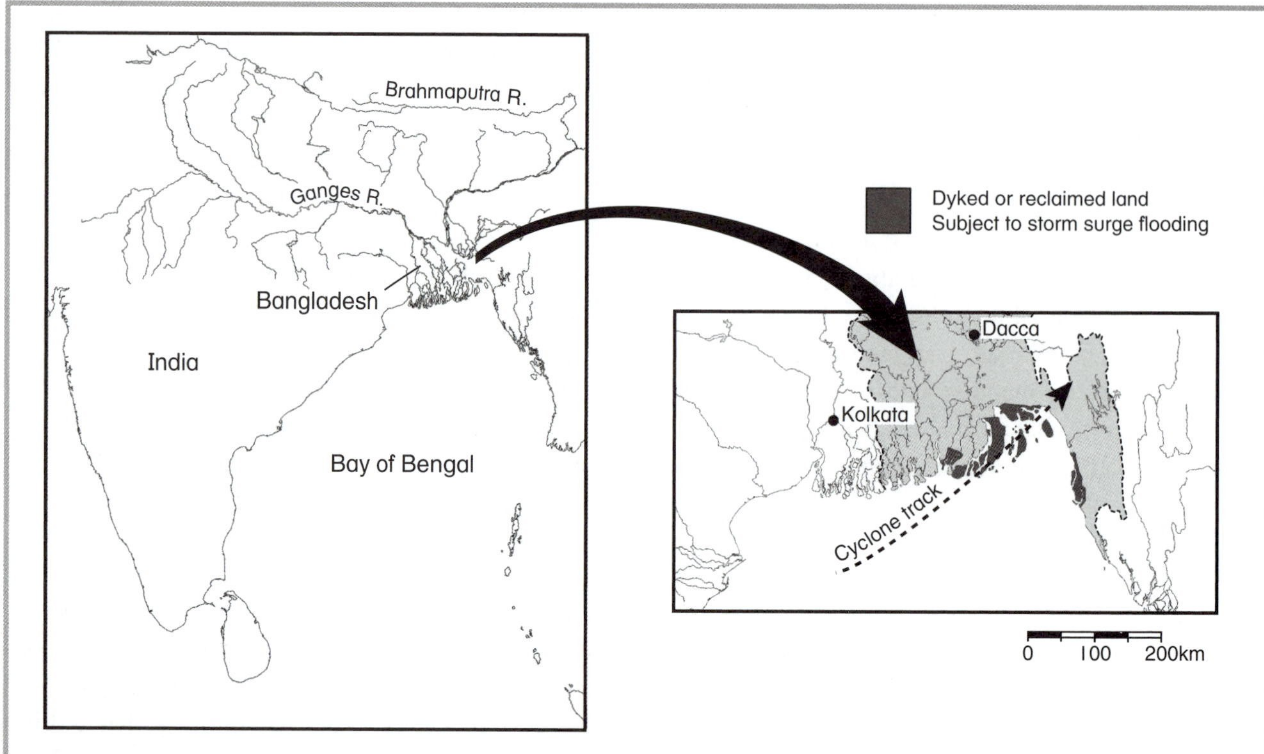

Fig. 3.10 Path of 13 November 1970 cyclone in the Bay of Bengal and location of embankments protecting low-lying islands in Bangladesh.

event. In the 1985 event, 100 000 people lost their lives.

The response in Bangladesh to tropical cyclones appears to be one of ignorance and acceptance of the disaster as the 'will of God'. Cyclones cannot be avoided and one must simply wait out the storm and see what happens. With a bit of luck a family that has experienced one cyclone storm surge in its lifetime will not experience another. In 1970, only 5 per cent of the people who survived had evacuated to specific shelters while 38 per cent had climbed trees to survive. Many view evacuation as unwise because homes could be looted. Even if evacuation had been possible, religious taboos in 1970 would have prevented Moslem women from leaving their houses because the cyclone struck during a month when women were forbidden by established religious convention from going outside. Since 1970, 1841 cyclone shelters have been built by Caritas Bangladesh and other non-government organizations. These shelters are built of reinforced concrete and raised on piers at least 4.6 m above historical storm surge levels. They can double as schools, family welfare centers, mosques, markets, and food stores during normal times. Each shelter is designed to give a person 0.5 m^2 of floor space, which is sufficient to allow them to sit comfortably on the floor during storms. When filled to capacity with people standing shoulder-to-shoulder, this personal space decreases to 0.14 m^2. Toilet facilities and potable water supplies pose problems under these latter circumstances. A shelter can hold 1000–1500 people and some livestock. Up to 50 per cent of the population can now be accommodated in cyclone shelters. Their effectiveness was proven on 28 May 1997 when tropical cyclone Helen hit the Chittagong region. A million people took refuge in 700 cyclone shelters for up to 20 hours. Winds of up to 230 km hr^{-1} damaged 600 000 houses in south-east Bangladesh. Instead of a death toll in the tens of thousands, only 106 people were killed.

EXTRA-TROPICAL CYCLONES

Extra-tropical cyclones or depressions are low-pressure cells that develop along the polar front. They also can develop over warm bodies of water outside the tropics, usually off the east coast of Australia, Japan, and the United States. This section will describe the formation of low-pressure storms due to upward forcing of warm air along the polar front, and the formation of *east-coast lows* that develop or intensify over bodies of warm water along limited sections of the world's coastline.

Polar-front lows

Formation

(Linacre & Hobbs, 1977; Whipple, 1982)

Low-pressure cells developing along the polar front have cold cores and an effective diameter that is much larger than that of a tropical cyclone (2000 km versus 400 km, respectively). Their intensity depends upon the difference in temperature between colliding cold and warm air masses. They always travel in an easterly direction, entrapped within the westerlies. Their location depends upon the location of Rossby waves in the polar front around 40°N and S of the equator. In the southern hemisphere, this occurs south of landmasses, but in the northern hemisphere, it coincides with heavily populated areas.

The theory for the formation of polar-front lows was formulated by Bjerknes and his colleagues at Bergen, Norway, in the 1920s. Outbreaks of cold polar air flow rapidly towards the equator and clash with much warmer air masses at mid-latitudes. Coriolis force dictates that this polar air has an easterly component while the movement of warmer air on the equatorial side of the cold front has a westerly component (Figure 3.11). At the same time, the warm air will be forced upwards by the advancing cold air. Winds on each side of the cold front blow in opposite directions and spiral upwards in a counterclockwise direction because of Coriolis force. This rotation establishes a wave or indentation along the front. Warm air begins to advance eastward and poleward into the cold air mass while cold air begins to encircle the warm air from the west. A V-shaped frontal system develops with warm and cold fronts at the leading and trailing edges, respectively. The lifting of warm, moist air generates low pressure and convective instability as latent heat of evaporation is released in the upper atmosphere. This uplift enhances the rotational circulation, increasing wind flow and producing precipitation over a wide area. The cold front at the back of the system pushes up warm air rapidly leading to thunderstorm formation. If the temperature difference at this location between the two air masses is great enough, tornadoes can be generated. The cold front, being backed by denser air, eventually advances faster than the warm front and begins to lift off the ground the warm air within the 'V', occluding the fronts. In the final stage of a depression, the polar front is re-established at the ground with decreasing rotational air movement stranded aloft. Without continued convection, cloud

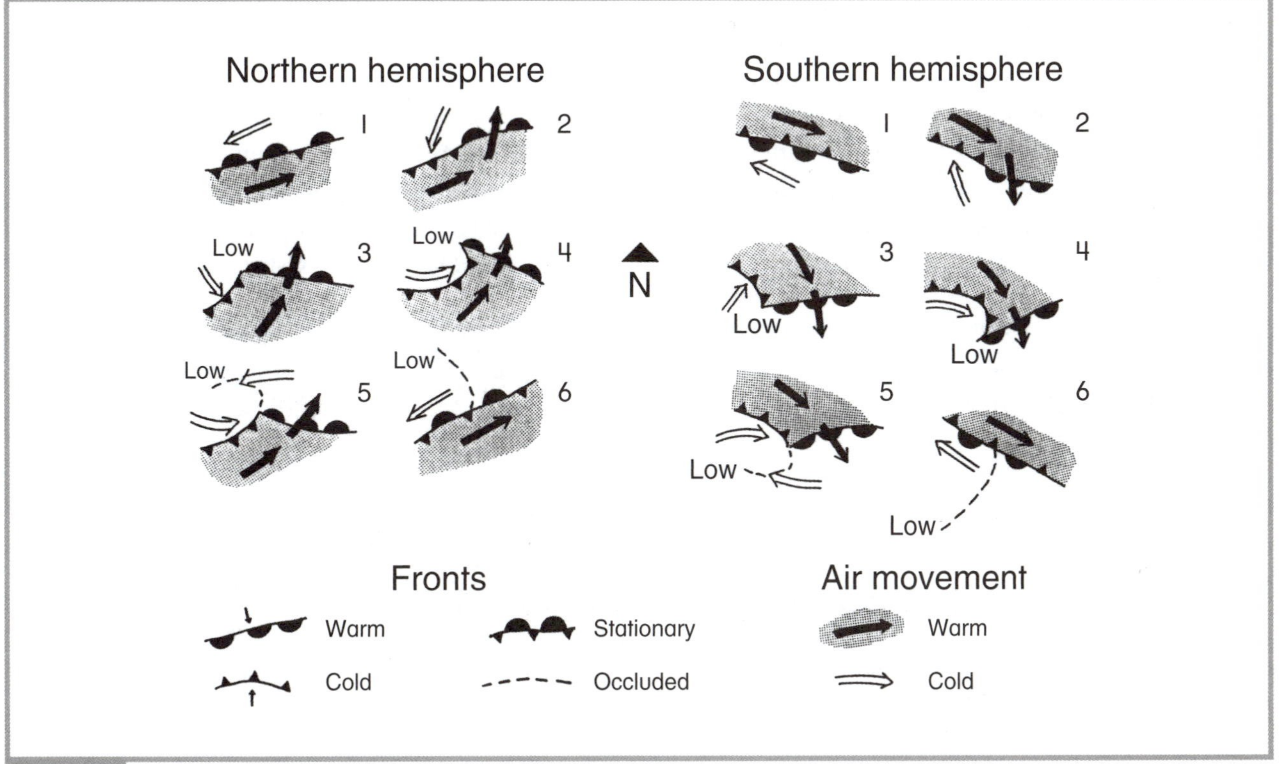

Fig. 3.11 Stages in the development of an extra-tropical depression as viewed in each hemisphere 1) and 2) initial development of wave form, 3) and 4) mature front development and full extent of low, 5) and 6) occlusion.

formation and precipitation slowly decrease and the storm cell disappears. Under some circumstances, the advancing cold front may entrap another lobe of warm air, generating a secondary low to the south-east of the original one that is now dissipating. It is even possible for the rotating air aloft to circle back to the original site of cyclogenesis, and trigger the formation of a new low-pressure wave along the polar front. These extra-tropical lows or depressions, because of their method of formation, have the potential to produce the same wind strengths and precipitation amounts as tropical cyclones. However, because they are larger and more frequent, they pose a greater hazard. Such extra-tropical storms also can generate ocean waves and storm surges similar in magnitude to those produced by tropical cyclones.

Historical events

(Wiegel, 1964; Lamb, 1982; Whipple, 1982; Milne, 1986; Bresch et al., 2000)

Mid-latitude depressions have also been as devastating as tropical cyclones. Their effects are dramatic in two areas, namely along the eastern seaboard of the United States and in the North Sea. Historical accounts for these areas document the strongest storms recorded. Figure 3.12 plots the number of recorded cyclonic depressions in northern Europe that have produced severe flooding over the past 2000 years. The Middle Ages was an unfortunate period of cataclysmic tempests that caused great loss of life and immense erosion. Four storms along the Dutch and German coasts in the thirteenth century killed at least 100 000 people each. The worst of these is estimated to have killed 300 000 people. North Sea storms on 11 November 1099, 18 November 1421 and in 1446 also killed 100 000 people each in England and the Netherlands. By far the worst storm was the All Saints Day Flood of 1–6 November 1570. An estimated 400 000 people were killed throughout western Europe. These death tolls rank with those produced in recent times in Bangladesh (East Pakistan) by storm surges from tropical cyclones. An English Channel storm on 26–27 November 1703, with an estimated central pressure of 950 hPa and wind speeds of 170 km hr^{-1}, sank virtually all ships in the Channel with the loss of 8000–10 000 lives (Figure 3.13). Other storms with similar death tolls occurred in 1634, 1671, 1682, 1686, 1694 and 1717 at the height of the Little Ice Age. Much of the modern-day coastline of northern Europe owes its origin to this period of storms. Erosion in the North Sea was exacerbated by 30 notable, destructive storm surges between 1000 and 1700 AD. North Sea storms reduced the island of Heligoland, 50 km into the German Bight, from a length of 60 km around the year 800 to 25 km by 1300 and to 1.5 km by the twentieth century. The Lucia Flood storm of 14 December 1287 in northern Europe was of immense proportions. Before that time, a continuous *barrier island* bordered the Netherlands coast. The storm breached this coastal barrier and formed the Gulf of Zuider Zee (Figure 3.13). It is only recently that the inlets at these breaches have been artificially dammed. Smaller embayments further east at Dollart and Jade Bay were also initiated and the Eider Estuary was widened to its present funnel shape. The same storm in northern France eroded several kilometres of marsh, making one headland a distant offshore island. In the Great Drowning Disaster storm of 16 January

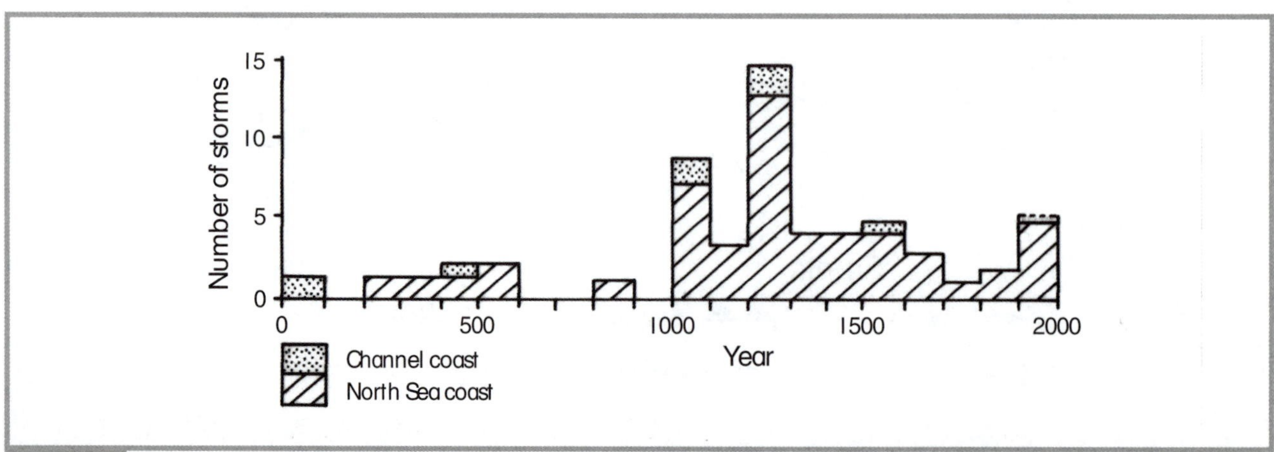

Fig. 3.12 Incidence of severe North Sea and English Channel storms per century for the last 2000 years (Lamb, ©1982; with permission from Methuen, London).

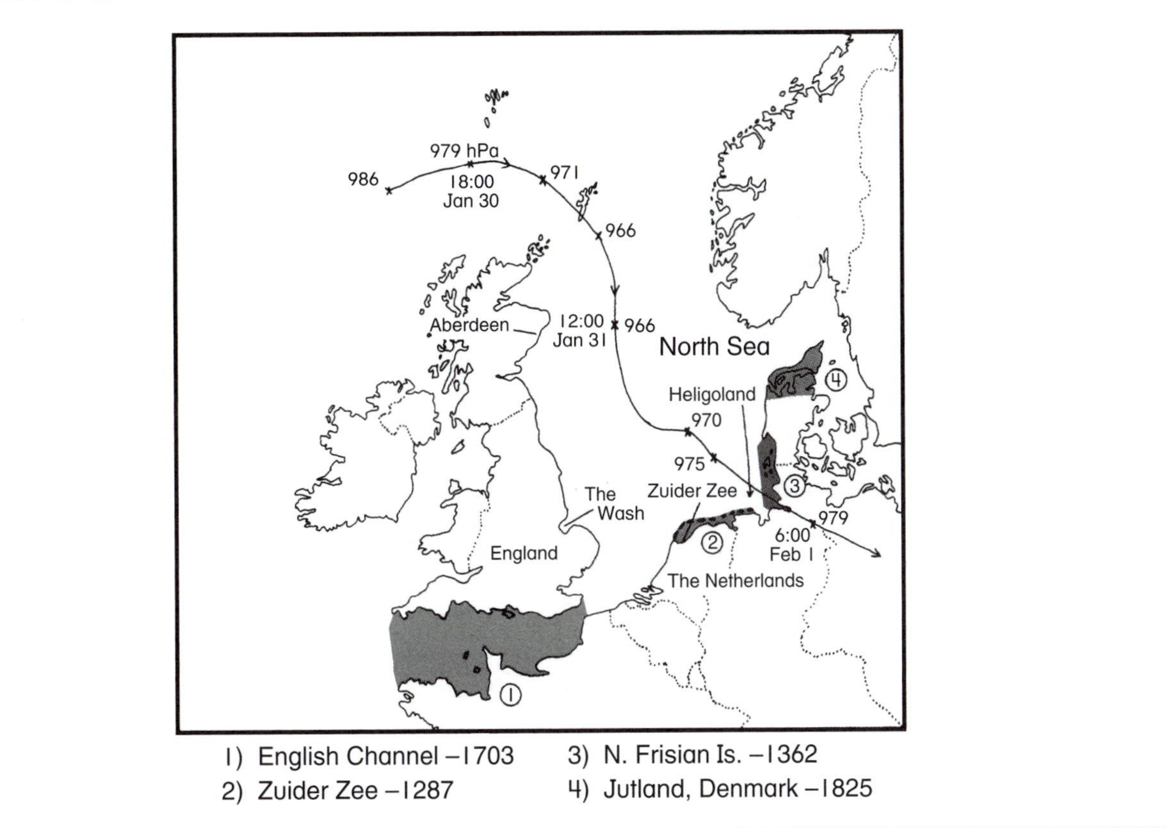

Fig. 3.13 Location of some historical, erosive North Sea storms including the path of the 1 February 1953 storm.

1362, parts of the North Sea coast of Schleswig–Holstein eroded 15 km landward, turning a low-lying estuarine coastline into one bordered by irregular barrier islands separated from the mainland by lagoons 5–10 km wide. Over 60 parishes, accounting for half the agricultural income in this area, were destroyed. Storms in 1634 severely eroded both the Danish and German coasts and, in 1825, a North Sea storm separated part of the northern tip of Denmark from the mainland, leaving behind several islands. These storms also directed the course of history. In 1588, the Spanish Armada of 130 ships, after being forced into the North Sea by marauding attacks from the outnumbered British fleet, was all but rendered ineffective by a five-day storm off the east coast of Scotland (Figure 3.14). What remained of the Armada was finished off by more of these storms as the fleet tried to escape west around Ireland. The summer of 1588 was characterized by an exceptional number of cyclonic depressions, corresponding to a polar jet stream diverted much further south and wind speeds much stronger than could be expected at present.

None of the extra-tropical storms mentioned above have been matched in the twentieth century for loss of life. However, there have been big storms. Most noteworthy was the North Sea storm of 1 February 1953. This storm developed as a low depression north of Scotland and then proceeded to sweep directly south-east across the North Sea (Figure 3.13). The lowest pressure recorded was 966 hPa. While this is not as low as pressures produced by tropical cyclones (see Table 3.1), it represents an extreme low pressure for a cold-core depression. Wind speeds at Aberdeen, 200 km from the center of the storm, exceeded 200 km hr^{-1}. These velocities are similar to those experienced in tropical cyclones. Tides in the North Sea were superimposed upon a seiching wave (also known as a Kelvin wave) that moved counterclockwise along the coast because of Coriolis force. The storm moved in the same direction as this Kelvin wave. In doing so, it set up a storm surge, exceeding 4 m in height, which flooded The Wash in England and breached the dams fronting the Zuider Zee. (The effect of this surge will be discussed in more detail at the end of this chapter.) In Britain, 307 people died, 32 000 inhabitants had to be evacuated, and 24 000 homes were damaged. Over 83 000 hectares of valuable agricultural land were submerged under salt

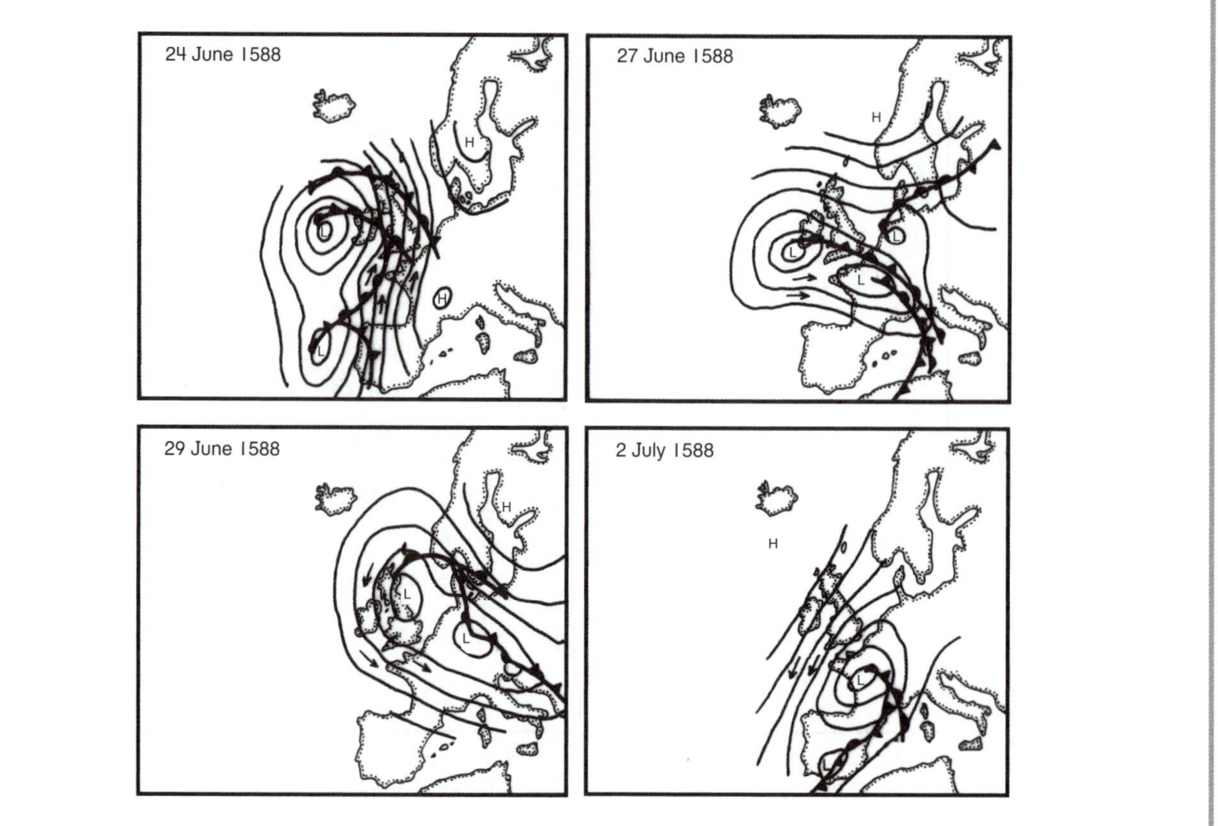

Fig. 3.14 Pressure pattern over northern Europe during the Spanish Armada-attempted invasion of Britain, 24 June–2 July 1588 (Lamb, ©1982; with permission from Methuen, London).

water. In the Netherlands, dykes were breached in over 100 places, flooding 1 600 000 hectares or one-sixth of the country. In this country, 2000 lives were lost and 250 000 head of livestock perished. The storm was estimated to have had a recurrence interval of 1:500 years, but does not rank as severe as some of those that occurred in the Middle Ages. It did become a benchmark for modern engineering design in the Netherlands and Britain and gave rise to renewed interest in coastal engineering in these countries.

The North Sea region still has the potential to spawn some of the most intense storms ever witnessed. The last years of the twentieth century have seen an increase in these intense mid-latitude storms, with some of the lessons of past events ignored. For example, on the night of 15 October 1987, very warm, humid air from the west of Africa clashed with cold Arctic air off the coast of France, triggering the formation of a very intense low-pressure cell off the Bay of Biscay. The storm, which swept up the English Channel and over the south of England, eventually developed a central pressure of 958 hPa, the lowest ever recorded in England. Wind gusts of 170–215 km hr^{-1} affected a wide area between the coast of France and England. Only 18 people were killed, but the storm virtually destroyed all large trees growing in the south of England. The French Meteorology Department had issued a very strong wind warning two days in advance. The same information was available to the British Meteorological Office, but went unheeded until the storm actually hit the English coastline. The storm was reported as the worst to hit England, breaking pressure and wind records, and was followed by similar storms in January–February 1990.

Two recent windstorms, on 26–27 December 1999, rank as intense as any medieval storm. Both storms were given names – *Lothar*, which crossed northern France, southern Germany and Switzerland on 26 December, and *Martin*, which crossed central and southern France, northern Spain and northern Italy a day later. Each developed in front of a rapidly moving mobile polar high and developed wind speeds of 180 km hr^{-1} and 160 km hr^{-1}, respectively. The storms caused 80 deaths and damaged 60 per cent of the roofs in Paris. *Lothar* and *Martin* blew over 120 and 80 electricity pylons, respectively. More than three million insurance claims were filed, bankrupting several insurance companies. The storms cost \$US12 and \$US6 billion damage, respectively, of which less than 50 per cent was insured. Destructive as these storms were, they had a theoretical return period of 5–10 years and were matched by similar magnitude events on at least eight other occasions, including the 16 October 1987 windstorm that devastated southern England. The 1:100 year storm event in Europe is currently estimated to cost \$US100 billion. Intense mid-latitude storms still plague Europe today as much as they did at any point over the past millennium.

East-coast lows or 'bombs'

Formation

(Sanders & Gyakum, 1980; Holland et al., 1988)

East-coast lows are storm depressions that develop, without attendant frontal systems, over warm bodies of water – usually off the east coasts of continents between 25 and 40° latitude. The development of such lows exhibits the forcing upon atmospheric instability by both topography and sea surface temperatures. The lows gain latent heat from warm ocean currents, which tend to form on the western sides of oceans. For some regions, where easterly moving low-pressure systems pass over a mountainous coastline and then out to sea over this warm poleward flowing current, intense cyclonic depressions can develop with structural features and intensities similar to tropical cyclones. This development is shown schematically in Figure 3.15. These regions include the east coasts of Japan, characterized by the Japanese Alps and the Kuroshio Current; the United States, characterized by the Appalachians and the Gulf Stream; and south-eastern Australia, characterized by the Great Dividing Range and the East Australian Current. Lows may also intensify off the east coast of South Africa, dominated by the Drakensburg Range and the Agulhas Current. The lows are not associated with any frontal structure, but tend to develop under, or downstream of, a cold-core low or depression that has formed in the upper atmosphere. The depressions are associated on the surface with a strong mobile polar high that may stall or block over the adjacent ocean. The initial sign that a low is developing is a poleward dip in surface isobars on the eastward side of these highs.

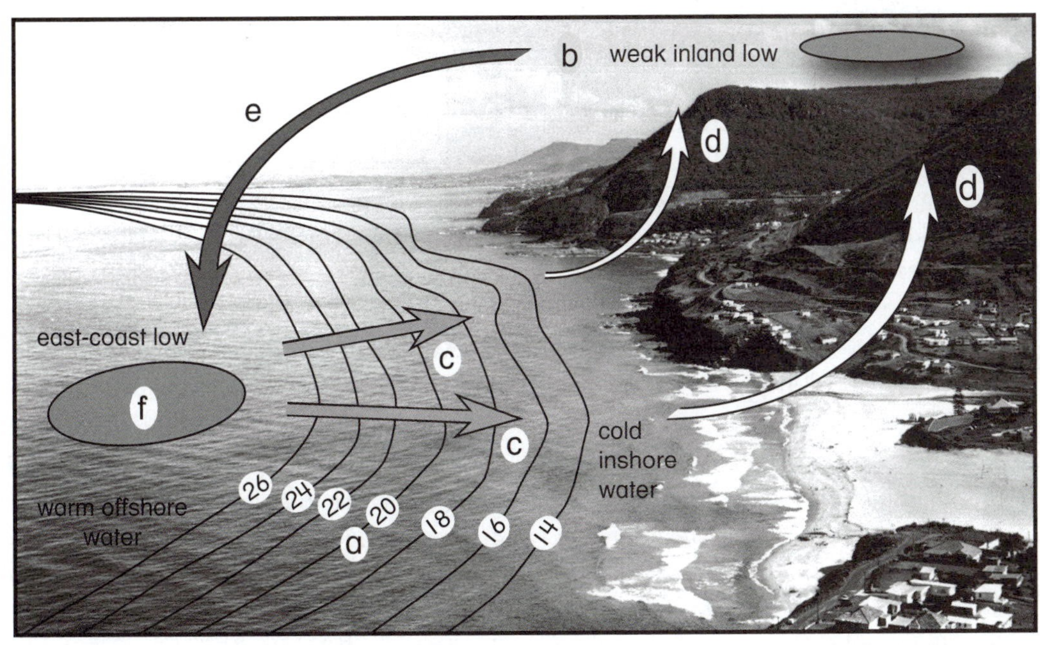

Fig. 3.15 Schematic development of an east-coast low.

The lows intensify and reach the ground near the coastline due to convection that is enhanced by onshore winds forced over steep topography at the coast. *Topographic control* also deflects low-level easterlies towards the equator, causing convergence and convection that can explode the formation of the cyclone. Some lows become 'bombs', developing rapidly within a few hours. This explosive development occurs when pressure gradients reach 4 hPa per 100 km. Convection of moisture-laden air is enhanced by cold inshore water that maximizes condensation and the release of latent heat. Along the east Australian coastline, the magnitude of the sea-surface temperature anomaly does not appear to be important in triggering a bomb. Rather, the temperature gradient perpendicular to the coast (the zonal gradient) is the crucial factor. This gradient must be greater than 4°C per 0.5° of longitude, within 50 km of the coastline, for a low to intensify.

Fig. 3.16 Infrared satellite image of the Halloween storm of October 1991 off the north-east coast of the United States. The storm developed as an east-coast low and, shortly after this photograph was taken, was reclassified as a tropical storm (source: McCowan, 2001).

East coast lows develop preferentially at night, at times when the maritime boundary layer is most unstable. East coast lows tend to form in late autumn or winter when steep sea-surface temperature gradients are most likely. However, they are not necessarily restricted to these times, and can occur in any month of the year. In Australia, a 4.5-year cycle is apparent in records of these storms over the last forty years, with a tendency for such lows to develop in transition years between ENSO and La Niña events. In extreme cases, the lows can develop the structure of a tropical cyclone. They may become warm-cored, obtain central pressures below 990 hPa, develop an 'eye' that appears on satellite images (Figure 3.16), and generate wind speeds in excess of 200 km hr^{-1}. These 'eye' structures have been detected off the east coast of both Australia and the United States. Once formed, the lows travel poleward along a coastline, locking into the location of warmest water. Such systems can persist off the coastline for up to one week, directing continual heavy rain onto the coast, producing a high storm surge, and generating waves up to 10 m in height. Heaviest rainfalls tend to occur towards the tropics, decreasing with increasing latitude. This rainfall pattern shifts towards the equator from summer to winter, reflecting the intensification and seasonal movement of high-pressure systems.

East coast lows have produced exceptional storms. The Ash Wednesday storm of 7 March 1962 in the United States and the 25 May 1974 storm in Australia are classic examples of this type of event. The most recent, significant, east-coast lows to occur in the United States were the Halloween storm of 28–31 October 1991 (The Perfect Storm), and the 21–22 December 1994 storm. The latter 'turned on' explosively, reaching a central pressure of 970 hPa, developed a warm core with attendant 'eye' structure, and generated winds of 160 km hr^{-1} and waves 11.9 m high. These events are further described below.

United States Ash Wednesday storm of 7 March 1962

(Dolan & Hayden, 1983; Nalivkin, 1983)

North America has experienced many severe, extra-tropical storms. For instance, in 1868–1869, cyclonic storms on the Great Lakes sank or damaged more than 3000 ships and killed over 500 people. However, the Ash Wednesday storm of 7 March 1962 is considered the worst storm recorded along the east coast of the United States. At no time did the storm reach land as it paralleled the coast some 100 km offshore. Its passage along the coast took four days, with waves exceeding 4 m in height for most of this time. In the decades leading up to the storm, housing development had taken place on barrier islands along the coast. The destruction of the storm was immense. Whole towns were destroyed as storm waves, superimposed on a storm surge of 1–2 m, overwashed barrier islands (Figure 3.17). The loss of life was minimal since most of the damage occurred to summer homes in the winter season. Coastal retreat was in the order of 10–100 m.

Fig. 3.17 Storm damage produced by the Ash Wednesday storm of 7 March 1962 along the mid-Atlantic United States coast.

However, this value varied considerably over short distances. The Ash Wednesday storm represented a turning point for coastal geomorphology and engineering in the United States. Afterwards, the federal government made research into coastal erosion and beach processes a priority. Since then, the concept of a coastal setback line has been developed and implemented in some states. This setback line is based on the probability, derived from historical evidence, of coastal flooding or erosion. Building development is prohibited seaward of the line, with the legality of such a prohibition being tested and upheld in the courts, specifically in North Carolina, where established insurance companies now will not insure buildings against wave damage if they are built seaward of the *setback line*.

The Halloween storm of October 1991 (The Perfect Storm)

(McCowan, 2001)

The Halloween storm of October 1991 was comparable to the Ash Wednesday storm. It began with the passage of an intense cold front preceding a mobile polar high that swept across warm seas off the north-east coast of the United States. In North America, these east coast lows are called 'nor'-easters'. The low intensified off Nova Scotia in the late evening of 28 October 1991. So strong was this extra-tropical low that Hurricane Grace, which had formed on 27 October and was moving northward along the east coast of the United States, swung dramatically towards the storm. Grace was already producing waves 3–5 m high. Grace merged with the low, which continued to deepen to a central pressure of 972 hPa by midday on 30 October, backed by a high-pressure cell with a central pressure of 1043 hPa. Within the next 24 hours, winds reached speeds of 120 km hr^{-1} and waves reached heights of 12 m. Storm conditions extended along the coast from North Carolina to Nova Scotia on a storm surge exceeding 2 m. This surge had been exceeded only during the Great Atlantic Hurricane of 1944 and the nor'-easter of March 1962. For hundreds of kilometers of coastline, sea walls, boardwalks, piers, and homes were reduced to rubble. Inland, high winds downed communication infrastructure, tore out trees, and blew roofs off buildings. Finally, in one last twist, the center of the storm passed over the warm Gulf Stream and acquired the characteristics of a hurricane.

Australian east-coast storms of May–June 1974

(Bryant & Kidd, 1975; Bryant, 1983)

East-coast lows off the Australian coast are as intense as tropical cyclones. However, the three May–June storms of 1974, while mainly confined to New South Wales, remain the most significant and widespread series of such lows to strike the east coast of Australia in 100 years. Each storm event consisted of a classic east-coast low that developed along the western edge of the Tasman Sea over warm pools of water. In each case, blocking highs over south-eastern Australia exacerbated the duration of the storms by directing south-east winds onto the coast for several days. Figure 3.18 presents the wave heights generated by the three storms and the elevation of the associated storm surge as shown by the difference between actual and predicted tide heights. Maximum wave heights reached 7 m on 25 May and again on 12 June. For most of the three weeks, wave heights were above 4 m. For the whole period, tide levels exceeded predicted ones, with the greatest deviation being 0.7–0.8 m during the 25 May storm. This is the highest recorded surge along the New South Wales coast. In embayments, this surge was accompanied by seiching and wave *set-up* such that water levels at the peak of each storm did not fall towards low tide but kept rising. In Sydney, many homes built close to the shoreline were threatened; however, only a few homes were destroyed (Figure 3.19). The sandy coastline suffered the greatest damage. In the three-month period from April to June, Stanwell Park beach high tide line retreated over 100 m. At Pearl Beach in Broken Bay, north of Sydney, a dune 7 m in height was overtopped by storm waves. Measured shoreline retreat amounted to 40 m on some beaches (Figure 2.11). During the third storm, Cudmirrah Beach south of Jervis Bay underwent dune

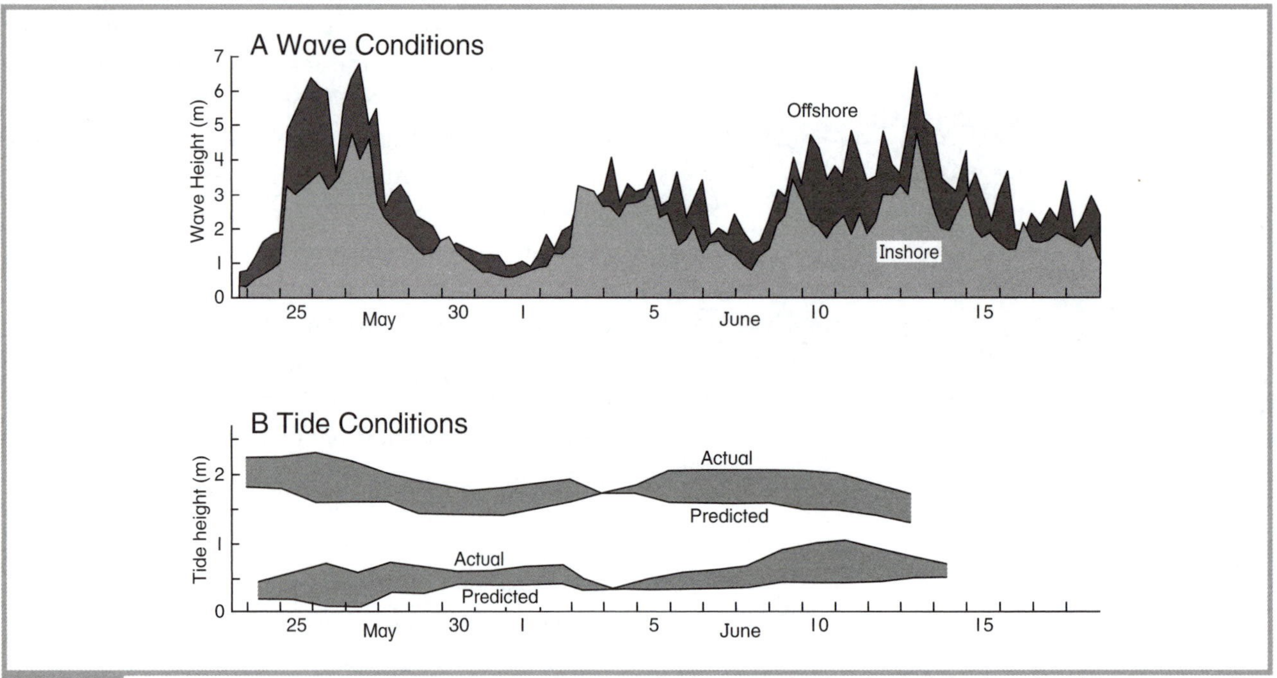

Fig. 3.18 Storms of 25 May–15 June 1974 at Sydney, Australia. A) wave heights B) actual and predicted tide heights (adapted from Bryant & Kidd, © 1975).

Fig. 3.19 A) Undermining and collapse of beachfront houses at Bilgola beach, Sydney following the 25 May 1974 storm.

Fig. 3.19 B) The same storm caused up to 40 m of erosion on other beaches. The collapsed structure shown here was a picnic table shelter situated about 5 m shoreward of the previous dune scarp at Whale Beach, Sydney.

scarping along a face 2 m in height at the rate of 1 m min^{-1}. Despite these large rates of retreat, coastal erosion was highly variable, with some beaches escaping the storm completely unscathed. Similar results were found for the Ash Wednesday storm along the east coast of the United States.

Since 1974, there have been similar east-coast lows just as intense as the May 1974 storm. The autumn storms of 1978 and 1985 both produced 10-meter waves at Newcastle. Historical documents also suggest that storms rivaling the 1974 storms for intensity have been common over the past century. The fact that the frequency and magnitude of storms appears to be constant over time is not unique to New South Wales. A similar effect has been observed along the east coast of North America. In the long term, there is no evidence that storms are solely responsible for coastal erosion in New South Wales. Additional causes of beach retreat will be described in more detail in Chapter 8.

The Sydney Hobart Yacht Race storm of 27–29 December 1998

(Australian Bureau of Meteorology, 2001)

Boxing Day in Sydney heralds the beginning of the Sydney Hobart Yacht Race that pits amateur and professional sailors against each other in a leisurely summer sail down the south-east coast of Australia. The deadly storm that beset the fleet in 1998 not only turned many off sailing forever, but was also astonishing in its intensity. A rescue helicopter provided one of the highest measurements of a wave ever recorded. As it hovered 30 m above a stricken yacht buffeted by high winds, a single wave, for a brief second, rose suddenly towards the helicopter and touched its skids. However, the worst failing surrounding the race was the fact that no warning was given at the final, pre-departure briefing – despite ominous signs that an unusual meteorological situation was developing.

The storm began with the passage of a cold front across the south of the continent on Christmas Day, the day before the race began. At the same time, an east-coast low had developed off the Queensland coast with a central pressure of 992 hPa. This latter low moved south-west and linked up with a low-pressure cell preceding the cold front. In the early hours of 27 December, both cells merged in Bass Strait between the mainland and Tasmania, to explosively produce a complex storm cell with a central pressure of 982 hPa. This cell then drifted east into the Tasman Sea and the path of most of the yachts sailing down the coast. Temporarily, the center of the maelstrom developed an eye structure characteristic of tropical cyclones. As it did so, the fleet was hammered by strong westerly winds gusting to 167 km hr^{-1}. At this time, average maximum wave heights reached 14 m, with isolated rogue waves theoretically reaching the 30 m height cited above. Many of the waves were over-steepened, probably by the East Australian current moving in the opposite direction. Not since the 1979 Fastnet Race off Britain – in which a freak storm killed 15 competitors – had the best intentions of organizers of a yacht race been so destroyed. Six people drowned in the Sydney Hobart storm, five boats sank, and fifty-five sailors were rescued under some of the most treacherous conditions ever experienced by helicopter rescue crews. The subsequent coroner's inquiry found that the Cruising Yacht Club of Australia, which organized the race, was blind to reports of impending disaster and had 'abdicated its responsibility to manage the race'.

SNOWSTORMS, BLIZZARDS AND FREEZING RAIN

Snowstorms

(Rooney, 1973; Whittow, 1980; Eagleman, 1983; Lott, 1993)

While many mid-latitude cyclonic depressions can give rise to exceptionally heavy rain and widespread flooding, conditions will always be worse if the precipitation falls in the form of snow. The fact that snow volume exceeds rainfall by a factor of 7–10 implies that even small amounts of precipitated water falling as snow can totally paralyze large sections of a continent. Even if the volume of snow in individual storms is small, it can accumulate over many falls to present a serious flood hazard lasting one to two weeks during the spring melting season. If snowstorms occur too frequently, then the effects of the previous storm may not be cleared away and urban transport systems can be slowly crippled. Whereas snow generally incapacitates transportation, freezing rain can cause severe and widespread damage to power transmission lines.

Most large snowstorms originate as mid-latitude depressions following the *meandering* path of the jet stream across the continents. In winter, the jet stream tends to be located across North America and Europe, so that the area affected by snowstorms is likely to persist over periods of several weeks. There is also the risk that abnormal or unseasonable jet stream paths can cause snowstorms to occur in areas unprepared for them. In North America, the preferred path for winter snowstorms follows jet stream looping down over the United States Midwest towards Texas and then northward parallel to the Appalachians, Great Lakes–St Lawrence River Valley, and eastward over Newfoundland. This jet stream path is topographically controlled by the Appalachian Mountains. Movement of the jet to the eastern side of the Appalachians can bring exceptionally heavy snowfalls to the east coast of the United States. In Europe, the jet stream tends to loop down over England and northern Europe. Movement southward is partially hampered by the Alps. Extended looping southward can bring very cold conditions to northern Europe, and unseasonable snowfalls to the Mediterranean region of Europe.

In winter, mid-latitude cyclonic depressions can affect very large areas of the continental United States. Figure 3.20 shows areas of intense snowfall and the development of a low-pressure cell. In this diagram, rain is falling in the warm part of the air mass. As the warm air is pushed up over the cold air, there is a small zone of freezing rain (to be discussed in detail later). As the air is forced higher along the warm front, temperatures drop below freezing point and precipitation falls as snow. The area in front, and to the poleward side of the advancing warm front, is the area of heaviest snowfall, which can cover a distance of 1000 km. Cold

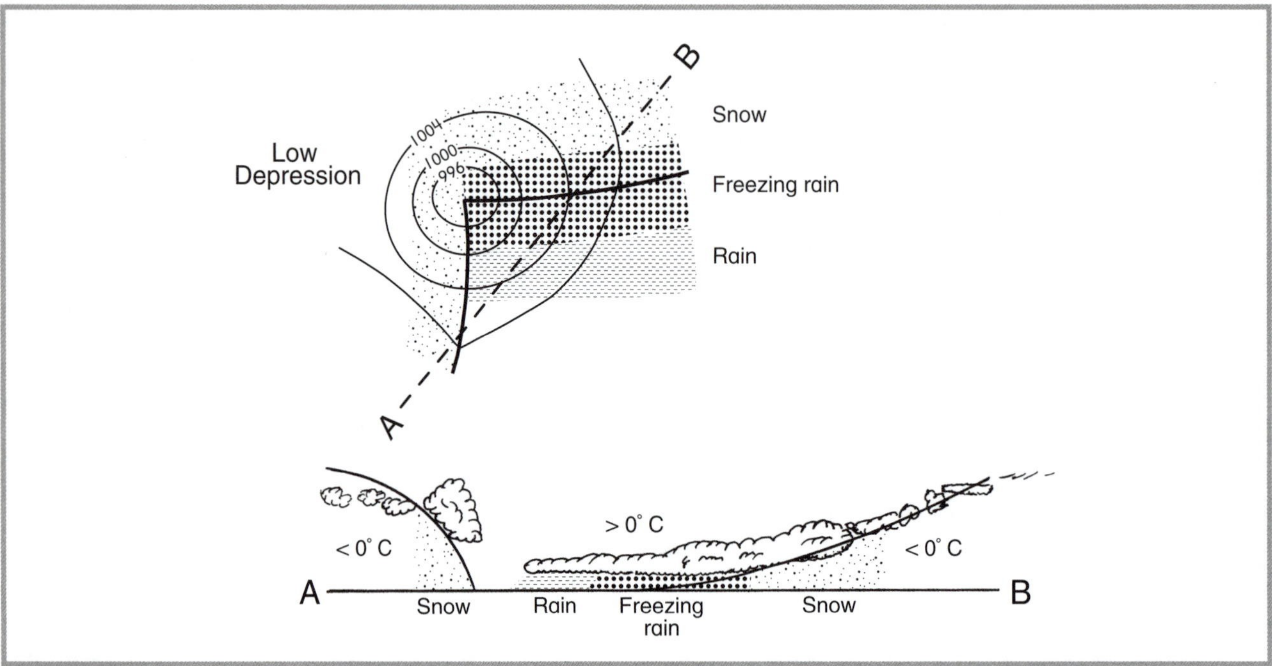

Fig. 3.20 Schematic representation of precipitation patterns for a mid-latitude cyclonic depression in winter in the northern hemisphere.

air is rapidly pushed into the low behind the warm air mass. In this region, there is rapid uplift of moist, warm air. However, condensing moisture is rapidly turned to snow. Here, snow may fall during thunderstorms, but the amounts are generally less than that accumulating in the path of the warm front. In exceptionally intense cases, another cyclonic frontal system can develop to the equatorial side of the first, forming twin low-pressure cells.

Figure 3.21 is typical of the big snowstorms to have affected eastern North America in the late twentieth century. It illustrates a cyclonic situation that brought snow to the eastern half of the United States at the end of February 1984, paralyzing transport. The eastern low was associated with very warm air from the Gulf of Mexico that brought widespread rain to the southern part of the United States. The western low was dominated by uplift of moist air along the polar front. Heavy snow extended across 1500 km, from New York to Chicago, and from Canada to the southern states. In some cases, as the cyclonic depression moves down the St Lawrence Valley and occludes, spiraling winds may turn the snow-bearing cloud back over parts of the continent already affected by the main storm, and prolong the fall of snow.

Notable events

The United States is most susceptible to large snowstorms because of the ability of mobile polar highs to interact with moisture-bearing air originating from the Gulf of Mexico. One of the worst snowstorms was 'The Great Snow of 1717', which resulted from four late-winter snowstorms that dumped 1.5 m of snow. New England's Blizzard of 1888 dropped between 1.0 and 1.5 m of snow over several days. It is most notable for the high winds that turned it into a *blizzard*. A century later the 12–15 March 1993 'Storm of the Century' replicated the havoc. This latter storm covered twenty-six states – from Texas to Maine – and the eastern provinces of Canada. The storm equated to a category 3 hurricane on the Saffir–Simpson scale. It began in the western Gulf of Mexico, tracked across Louisiana, and then rapidly intensified up the eastern seaboard before moving north-east over eastern Canada. It generated maximum winds of 232 km hr^{-1} at Mt Washington, waves 19.8 m high off Nova Scotia, a 3.7 m storm surge in Florida and a minimum pressure of 961 hPa over New York State. The snowfalls were most notable, reaching 1.4 m in Tennessee. These occurred with record-breaking cold temperatures in conjunction with power failures that left three million people without heating. The storm killed 270 people, paralyzed communications and air traffic, and impaired people's ability to get to work. Fifteen tornadoes formed over Florida, accounting for 44 of those killed. The storm cost $US6 billion, making it the most expensive extra-tropical storm in United States history. Three years later, on 6–8 January 1996, a similar storm

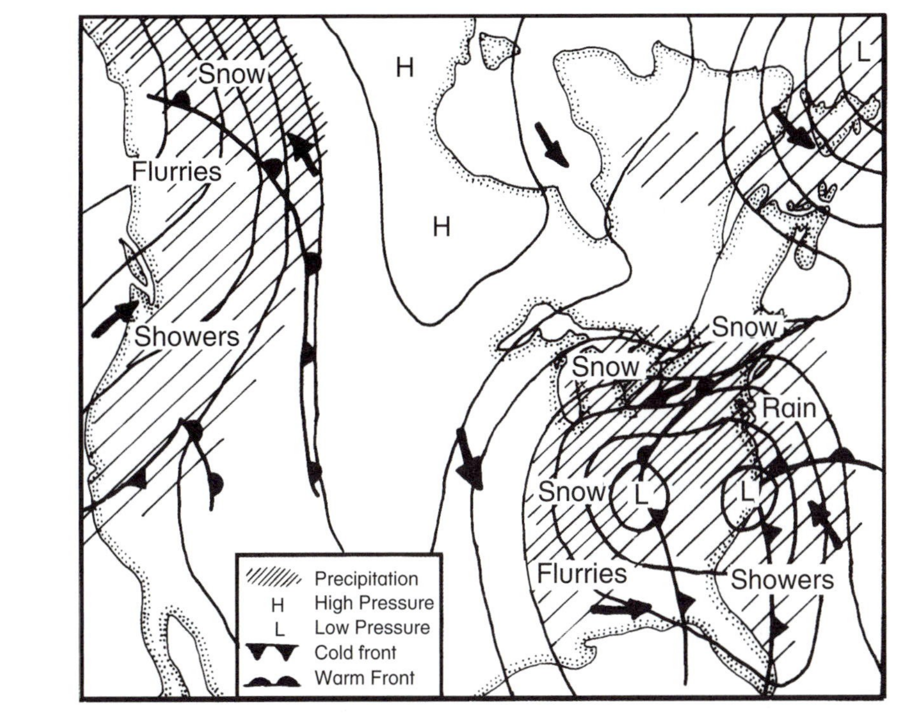

Fig. 3.21 Pressure patterns for February 1984 snowstorm that paralyzed the eastern United States.

sweeping along the same track produced heavier snowfalls in the eastern United States.

Impact of snowstorms

Heavy, northern hemisphere snowstorms have substantial effects. As soon as snow begins to fall, transportation networks in the United States are disrupted. Local roads can be completely closed off within hours of a storm striking and, depending upon the severity of the storm, closed for several days (Figure 3.22). Accident rates increase 200 per cent above average within a few hours of a storm. Geographically, motor vehicle insurance rates reflect the likely incidence of storms. During moderate falls, vehicles caught on roads can become stranded and impede snow removal. In extreme cases, stranded vehicles can become death traps. Motors left idling for warmth can suffocate occupants with carbon monoxide, while accompanying cold temperatures can freeze people to death. In the severe Illinois winter of 1977–1978, 24 people died in motor cars from these causes.

Fig. 3.22 The effect of a heavy snowstorm on roadways in St John's, Newfoundland (photograph from Canadian Atmospheric Environment Service).

Airports are also affected within hours, and can be kept closed for several days because of the time required to clear long, wide runways. Closure of major airports such as Chicago can disrupt airline schedules nationwide. Blocking of roads leads to closure of schools, industry and retail trade. The effect on retail trade, however, is often short-term because shoppers usually postpone purchases until after the storm. In fact, the seasonal onset of snow is seen in many northern states as a blessing, because it triggers consumer spending for Christmas. Purchase of winter clothes, snow tires, and heaters can also increase.

Motel business can boom during major storms. Surprisingly, garage stations record heavier business, too, because people unable to reach work can still manage to travel to the local service station to complete delayed vehicle maintenance. For most industries, a major storm means lost production. Contingencies must be made by certain industries – such as steelworks – to begin shutting down blast furnaces before a storm affects the incoming shift of workers. Unless production is made up, industry can lose profits and contracts, workers can lose pay and governments, income-tax revenue. Worker and student absenteeism can overload domestic power demand, as these people are at home during the day utilizing heating and appliances not normally used at that time. Severe, rapidly developing, or unpredicted snowstorms may even prevent the workers required for snow removal from reaching equipment depots.

Annually, many cities in North America budget millions of dollars to remove snow from city streets. Most cities have programs of sanding and salting roads and expressways to keep the snow from being compacted by traffic and turning into ice. For the cities surrounding Lake Ontario, the volume of salt used on roads is so large that increased salinity now threatens freshwater life. Snow-removal programs necessitate a bureaucracy that can plan effective removal of snow at any time of day or night during the winter months. Inefficiency in this aspect has severe economic repercussions on industry and retail trading.

All of these disruptions depend upon the perception of the storm. Cities experiencing large annual snowfalls are generally more prepared to deal with them as a hazard. Snowfalls under 10 cm – considered minor in the north-east of the United States – will paralyze southern states. However, an awareness of snowstorms can be counterproductive. In the western United States – where on a personal level snowfall is considered an everyday element of winter life – large snowfalls can cripple many major cities because they do not prepare for these higher magnitude events. In the eastern states, large storms are expected, but continuous small falls are not. Here, snow clearing operations and responses are geared to dealing with larger storms.

The variation in perception and response to the snow hazard is best illustrated around the Great Lakes. Because of their heat capacity, the Great Lakes remain unfrozen long into the winter. Cool air blowing over these bodies of relatively warm water can cause them to reach their saturation point and lead to rapid accumulation of snow downwind. As the prevailing winds are westerly in the southern lakes, and northerly around Lake Michigan and Lake Superior, snowbelts develop on the eastern or southern sides of the Lakes. In these snowbelts, average annual snowfalls can exceed 3 m – double the regional average. Often this snow falls during a few events, each lasting several days. Cities such as Buffalo on the eastern side of Lake Erie are prepared for the clearing of this type of snowfall, on top of which may be superimposed major mid-latitude storms. Cities such as Hamilton at the western end of Lake Ontario have to cope only with the irregular occurrence of mid-latitude storms. Each city has built up a different response to snow as a hazard, and each city allocates different sums of money for snow-removal operations. Buffalo removes its snow from roads and dumps it into Lake Erie; if it did not, snow would soon overwhelm roadsides. Hamilton uses sanding and salting operations to melt smaller amounts of snow, and only resorts to snow removal for the larger storms. These two cities, within 60 km of each other, thus have very different responses to snow as a hazard.

However, these average winter conditions are not guaranteed. Too many mid-latitude cells – even of low intensity – tracking across the Great Lakes will diminish the duration of westerly winds and increase the duration of easterly winds. Hamilton then becomes part of a localized snowbelt at the western end of Lake Ontario, while Buffalo receives little snowfall. Buffalo may end the winter with millions of unspent dollars in its snow-removal budget, while Hamilton has to request provincial government assistance, or increase tax rates to cope with the changed conditions of one winter's snowfall.

The winter of 1976–1977 was a dramatic exception for Buffalo. Abnormally cold temperatures set records throughout the north-east. Buffalo's well-planned snow removal budget was decimated by a succession of blizzards. On 28 January, one of the worst blizzards to hit an American city swept off Lake Erie. The storm raged for five days and occurred so suddenly that 17 000 people were trapped at work. Nine deaths occurred on expressways cutting through the city, when cars became stalled and people were unable to walk to nearby houses before freezing to death. Over 3.5 m of snow fell. Winds formed drifts 6–9 m high. For the first time in United States history, the federal government proclaimed a disaster area because of a blizzard. A military operation had to be mounted, with 2000 troops digging out the city. Not only was Buffalo cut off from the rest of the world, it was also completely immobilized within. Over one million people were stranded in streets, offices, their own homes, or on isolated farms. It took two weeks to clear the streets. To minimize the local spring flooding risk, snow had to be removed to other parts of the country using trains. The final clearance bill totalled over $US200 million, more than five times the budgeted amount for snow removal. In contrast, the city of Hamilton was relatively unaffected by this blizzard, and ended up loaning snow removal equipment to Buffalo as part of an international disaster response.

The Buffalo blizzards of 1976–1977 illustrate the additional hazard caused by abnormal amounts of snow: flooding in the subsequent spring melting period. While snow accumulation can be easily measured, the rate of spring melting is to some degree unpredictable. If rapid spring warming is accompanied by rainfall, then most of the winter's accumulation of snow will melt within a few days, flooding major rivers. The historic flooding of the Mississippi River in the spring of 1973 occurred as the result of such conditions, after a heavy snowfall season within its drainage basin. Snowmelt flooding is a major problem in the drainage basins of rivers flowing from mountains, and in the interior of continents with significant winter snow accumulation. This aspect is particularly severe in the Po Basin of northern Italy, along the lower Rhine, and downstream from the Sierra Nevada Mountains of California. On a continental scale, the flooding of the Mississippi River and its tributaries is well known. However, for river systems draining north to the Arctic Ocean, snowmelt flooding can be a catastrophic hazard. Because the headwaters of these rivers lie in southern latitudes, melt runoff and ice thawing within the river channel occur first upstream and then downstream at higher latitudes. The ice-dammed rivers, swollen by meltwater, easily flood adjacent floodplains and low-lying topography. In Canada, spring flooding of the Red River passing through Winnipeg, and the Mackenzie leading to the Arctic Ocean, is an annual hazard. Nowhere is the problem more severe than on the Ob River system in Russia. All northward flowing rivers in Russia experience an annual spring flooding hazard; but on the Ob River, the problem is exacerbated by the swampy nature of the low-lying countryside.

Blizzards

(Orville, 1993; Environment Canada, 2002; Henson, 2002)

There is one special case of snowfall which does not depend upon the amount of snow falling but upon the strength of the wind. The term 'blizzard' is given to any event with winds exceeding 60 km hr^{-1}, visibility below 0.4 km for more than three hours because of blowing snow, and temperatures below –6°C. The word derives from the description of a rifle burst and was first used to describe a fierce snowstorm on 14 March 1870 in Minnesota. The term 'purga' is used in Siberia. A severe blizzard occurs if winds exceed 75 km hr^{-1} with temperatures below –12°C. Strong mid-latitude depressions with central pressures below 960 hPa (less than some tropical cyclones), and accompanied by snow, will produce blizzard conditions. On the prairies of North America strong, dry north-westerly winds without snow are a common feature of winter outbreaks of cold Arctic air. These winds can persist for several days. Snow is picked up from the ground and blown at high velocity, in a similar manner to dust in dust storms. Surprisingly, the death toll annually from blizzards in North America is the same as that caused by tornadoes. The 28 January 1977 blizzard at Buffalo, described above, resulted in over 100 deaths across the eastern United States, and produced winds in excess of 134 km hr^{-1}. In March 1888, the late New England winter snowstorm mentioned above turned into a blizzard as winds gusted 128 km hr^{-1}. Temperatures in New York City dropped to –14.5°C – the coldest on record for March.

The main hazard of blizzards is the strong wind, which can drop the *wind-chill factor*, the index measuring equivalent still-air temperature due to the combined effects of wind and temperature. Wind chill quantifies terms such as 'brisk', 'bracing' and 'bone-chilling'. Since 1973, wind chill has been measured using the Siple–Passel Index based upon how long it takes to freeze a water-filled plastic cylinder positioned 10 m above the ground. It is assumed that, at temperatures of less than –35°C, lightly clothed skin or bare flesh will freeze in 60 seconds. If body temperature drops by more than 5°C, hypothermia and death will result. However, water in a plastic cylinder bears no relationship to reality. In addition, body heat forms a layer of warm air adjacent to exposed flesh. This layer is dispersed faster by increasing wind speeds. Finally, there is biofeedback such that flesh exposed to wind feels less cold as time progresses. A new index has been constructed that is referenced to a human walking at 5–6 km hr^{-1} with their face exposed. This index was established using volunteers walking on a treadmill in a wind tunnel. Figure 3.23 presents this version of the wind chill index. For example, a face exposed to a 50 km hr^{-1} wind at a temperature of –25°C experiences the equivalent, still-condition temperature of –42°C. For this scenario, the old index gave a greater wind chill of –51°C. The chart also indicates the temperatures and wind speeds at which human flesh will freeze after 2 minutes, 10 minutes, and prolonged exposure to the elements. For example, if the air temperature near the ground is –25°C and the wind speed is 20 km hr^{-1}, then frostbite will occur within 10 minutes. Other factors are being incorporated into this schema. One's flesh cools more quickly at night than in sunshine, an effect that is, surprisingly, exacerbated by precipitation but not by increased humidity. In addition, nothing substitutes for acclimatization. Hence, the original inhabitants of Tierra del Fuego at the tip of Patagonia were discovered walking around naked – by Europeans who found the cold unbearable.

Freezing rain

(Environment Canada, 1998; Jones & Mulherin, 1998; Fell, 2002–2003)

In some cases in winter or early spring, mid-latitude depressions will develop rain instead of snow along the uplifted warm front. To reach the ground, this rain must fall through the colder air mass (Figure 3.20). If this colder air is below freezing and the rain does not freeze before it reaches the ground, then it will immediately turn to ice upon contacting any object with a temperature below 0°C. If the warm front becomes stationary, then continual freezing rain will occur over a period of several days, and ice can accumulate to a thickness of tens of millimetres around objects. The added weight on telephone and power lines may snap the lines where they connect to a pole. Collapsing tree branches exacerbate the problem and disrupt road transport (Figure 3.24). The area most susceptible to crippling ice storms occurs in a band from central Texas, north over the Appalachian Mountains and along the north-eastern United States–south-eastern Canada border. This is the region where low pressure can develop at the leading edge of mobile polar highs and stall against the Appalachians or a preceding high-pressure cell. Severe ice storms occurred in Kentucky and Tennessee in January 1951, Georgia and South Carolina in December 1962, Maine in December 1964 and January 1979, Texas and western New York in March 1976, Iowa in March 1990, Mississippi in February 1994, and the US Midwest in December 2000.

The worst ice storm occurred on 4–10 January 1998 along the Canadian–US border east of the Great Lakes. In the United States, federal disaster areas were declared in 37 counties, 11 people died and 18 million acres of forest worth $US1–2 billion was damaged by ice. Analysis indicates that the storm has a return

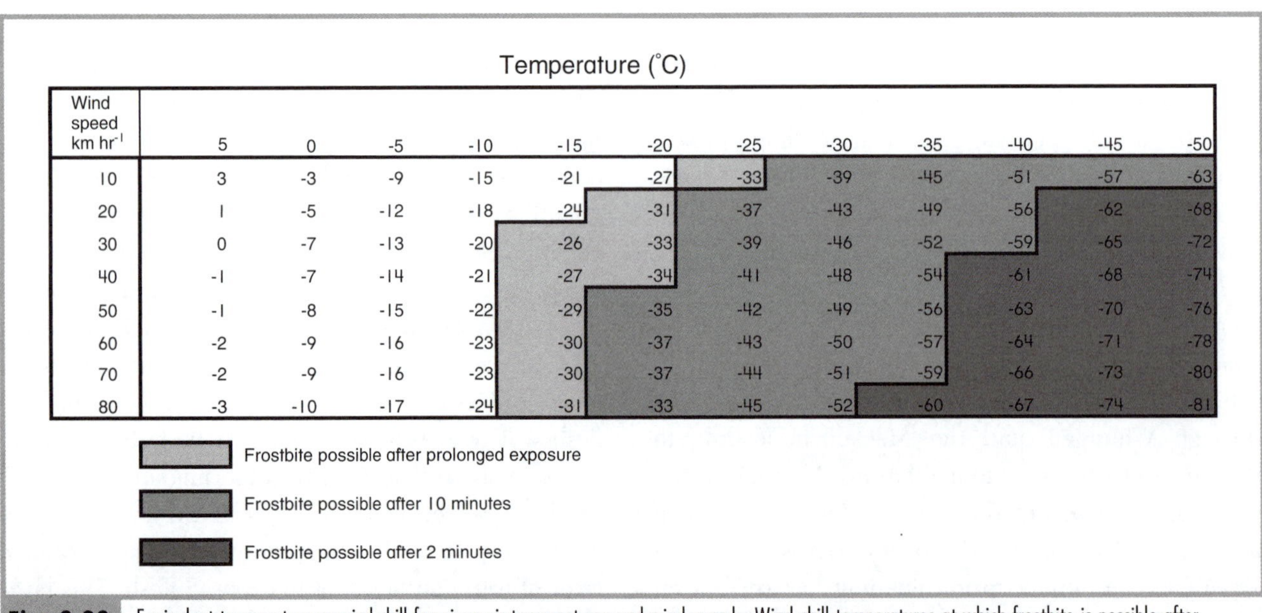

Wind speed km hr^{-1}	Temperature (°C) 5	0	-5	-10	-15	-20	-25	-30	-35	-40	-45	-50
10	3	-3	-9	-15	-21	-27	-33	-39	-45	-51	-57	-63
20	1	-5	-12	-18	-24	-31	-37	-43	-49	-56	-62	-68
30	0	-7	-13	-20	-26	-33	-39	-46	-52	-59	-65	-72
40	-1	-7	-14	-21	-27	-34	-41	-48	-54	-61	-68	-74
50	-1	-8	-15	-22	-29	-35	-42	-49	-56	-63	-70	-76
60	-2	-9	-16	-23	-30	-37	-43	-50	-57	-64	-71	-78
70	-2	-9	-16	-23	-30	-37	-44	-51	-59	-66	-73	-80
80	-3	-10	-17	-24	-31	-33	-45	-52	-60	-67	-74	-81

Fig. 3.23 Equivalent temperature or wind chill for given air temperatures and wind speeds. Wind chill temperatures at which frostbite is possible after 10 minute and 2 minute exposures are shaded (based on Environment Canada, 2002).

Fig. 3.24 The effect of freezing precipitation and wet snow on wires and tree branches in Ontario, Canada (photograph from Canadian Atmospheric Environment Service).

period of between 35 and 85 years, and that the damage was comparable to a storm that struck the same region in December 1929. In Canada, the storm paralyzed the city of Montreal and its population of four million for several weeks. Here, the accumulation of ice around objects was unprecedented and amounted to 100 mm at some locations – double any previous storm. Around Montreal, more than 1000 power transmission towers and 30 000 wooden utility poles crumbled under the weight. Over 120 000 km of power lines and telephone cables were downed. Close to 1.4 million people in Quebec and 230 000 in Ontario were without electricity for over a week. Temperatures dropped to –40°C afterwards and 100 000 people had to be evacuated to shelters for a month because of a lack of home heating. Twenty-five people died – mainly from hypothermia, 20 per cent of Canada's work force missed work, 25 per cent of its dairy industry was suspended and five million sugar maples lost the capacity to produce syrup for 30 to 40 years. Over 16 000 Armed Forces personnel had to be brought in to assist overwhelmed utility workers. The total coast to the economy was in excess of $CAD2 billion.

Two recent innovations may go a long way towards minimizing the downing of power lines by ice. The first utilizes a property of ice whereby the top layer of water molecules exposed to air is quasi–liquid and conducts electricity. Ice can be melted from power lines simply by passing a high-frequency electrical current at 50 watts m^{-1} along the cables. The second innovation completely does away with power lines. The French are developing a system whereby electricity can be transferred between towers using microwaves, thus eliminating cables.

STORM SURGES

(Wiegel, 1964; Coastal Engineering Research Center, 1977; Anthes, 1982)

Introduction

Storm surge, as alluded to above, plays a significant role in tropical and extra-tropical cyclone damage. Storm surge was the main cause of death in Bangladesh (East Pakistan) in the 1970 and 1985 cyclones; the main cause of destruction in the February 1953 North Sea storm; and the main reason that waves were so effective in eroding beaches in the May–June 1974 storms in New South Wales. In the United States storm surge is considered the main threat necessitating evacuation of residents from coastal areas preceding tropical cyclone landfall. Here, evacuation was instituted in response to the Galveston hurricane disaster of September 1900, during which 6000 people died because of storm-surge inundation. The phenomenon is also a recurring hazard in the embayments along much of the Japanese and south-east Chinese coastlines. For example, approximately 50 000 people lost their lives around Shantou (Swatow), China, on 3 August 1922 and, as recently as September 1959, 5500 people lost their lives in Ise Bay, Japan, because of storm surges. In this concluding section, the causes of storm surge and the concept of probability of occurrence will be discussed.

Causes

Storm surge is generated by a number of factors including:

- wind set-up,
- decreases in the atmospheric weight on a column of water,
- the direction and speed of movement of the pressure system,
- the shallowness of the continental shelf, bay or lake, and
- the shape of the coastline.

This discussion will ignore any addition effects due to river runoff, direct rainfall, wave set-up inside the surf zone, or Coriolis force.

The main reason for storm surge would appear to be the piling up of water by wind. The exact amount of water piled up depends upon the speed of the wind, its duration, and its location relative to the center of a cyclone. Wind set-up is difficult to calculate, but Wiegel (1964) gives an equation that shows the magnitude of wind set-up in a channel as follows:

$$h^2 \sim [(2.5\,\Pi\rho^{-1}\,g^{-1})\,(x + C_1)] - d \qquad (3.1)$$

where
h = height of wind set-up in metres
d = water depth of channel
$\Pi = 0.0025U_o^{\,2}$
ρ = density of salt water
g = gravitational constant
$U_o^{\,2}$ = wind speed
$(x + C_1)$ = the *fetch* length of the surge

Equation 3.1 does not help one to easily conceptualize wind-induced surge heights, because the actual surge depends upon where you are relative to the movement of the cyclone and its center. If the wind is moving away from a coast, it is possible to get set-down. In general, for a cyclone with 200 km hr^{-1} winds, one could expect a wind set-up of at least 2 m somewhere around the cyclone. Sea level elevation depends more upon the weight of air positioned above it. As air pressure drops, sea level rises proportionally. This is known as the *inverted barometer effect*. It must be remembered that some tropical cyclones have a pressure reduction of up to 13 per cent. Equation 3.2 expresses simply the relationship between sea level and atmospheric pressure:

$$h_{max} = 0.0433(1023 - P_o) \qquad (3.2)$$

where
h_{max} = the height of the storm surge due to atmospheric effects
P_o = the pressure at the center of the hurricane in hPa

From Equation 3.2, it can be seen that the lowest central pressure of 870 hPa ever recorded for a tropical cyclone could have produced a theoretical surge height, due to atmospheric deloading, of 6.6 m. The theoretical surge height for the North Sea storm of 1953 is 2.47 m, while for the 25 May 1974 storm in New South Wales it is 1.2 m. The last value is higher than the recorded tide level difference and reflects cyclone movement.

If a storm moves in the direction of its wind speed, then it will tend to drive a wall of water ahead of it. This wall behaves as a wave and travels with a speed similar to that of the storm. The height of the wave is related to the size of the disturbance. Along the United States east coast, this long wave can have a height of several metres, becoming highest where land juts out into the path of the tropical cyclone. Figure 3.25 shows the track of Hurricane Carol in 1954 along the United States east coast, together with records of sea level elevation at various tide gauges. Note that as the cyclone approaches land, the surge height increases, but it increases most where land at Long Island and Cape Cod intercepts the storm path. At these locations the long wave, which has been moving with the storm, piles up against the shoreline. The 21 September 1938 east-coast hurricane, moving along a similar path, sent a wall of water 6 m in height plowing into the Long Island coastline, where it made landfall. Along the Cape Hatteras barrier island chain that sticks out into the Atlantic Ocean, and along the Gulf of Mexico coastline of Florida, planning dictates that houses must be constructed above the 6.5 m storm-surge flood limit. In New South Wales, Australia, most storms move offshore; so, the long wave is in fact traveling away from the coast, and there is minimal storm surge felt at the shoreline. This is the reason storm-surge elevation during the May 1974 storm reached only 0.7–0.8 m instead of the theoretically possible value of 1.2 m due to atmospheric deloading.

Figure 3.25 also reflects two other factors affecting surge. As Hurricane Carol moved along the coast, it began to cross a shallower shelf. The storm-surge wave underwent shoaling (shallowing) as it moved shoreward. This raised its height. Figure 3.26 illustrates the effect of shoaling on a long wave. The movement of the storm-surge wave throughout the water column is dictated by the speed of the storm. As this wave moves through shallower water, its speed decreases and, because the energy flux is conserved through a decreasing water depth, wave height must increase. Any shallow body of water can generate large surges for this reason. The Great Lakes in North America are very susceptible to surges of the order of 1–2 m. In December 1985, a major storm struck Lake Erie, producing a surge of 2.5 m at Long Point, Ontario, coincident with record high lake levels. Lakeshore cottages were floated like corks 0.5–1.0 km inland. In the shallow Gulf of Finland towards St Petersburg, surges 2–4 m in height can be generated. Since 1703, there have been at least 50 occasions when surges greater than 2 m have flooded this city.

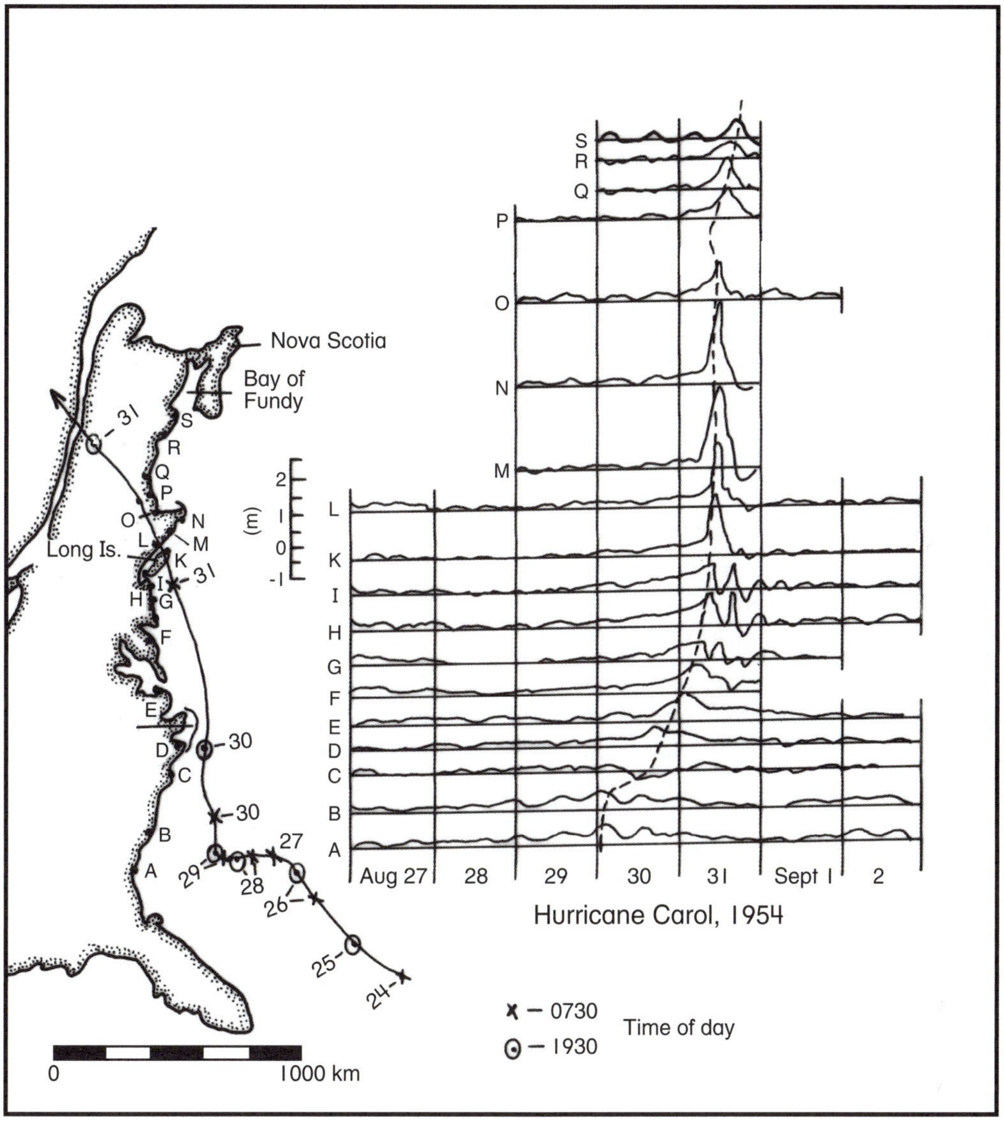

Fig. 3.25 Track of Hurricane Carol, 1954, and tide records along the United States east coast (after Harris, 1956).

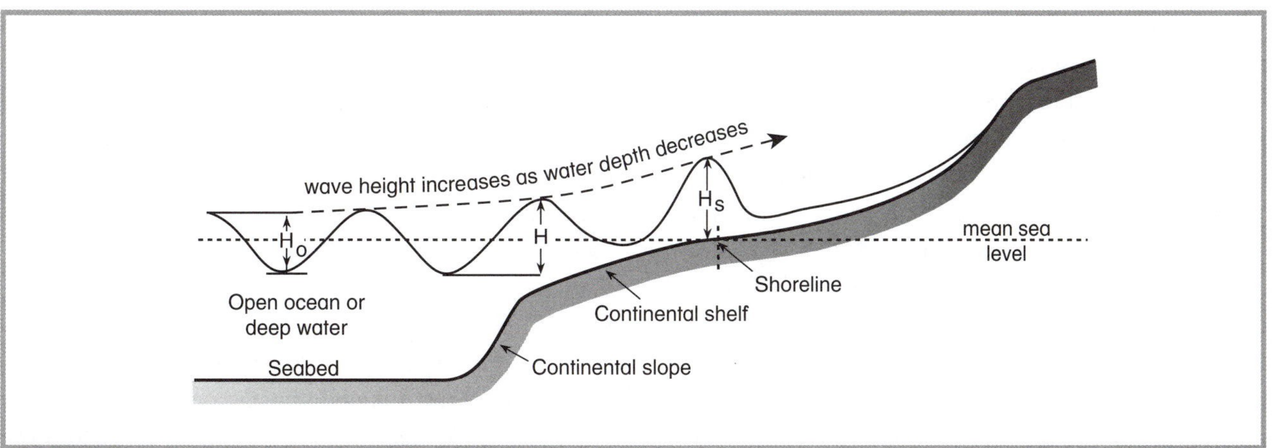

Fig. 3.26 Effect of wave shoaling on the height of long waves crossing the continental shelf.

Finally, the shape of the coastline accounts for the largest storm surge. The Bay of Bengal, where storm surges have drowned so many people, is funnel-shaped (Figure 3.10). Any surge moving up the funnel, as the 1970 cyclone storm surge did, will be laterally compressed. In addition, the size of the basin can lead to resonance if its shape matches the period of any wave entering it. A similar enhancement effect happened in 1770 to the tidal bore on the Qiantang River, south of Shanghai, China. A storm surge lifted the tidal bore on the funnel-shaped estuary to heights in excess of 4 m. The protecting dykes on each side of the river were overwashed and over 10 000 people drowned within minutes. The highest tides in the world are recorded in the Bay of Fundy in Canada because the basin shape matches, within six minutes, the diurnal tidal period of 12.42 hours. On 4 October 1869, a cyclone, called the Saxby Gale, moved up the United States coast in a similar fashion to Hurricane Carol in 1954, but slightly eastward. The storm traveled up the Bay of Fundy, moving a mass of water at about the resonance frequency (13.3 hours) of the Bay of Fundy–Gulf of Maine system. The resulting storm surge of 16 m was superimposed on a tide height of 14 m. In all, water levels were raised 30 m above low tide level, almost overwashing the 10 km isthmus joining Nova Scotia to the mainland.

PROBABILITY OF OCCURRENCE

(Leopold et al., 1964; Wiegel, 1964)

The probability of occurrence of a surge height is highly dependent upon the physical characteristics of a coastal site. To define this probability, knowledge of the size of past events and how often they have occurred over time (magnitude–frequency) is also required. This information is usually obtained from tide gauges. The maximum storm-surge values have been calculated for many tide stations around the United States Gulf and east Atlantic coast. This type of information can be used for planning; however, it is fraught with the danger that one may not have records of the most extreme events. Take the example of the 1953 storm surge in the Netherlands. Engineers correctly designed dykes to withstand the 100–200-year storm-surge event; but the 1953 event exceeded those limits. The probability of occurrence and magnitude of this event, but not its timing, could have been foreseen from a simple analysis of past historical events.

Recurrence intervals

The probability of rare events of high magnitude can be ascertained in one of two ways: by determining the recurrence interval of that event or by constructing a *probability of exceedence* diagram. In both methods, it is assumed that the magnitude of an event can be measured over discrete time intervals – for example, daily in the case of storm waves, or yearly in the case of storm surges. All the events in a time series at such intervals are then ranked in magnitude from largest to smallest. The recurrence interval for a particular ranked event is calculated using the following equation:

$$\text{Recurrence interval} = (N + 1)\,M^{-1} \qquad (3.3)$$

where N = the number of ranks
M = the rank of the individual event (highest = 1)

The resulting values are then plotted on special logarithmic graph paper. This type of plot is termed a *Gumbel distribution*. An example of such a plot is shown in Figure 3.27 for 70 years of maximum fortnightly tide heights in the Netherlands before the 1953 storm surge event. Note that the recurrence interval is plotted along the x-axis, which is logarithmic, while the magnitude of the event is plotted along the y-axis. Often the points plotting any natural hazard time series will closely fit a straight line. The extrapolation of this line beyond the upper boundary of the data permits the recurrence interval of unmeasured extreme events to be determined. For example, in the century before the 1953 storm surge in the Netherlands, the greatest recorded surge had a height of 3.3 m. This event occurred in 1894 and had a recurrence interval of one in 70 (1:70) years. The ranked storm-surge data for the Netherlands fit a straight line which, when extended beyond the 70-year time span of data, permits one to predict that a 4 m surge event will occur once in 800 years. The 1953 surge event fits the straight line drawn through the existing data and had a recurrence interval of once in 500 (1:500) years. If the dykes had been built to this elevation, they would be expected to be overtopped only once in any 500-year period. Note that the exact timing of this event is not predicted, but just its elevation. The engineers who designed the dykes in the Netherlands before the 1953 North Sea storm cannot be blamed for the extensive flooding that occurred because they had built the dykes to withstand only the 1:200 year event. At present, dykes in

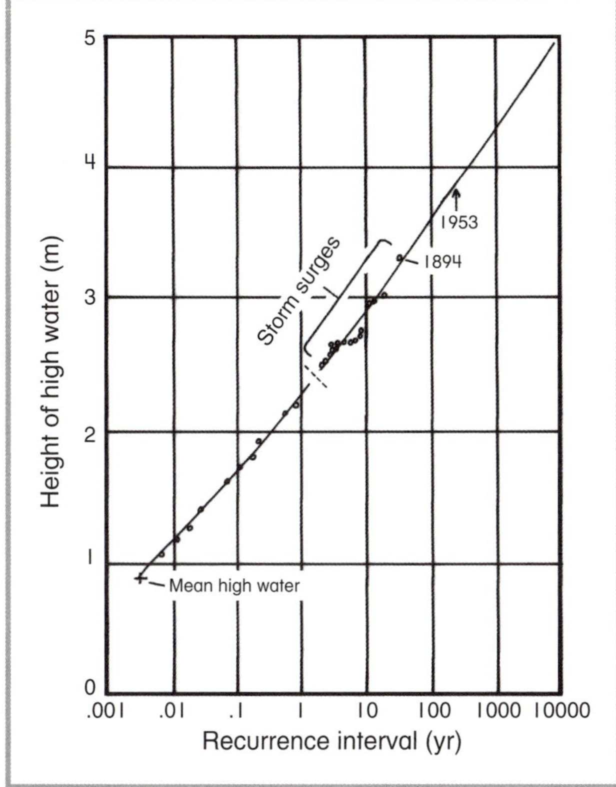

Fig. 3.27 Recurrence intervals of storm surges in the North Sea along the Netherlands coast for the 70 years before the 1953 February surge event (after Wemelsfelder, 1961).

the Netherlands have been elevated to withstand the 1:10 000 year event with a predicted surge height of 5 m.

Probability of exceedence diagrams

Recurrence intervals can be awkward to interpret, especially if the period of measurement is limited. In this case, a probability of exceedence diagram is favored because it expresses the rarity of an event in terms of the percentage of time that such an event will be exceeded. Probability of exceedence diagrams are constructed by inverting Equation 3.4 as follows:

$$\text{Exceedence probability} = M(N + 1)^{-1}\ 100\% \qquad (3.4)$$

Data in this form cannot be plotted on logarithmic paper, but instead are plotted on normal probability paper. This paper has the same advantage as logarithmic paper, in that the probability of occurrence of rare events beyond the sampling period can be readily determined, because the data tend to plot along one or two straight lines. This is shown in Figure 3.28, a probability plot for the Netherlands storm-surge data plotted in Figure 3.27. For these data, two lines fit the data well. Coincidentally, where the two lines intersect at a high-water height of 2.4 m is the point separating storm surges from astronomical tides. The advantage of a probability plot is that the frequency of an event can be expressed as a percentage. For example, a storm surge of 3 m – which in Figure 3.27 plots with a recurrence interval of 1:30 years – has a probability of being exceeded only 3.8 per cent of the time, or once every 26 years (1:26) (see Figure 3.28). Engineers find this type of diagram convenient and tend to define an event as rare if it is exceeded 1 per cent of the time. Probability plots have an additional advantage in that the probability of low-magnitude events can also be determined. For instance, annual rainfalls, when ranked and plotted as recurrence intervals, will not provide any information about the occurrence of drought. If the same data are plotted on probability paper, both the probability of extreme rainfall as well as deficient rainfall can be determined from the same graph.

Both types of plots have limitations. Firstly, an extreme event such as a 1:10 000 year storm surge, while appearing rare, can occur at any time. Secondly, once a rare event does happen, there is nothing to preclude that event recurring the next day. This constitutes a major problem in people's perception of hazards. Once high-magnitude, low-frequency events have been experienced, people tend to regard them as beyond their life experience again. The farmers of the Chars, having lived through the 1970 storm-surge event that killed 500 000 people, came back and re-established their flooded farms with the notion that they were now safe because such an event was so infrequent that it would not recur in their lifetimes. Only 15 years later, a storm surge of similar magnitude happened again. The 1973 flooding of the Mississippi River was the greatest on record in 200 years. Within two years that flood had been exceeded. High-magnitude events in nature tend to cluster over time regardless of their frequency. Such *clustering* is related to the persistence over time of the forcing mechanism causing the hazard. In the case of storm surges in the Bay of Bengal, the climatic patterns responsible for the cyclones may be semi-permanent over the span of a couple of decades. A third drawback about probability diagrams is that the baseline for measurements may change over time. For example, deforestation of a major drainage basin will cause rare flood events to become more common. In the case of the Netherlands, storm surges in the Middle Ages were certainly

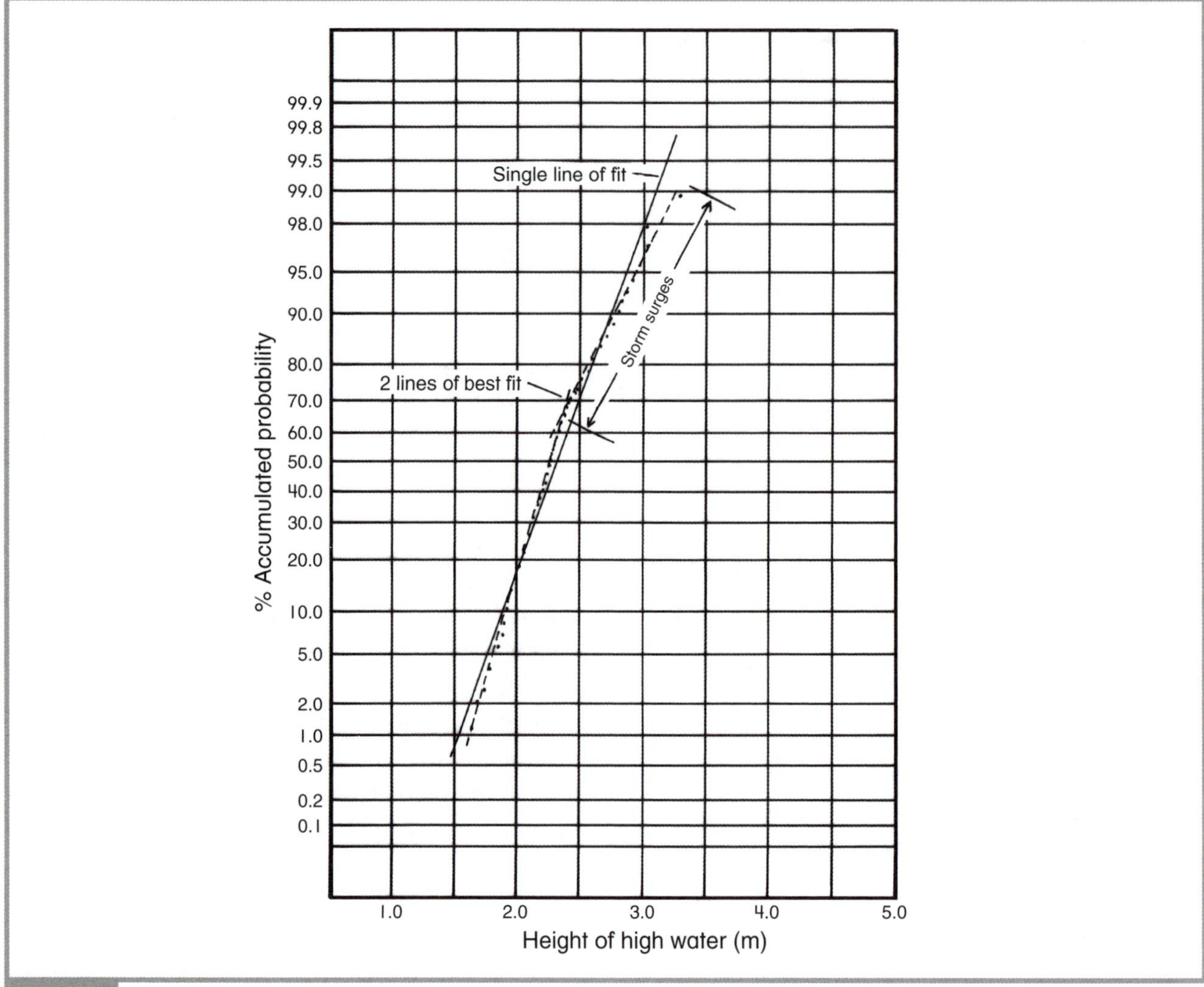

Fig. 3.28 Probability of exceedence diagram for the same data set as depicted in Figure 3.27 (data from Wemelsfelder, 1961).

more frequent and to higher elevations than at present. The 1:10 000 storm-surge elevation planned for today could also be exceeded by an event smaller than 5 m if mean sea level were to rise. Finally, neither diagram may be the most appropriate method in engineering design to determine the frequency of rare events. In the case where high-magnitude events tend to cluster over time, engineers will resort to frequency distributions that are more complex to permit the recurrence of clustered events to be evaluated more realistically.

DUST STORMS

(Lockeretz, 1978; Goudie, 1983; Middleton et al., 1986; Pearce, 1994; Tegen et al., 1996; Griffin et al., 2002)

Introduction

Dust storms are windstorms accompanied by suspended clay and silt material, usually but not always without precipitation. Presently, between 130 and 800 million tonnes of dust, with extremes as high as 5000 million tonnes, are entrained by winds each year. The average annual amount of clay-sized particles moved is about 500 million tonnes. Dust storms are responsible for most of the terrigenous material found in ocean basins, contributing over 75 million tonnes of material per year to the Atlantic Ocean alone. At distances 5000 km out into the Atlantic Ocean, fallout from the Sahara is still deposited at a rate of 3000 tonnes km^{-2} yr^{-1}. Dust from the Sahara is frequently deposited 7000 km away on Caribbean islands and in the North Amazon Basin. In the latter location, it accumulates at the rate of 13 million tonnes every year. Saharan dust often falls over western Europe with single storms being used as marker horizons in glaciers in the Alps. Dust from China exceeds these statistics. Dust from the Gobi Desert has been recorded in Hawaii and Alaska, a distance in excess of 10 000 km. A dust plume, which left China around 25 February

1990, was tracked across North America using satellite imagery. It arrived in the French Alps ten days later – having traveled a distance of 20 000 km.

Formation

Dust storms commonly form as the result of the passage of cold fronts linked to mobile polar highs across arid or drought-affected plains. The passage of these fronts can give rise to dust storms lasting for several days. In the northern Sahara region, dust storms are mainly produced by complex depressions associated with the westerlies. The depressions originate in winter in the eastern Mediterranean Sea or the Atlantic Ocean. In the southern Sahara, low-pressure fronts are responsible for *harmattan* winds. These depressions track south during the northern hemisphere winter and easterly during the summer. A particularly common feature of dust storms in North Africa is the *haboob*, generated by cold downbursts associated with large convective cells, and linked to the advance of the intertropical convergence. These types of winds also account for many of the dust storms in the Gobi and Thar deserts in central Asia, in the Sudan, and in Arizona. The seasonal movement of the monsoon into the Sudan region of the Sahara is responsible for dust storms in eastern Africa, while the Indian monsoon controls the timing of dust storms on the Arabian Peninsula. Dust transport out of northern Africa to the Mediterranean Sea and the Atlantic Ocean is correlated to stages in the North Atlantic Oscillation (NAO) described in the previous chapter. When the NAO index is high in winter, dust transport out of Africa is greater. East of the Mediterranean, depressions moving across Turkey and northern Iraq generate most dust storms. In Australia, dust storms are generated by the passage of intense, cold fronts across the continent, following the desiccating effects of hot, dry winds produced by the subtropical jet stream in the lee of high-pressure cells. The Victorian dust storm of 8 February 1983 (Figure 3.29), and the devastating bushfires of Ash Wednesday exactly one week later, both originated via this mechanism. In mountainous regions such as western North America, katabatic winds can generate dust storms in places such as California and along Colorado's Front Ranges. Similarly, cold, dense winds flowing off the mountains in the Himalayas and the Hindu Kush generate dust storms southward over the Arabian Sea. Such localized winds also produce dust storms in the valleys leading

Fig. 3.29 Dust storm from the mallee country of western Victoria about to bear down on the city of Melbourne for only the second time in a century, 8 February 1983 (photograph courtesy the Australian Bureau of Meteorology).

from the Argentine foothills. There is a correlation worldwide between dust storms and rainfall. Where rainfall exceeds 1000 mm yr^{-1}, dust storms occur on less than one day per year, mainly because vegetation prevents silt entrainment by winds into the atmosphere, and convective instability tends to give rise to thunderstorms rather than dust storms. Dust storm frequency is greatest where rainfall lies between 100 and 200 mm. Except under unusual conditions, dust storms in both hemispheres tend to have their greatest frequency of occurrence in late spring and early summer. This pattern corresponds to the changeover from minimum rainfall in early spring to intense evaporation in summer. In Australia, however, the pattern also represents the most likely period of drought associated with the Southern Oscillation.

Where cold fronts form dust storms, cold air forces warm air up rapidly, giving a convex, lobed appearance to the front of the dust storm. As the air is forced up, it sets up a vortex that tends to depress the top of the cold front and further enhance the lobe-like appearance of the storm. The classic image is one of a sharp wall of sediment moving across the landscape with turbulent overturning of dense clouds of dust (see Figure 3.29). Wind speeds can obtain consistent velocities exceeding 60 km hr^{-1} or 30 m s^{-1}. Material the size of coarse sand can be moved via *saltation* in the first tens of centimeters above the ground surface. Fine sand can be moved within 2 m of the ground, while silt-sized particles can be carried to heights in excess of 1.5 km. Clay-sized particles can be suspended throughout the depth of the troposphere and carried thousands of kilometers. Most of the dust is suspended in the lower 1 km of the atmosphere. Concentrations can be such that noonday visibility is reduced to zero. Gravel-sized material cannot be transported by winds, so it is left behind forming a stony or *reg* desert. Sand-sized particles in saltation are also left behind by the removal of silt and clay. These sand deposits become non-cohesive and mobile. Once the fine material has been removed from topsoil its cohesiveness and fertility cannot easily be re-established. The removal of fines often means the removal of organics as well. This is why the initial occurrence of dust storms in an area signifies the removal of soil fertility and, over extended periods, the *desertification* of marginal semi-arid areas.

The role of dust

Dust in storms diffuses into the upper troposphere where it absorbs heat during the day, but blocks incoming solar radiation. At the top of the atmosphere, the dust blocks incoming solar radiation by –0.25 W m^{-2} and absorbs outgoing long-wave radiation by +0.34 W m^{-2}. The global mean cooling of dust, at the ground, is –0.96 W m^{-2}, rising above –8.00 W m^{-2} over such arid locations as the Arabian Peninsula and adjacent sea (Figure 3.30). Because of dust, the ground surface is slightly cooler than normal during daytime, resulting in less convection. At night, high-altitude dust radiates long wave radiation in the upper atmosphere, cooling air and causing it to sink. At the ground, however, the dust traps in long-wave radiation that would normally escape from the Earth's surface, causing the air above the surface to remain warmer than normal, thus preventing dew formation. The sinking air aloft leads to conditions of stability, while the lack of dew keeps the ground surface dry and friable. These conditions favor aridity.

Human activity is also a major contributor of dust to the atmosphere. This effect was espoused in the 1970s during a debate about global cooling. At this time, it was postulated that humans were increasing the dust content of the atmosphere at an accelerating rate through industrial and agricultural activity. In the 1970s, archaeological evidence was used to show that some of the earliest civilizations and societies reliant upon agriculture in semi-arid regions collapsed because their poor land management practices led to increased atmospheric dust and aridity. Insidiously, arable topsoil was blown away, and the slow process of marginalization or desertification of semi-arid land took place. The long-term social consequences involved the dislocation of communities, famine, and ultimately the destruction of civilizations dependent upon such areas for their existence. This was the fate of

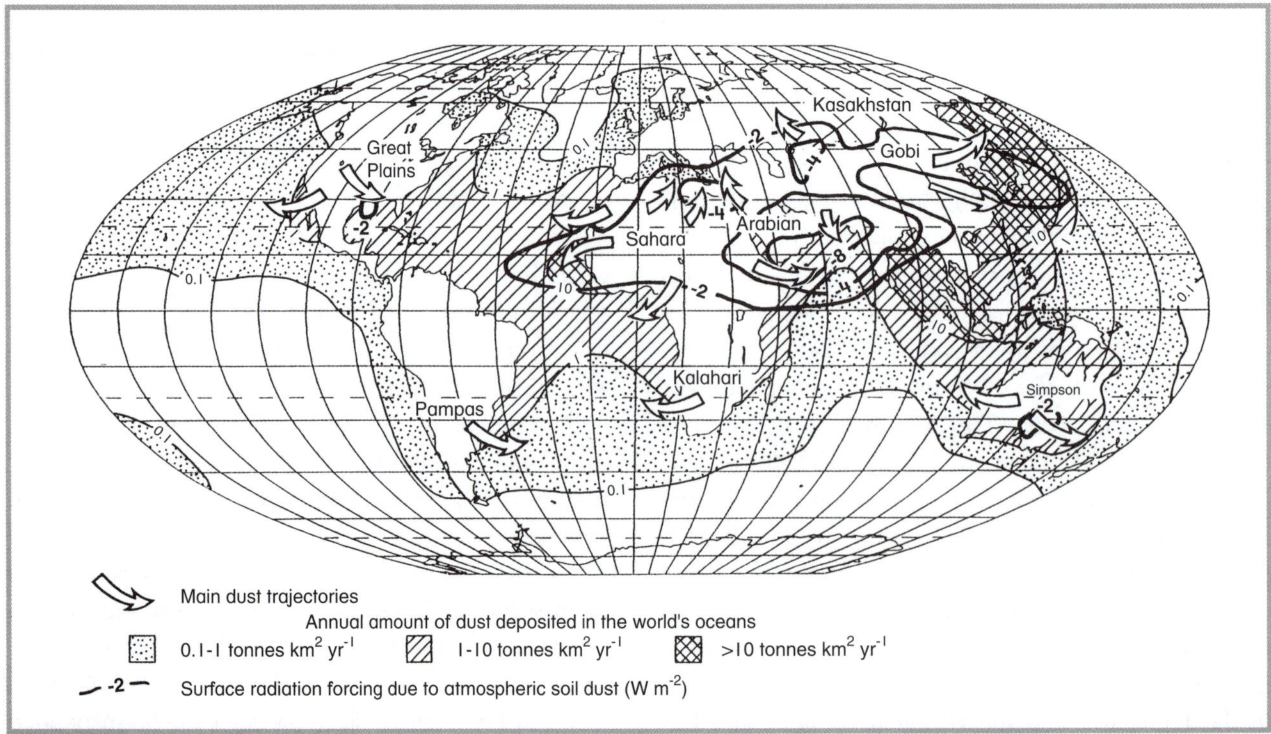

Fig. 3.30 Trajectories of dust storms and annual amounts of atmospheric dust transported to the world's oceans (based on Middleton et al., 1986; Goudie, 1983; and Pearce, 1994). Also mapped is the radiative cooling effect at the Earth's surface due to soil dust (from Tegen et al., 1996).

early agricultural civilizations in Mesopotamia, dependent upon the Euphrates and Tigris Rivers for their water. Deserts now occur in humid regions and are maintained by this human-induced dust effect. The Rajputana Desert on the Indian subcontinent exemplifies this best. Historically, the Rajputana was one of the cradles of civilization, with a well-developed agrarian society based on irrigation around the Indus River. Successive cultures occupied the desert, but each time they collapsed. Monsoons affect the Rajputana area; however, rainfall generally amounts to less than 400 mm yr^{-1}. At the time of the development of civilization by the Harappan culture, 4500 BP, rainfall exceeded 600 mm yr^{-1}. The dust put into the atmosphere by agricultural activities inhibited convection and rainfall, leading to collapse of this, and successive, civilizations. Extrapolated to the modern world, these case studies imply that anthropogenic dust from industrialization and intensifying agriculture is creating an enhanced global cooling effect.

Figure 3.30 also maps the tracks of the main dust storms worldwide and the annual amount of dust currently being deposited in the world's oceans. A broad band of dust, amounting to over 10 tonnes km^{-2}, occurs up to 1000 km off the east coast of Asia. A similar swathe extends south-east of the Indian subcontinent and through South-East Asia. High dust transport rates also occur off the west coast of northern Africa, associated with thunderstorms generated in front of mobile polar highs as these sweep along 1000 km-wide fronts. The above areas are not necessarily downwind of deserts or cultivated semi-arid regions. They represent the global signature of dust output due to both industrial and agricultural activities. Carried with this dust are substantially increased amounts of iron, both from industrial activity and soil weathering. In addition to atmospheric cooling caused directly by dust, windblown material may be indirectly cooling the atmosphere by 'fertilizing' the growth of phytoplankton and, with it, increasing dimethylsulphides and cloudiness in the marine atmosphere. Dust fertilization of algae may be responsible for red tides off Florida – with the incidence increasing concomitantly with droughts in northern Africa.

Dust storms also carry heavy metals, fungi, bacteria, and viruses. These particles, along with toxic pollutants from the burning of rubbish, and pesticides and herbicides used in semi-arid regions, are absorbed on dust particles and transported across the oceans. The problem is being exacerbated by the exposure of sediments on drying lake beds such as the Aral Sea in central Asia and Lake Chad in northern Africa. Most important is the transport of pathogens. Following major African dust storms, sugar cane rust, coffee rust and banana leaf spot originating in Africa have been found in the air over the Caribbean. The fungus *Aspergillus sydowii*, found in African dust, is also linked to sea fan disease that kills corals in the Mediterranean and Caribbean seas. Airborne dust may also be responsible for asthma, its increased incidence in some countries possibly following the increased volumes of dust originating from semi-arid countries due to human activities.

Frequency of dust storms

The frequency of dust storms cannot be overemphasized. Between 1951 and 1955, central Asia recorded 3882 dust storms. Turkmenistan in central Asia recorded 9270 dust storms over a 25-year period. In the United States Midwest, as many as ten dust storms or more per month occurred between 1933 and 1938. Dust storms occur, geographically, in the southern arid zones of Siberia, European Russia, Ukraine, western Europe, western United States, north-western China, and northern Africa. Frequencies in the Sahara range from 2–4 per year in Nigeria to 10–20 per year in the Sudan. In Iraq, the incidence of dust storms rises to over 30 per year. Many parts of central Asia average about 20 storms per year, with the frequency rising to over 60 per year in Turkmenistan on the eastern side of the Caspian Sea. North-western India, adjacent to the Thar Desert, is affected by as many as 17 events per year, with the frequency increasing northward into Afghanistan to over 70 storms per year. The highest frequency of dust storms occurs in the Gobi Desert of China, which averages 100–174 dust storms annually under the influence of the Mongolian–Siberian high-pressure cell. Dust has been blown as far as Beijing and Shanghai, where as many as four storms per year have been recorded. In Australia, more than five dust storms per year are recorded around Shark Bay, in Western Australia, and Alice Springs, north of the Simpson Desert.

It is debatable whether the incidence of dust storms is increasing because of growing populations. Chinese data show no such increase; but intense cultivation of the United States Great Plains in the 1920s, the opening up of the 'Virgin Lands' in the former Soviet Union in the 1950s, and the Sahel drought of the

1960s–1970s, all led to an increase in the frequency of dust storms in these locations. While the latter two regions have produced the most dramatic recent increase in dust storm activity because of human activities, the best-documented effect has been on the United States Great Plains during the *dust bowl* years of the 1930s Great Depression. Here, it can be shown that dust storms have a cyclic occurrence. In the 1870s, dust storms during drought drove out the first group of farmers. While no major dust storm period was recorded until the 1930s, drought has tended to occur synchronously with maxima in the 18.6-year M_N lunar cycle. Dust storm degradation of arable land also occurred in the mid-1950s and mid-1970s. While the dust bowl years of the 1930s have become part of American folklore, the drought years of the mid 1950s and 1970s actually produced more soil damage. This is despite the fact that reclamation and rehabilitation programs had been initiated by the federal government in response to the 1930s dust bowl. These programs included the return of marginal land to grassland, and the use of *strip farming*, crop rotation, and mulching practices. These techniques build up soil moisture and nutrients while protecting the surface soil from wind *deflation*. In most cases, the dust storms followed a period of rapid crop expansion during favorable times of rainfall and commodity prices.

Major storm events

There have been many notable dust storms. For instance, in April 1928 a dust storm affected the whole of the Ukrainian steppe, an area in excess of 1 million km^2. Up to 15 million tonnes of black *chernozem* soil were removed and deposited over an area of 6 million km^2 in Romania and Poland. In the affected area, soil was eroded to a depth of 12–25 cm in some places. Particles 0.02–0.5 mm in size were entrained by the wind. In March 1901, 1 million tonnes of red dust from the Sahara were spread over an area from western Europe to the Ural Mountains in Russia. In the 1930s, the dust bowl of the United States Midwest produced some of the most dramatic dust storms of the twentieth century. Their effect was enhanced by the fact that the area had not previously experienced such extreme events. Most of the storms occurred in late winter–early spring, just as the snow cover had melted, and the frozen ground inhibiting soil deflation had thawed. However, at this time of year, the polar westerlies were still strong. The most severe storms occurred between 1933 and 1938 on the southern Plains and between 1933 and 1936 on the northern Plains. The number and extent of storms during the worst period of March 1936 is shown in Figure 3.31. These storms covered an area from the Gulf of Mexico

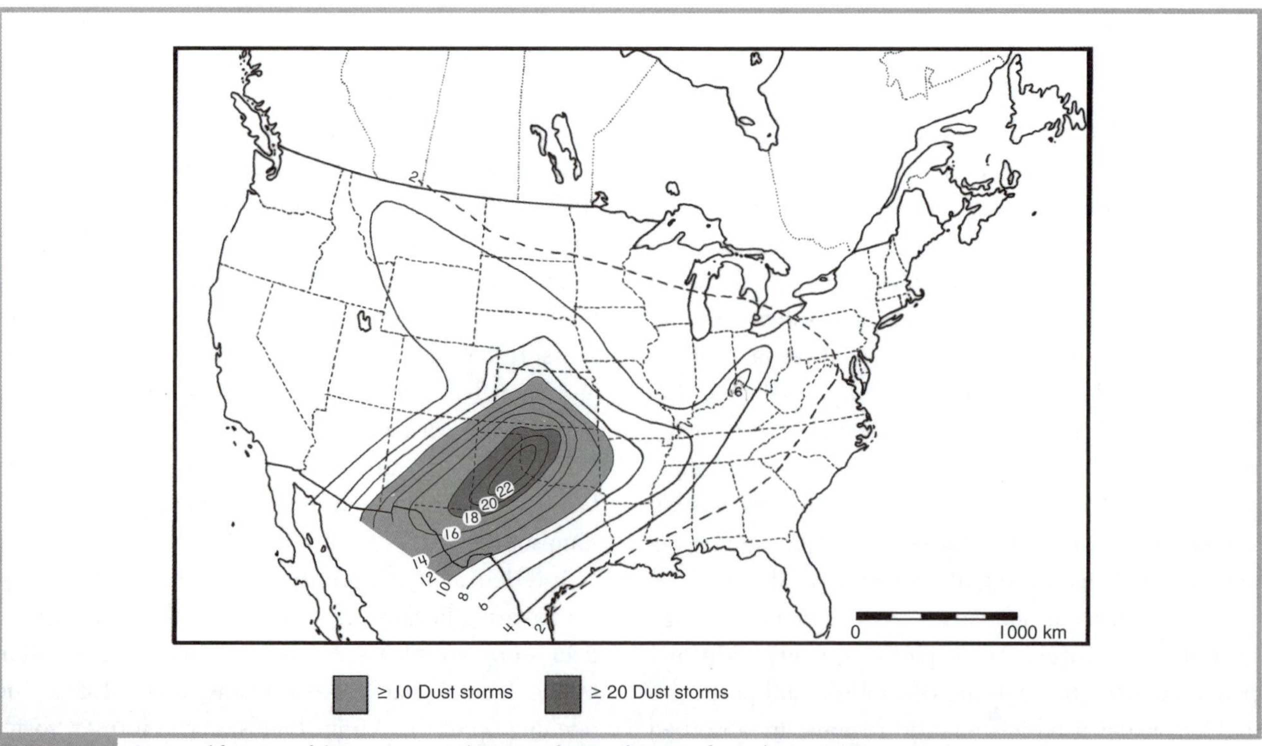

Fig. 3.31 Extent and frequency of dust storms in March 1936 in the United States (after Lockeretz, 1978).

north into Canada. During this month, the panhandle of Texas and Oklahoma experienced dust storms on 28 out of 31 days. At one point, as the United States Congress debated the issue in Washington, dust originating 2000 km to the west drifted over the capital. By 1937, 43 per cent or 2.8 million hectares of the arable land at the center of the dust bowl had been seriously depleted. In some counties over 50 per cent of the farm population went bankrupt and onto relief.

CONCLUDING COMMENTS

The phenomena described in this chapter account for most of the deaths, economic disruption, and societal change due to climatic hazards. While it is relatively easy today to model the global pressure changes described in the previous chapter, the storm systems in this chapter pose a problem because they contain elements of chaotic systems. East coast bombs are one of the best examples of this, while the Great Ice Storm of 1998 in eastern North America illustrates the unpredictability of many storms. It is thus relatively simple to predict temperature increases in a 'Greenhouse'-warmed world fifty years from now using sophisticated computer modeling; but it is impossible to predict subsequent changes in the magnitude and frequency of tropical cyclones, mid-latitude depressions or ice storms stalled at the boundary between air masses. The Great Ice Storm of 1998 also illustrates how vulnerable major cities are to climatic phenomenon. The world's largest cities of economic importance operate on electricity. Cut off power supplies and these cities cease to operate at any economically viable level for days, weeks or months. However, such disruptions are amenable to engineering solutions. Storms similar to those that struck northern Europe in the Middle Ages are not. Were a storm the size and intensity of the All Saints Day storm of November 1570 to recur today, the effects on modern European economies would be catastrophic. At present, the types of large storms described in this chapter are not so deadly as they have been historically. One of the main reasons for this is the fact that emergency services in many countries are geared to respond and militate against extreme death tolls. This has been a trend in developed countries such as the United States as well as impoverished ones such as Bangladesh. High death tolls are now more likely to occur during localized events such as thunderstorms. These localized storms will be discussed in the following chapter.

REFERENCES AND FURTHER READING

American Meteorological Society 1986. Is the United States headed for hurricane disaster?. *Bulletin American Meteorological Society* 67(5): 537–538.

Ananthaswamy, V. 2003. Historic storms live again. *New Scientist* 27 September, pp. 14–15.

Anthes, R. A. 1982. Tropical cyclones: their evolution, structure and effects. *American Meteorological Society Meteorological Monograph* 19 No. 41.

Australian Bureau of Meteorology 1977. *Report on Cyclone Tracy, December 1974*. AGPS, Canberra.

Australian Bureau of Meteorology 2001. Preliminary report on meteorological aspects of the 1998 Sydney to Hobart yacht race. <http://www.bom.gov.au/inside/services_policy/marine/sydney_hobart/contents.shtml>

Australian Bureau of Meteorology 2003. *Tropical cyclone severity categories*. <http://www.bom.gov.au/info/cyclone/#severity>

Blong, R.J. and Johnson, R.W. 1986. Geological hazards in the southwest Pacific and southeast Asian region; identification, assessment, and impact. *Bureau Mineral Resources Journal Australian Geology and Geophysics* 10: 1–15.

Bresch, D.N., Bisping, M. and Lemcke, G. 2000. *Storm over Europe: An Underestimated Risk*. Swiss Re, Zurich.

Bryant, E.A. 1983. Coastal erosion and beach accretion Stanwell Park beach, N.S.W., 1890–1980. *Australian Geographer* 15: 382–390.

Bryant, E.A. and Kidd, R.W. 1975. Beach erosion, May–June, 1974, Central and South Coast, NSW. *Search* 6(11–12): 511–513.

Burton, I., Kates, R.W. and White, G.F. 1978. *The Environment as Hazard*. Oxford University Press, NY.

Carter, R.W.G. 1987. Man's response to sea-level change. In Devoy, R.J.N. (ed.) *Sea Surface Studies: A Global Review*. Croom Helm, London, pp. 464–498.

Chang, C.P., Liu, C.H. and Kuo, H.C. 2003. Typhoon Vamei: An equatorial tropical cyclone formation. *Geophysical Research Letters* 30(3) 10.1029/2002GL016365.

Coastal Engineering Research Center 1977. *Shore Protection Manual* (3 vols). United States Army, Washington.

Cornell, J. 1976. *The Great International Disaster Book*. Scribner's, NY.

Dolan, R. and Hayden, B. 1983. Patterns and prediction of shoreline change. In Komar, P.D. (ed.) *CRC Handbook of Coastal Processes and Erosion*. CRC Press, Boca Raton, Florida, pp. 123–150.

Donnelly, J.P., Roll, S., Wengren, M., Butler, J., Lederer, R. and Webb, T. III 2001. Sedimentary evidence of intense hurricane strikes from New Jersey. *Geology* 29: 615–618.

Dvorak, V.F. 1975. Tropical cyclone intensity analysis and forecasting from satellite imagery. *Monthly Weather Review* 103: 420–430.

Eagleman, J.R. 1983. *Severe and Unusual Weather*. Van Nostrand Reinhold, NY.

Environment Canada 1984. Atlantic seasonal hurricane frequency Part I: El Niño and 30 mb quasi-biennial oscillation influences. *Monthly Weather Review* 112: 1649–1667.

Environment Canada 1998. *The worst ice storm in Canadian history* <http://www.msc-smc.ec.gc.ca/media/icestorm98/icestorm98_the_worst_e.cfm>

Environment Canada 2002. *Wind chill calculation chart*. <http://www.msc-smc.ec.gc.ca/education/windchill/windchill_calculator_e.cfm>

Fell, N. 2002–2003. Piste lightning. *New Scientist* 21 December 2002: 42; 25 January 2003: 27.

Gore, R. 1993. Andrew aftermath. *National Geographic* 183(4): 2–37.

Goudie, A.S. 1983. Dust storms in space and time. *Progress in Physical Geography* 7: 503–530.

Gray, W.M. 1975. Tropical Cyclone Genesis. Atmospheric Science Paper No. 4, Department of Atmospheric Science, Colorado State University, Fort Collins, Colorado.

Gray, W.M. 1984. Atlantic seasonal hurricane frequency Part 1: El Niño and 30 mb quasi-biennial oscillation influences. *Monthly Weather Review* 112: 1649–1667.

Gray, W.M., Landsea, C.W., Mielke, P.W. and Berry, K.J. 1994. Predicting Atlantic Basin seasonal tropical cyclone activity by 1 June. *Weather and Forecasting* 9: 103–115.

Griffin, D.W., Kellogg, C.A. Garrison, V.H., and Shinn, E.A. 2002. The global transport of dust. *American Scientist* 90: 227–235.

Harris, D.L. 1956. Some problems involved in the study of storm surges. *United States Weather Bureau, National Hurricane Research Project Report*, No. 54. Washington.

Hawkins, H.F. and Imbembo, S.M. 1976. The structure of a small, intense hurricane, Inez 1966. *Monthly Weather Review* 104: 418–425.

Henson, R. 2002. Cold rush scientists search for an index that fits the chill. *Weatherwise* 55 January/February pp. 14–20.

Holland, G.J., Lynch, A.H. and Leslie, L.M. 1988. Australian east-coast cyclones. Part I: Overview and case study. *Monthly Weather Review* 115: 3024–3036.

Holthouse, H. 1986. *Cyclone: A Century of Cyclonic Destruction*. Angus and Robertson, Sydney.

Jones, K.F. and Mulherin, N.D. 1998. *An evaluation of the severity of the January 1998 ice storm in northern New England*. US Army Cold Regions Research and Engineering Laboratory <http://www.crrel.usace.army.mil/techpub/CRREL_Reports/reports/IceStorm98.pdf>

Lamb, H.H. 1982. *Climate, History and the Modern World*. Methuen, London.

Leopold, L.B., Wolman, M.G. and Miller, J.P. 1964. *Fluvial Processes in Geomorphology*. Freeman, San Francisco.

Linacre, E. and Hobbs, J. 1977. *The Australian Climatic Environment*. Wiley, Brisbane.

Lockeretz, W. 1981. The lessons of the dust bowl. In Skinner, B.J. (ed.) 1981 *Use and Misuse of Earth's Surface*. Kaufmann, Los Altos, California, pp. 140–149.

Lott, N. 1993. *The big one! A review of the March 12–14, 1993 'Storm of the Century'*. National Climatic Data Center Research Customer Service Group Technical Report 93-01 <ftp://ftp.ncdc.noaa.gov/pub/data/techrpts/tr9301/tr9301.pdf>

Lott, N., McCown, S., Graumann, A. and Ross, T. 1999. *Mitch: The deadliest Atlantic hurricane since 1780*. United States National Climatic Data Center <http://www.ncdc.noaa.gov/oa/reports/mitch/mitch.html#DAMAGE>

McCowan, S. 2001. *The Perfect Storm*. US National Climatic Data Center, <http://www.ncdc.noaa.gov/oa/satellite/satelliteseye/cyclones/pfctstorm91/pfctstorm.html>

Middleton, N.J., Goudie, A.S. and Wells, G.L. 1986. The frequency and source areas of dust storms. In Nickling, W.G. (ed.) *Aeolian Geomorphology*. Allen and Unwin, London, pp. 237–260.

Milne, A. 1986. *Floodshock: The Drowning of Planet Earth*. Sutton, Gloucester.

Nalivkin, D.V. 1983. *Hurricanes, Storms and Tornadoes*. Balkema, Rotterdam.

NOAA 2002. *NOAA composite satellite image of Hurricane Andrew's path August 22–25, 1992*. <http://www.noaa.gov/images/andrew-comp0822-2592.jpg>

NOAA 2003. *Image ID: wea00544, historic NWS collection*. http://www.photolib.noaa.gov/historic/nws/wea00544.htm

Orville, R.E. 1993 Cloud-to-ground lightning in the Blizzard of '93. *Geophysical Research Letters* 20: 1367–1370.

Pearce, F. 1994. Not warming, but cooling. *New Scientist* 9 July: 37–41.

Rooney, J. F. 1973. The urban snow hazard in the United States: an appraisal of disruption. In McBoyle, G. (ed.) *Climate in Review*. Houghton Mifflin, Boston, pp. 294–307.

Sanders, F. and Gyakum, J.R. 1980. Synoptic-dynamic climatology of the 'Bomb'. *Monthly Weather Review* 108: 1589–1606.

Santana, S. 2002. Remembering Andrew. *Weatherwise* 55 July/August: 14–20.

Sheets, R.C. 1980. Some aspects of tropical cyclone modification. *Australian Meteorological Magazine* 27: 259–286.

Simpson, R. (ed.) 2003. *Hurricane! Coping with Disaster*. American Geophysical Union, Washington.

Simpson, R.H. and Riehl, H. 1981. *The Hurricane and Its Impact*. Louisiana State University Press, Baton Rouge.

Tegen, I., Lacis, A.A., and Fung, I. 1996. The influence on climate forcing of mineral aerosols from disturbed soils. *Nature* 380: 419–422.

The Disaster Center 1998. *Hurricane Mitch reports*. <http://www.disastercenter.com/hurricmr.htm>

Wemelsfelder, P.J. 1961. On the use of frequency curves of storm floods. *Proceedings of the Seventh Conference on Coastal Engineering*, The Engineering Foundation Council on Wave Research, ASCE, NY, pp. 617–632.

Western, J.S. and Milne, G. 1979. Some social effects of a natural hazard: Darwin residents and cyclone 'Tracy'. In Heathcote, R.L. and Thom, B.G. (eds) *Natural Hazards in Australia*. Australian Academy of Science, Canberra, pp. 488–502.

Whipple, A.B.C. 1982. *Storm*. Time-Life Books, Amsterdam.

Whittow, J. 1980. *Disasters: The Anatomy of Environmental Hazards*. Penguin, Harmondsworth.

Wiegel, R. L. 1964. *Oceanographical Engineering*. Prentice Hall, Englewood Cliffs, N.J.

Localized Storms

CHAPTER 4

INTRODUCTION

The previous chapter was mainly concerned with the effects of strong winds generated by secondary features of general air circulation. These winds were associated with the development of low-pressure cells spanning areas of 10 000–100 000 km^{-2}. While tropical cyclones and extra-tropical depressions produce some of the strongest winds and highest amounts of precipitation over the widest areas, they are by no means the only source of high winds or heavy precipitation. These values can be matched by thunderstorms, which cover no more than 500 km^2 in area and rarely travel more than 100–200 km before dissipating. They can produce high-magnitude, short-period rainfalls leading to flash flooding. Thunderstorms are also associated with a wide range of climatic phenomena such as lightning, hail and *tornadoes* that bring death and destruction. Tornadoes generate the highest wind speeds and can produce localized wind damage just as severe as that produced by a tropical cyclone. This chapter will examine first the development and structure of thunderstorms resulting in lightning and hail. This is followed by a description of tornadoes and the major disasters associated with them. The chapter concludes with a discussion of warning and response to the tornado threat, an aspect that has been responsible for decreasing death tolls throughout the twentieth century.

THUNDERSTORMS, LIGHTNING AND HAIL

Thunderstorms

(Whipple, 1982; Eagleman, 1983)

Thunderstorms are a common feature of the Earth's environment. There are about 1800–2000 storms per hour or 44 000 per day. In tropical regions, they occur daily in the wet season. However, in these regions thunderstorms may not represent a hazard because they do not intensify. As thunderstorms represent localized areas of instability, their intensity is dependent upon factors that increase this instability. On a world scale, instability is usually defined by the rate at which the base of the atmosphere is heated by incoming solar radiation, especially where evaporation at the ground and condensation in the atmosphere both occur. In this case, the saturated *adiabatic lapse rate* prevails. Under these conditions, large quantities of heat energy (2400 joules gm^{-1} of liquid water that condenses) are released into the atmosphere. This process causes convective instability, which terminates only when the source of moisture is removed.

On a localized scale, the degree of instability is also dependent upon topography and atmospheric conditions such as convergence. If air is forced over a hill, then this may be the impetus required to initiate convective instability. Convergence of air masses by topography, or by the spatial arrangement of pressure patterns, can also initiate uplift. The most likely occurrence of instability takes place along cold fronts, mainly the polar front where it intrudes into moist tropical air. Such intrusions are common over the United States Midwest and along the south-eastern part of Australia.

Figure 4.1 illustrates all of the above processes over the United States. The highest incidence of thunderstorms – greater than 60 per year – occurs in Florida, which is dominated by tropical air masses, especially in summer. A smaller area of high thunderstorm frequency over the Great Plains corresponds to an area affected by topographic uplift. From here across to the east coast, thunderstorm activity occurs because of the interaction of polar and tropical air. In Australia, the distribution of thunderstorms follows a similar pattern (Figure 4.2). The greatest intensity of storms – greater than 30 per year – occurs in the tropics and along the eastern Divide, where topographic uplift is favored. In terms of forcing thunderstorm development, the passage of cold fronts plays a minor role in Australia compared to the United States. While the incidence of storms in Australia is less than in the United States, thunderstorms over tropical Australia are some of the most intense in the world. They have been referred to as stratospheric fountains because their associated convection is strong enough to pierce the tropopause and inject air from the troposphere into the normally isolated stratosphere. This process is highly effective during periods when Walker circulation is 'turned on'.

Thunderstorms globally generate electricity between a negative ground surface, and a more positively charged *electrosphere* at about 50 km elevation. Tropical thunderstorms are responsible for most of this upward current of electricity, each averaging about 1000 amperes (A). This current creates a positive charge in the ionosphere reaching 250 kV with respective to the ground. There is, thus, a voltage gradient with altitude. A steepening of this gradient enhances thunderstorm activity. This can be accomplished by increasing the amount of atmospheric aerosols such as smoke from bushfires, pollution, or thermonuclear devices. The gradient can also be increased by solar activity through the interference of the solar wind with the Earth's magnetic field. During sunspot activity, the electrosphere becomes more positively charged, increasing the frequency of thunderstorms. Thus thunderstorm

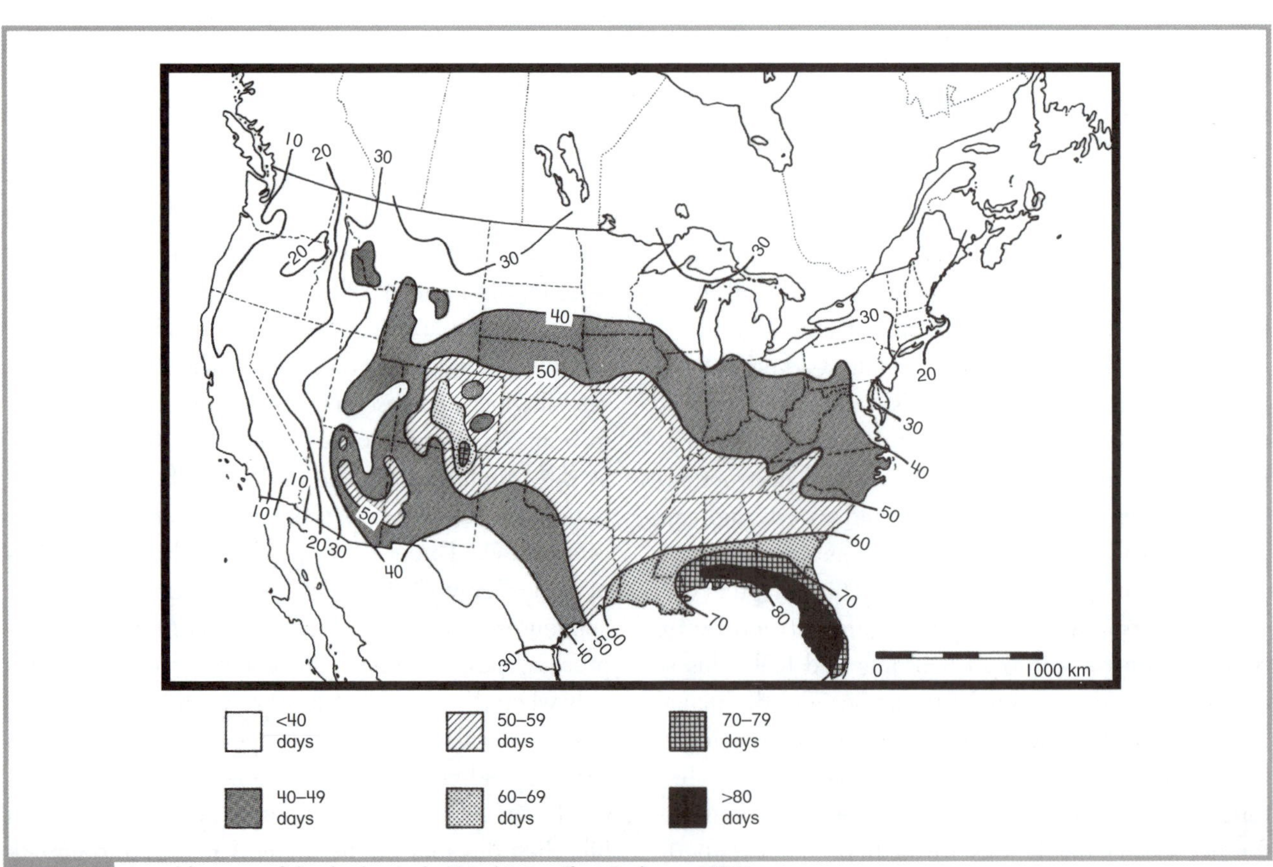

Fig. 4.1 Annual frequency of thunder days in the United States (from Eagleman, 1983).

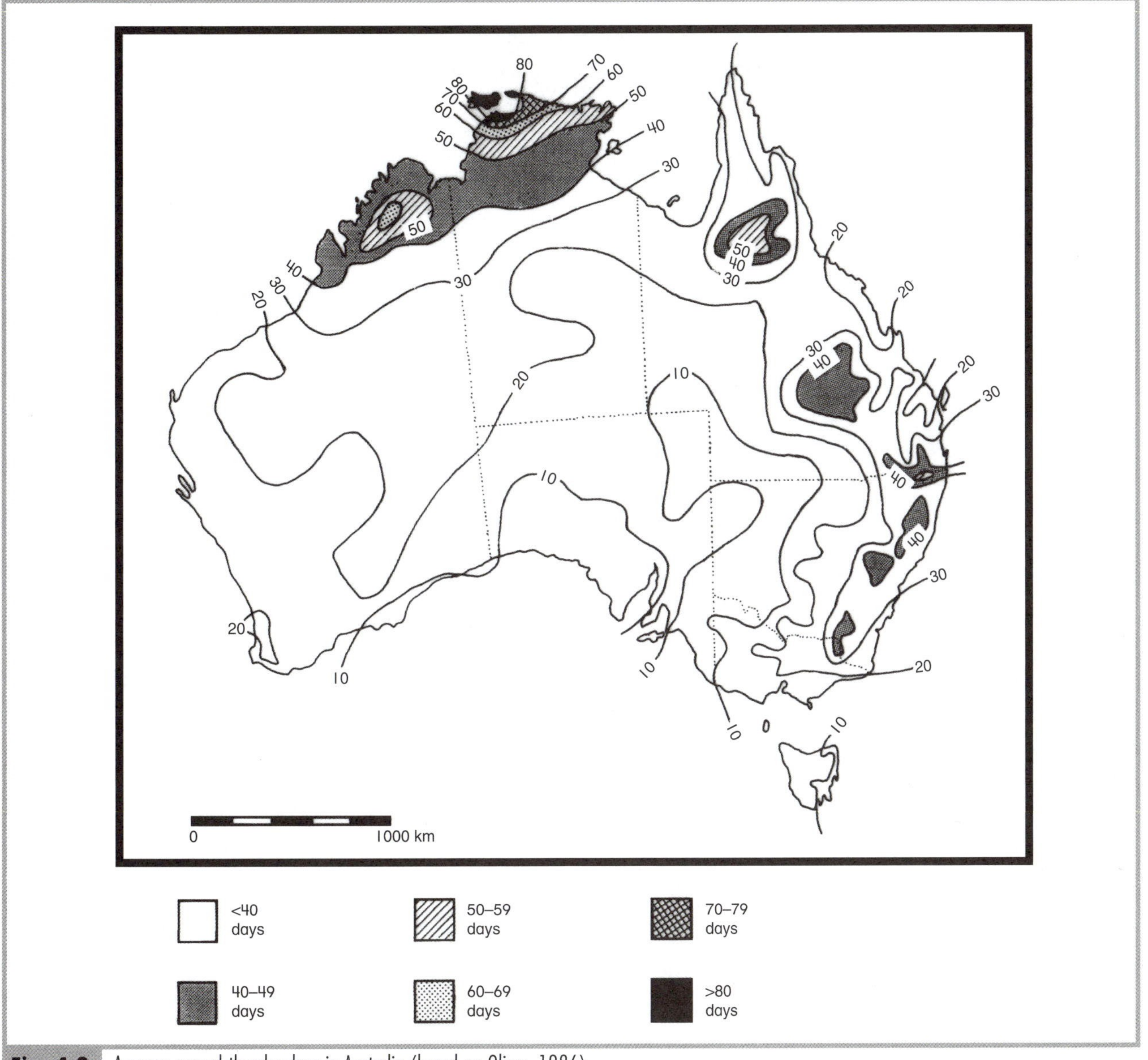

Fig. 4.2 Average annual thunder days in Australia (based on Oliver, 1986).

frequency strongly parallels the 11-year geomagnetic cycle.

Because thunderstorms in the United States are so closely linked to tornado occurrence, the threat of thunderstorms is forecast daily by evaluating the degree of atmospheric instability. This is calculated at 93 stations using air temperature measured at the 500 hPa level, and comparing this to a theoretical value assuming that air has been forced up at the forecast temperature and humidity for that day. A difference in temperature of only 4°C has been associated with tornado development. The largest tornado occurrence in the United States (when over 148 tornadoes were recorded in 11 states on 3–4 April 1974) was associated with a predicted temperature difference of 6°C in the Mississippi Valley. As well, the movement of thunderstorms is predictable. Most move in the direction of, and to the right of, the mean wind. The effect is due mainly to the strength of cyclonic rotation in the thunderstorm. In the United States Midwest, thunderstorm cells also tend to originate in the same place because of topographic effects. Thunderstorms, like tornadoes, tend to follow lower lying topography. Cities, which develop an urban heat island, also attract thunderstorms. The urban heat island initiates updrafts, which then draw in any thunderstorms developing in the area. This phenomenon has been observed, for example, in both Kansas City, United States, and London, England (see Figure 4.3 for the location of major placenames mentioned in this

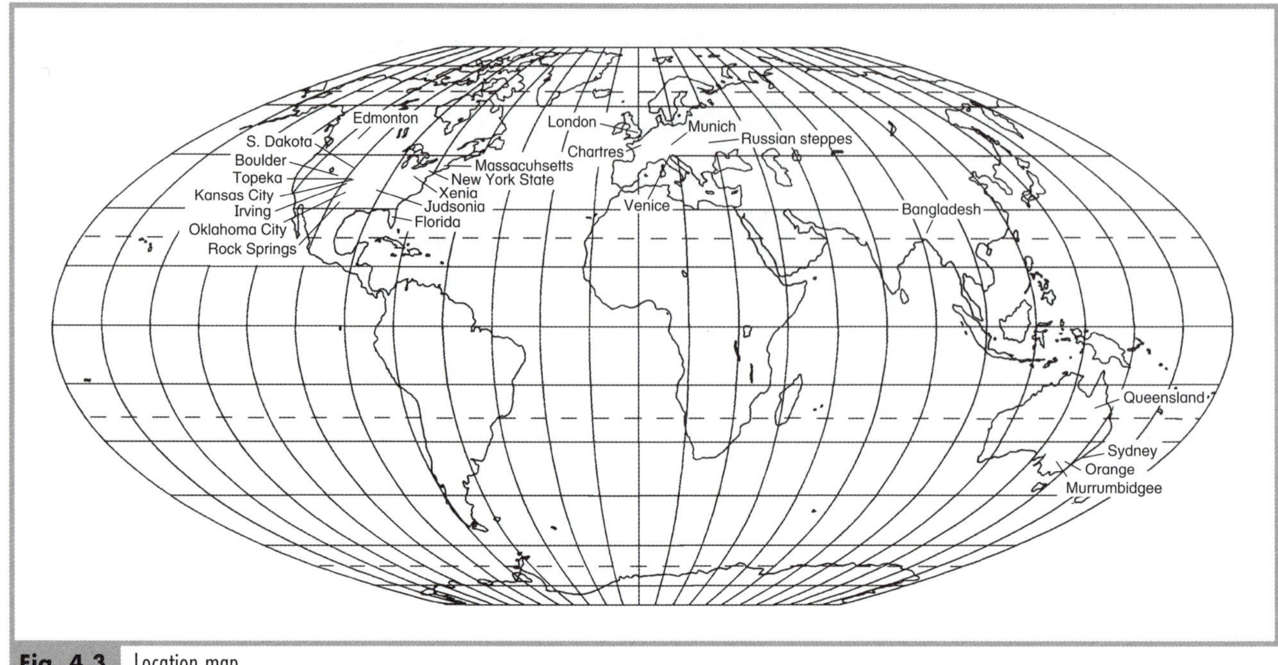

Fig. 4.3 Location map.

chapter). Thunderstorms can also be enhanced by the warmer temperatures and increased surface roughness found in urban areas. For example, the amount of rainfall from thunderstorms in Sydney, Australia, increases with increase in the percentage of land covered by concrete and buildings. The highest rainfalls occur over Sydney's central business district where turbulence increases over a regular grid of skyscrapers and streets.

Thunderstorms accompanied by their attendant phenomena have produced some colossal damage. In Sydney, on 21 January 1991, a thunderstorm swept uphill through the northern suburbs of the city accompanied by wind gusts of 230 km hr^{-1}, hail up to 7 cm in diameter and rainfall intensities up to 205 mm hr^{-1}. In twenty minutes, 50 000 mature trees were destroyed, uprooting gas and water mains and bringing down 140 km of power lines concentrated in an area of only 100 km^2. Over 7000 homes were damaged by wind or fallen trees. The storm cost $A560 million and the clean-up involved 4000 utility personnel and emergency workers.

Lightning

(Geophysics Study Committee, 1987; Black & Hallett, 1998; Anagnostou et al., 2002)

Thunder is generated when lightning briefly heats the air to temperatures of 30 000 K or five times the surface temperature of the sun. As a result, the width of the conducting channel expands from a few millimetres to a few centimetres in a few millionths of a second. This expansion produces a shock wave and noise. A build-up of charged ions in the thundercloud causes lightning (Figure 4.4). The process is known as charge separation or polarization and is aided by updrafts and the presence of ice or *graupel*. Graupel consists of aggregated ice or hail that is coated in supercooled water. If this freezes, expansion causes the outer layer of the graupel to shatter into small pieces that are carried in updrafts at speeds of 50 m s^{-1}. Shattering occurs at temperatures between –3 and –10°C. These shattered particles tend to carry a positive charge away from the larger graupel particle leaving behind a negative charge. Thus, positive charges migrate to the top of the cloud leaving negative charges at the bottom. Positive charges can also build up in a shadow zone at the ground under the thundercloud. The latter build-up, with its concomitant depletion of negative charge, has been postulated as the reason for an increase in suicides, murders and traffic accidents before the occurrence of thunderstorms. This distribution in charge produces an electrical gradient, both within the cloud, and between the base of the cloud and the ground. If the gradient reaches a critical *threshold*, it is reduced by sparking between the area of negative charge and the two areas of positive charge build-up. The amount of energy involved in a lightning discharge varies greatly, but is quite small compared to the

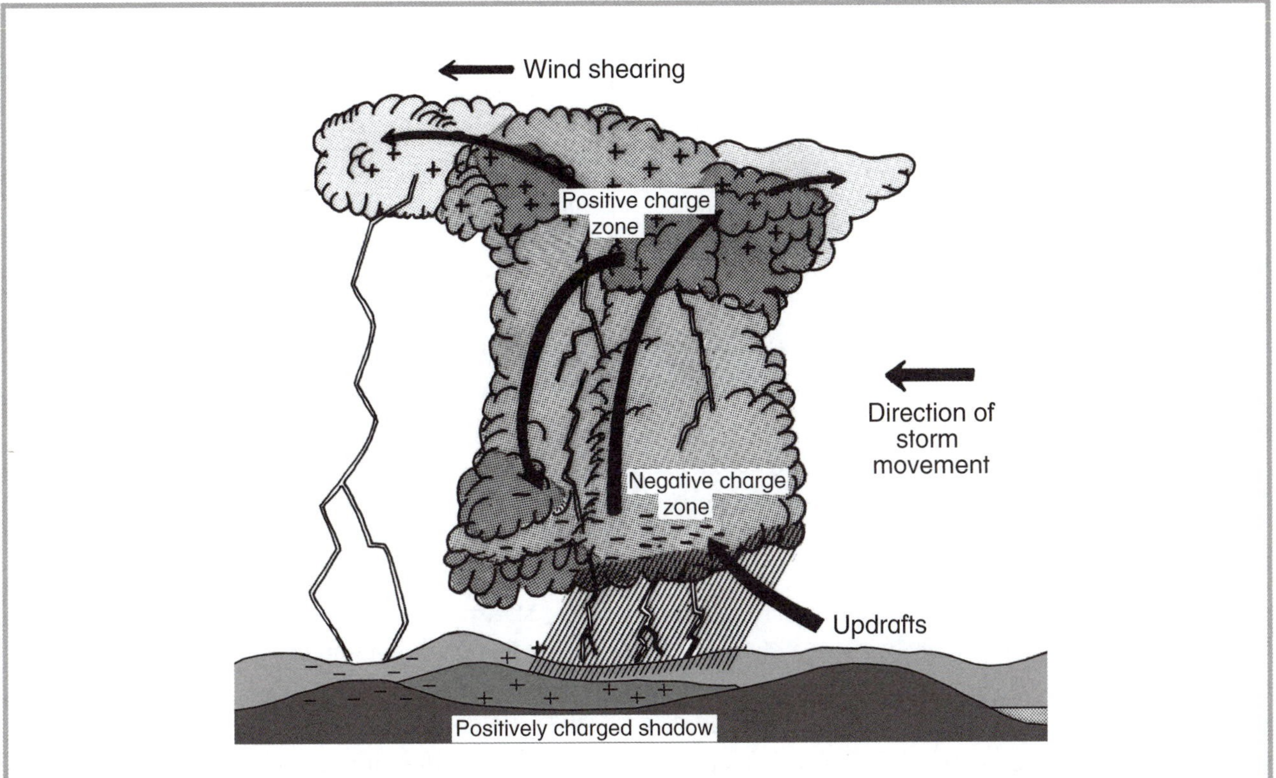

Fig. 4.4 Schematic representation of the separation of charged water droplets by updrafts in a thundercloud, giving rise to lightning.

volume of air included in the thunderstorm. For instance, the energy density of the charge is 100 million times smaller than that produced in the same volume of the air by convection or latent heat due to condensation. However, the total energy of the bolt, because it is concentrated in such a small channel, can reach 9×10^8 J. Under extreme *wind shear*, the area of positive charge at the top of the atmosphere may move downwind faster than the central core of negative charge. If negative charge builds up at the Earth's surface, a discharge can take place from the top of the atmosphere to the ground with ten times the energy of a normal lightning flash. Often the imbalance in charge adds an attractive force to moisture in the cloud. The opposing charges tend to hold moisture in the cloud. A lightning spark may weaken this hold, and lightning flashes are often accompanied by localized increases in precipitation.

Initially, the sparking originates in the area of negative charge at the base of the cloud, and surges in jerky steps towards the top of the cloud or to the ground at the rate of 500 km s^{-1}. Once a leader channel (no more than a few millimeters wide but up to 100 km long) has been established, a current races along this channel at a speed of 100 000 km s^{-1}. Currents in the channel can reach 300 000 A and be driven by a voltage potential of hundreds of millions of volts between the ground and the cloud. On average, a channel, once established, lasts long enough to carry three distinct pulses of lightning between the positive and negative ionized areas. There are on average 100–300 flashes per second globally, representing a continuous power flow of about 4000 million kilowatts in the global electricity circuit. On average, in the United States, the greatest number of cloud-to-ground lightning flashes per year – more than 10 km^{-2} – occurs in Florida; however, the greatest daily incidence of discharges to date happened in June 1984, when 50 836 flashes were recorded – mainly in and around New York State. The 'Storm of the Century' of 12–15 March 1993 mentioned in the previous chapter generated 59 000 flashes across the Atlantic seaboard, with a peak rate exceeding 5100 flashes per hour. Finally, lightning frequency relates to increases in temperature and pollutants. Urban areas with large heat islands, more dust, and increased pollution have more cloud and lightning. This implies that regions subject to recent global warming will be subject to more lightning during thunderstorms.

Lightning also generates high-frequency radio noise, gamma rays and X-ray bursts. The radio noise can be monitored by satellites, and located to a high degree of accuracy. A network of automatic direction-finding, wide-band *VHF*, magnetic receivers has been established, across North America and other continents, to locate and measure the direction, polarity, intensity, and stroke patterns of lightning. In the United States, 100 stations in the National Lightning Detection Network measure the exact time and direction of electromagnetic energy bursts produced by lightning. Lightning can also be located by measuring the difference in the time of arrival of the electromagnetic wave between two sensors. The latter method uses radio noise produced by lightning in the 7–16 kHz very low frequency (VLF) spectrum. In Europe, a network of six very low frequency receivers has been established in conjunction with global positioning systems and signal processing algorithms to locate lightning flashes globally. This network, in real time, can detect lightning within severe storms resulting in flash flooding. This can be verified using the Lightning Imaging Sensor (LIS) on the Tropical Rainfall Measuring Mission (TRMM) satellite.

Lightning is a localized, repetitive hazard. Because the lightning discharge interacts with the best conductors on the ground, tall objects such as radio and television towers are continually struck. When lightning strikes a building or house, fire is likely as the lightning grounds itself. However, this effect, while noticeable, is not severe. In the United States, about $US25 million in damage a year is attributable to lightning. Insurance companies consider this a low risk. However, because most of the damage can be avoided by the use of properly grounded lightning rods on buildings, insurance companies in North America either will not insure, or will charge higher rates for, unprotected buildings. Lightning, besides igniting forest fires, is also a severe hazard to trees and other vegetation. In North America, tree mortality in some areas can reach 0.7–1.0 per cent per year because of lightning strikes, and up to 70 per cent of tree damage is directly attributable to lightning. Scorching is even more widespread. A single lightning bolt can affect an area 0.1–10 hectares in size, inducing physiological trauma in plants and triggering die-offs in crops and stands of trees. Lightning can also be beneficial to vegetation, in that oxygen and nitrogen are combined and then dissolved in rain to form a nitrogen fertilizer.

The greatest threat from lightning is the fact that the high voltage kills people. Lightning enters a person through the body's orifices, flowing along blood vessels and the cerebral spinal fluid pathways. The current builds up in the brain and then discharges along the skin, leading to a dramatic drop in charge inside the body. This causes failure of the cardiac and respiratory systems linked to the brainstem. Victims of lightning die from suffocation because breathing stops. About 40 per cent more people die in the United States and Australia from lightning than from any other meteorological hazard. This incidence is decreasing over time despite a growth in population. Decreased fatalities reflect increased urbanization where protection is provided by tall buildings, an increase in safety education, and the improved availability of medical services. Despite the misconception that standing under a tree during a lightning storm is playing with death, statistics indicate that the number of people who die from lightning-induced charges inside houses or barns is equal to the number of those who die outside standing under trees. This statistic may merely reflect the fact that most people know about the risks of standing under trees during a storm, but not about other hazardous activities. In Australia, Telecom had to mount a major campaign in 1983 warning subscribers about the risk of using telephones during thunderstorms. Risky behavior also includes standing near electrical wiring or metal piping (including toilets) inside homes, or standing at the edges of forests, near mountains or on beaches during thunderstorms. In the latter cases, an observer may be the first tall object available for lightning discharge for some kilometers.

Hail

(Eagleman, 1983; Yeo et al., 1999)

The formation of hail depends upon the strength of updrafts, which in turn depends upon the amount of surface heating. The likelihood of hail increases with more intense heating at the Earth's surface and cooler temperatures aloft. The surface heating produces the updrafts, while the cooling ensures hail formation. Almost all hailstorms form along a cold front shepherded aloft by the jet stream. The jet stream provides the mechanism for creating updrafts. Wind shear, which represents a large change in wind speed over a short altitude, also facilitates hail formation. The degree of uplift in a thunderstorm can be assessed by the height to which the storm grows. In colder climates, however, this height is less than in more temperate or subtropical latitudes.

Locations receiving hail do not coincide with areas of maximum thunderstorm or tornado development. Figure 4.5 illustrates the annual frequency of hailstorms in the United States. Areas with the highest incidence of hailstorms on this map do not correspond well to those areas of maximum thunderstorm activity shown on Figure 4.1. The middle of the Great Plains, underlying the zone of seasonal migration of the jet stream, is the most frequently affected region. Most hailstorms in the United States occur in late spring and early summer as the jet stream moves north. A similar seasonal tendency occurs in Australia.

The size of hail is a direct function of the severity and size of the thunderstorm. General physics indicates that a hailstone 2–3 cm in diameter requires updraft velocities in excess of 96 km hr^{-1} to keep it in suspension. An 8 cm hailstone would require wind speeds in excess of 200 km hr^{-1}, and the largest stones measured (>13 cm) require wind velocities greater than 375 km hr^{-1}. If graupel is present, hailstones can grow quickly to large sizes as pieces freeze together (Figure 4.6a). Multiple-layer hailstones indicate that the stone has been repetitively lifted up into the thunderstorm (Figure 4.6b). Ice accumulates less when the stone is being uplifted because it often passes through air cooler than the freezing point. The ice in this process can be recognized from its milky appearance. When the stone descends again, water freezes rapidly to the surface forming a clear layer.

Because hail can do tremendous damage to crops, there has been substantial research on hail suppression in wheat- and corn-growing areas. Supercooled water (water cooled below the freezing point but still liquid) is very conducive to hail formation, and attempts have been made to precipitate this water out of the atmosphere before large hailstones have a chance to form. This process is accomplished by seeding clouds containing supercooled water with silver iodide nuclei, under the assumption that the more nuclei present in a cloud, the greater will be the competition for water and the smaller the resulting hailstones. In the United States seeding is carried out using airplanes, while in Russia and Italy the silver iodide is injected into clouds using rockets. In the former Soviet Union, seeding reduced hail damage in protected areas by a factor of 3–5 at a cost of 2–3 per cent of the value of the crops saved. In South Africa, crop loss reductions of about 20 per cent have been achieved by cloud seeding. In

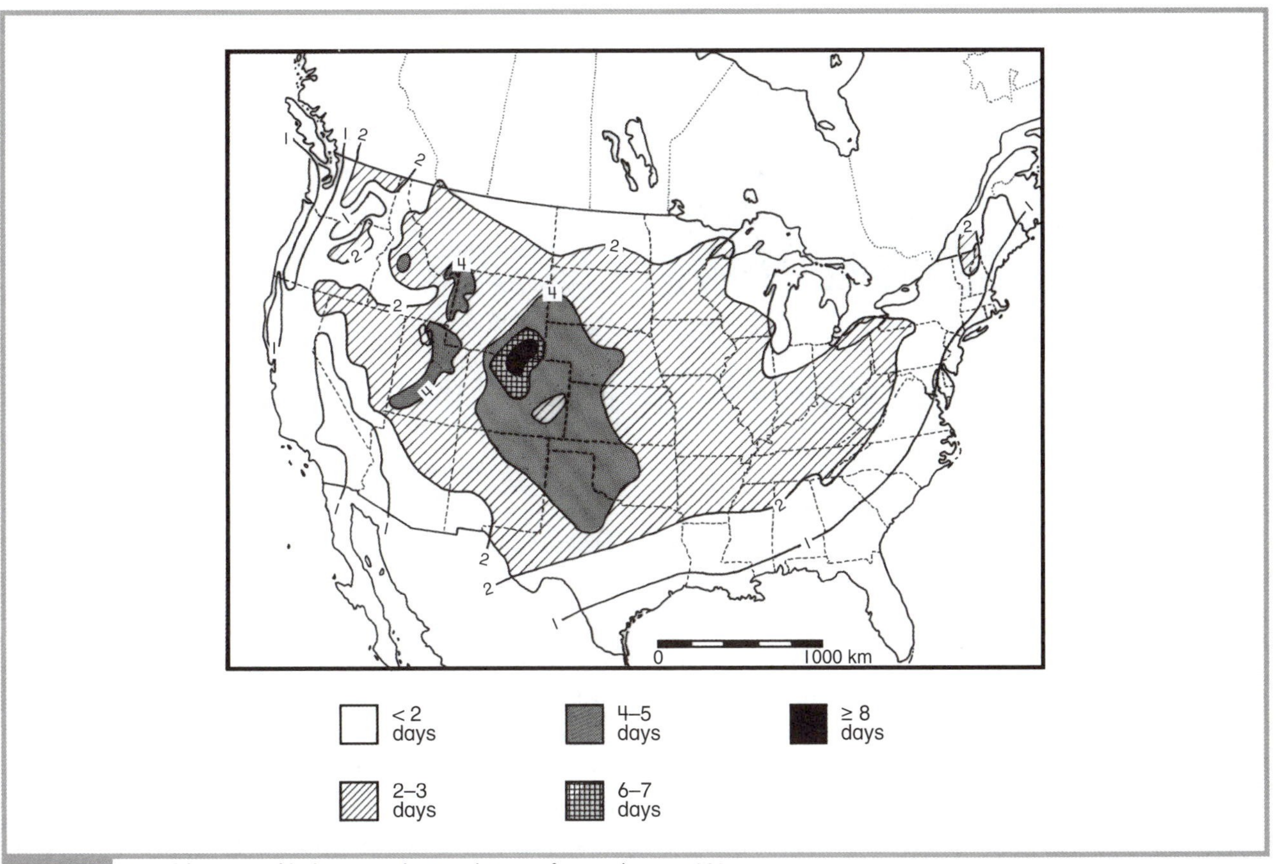

Fig. 4.5 Annual frequency of hailstorms in the United States (from Eagleman, 1983).

Fig. 4.6 A) Large hailstones that have built up quickly from the aggregation of graupel particles (photo credit: Eric Beach, 28 April 2002, Pomfret, MD http://www.erh.noaa.gov/er/lwx/publicpix/index.htm).

Fig. 4.6 B) A layered hailstone created by hail passing through the melting–freezing boundary at least five times (photo credit: http://www.crh.noaa.gov/mkx/slideshow/tstm/images/slide22.jpg).

the United States, hail suppression programs were terminated at the end of the 1970s because of political controversy. The main reason was marginal evidence that cloud seeding experiments actually increased the amount of hail, especially if thunderstorms had reached a mature or supercell stage. However, data collected over a three-year period indicated that crop losses had been reduced by 48 per cent in western Texas and by 20 per cent in South Dakota. In Australia, major hail suppression programs have not progressed beyond the experimental stage. There is some question of cost-effectiveness, given the lower crop yields over large areas. However, in the Griffith and Murrumbidgee irrigation areas, the cost of hail suppression would certainly be less than the losses experienced. To date, the purpose of cloud seeding in Australia is usually not for hail suppression, but for rain enhancement to relieve drought in New South Wales, and to prevent bushfires in Victoria.

Hail can be deadly. On two occasions since 1960, hail has killed over 300 people in Bangladesh, which experiences the most frequent and intense hailstorms of any location in the world. Nothing, however, has eclipsed the disaster that struck the army of Edward III in 1359 near Chartres, France. In a matter of minutes hail the size of goose eggs killed 1000 men and 6000 horses, effectively ending his campaign to take France in the early years of the Hundred Years' War. Hail is also destructive to crops, animals, buildings, and vehicles. About 2 per cent of the United States' crop production is damaged by hail each year. On the Great Plains of the United States, losses in some years have amounted to 20 per cent of the value of the crop. In Australia in 1985, 20 per cent of the apple crop was wiped out by a single hailstorm around Orange, New South Wales. Following the drought of 1982–1983, some wheat farmers in eastern Australia suffered the misfortune of seeing bumper wheat crops destroyed by hail. Property damage is less extensive than crop losses; however, hailstorms are the major cause of window breakage in buildings, and the major natural cause of automobile damage. The largest single insurance payout – made by the major international insurance underwriter Munich Reinsurance Corporation – occurred following the 12 July 1984 hailstorm that struck the city of Munich, West Germany. This company paid out $US500 million, mainly on damage to automobiles.

Sydney, Australia, has the misfortune of being recognized as the hail capital of the world. Insurance brokers in Australia call hailstorms 'glaziers' picnics' because of the damage they do to windows. Here, there is an additional risk to property because homes are traditionally roofed with clay tiles that can be broken by hail. Worse, low-cost public housing uses fibrocement sheeting as exterior wall cladding. Between 1980 and 1990, three or four localized storms in the Sydney metropolitan area led to claims each amounting to $A40–60 million. This was accepted as the norm until two devastating events in the 1990s.

The first occurred on 18 March 1990 and dropped hail up to 9 cm in diameter, breaking windows, roofing tiles and fibro cladding. The damage bill reached $A500 million with houses completely smashed beyond repair. This event paled in significance compared to the damage done during the Great Hailstorm of 14 April 1999.

The storm reached the outskirts of Sydney around 7:20 pm, just after the Meteorological Bureau's storm-watch roster changed and the evening television stations issued weather forecasts for plenty of sunshine lasting the next week. The forecasts were ironclad because the Bureau had already dismissed volunteers' phoned reports of the storm's intensity up to 200 km south of the city hours beforehand. Over a distance of 1 km, between pristine bushland at the edge of the urban heat island and the first row of houses in southern Sydney, the storm unleashed its fury. The lack of warning led to 30 aircraft at Sydney International airport being damaged to the tune of $A100 million. Hailstones reaching 9 cm in diameter punched through car roofs and houses at velocities of 200 km hr^{-1}. Hailstones in Sydney had reached this size only twice before: in January 1947 and March 1990. So intense was the hail that flocks of seagulls, cormorants, and flying foxes were killed in the open where they stood. By the end of the evening, 2500 km^2 of urban Sydney had been declared a natural disaster area with many homes uninhabitable as rain poured in through broken roofs. In the hail zone, 34 and 62 per cent of homes had broken windows and roofing tiles, respectively, while 53 per cent of cars were damaged. The storm statistics were massive: 500 000 tonnes of hail fell, 20 415 homes and 60 000 cars were damaged, 25 000 calls were made for assistance, 170 000 tarpaulins were fitted to roofs, 55 000 m^2 of slate and 11 000 000 terracotta tiles were replaced, and 32 000 home and 3000 commercial insurance claims were processed. It was the costliest natural disaster in Australian history, resulting in a total insurance bill ($A1.5 billion rising to $A3 billion when non-insured damage was included) that surpassed the damage caused by the Newcastle earthquake of 1989 and by Cyclone Tracy in Darwin in 1974. Despite this damage, the storm does not even rank in the top 40 most costly natural disasters in the world since 1970. It ranks equal fourth with losses due to tornadoes in the United States over the same period. The storm actually stimulated the regional economy, providing work for home renovators and car salespeople. In the weeks following the storm, the supply of roofing tiles, tarpaulins, and household windows was exhausted, and the reputation of the State Emergency Services was in tatters. Untrained to climb to the top of 3–4 storey blocks of flats to fix tarpaulins, and understaffed by the extent of the disaster, the emergency services had to defer operations to the Rural Fire Brigade which at least had extension ladders and professionals trained to skip across roofs. A week later, the army relieved everyone as the enormity of the task unfolded. In the end, the major occupational hazard for emergency services personnel – besides falling off buildings – was being stung by the thousands of wasps nesting on the sunny sides of damaged buildings.

TORNADOES

Introduction

(Miller, 1971; Eagleman, 1983; Nalivkin, 1983)

A tornado is a rapidly rotating vortex of air protruding funnel-like towards the ground from a *cumulonimbus* cloud. Most of the time, these vortices remain suspended in the atmosphere, and it is only when they connect to the ground or ocean surface that they become destructive. Tornadoes are related to larger vortex formation in clouds. Thus, they often form in convective cells such as thunderstorms, or in the right-forward quadrant of a hurricane at large distances (> 200 km) from the area of maximum winds. In the latter case, tornadoes herald the approach of the hurricane. Often, the weakest hurricanes produce the most tornadoes. Tornadoes are a secondary phenomenon, in which the primary process is the development of a vortex cloud. Given the large number of vortices that form in the atmosphere, tornadoes are generally rare; however, because vortices can be generated by a myriad of processes, tornadoes have no one mechanism of formation. On average, 850 tornadoes are reported annually, of which 600 originate in the United States. This frequency is increasing in the United States; the increase, however, is due mainly to increased monitoring. The width of a single tornado's destructive path is generally less than 1 km. They rarely last more than half-an-hour and have a path length extending from a few hundred meters to tens of kilometres. In contrast, the destructive path of a tropical cyclone can reach a width of 200 km, spread over thousands of kilometers, and last for up to two weeks. However, tornadoes move faster than cyclones and generate higher wind velocities. Tornadoes move at speeds of 50–200 km hr^{-1} and generate internal winds at speeds exceeding 400–500 km hr^{-1}. Tornadoes, similar to tropical cyclones, are almost always accompanied by heavy precipitation.

Form and formation

(Nalivkin, 1983; Grazulis, 1993)

Supercell tornado formation

By far the best-studied and most common process of tornado formation is that associated with supercell thunderstorms. The *parent cloud* may be 15–20 km in diameter with strong rotational uplift that pierces the stratosphere as a dome cloud capping the storm (Figure 4.7). This mesocyclone links to the jet stream producing an anvil shape that is skewed in the direction of movement. As precipitation falls from the storm it cools air, which sinks to the ground in front of the storm producing a gust front that precedes the thunderstorm. Destructive winds in this zone are known as a downburst and when they hit the ground can produce dangerous wind shear. Drier air behind the thunderstorm is drawn into the mesocyclone in the middle atmosphere. This evaporates cloud, giving a sharply defined outline to the storm. More importantly, evaporation causes the air to cool, become unstable and cascade to the ground as a downdraft. Tornadoes develop between this rear downdraft and the main updraft. As the downdraft expands, it collapses the supercell and at this point the tornado descends to the surface.

Dust devils, mountainadoes, fire tornadoes and waterspouts

There are also a number of small vortex phenomena that do not originate under the above conditions. Horizontal vortices tend to form over flat surfaces because air at elevation moves faster and overturns without the influence of frictional drag at the Earth's surface. If the horizontal vortex strikes an obstacle, it can be split in two and turned upright to form a dry tornado-like vortex (Figure 4.8). Because of its small size, Coriolis force does not affect these vortices, so they can rotate either clockwise or counterclockwise. On very flat ground, where solar heating is strong, thermal convection can become intense, especially in low latitudes. If the rate of cooling with altitude of air over this surface exceeds the usual rate of –1°C per 100 m (the normal adiabatic lapse rate), then the air can become very unstable, forming intense updrafts that can begin to rotate. These features are called 'dust devils' or, in Australia, 'willy-willys'. While most dust devils are not hazardous, some can intensify downwind to take on tornado-like proportions. Tornado-like vortices can also be generated by strong winds blowing over topography. Strong wind *shearing* away from the top of an obstacle can suck air up rapidly from the ground. Such rising air will undergo rotation, forming small vortices. Around Boulder, Colorado, strong winds are

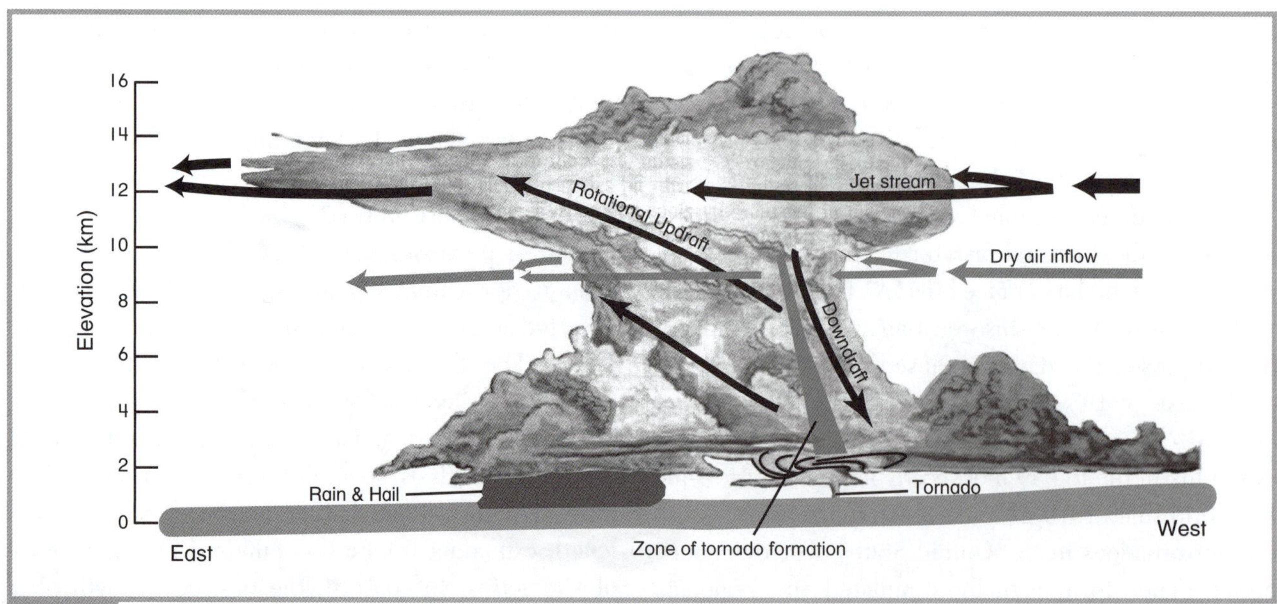

Fig. 4.7 Generation of a tornado within a supercell thunderstorm (after Grazulis, 1993).

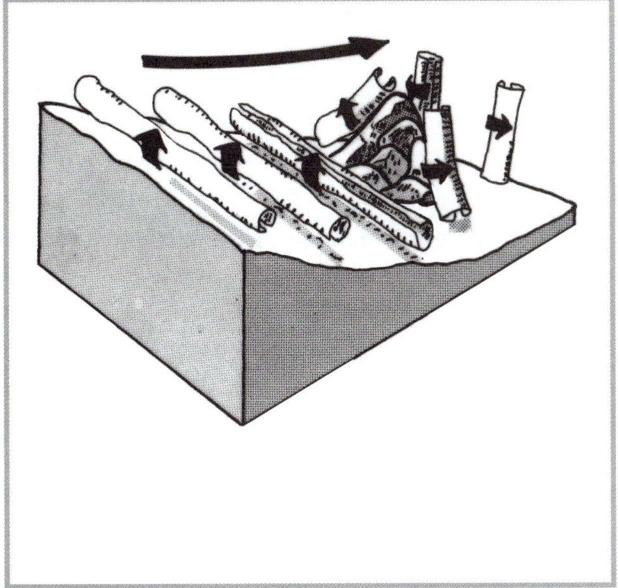

Fig. 4.8 Development of a mountainado from the splitting of a horizontal vortex rolling along the ground (after Idso, 1976).

a common feature in winter, blowing across the Rocky Mountains. Vortices have often been produced with sufficient strength to destroy buildings. Such vortices behave like small tornadoes and are called 'mountainadoes'.

Strong updrafts can also form because of intense heating under the influence of fires. Most commonly, this involves the intense burning of a forest; however, the earthquake-induced fires that destroyed Tokyo on 2 September 1923, and the fires induced by the dropping of the atomic bomb on Hiroshima on 6 August 1945, also produced tornadoes. Experiments, where the atmosphere has been heated from below using gas burners, commonly have generated twin vortices on either side of the rapidly rising columns of air. These fire whirlwinds can vary from a few metres in height and diameter to fire tornadoes hundreds of metres high. The most intense and hazardous fire tornadoes are generated by forest fires, and usually form on the lee side of hills or ridges during extreme bushfires or forest fires such as that of the Ash Wednesday disaster in southern Australia in February 1983. Rising wind velocities can exceed 250 km hr^{-1}, with winds inflowing into the maelstrom reaching velocities of 100 km hr^{-1}. Such *firestorms* pose a serious, life-threatening hazard for firefighters on the ground. Fire whirlwinds have been known to lift flaming debris high into the sky and have snapped off fully grown trees like matchsticks. In many cases, they are linked to the convective column above the fire, and are fuelled by massive gaseous explosions within this rising thermal.

Finally, vortices can easily develop over warm seas, bays, or lakes under conditions of convective instability because of the lower friction coefficient of water. The waterspouts that develop can be impressive (Figure 4.9); however, wind velocities are considerably lower than their land counterparts – around 25–25 m s^{-1} – and damage is rare. They often form as multiple vortices (Figure 4.9). Florida has 50–500 sightings each year and they are probably just as frequent in other tropical environments, especially on the eastern sides of continents where warm water accumulates under easterly trade winds. Waterspouts that move inland and kill people are probably tornadoes that have simply developed first over water. These events are rare. One of the most notable waterspouts occurred during Australia's 2001 Sydney Hobart Yacht Race. The crew of the 24 metre maxi-yacht *Nicorette* video-taped the approach of an enormous waterspout. Despite efforts to avoid it, the yacht sailed through the waterspout, was buffeted by 130 km hr^{-1} winds and hail, but came out the other side with only its mainsail shredded.

Fig. 4.9 An artist's impression of the sailing vessel *Trombes* being struck by multiple waterspouts. Sketch first appeared in *Les Meteores*, Margolle et Zurcher, 3rd edition, 1869, page 126 (NOAA, 2003).

Structure of a tornado

Simple tornadoes consist of a funnel, which is virtually invisible until it begins to pick up debris. Many tornadoes are associated with multiple funnels. In June 1955 a single tornado cloud in the United States produced 13 funnels that reached the ground and left still more suspended in midair. Generally, the larger the parent cloud, the greater the number of tornadoes that can be produced. The funnel has horizontal pressure and wind profiles similar to a tropical cyclone (Figure 3.5), except that the spatial dimensions are reduced to a few hundred meters. In addition, tornadoes develop greater depressed central pressure and more extreme wind velocities at the funnel boundary. The sides of the funnel enclose the eye of the tornado, which is characterized by lightning and smaller, short-lived mini-tornadoes. Air movement inside the funnel is downward. Vertical air motion occurs close to the wall with upward speeds of 100–200 m s^{-1}. The most visible part of a tornado, the funnel, is actually enclosed within a slower spinning vortex. The air moving down through the center of the funnel can kick up a curtain of turbulent air and debris at the ground – in a process similar to that of a hydrofoil moving over water (Figure 4.10). This feature – termed a 'cascade' – can also form above the ground before the main funnel actually develops. If the cascade spreads laterally far enough, its debris can be entrained by the outer vortex, and if wind velocities in this outer vortex are strong enough, the main funnel may slowly be enshrouded by debris from the ground upwards to form an 'envelope'. Envelopes are more characteristic of waterspouts than tornadoes that develop over land.

Often tornado structure is more complex than a single funnel. Air may descend through the center of the funnel at great speed. At the surface, this air spreads out and causes the vortex to break down. Air spinning into the vortex undergoes a rapid change in direction or shearing where it meets this downdraft, leading to the formation of mini-vortices around the main zone of rotating air. Up to half a dozen mini-vortices may form, rotating in a counterclockwise direction around the edge of the tornado itself. These smaller vortices are unstable, often appearing as wispy spirals. However, their presence can be devastating. Mini-vortices can rotate at speeds of 20–40 m s^{-1}. On the inner side of the mini-vortex, rotation opposes that of the main funnel leading to wind reversal or shear over very small distances (T-V in Figure 4.11). On the outer edge of the mini-vortex, air moves in the same direction as that of the main funnel and enhances wind speeds (T+V in Figure 4.11). The former condition tears structures apart, while the latter destroys structures that might resist lower wind speeds.

Fig. 4.10 The first tornado captured by the NSSL Doppler radar and NSSL chase personnel. Note the defined funnel shape and the cascade formed at the ground by the incorporation of debris. Union City, Oklahoma, 24 May 1973. Image ID: nssl0062, National Severe Storms Laboratory (NSSL) Collection, NOAA Photo Library, NOAA.

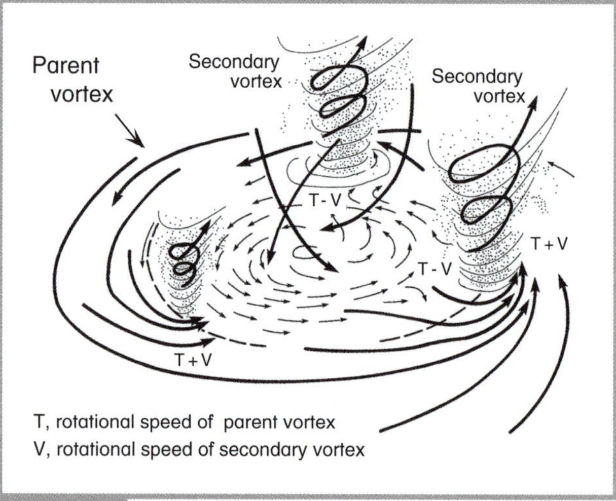

Fig. 4.11 Model for multiple-vortex formation in a tornado (based on Fujita, 1971 and Grazulis, 1993).

The highest wind velocities generated within tornadoes have not been measured; but they can be determined theoretically. For example, eggs or windows cleanly pierced by a piece of debris without shattering or a wooden pole pierced by a piece of straw requires wind speeds in excess of 1300 km hr^{-1}. However, this wind speed exceeds the speed of sound, a phenomenon that has not been reported during the passage of a tornado. The maximum wind speed measured using photogrammetry is 457 km hr^{-1}. These wind speeds can singe the skin of defeathered chickens, move locomotives off railway tracks and shift bridges. Because of the lack of measurement, a visual scale of tornado intensity has been developed; called the Fujita Tornado Intensity Scale, it is based on wind speeds (Table 4.1). The most intense tornado ranks as an F5 event on this scale. Such tornadoes tend to completely destroy houses, leaving nothing but the concrete slab, and badly damage steel-reinforced concrete structures. Wind speeds during such maelstroms range from 419–510 km hr^{-1}. Note that winds during Cyclone Tracy (mentioned in the previous chapter) not only twisted steel girders but also transported them through the air. These winds must have exceeded F5 on the Fujita scale. Only 0.4 per cent of tornadoes in the United States reach category F5; however with 1200 tornadoes observed each year, this statistic equates to five extreme events annually. Over 50 per cent of tornadoes produce only minor damage to homes.

Mini-vortices are responsible for the haphazard and indiscriminate damage caused by tornadoes. People driving a horse and cart have witnessed a tornado taking their horse and leaving the cart untouched. One tornado that swept through a barn took a cow, and left the girl milking it and her milk bucket untouched. Numerous photographs exist that show houses and large buildings cleanly cut in half with the remaining contents undisturbed.

Occurrence

(Miller, 1971; Nalivkin, 1983; Grazulis, 1993)

Tornadoes do not form near the equator because Coriolis force is lacking. Nor do they form in polar or subpolar regions where convective instability is suppressed by stable, cold air masses. About 80 per cent of all tornadoes occur in the United States, particularly on the Great Plains. Figure 4.12 illustrates the density of tornadoes and the preferred path of travel in this region. The area with five or more tornadoes per year, which is referred to as 'tornado alley', takes in Kansas, Oklahoma, Texas, Missouri, and Arkansas. Tornadoes occur commonly in this region, because here cold Arctic air meets warm air from the Gulf of Mexico, with temperature differences of 20–30°C. The flat nature of the Plains enhances the rapid movement of cold air, while the high humidity of Gulf air creates optimum conditions of atmospheric instability. Tornadoes in the United States tend to occur in late spring, when this temperature difference between air masses is at its maximum. At this time of year, the polar jet stream still can loop far south over the Great Plains.

Between 1950 and 1999 there were 40 522 tornadoes recorded in the United States, an average of 810 events per year. Of these, approximately 43 per cent were F2–F5 events. Since 1950, the frequency of tornadoes has been increasing from 600 tornadoes per year to a current average of 1200 per year (Figure 4.13). While Texas has the greatest number of tornadoes, the greatest density of tornadoes occurs in Florida. Kansas reports the most tornadoes in North America by population: 3.4 per 10 000 people. Surprisingly, the greatest number of deaths per unit area occurs in Massachusetts. The greatest number of tornadoes reported in a single week occurred 4–10 May 2003 (384 reports), while the

Table 4.1 Fujita Tornado Intensity Scale and frequency of tornadoes on the Fujita Scale in the USA (from Fujita, 1981; Grazulis, 1993).

F-scale	Wind Speed	Type of Damage	Frequency
F0	64–116 km/h	MINIMAL DAMAGE: Some damage to chimneys, TV antennas, roof shingles, trees, and windows.	82.0%
F1	117–180 km/h	MODERATE DAMAGE: Automobiles overturned, carports destroyed, trees uprooted.	11.0%
F2	181–253 km/h	MAJOR DAMAGE: Roofs blown off homes, sheds and outbuildings demolished, mobile homes overturned.	4.0%
F3	254–332 km/h	SEVERE DAMAGE: Exterior walls and roofs blown off homes. Metal supported buildings severely damaged. Trees uprooted.	1.8%
F4	333–418 km/h	DEVASTATING DAMAGE: Well-built homes levelled. Large steel and concrete missiles thrown far distances.	0.9%
F5	419–512 km/h	INCREDIBLE DAMAGE: Homes leveled with all debris removed. Large buildings have considerable damage with exterior walls and roofs gone. Trees debarked.	0.4%

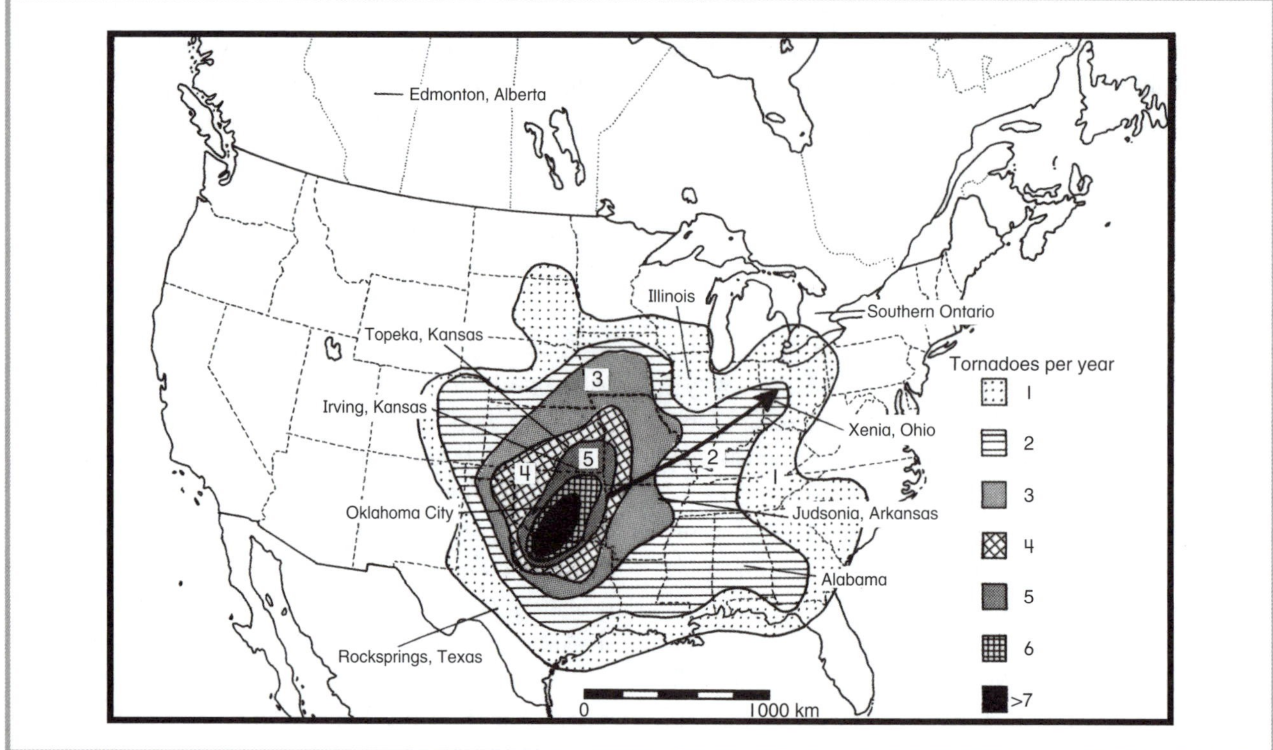

Fig. 4.12 Average number of tornadoes per year and preferred path of movement in the continental United States (adapted from Miller, 1971).

greatest number in one day occurred on 29 March 1974 (274 reports). However, these trends must be set within the context of variability from year to year. The 2003 season mentioned above occurred following a year that witnessed the fewest tornadoes – adjusted for increased report trends – in 50 years.

Canada now experiences the second largest number of tornadoes: 100–200 events annually as far north as

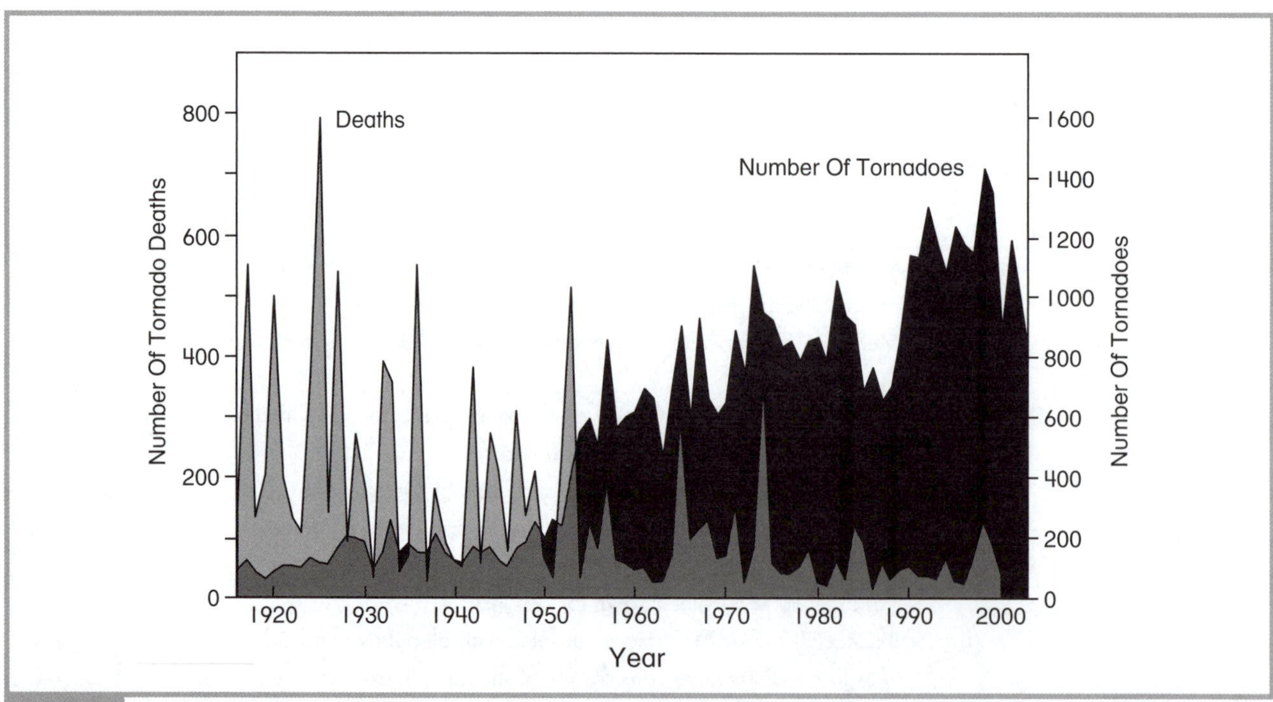

Fig. 4.13 Number of tornadoes and deaths due to tornadoes in the United States between 1916 and 2002 (U.S. Severe Weather Meteorology and Climatology, 2002a, b).

Edmonton, Alberta, at 53.5°N. This is far north of the zone where sharp temperature differences favor tornado development. Here, in 1987, a F4 tornado killed 27 people. Other areas of the world are susceptible to tornadoes. Surprisingly, a region such as Queensland, Australia, reports 4.0 tornadoes per 10 000 people – more than Kansas. The Russian Steppes are similar in relief to the Great Plains. Here, however, there is no strong looping of the jet stream or any nearby warm body of water supplying high humidity. Great Britain and the Netherlands average 50 and 35 tornado-like vortices a year, respectively. In Europe overall, about 330 vortices are reported each year, of which half occur as waterspouts and fewer than ten cause noteworthy damage. One of the greatest tragedies on this continent occurred in Venice on 11 September 1970 when a tornado sank a waterbus near St Mark's Square and swept later through a campground, killing 35 people. Surprisingly, tornadoes are rare in Russia, Africa and the Indian subcontinent. In the case of the last, tornadoes are rarely reported outside the northern part of the Bay of Bengal where they are associated with tropical cyclones. However, the deadliest tornado occurred in Bangladesh on 26 April 1989, killing 1300 people and injuring 12 000 others.

Tornado destruction

(Whipple, 1982; Nalivkin, 1983)

Tornadoes are one of the most intense and destructive winds found on the Earth's surface. The fact that they most frequently form along cold fronts means that they can occur in groups over wide areas. Their destructive effects are due to the lifting force at the funnel wall, and to the sudden change in pressure across this boundary. The lifting force of tornadoes is considerable, both in terms of the weight of objects that can be moved, and the volume of material that can be lifted. Large objects weighing 200–300 tonnes have been shifted tens of metres, while houses and train carriages weighing 10–20 tonnes have been carried hundreds of meters. Tornadoes are also able to suck up millions of tonnes of water, which can be carried in the parent cloud. Tornadoes have been known to drain the water from rivers, thus temporarily exposing the bed.

The most destructive mechanism in a tornado is the dip in barometric pressure across the wall of the funnel. It is not how low the pressure gets in the eye (about 800 hPa), but how rapidly this change in pressure occurs. The fact that the tornado is so compact and has a sharp wall means that a pressure reduction of tens of hPa s^{-1} could occur, although this dip has been difficult to substantiate by measurement. Air pressure in buildings affected by a tornado ends up being much higher than the outside air pressure, and cannot escape from the structure quickly enough without causing an explosion. This exploding effect is the main reason for the total devastation of buildings, whereas adjacent structures or remaining remnants appear untouched. The pressure reduction can be so rapid that live objects appear burnt because of dehydration. One of the major causes of death from a tornado is not due to people being struck by flying debris or tossed into objects, but to their receiving severe burns. It is even possible that the defeathered chickens described above could be cooked because of this phenomenon.

Between 1916 and 2000, tornadoes killed 12 766 people in the United States. This death toll has decreased by 200 people per decade because of increased warning and awareness (Figure 4.13). In April 1965, 257 people lost their lives as 47 tornadoes developed along a cold front in the central United States. The 'Three States' tornado of March 1925 created a path of destruction over 350 km long. Over 700 people were killed by the tornadoes spawned from the movement of this single parent cloud. One tornado in March 1952 destroyed 945 buildings in Judsonia, Arkansas. Another tornado – in Rock Springs, Texas, in April 1927 – killed 72 people, injured 240 others, and wiped out most of the town of 1200 people in 90 seconds. In all, 26 per cent of the population were either killed or injured. Cities are not immune, although high-rise buildings tend to survive better than single-unit dwellings. For example, in June 1966, a tornado went through Topeka, Kansas, destroying most homes and shops in its path but leaving multistoried flats and office buildings undamaged. The worst tornado disaster in the United States occurred during a 16-hour period on 3–4 April 1974. A total of 148 tornadoes was recorded from Alabama to Canada, killing 315 people and causing damage amounting to $US500 million. Six of the tornadoes were amongst the largest ever recorded. Complete towns were obliterated, the worst being Xenia, Ohio, where 3000 homes were destroyed or damaged in a town with only 27 000 residents.

Warning

(Golden & Adams, 2000)

Tornado prediction in the United States has dramatically improved since 1970 with the explosion in private meteorological forecasting and improved federally funded technology. In the 1950s, the US Air Force instigated tornado forecasting. A radar network was established in which tornadoes could be detected based upon their 'hock echo' signature on a radar map. However, in most cases, these identified only tornadoes that had already touched down. In the 1970s, lightning detectors were established to detect tornado-bearing storms. In the 1990s, infrasound and seismic ground detector systems augmented a range of remote sensing systems. The latter included precursor Geostationary Operational Environmental Satellites (GOES) 8 and 10 forecasting in tandem with 161 WSR88-Doppler radar installations, all of which were backed up by improved personnel training, increased numbers of local storm spotters and enhanced public awareness. The satellite monitoring permits severe convective systems to be detected days in advance. No longer is a simple vortex approach used to model tornadoes. Instead, satellite detection and computer modelling provides, hours in advance, evidence of mesocyclones favorable to tornadoes. The Doppler radar network monitors the instantaneous development of severe thunderstorms, large hail, heavy rain, and high winds associated with tornadoes. These signals are processed locally using sophisticated algorithms at over 1200 sites using the Automated Surface Observing System. In 1978, the probability of detecting a tornado was 22 per cent with a lead time of three minutes. In 1998 this probability had risen to 65 per cent with a lead time of 11 minutes. The events that occurred on 3 May 1999 around Oklahoma City provide an example of how effective recent procedures have become. That night 116 warnings, with lead times of 20–60 minutes, were issued for 57 tornadoes. No warning is useful unless it can be communicated effectively to the public. Radio messages can be transmitted by a network of 504 transmitters to 80–85 per cent of the US population. About 25 per cent of households can automatically receive these signals. Many TV stations also receive these automatic warnings and display them on television screens as either a text crawler or a flashing icon. In the United States the fear of tornadoes occurring without any warning has virtually disappeared.

Response

(Sims & Baumann, 1972; Oliver, 1981)

For most of the world, tornadoes can be put into the 'act of God' category. For instance, in Australia, it is just bad luck if a tornado strikes you or your property. In the United States, especially in 'tornado alley', tornadoes are a frequent hazard. While it is difficult to avoid their occurrence, there are many ways to lessen their impact. For instance, in the United States Midwest, schoolchildren are taught at an early age how to take evasive action if they see, or are warned about, the approach of a tornado. I can remember, as a schoolchild living in southern Ontario, being sent home from school one day when tornadoes were forecast. Southern Ontario, then, was not considered highly vulnerable; but the school authorities were aware that a tornado hitting a school full of children would generate a major disaster. We were sent home to minimize the risk. We were also drilled in procedures to take if a tornado did strike during school hours – the same procedures we would have taken if a nuclear attack were imminent! We knew that tornadoes traveled mainly north-east; so, if we could see a tornado approaching and were caught in the open, we were instructed to run to the north-west. In the United States Midwest, most farms or homes have a basement or cellar separate from the house that can be used as shelter. All public buildings have shelters and workers are instructed in procedures for safe evacuation. Safety points in buildings, and in the open, are well sign-posted. In the worst threatened areas such as 'tornado alley', tornado alerts are broadcast on the radio as part of weather forecasts. The actual presence of a tornado can be detected using Doppler radar or tornado watches maintained by the Weather Service. The weather watch consists of a network of volunteer ham radio operators known as the Radio Emergency Associated Citizen Team (REACT), who will move to vantage points with their radio equipment on high-risk days. Immediately a tornado is detected, local weather centers, civil defense and law enforcement agencies are alerted and warnings then broadcast to the public using radio, television, and alarm sirens.

For many rural communities in the Midwest, a tornado is the worst disaster that can strike the area. State governments will declare a state of emergency and, if outbreaks of tornadoes occur over several states, federal assistance may be sought through the Federal

Emergency Management Agency (FEMA). In rural areas, community response to a tornado disaster is automatic and there are government agencies, in addition to police and fire services, which have been specially created for disaster relief. Generally, however, tornadoes are accepted as a risk that must be lived with. The above observations are not universal; they may be modified by the cultural attitudes of people living in various threatened areas. Sims and Baumann (1972) analyzed the attitudes of people to the tornado risk between Illinois and Alabama. The death rate from tornadoes apparently was higher in the latter state. They concluded, based upon interviews, that there was a substantial attitude difference between the two states. People in Illinois believed that they were in control of their own future, took measures to minimize the tornado threat, and participated in community efforts to help those who became victims. In Alabama, however, people did not heed tornado warnings, took few precautions, were more likely to believe that God controlled their lives, and were less sympathetic towards victims. Oliver (1981) has analyzed this study and pointed out some of its fallacies. For instance, the death toll from tornadoes is higher in Alabama because the tornadoes there are more massive than in Illinois. However, the Sims and Baumann (1972) study does highlight the fact that not all people in the United States have the same perception of the tornado hazard or response to dealing with its aftermath. These differences will be elaborated further in Chapter 13.

Your attitude to tornadoes may make little difference to your survival; rather, where you live in the Midwest dictates your chances of survival. Inadvertently, Europeans settling the Midwest may have chosen the most susceptible town sites. For example, Oklahoma City on the Canadian River has been struck by more than 25 tornadoes since 1895, even though the State of Oklahoma ranks seventh in the United States for deaths per unit area due to tornadoes – 7.3 deaths per 10 000 km^2 compared to 20.5 per 10 000 km^2 for Mississippi, which ranks first. Indian tribes in the Midwest rarely camped in low-lying areas such as the bottoms of ravines, or along river courses, because they were aware that tornadoes tend to move towards the lowest elevation once they touch ground. This makes sense since the tornado vortex is free-moving and will tend to follow the path of least resistance. Figure 4.14 dramatically illustrates this tendency. The first tornado to pass through Irving, Kansas, on 30 May 1879, ended up being trapped in the valley of the Big Blue River, and eventually in a small gully off to the side. Tornadoes do not always adhere to this pattern, as illustrated in the same figure by the path of the second tornado that came through the town minutes later. For historical reasons such as access to transport and water, many towns in the United States Midwest are built near rivers or in low-lying areas. The fact that tornadoes tend to travel along these paths may be one of the reasons why they have been, and will continue to be, a major hazard in the United States Midwest.

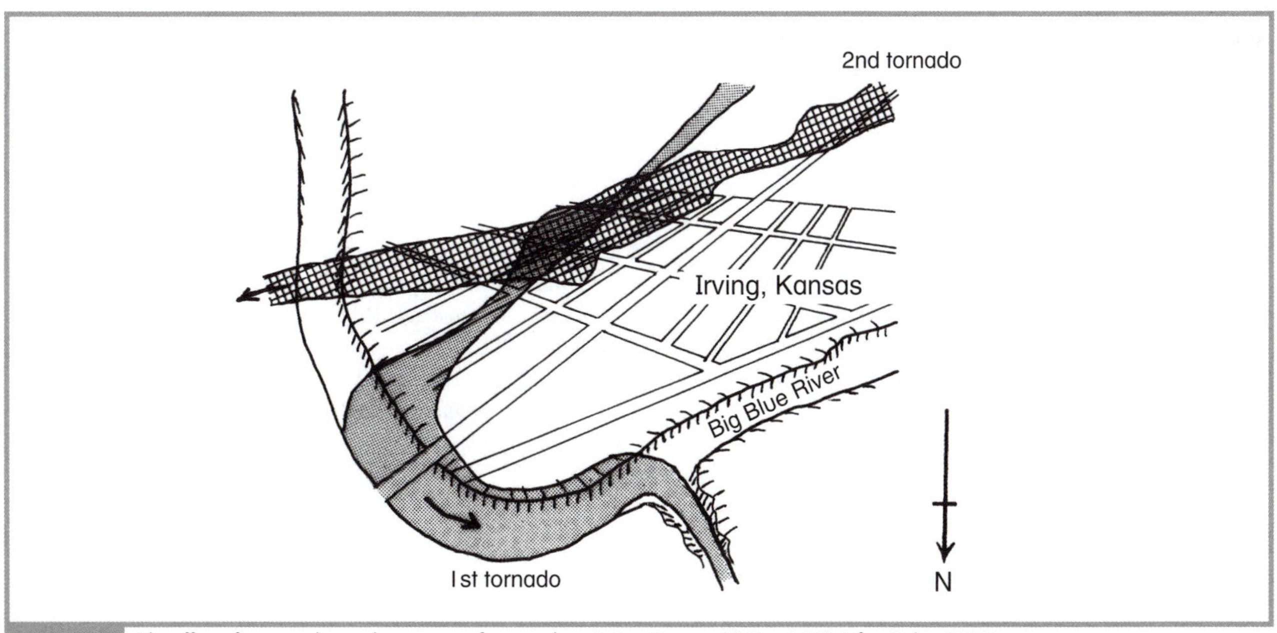

Fig. 4.14 The effect of topography on the passage of a tornado at Irving, Kansas, 30 May 1879 (after Finley, 1881).

CONCLUDING COMMENTS

The thought that a series of tornadoes – such as those that struck the United States in April 1974 or a hailstorm as devastating as the Sydney hailstorm of 1999 – could strike again demolishes complacency that our present climate is stable or that humans can prevent extreme climatic events. Thunderstorms will continue to drop rain at the upper level of physical possibility, produce hail that can kill crowds in the open, and spawn tornadoes that are certain to set records for damage and loss of life. The uncertainty is not if, but when, such events will recur. While scientists are tempted to invoke global warming as the cause of an increase in the severity of thunderstorms, thunderstorm magnitude simply represents the natural occurrence of random natural events. King Edward III's defeat due to a hailstorm at Chartres, France, in 1359 was not a signature of global warning. History is replete with similar events. Today, were such a disaster to strike the final match of the baseball World Series or the soccer World Cup, the blame would certainly be directed towards global warming as the scapegoat.

While short-lived events such as hailstorms and tornado swarms are easily dealt with in modern societies, the effects of flash flooding, catastrophic regional rains, and droughts always will test our climatic modelling, our societal altruism and, ultimately, the capacity of the international community to respond in any meaningful way. The following chapters examine the nature of droughts and regional floods, and our attempts to mitigate, and respond to, such large-scale disasters.

REFERENCES AND FURTHER READING

Anagnostou, E.N., Chronis, T., and Lalas, D.P. 2002. New receiver network advances in long-range lightning monitoring. *EOS, Transactions American Geophysical Union* 83: 589, 594–595.

Black, R.A. and Hallett, J. 1998. The mystery of cloud electrification. *American Scientist* 18: 526–534.

Eagleman, J.R. 1983. *Severe and Unusual Weather*. Van Nostrand Reinhold, NY.

Finley, J.P. 1881. Report on the tornadoes of May 29 and 30, 1879, in Kansas, Nebraska. *Professional Paper of the Signal Service* No. 4, Washington.

Fujita, T.T. 1971. Proposed mechanism of suction spots accompanied by tornadoes. Preprints, Seventh Conference on Severe Local Storms, American Meteorological Society, Kansas City, pp. 208–213.

Fujita, T.T. 1981. Tornadoes and downbursts in the context of generalized planetary scales. *Journal of Atmospheric Research* 38: 1511–1534.

Geophysics Study Committee 1987. *The Earth's Electrical Environment*. United States National Research Council, National Academy Press.

Golden, J.H. and Adams, C.R. 2000. The tornado problem: forecast, warning, and response. *Natural Hazards Review* 1(2): 107–118.

Grazulis, T.P. 1993. *Significant Tornadoes 1680–1991*. Environmental Films, St. Johnsbury.

Idso, S. 1976. Whirlwinds, density current and topographical disturbances: A meteorological mélange of intriguing interactions. *Weatherwise* 28: 61–65.

Miller, A. 1971. *Meteorology* (second edn). Merrill, Columbus.

Nalivkin, D.V. 1983. *Hurricanes, Storms and Tornadoes*. Balkema, Rotterdam.

NOAA 2003. *Image ID: wea00300, historic NWS collection*. <http://www.photolib.noaa.gov/historic/nws/wea00300.htm>

Oliver, J. 1981. *Climatology: Selected Applications*. Arnold, London.

Oliver, J. 1986. Natural hazards. In Jeans, D.N. (ed.) *Australia: A Geography*. Sydney University Press, Sydney, pp. 283–314.

Sims, J.H. and Baumann, D.D. 1972. The tornado threat: coping styles of the North and South. *Science* 176: 1386–1392.

U.S. Severe Weather Meteorology and Climatology 2002a. *United States tornadoes, 1916–2000*. <http://www.hprcc.unl.edu/nebraska/ustornadoes1916-2000.html>

U.S. Severe Weather Meteorology and Climatology 2002b. *United States tornado deaths* 1916–2000. <http://www.hprcc.unl.edu/nebraska/ustornadoes1916-2000.html>

Whipple, A.B.C. 1982. *Storm*. Time-Life Books, Amsterdam.

Yeo, A., Leigh, R., and Kuhnel, I. 1999. The April 1999 Sydney Hailstorm. *Natural Hazards Quarterly* 5(2), <http://www.es.mq.edu.au/nhrc/web/nhq/nhq5-2tables.htm>

Drought as a Hazard

CHAPTER 5

INTRODUCTION

Drought and famine have plagued urban–agricultural societies since civilizations first developed. While many definitions exist, drought can be defined simply as an extended period of rainfall deficit during which agricultural *biomass* is severely curtailed. In some parts of the world, such as the north-east United States and southern England, a drought may have more of an effect on urban water supplies than on agriculture. The definition of drought, including the period of rainfall deficit prior to the event, varies worldwide. In southern Canada, for instance, a drought is any period where no rain has fallen in 30 days. Lack of rain for this length of time can severely reduce crop yields in an area where crops are sown, grown, and harvested in a period of three to four months. In Australia, such a definition is meaningless, as most of the country receives no rainfall for at least one 30-day period per year. Indeed, in tropical areas subject to monsoons, drought conditions occur each dry season: most of tropical Australia, even in coastal regions, endures a rainless dry season lasting several months. In Australia, drought is usually defined as a calendar year in which rainfall registers in the lower 10 per cent of all the records. Unfortunately, in the southern hemisphere, a calendar year splits the summer growing season in two. A more effective criterion for drought declaration should consider abnormally low rainfall in the summer growing season.

Drought onset is aperiodic, slow, and insidious. As a result, communities in drought-prone areas must be constantly prepared to insulate themselves from the probability of drought and finally, when overwhelmed, to ensure survival. However, a community's response to drought varies depending upon its social and economic structure. This chapter examines first the pre-colonial response of societies to drought. This is followed by an examination of drought response in modern societies. While westernized countries have generally fared better than underdeveloped countries, it will be shown that national perception and policy ultimately determine success in drought mitigation. The chapter concludes by comparing the success of drought relief organized by world organizations, mainly the United Nations, to that of private responses such as the Band Aid appeal headed by Bob Geldof in response to the Ethiopian drought of 1983–1984.

PRE-COLONIAL RESPONSE TO DROUGHT

(Garcia, 1972; Morren, 1983; Watts, 1983; Scott, 1984)

In nomadic settings, there is a heavy reliance upon social interaction as drought develops. The !Kung of the Kalahari Desert of south-west Africa congregated around permanent water holes only during the dry season (see Figure 5.1 for the location of major African placenames). If conditions became wetter and more

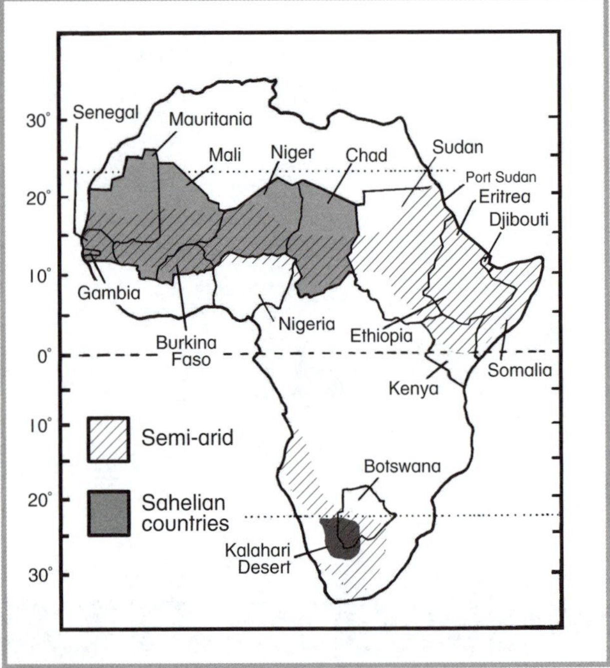

Fig. 5.1 Map of semi-arid regions and the Sahel in Africa.

favorable for food gathering and hunting, !Kung clans split into progressively smaller groups with as few as seven people. During severe droughts, however, the !Kung gathered in groups of over 200 at two permanent water holes, where they attempted to sit out the dry conditions until the drought broke.

In societies that are more agrarian, there is a critical reliance upon agricultural diversification to mitigate all but the severest drought. Pre-colonial Hausa society in northern Nigeria (on the border of the Sahara Desert) exemplifies this well. Before 1903, Hausan farmers rarely mono-cropped, but planted two or three different crops in the same field. Each crop had different requirements and tolerances to drought. This practice minimized risk and guaranteed some food production under almost any conditions. Water conservation practices were also used as the need arose. If replanting was required after an early drought, then the spacing of plants was widened or fast-maturing cereals planted in case the drought persisted. Hausan farmers also planted crops suited to specific microenvironments. Floodplains were used for rice and sorghum, interfluves for tobacco or sorghum, and marginal land for dry season irrigation and cultivation of vegetables. Using these practices, the Hausan farmer could respond immediately to any change in the rainfall regime. Failing these measures, individuals could rely upon traditional kinship ties to borrow food should famine ensue. Requests for food from neighbors customarily could not be refused, and relatives working in cities were obligated to send money back to the community or, in desperate circumstances, were required to take in relatives from rural areas as refugees.

A resilient social structure had built up over several centuries before British colonial rule occurred in 1903. Hausan farmers lived in a feudal Islamic society. Emirs ruling over local areas were at the top of the social 'pecking order'. Drought was perceived as an aperiodic event during which kinship and descent grouping generally ensured that the risk from drought was diffused throughout society. There was little reliance upon a central authority, with the social structure giving collective security against drought. Taxes were collected in kind rather than cash, and varied depending upon the condition of the people. The emir might recycle this wealth, in the form of loans of grain, in a local area during drought. This type of social structure characterized many pre-colonial societies throughout Africa and Asia. Unfortunately, colonialism made these agrarian societies more susceptible to droughts.

After colonialism occurred in Nigeria, taxes were collected in cash from the Hausan people. These taxes remained fixed over time and were collected by a centralized government rather than a local emir. The emphasis upon cash led to cash cropping, with returns dependent upon world commodity markets. Production shifted from subsistence grain farming, which directly supplied everyday food needs, to cash cropping, which provided only cash to buy food. If commodity prices fell on world markets, people had to take out loans of money to buy grain during lean times. As a result, farmers became locked into a cycle of seasonal debt that accrued over time. This indebtedness and economic restructuring caused the breakdown of group and communal sharing practices. The process of social disruption was abetted by central government intervention in the form of food aid to avert famine. Hausan peasant producers became increasingly vulnerable to even small variations in rainfall. A light harvest could signal a crisis of famine proportions, particularly if it occurred concomitantly with declining export prices. The Hausan peasantry now lives under the threat of constant famine.

During drought, a predictable sequence of events unfolds. After a poor harvest, farmers try to generate income through laboring or craft activity. As the famine intensifies, they seek relief from kin and subsequently begin to dispose of assets. Further steps include borrowing of money or grain and, if all else fails, either selling out or emigrating from the drought-afflicted region. With outward migration, governments notice the drought and begin organizing food relief. After the drought, the farm household has few resources left to re-establish pre-existing lifestyles. Rich households in the city have the resources (the farmer's sold assets) to withstand poor harvests. The natural hazard of drought now affects exclusively the Hausan farmer. In this process, society has become differentiated such that the wealthy beneficially control their response to drought, while the Hausan farmer becomes largely powerless and subjected to the worst effects of a drought disaster. This sequence of events now characterizes drought response throughout Africa and other Third World countries, even though colonialism has disappeared. Dependency upon cash cropping, export commodity prices or the vagaries of war have repeatedly brought large segments of various societies to crisis point during the major droughts that have occurred towards the end of the twentieth century.

POST-COLONIAL RESPONSE

(Garcia, 1972; French, 1983)

In contrast, westernized societies rely heavily upon technology to survive a drought. This is exemplified in Australia where individual farmers – who withstand the worst droughts – have developed one of the most efficient extensive agricultural systems in the world. Self-reliance is emphasized over kin or community ties, and it is efficiency in land management that ultimately determines a farmer's ability to survive a drought. As a drought takes hold, these management practices are subtly altered to maximize return and minimize permanent damage to the land and to the farming unit. Where land is cultivated, land management practices are directed towards moisture conservation and crop protection. Practices promoting the penetration of rain into the subsoil include mulching, tillage, and construction of contour banks and furrows that minimize runoff. In certain cases, catchments will be cleared and compacted to increase runoff into farm dams that can be used for irrigation during drought. Because most crops mature at times of highest moisture stress, the time of sowing is crucial to the final crop yield. Early sowing in winter ensures grain flowering and maximum growth before high water-stress conditions in summer. In addition, application of phosphate fertilizers can double crop yield for each millimetre of rainfall received. Where land is used for grazing, some pasture is set aside in spring so that hay can be cut to use as feed over a dry summer period. In mixed farming areas, grain must also be held over from the previous year's harvest to supplement sheep and cattle feed. When a drought persists for several years, stock are either dramatically culled or breeding stock are shipped considerable distances and grazed at a cost (agistment) in drought-free areas.

Technological advances also play a key role in farming drought-prone, semi-arid areas. These techniques include aerial spraying of herbicides to control weed growth, and aerial sowing of seed to prevent soil compaction by heavy farm machinery. Aerial monitoring is used to track livestock and give early warning of areas that are threatened by overgrazing. Computer management models are also utilized to maximize crop yields, given existing or forecast climate, moisture, and crop conditions. Finally, efficient financial management becomes crucial as much in drought survival as in recovery. Most farmers aim for three profitable years per decade and must be able to sustain the economic viability of the farm for up to seven consecutive drought years. Such a mode of operation requires either cash reserves or a line of credit with a bank to permit breeding stock to be carried for this period of time. Until the rural crisis of the 1980s, fewer than 15 per cent of farmers in Australia entered financial difficulty following a major drought.

The above management practices set the westernized farmers apart from those in nomadic and underdeveloped countries. Consequently, the Australian farmer can survive droughts as severe as those that have wracked the Sahel region of Africa in recent years. However, the Australian farm does not have to support the large number of people that units in Africa do. As a result, drought-induced famine in Africa can detrimentally disturb the entire social and economic fabric of a country, whereas in Australia or the United States, much of the population may be unaware that a major drought is occurring. In the 1970s and 1980s, the intensity of drought worldwide not only crippled many African countries but, for the first time since the depression years of the 1930s, it also perturbed many westernized societies.

DROUGHT CONDITIONS EXACERBATED BY MODERN SOCIETIES

(Bryson & Murray, 1976; Glantz, 1977; Lockwood, 1986)

Despite these technological solutions, human activity can greatly exacerbate droughts through over-cropping of marginalized land, massive vegetation clearing, and poor soil management. These activities have affected all semi-arid regions of the world. In the Rajputana Desert of India, which was a cradle of civilization, successive cultures have flourished, but each time collapsed. The region is humid and affected seasonally by the monsoon. However, the atmosphere is extremely stable because overgrazing leads to dust being continually added to it. Dust in the upper troposphere absorbs heat during the day, lessening the temperature gradient of the atmosphere and promoting atmospheric stability. At night, dust absorbs long wave radiation, keeping surface temperatures elevated above the dew point. An artificial desert has been produced that is self-maintaining as long as dust gets into the atmosphere.

Even without massive dust production, clearing of vegetation can itself establish a negative, bio-geophysical feedback mechanism, locking a region into aridity. In the Sahel of Africa, decreasing precipitation since 1960 has reduced plant growth, leading to reduced *evapotranspiration*, decreased moisture content in the atmosphere, and a further decrease in rainfall. Over time, soil moisture slowly diminishes, adding to the reduction in evaporation and cloud cover. As the soil surface dries out and vegetation dies off, the surface *albedo* – the degree to which short wave solar radiation is reflected from the surface of plants – is reduced, leading to an increase in ground heating and a rise in near-ground air temperatures. This process also reduces precipitation. The destruction of vegetation exposes the ground to wind, permitting more dust in the atmosphere.

Drought in the Sahel occurred concomitantly with an increase in population and deleterious economic conditions. In the 1970s, world fuel prices soared and people switched from kerosene to wood for cooking and heating. Such a change led to rapid harvesting of shrubs and trees. With diminishing crop yields, land was taken out of fallow, further reducing soil moisture. Western techniques of plowing – introduced to increase farming efficiency – destroyed soil structure, leading to the formation of surface crusts that increased runoff and prevented rain infiltration. All of these practices have enhanced the negative feedback mechanisms, favoring continued drought and desertification.

Similar physiographic conditions exist at present in Australia, but without the social outcomes. Here, humans have interfered more broadly with the ecosystem. The reduction of the dingo population allows the kangaroo population to increase, competing with domestic stock for food and water during drier conditions. Construction of large numbers of dams and ponds for stock watering allows native animals to survive into a drought; and introduction of feral animals – such as rabbits, pigs, goats, cats, cattle, water buffalo and donkeys – means passive native species have been replaced with more aggressive species. Many of these introduced species are ungulates (hoofed animals) and cause soil compaction. Wholesale ringbarking of trees has occurred, to such an extent that in many areas there are not enough mature trees left to naturally re-seed a region. Dieback is *endemic*. Without deep-rooted trees and evapotranspiration, watertables have risen to the surface, flushing salt from the ground to be precipitated in insoluble form at the surface. This salt scalds vegetation, leaving bare ground susceptible to wind erosion. Soil structure is altered and surface crusting takes place with the result that wetting and water penetration of the ground during rain becomes more difficult. In Victoria, salinity affects about 5200 km^2 of land and is increasing by 2–5 per cent per year, while in Western Australia, 3000 km^2 of arable land has been abandoned. Large sections of the Murray River drainage basin are threatened by this increasing problem, which is exacerbated by irrigation. As a result, arable land in some agricultural river basins within Australia is being ruined at a rate ten times greater than Babylonian civilization's degradation of the Euphrates–Tigris system, and five times faster than it took to change the Indus River floodplain into the Rajputana Desert.

MODERN RESPONSE TO DROUGHTS

Societies that expect drought: the United States

(Rosenberg & Wilhite, 1983; Warrick, 1983)

The above examples of national response to drought would imply that underdeveloped countries have great difficulty in coping with drought conditions. Some countries – such as Ethiopia in the 1980s – even go so far as to ignore the plight of victims and deny to the outside world that they are undergoing a major calamity. The western world should not sit smug just because our societies are developed to a higher technical level. Often, by delving into history, it becomes evident that western countries have responded to drought in exactly the same manner that some Third World countries do today.

The national response to drought is well-documented in the United States. In Chapter 2, it was pointed out that there was a well-defined cyclicity to drought occurrence on the Great Plains, synchronous with the 18.6-year lunar cycle. Major droughts have occurred in the 1890s, 1910s, 1930s, 1950s, and 1970s. Most people have a vivid picture of the 1930s drought with mass outward migration of destitute farmers from a barren, windswept landscape. However, the 1930s drought was not the worst to occur. Before the 1890s, the Great Plains were opened up to cultivation on a large scale. Settlement was encouraged via massive advertising campaigns sponsored by governments and private railway companies in the eastern cities and abroad. During the 1890s drought, widespread starvation and malnutrition occurred in the central and southern High Plains. Similar conditions accompanied the drought in the 1910s in the Dakotas and eastern Montana. There was no disaster relief in the form of food or money from the federal government. During the 1890s drought, many state governments refused to acknowledge either the drought, or the settlers' plight, because government was trying to foster an image of prosperity to attract more migrants. In both the 1890s and 1910s, mass outward migration took place with settlers simply abandoning their farms. In some counties, the migration was total. Those that stayed behind had to shoulder the burden individually or with modest support from their own or neighboring states. The similarity between responses to these two droughts and to the 1980s Ethiopian one is striking.

The 1930s drought on the Great Plains is notable simply because the government began to respond to drought conditions. Outward migration was not so massive as some people think, but merely followed a pattern that had been established in the preceding years. Many over-capitalized farmers, as their finances collapsed, simply liquidated or signed over their farms to creditors. At a national level, the 1930s drought had little impact. It received massive federal aid but this was minor in comparison to the money being injected into the depressed American economy. There were no food shortages (in fact, there was a surplus of food production), no rising costs and few cases of starvation or malnutrition. For the first time, the government assisted farmers so that by 1936, 89 per cent of farmers in some counties were receiving federal funds. The aid was not in the form of food, but cash. The government undertook a number of programs to aid farmers' staying on the land. Irrigation projects were financed, government-sponsored crop insurance initiated, and a Soil Conservation Service established. Programs were designed to rescue farmers, re-establish agricultural land, and mitigate the effect of future droughts.

The 1950s drought was very different. The programs begun in the 1930s were accelerated. There was a six-fold increase in irrigated land between the 1930s and the 1980s. The amount of land under crop insurance increased, while marginal land was tucked away under the Soil Bank program. In the 1950s, few farms were abandoned: they were sold off to neighbors to form larger economies of scale. People selling their properties retired into adjacent towns. Government aid poured into the region, but at a lesser scale than in the previous drought. Again, there was little effect upon the national economy. The government responded by initiating more technical programs. Irrigation and river basin projects were started, and weather prediction, control and modification programs researched and carried out. The Great Plains Conservation Program established mechanisms for recharging groundwater, increasing runoff, minimizing evaporation, desalinizing soil, and minimizing leakage from irrigation canals. Since the 1950s, these programs have increased in complexity, to the point that drought now has little impact on the regional or national economies.

It would appear that, in the United States, the effect of drought on the Great Plains has been dramatically reduced to a level unachievable at present in most African countries. The evolution of drought-reduction strategies indicates an increasing commitment to greater social organization and technological sophistication. Hence, the local and regional impacts of recurrent droughts have diminished over time; however, the potential for catastrophe from rarer events may have increased. The latter effect is due to the reliance placed on technology to mitigate the drought. If the social organization and technology cannot cope with a drought disaster, then the consequences to the nation may be very large indeed. Whereas droughts at the turn of the century had to be borne by individuals, and had little effect outside the region, society today in the United States is structured so that the failure of existing technology to minimize a drought will severely disrupt the social and economic fabric of that nation. A ripple effect will also permeate its way into the international community, because many Third World countries are reliant upon the Great Plains breadbasket to make up shortfalls should their own food production fail. This international effect will only be exacerbated if droughts occur worldwide at the same time. Until the 1990s, this appeared unlikely because the timing of drought on the Great Plains was out of sequence with that in other continents. However, since the early 1990s, droughts have occurred globally at the same time. This is sobering considering the international community's inability to respond to more than one major drought at a time.

Societies that don't expect drought: the United Kingdom

(Whittow, 1980; Morren, 1983; Young, 1983)

The United Kingdom is not usually associated with drought; however, the south and south-east sections of England have annual rainfalls of less than 500 mm. In 1976 England was subjected to 16 months of drought, the worst recorded in its history. For the previous five years, conditions had become drier as the rain-bearing, low-pressure cells, which usually tracked across Scotland and Norway, shifted northward into the Arctic. The low-pressure cells were replaced with unusually persistent high-pressure cells that often blocked over the British Isles, and directed the northward shift of the lows. By the summer of 1976, poor rainfalls affected many farmers and temperatures uncharacteristically soared over 30°C. Domestic water supplies were hard hit as wells dried up. By the end of July, water tankers were carrying water to towns throughout the United Kingdom, and water restrictions were brought into force. In some cases, municipal water supplies were simply turned off to conserve water. In mid-August, at the height of the drought, the Thames River ceased to flow above tide limits. By September, there were calls for national programs to mitigate the worsening situation. When the government finally shifted into action, the drought had broken. By November of 1976, most of the country had returned to normal rainfall conditions.

The drought was severe, not because of a lack of rain, but because of increased competition for available water supplies that in some parts of the country had traditionally been low. Since the Industrial Revolution, public priority ensured water supplies to cities and industry. Commercial water projects were common and, where they were not available, councils were empowered to supply water. The *Water Act* of 1945 encouraged amalgamation of local authorities and nationalization of private companies. A severe drought in 1959 resulted in the *Water Resources Act* of 1963; this led to the establishment of river authorities to administer national water resources, and a Water Resources Board to act in the national interest. By 1973 there were still 180 private companies supplying water, and through the *Water Act* of 1973, ten regional water authorities were set up to take control of all matters relating to water resources. In addition, a centralized planning unit and research center were set up. It would appear that the United Kingdom should have been well-prepared for the drought; but, despite this legislation, per capita consumption of water had increased as standards of living increased. Water mains had been extended into rural, as well as suburban areas, so that agriculture was reliant upon mains supply for irrigation and supplementary watering.

As the drought heightened in June of 1976, regional water authorities complained that their legal responses were either too weak or too extreme. Either they could ban non-essential uses of water such as lawn watering by ordinary citizens, or they could overreact and cut off domestic water supplies and supply water from street hydrants. They had no control whatsoever over industrial or agricultural usage. Except in south Wales, industry at no time had water usage drastically curtailed. On the other hand, water mains in many places were shut down, with special wardens appointed to open water mains when a fire broke out. In August, the national government gave the water authorities more powers through the *Drought Act*. The authorities were free to respond to local conditions as they saw fit. Publicity campaigns managed to get domestic consumers, who were collectively the largest water users, to decrease consumption by 25 per cent. Because there were virtually no water meters in the United Kingdom, it was not possible to reduce consumption by imposing a limit on the volume of water used at the individual household level and charging heavily for usage beyond that limit. Smaller towns had their water shut off, but at no time was it suggested that London should have its water supply reduced. When canals and rivers dried up, there was concern that the clay linings would desiccate and crack, leading to expensive repairs to prevent future leakage or even flooding. Ground subsidence on clay soils became a major problem, with damage estimated at close to £50 million. This latter aspect was one of the costliest consequences of the drought.

The agriculture sector was badly affected and poorly managed. From the onset, grain and livestock farmers were disadvantaged by the drier weather, receiving little assistance. Just when they needed water most – at the time of highest moisture stress – irrigation farmers found their water supplies cut by half. Britain's agriculture had become so dependent upon mains water supply that it virtually lost all ability to respond to the drought. This was not only a problem with the 1976 drought – it had been a growing concern throughout the 1960s. Fortunately, because of Britain's entry into the European Common Market, the agricultural sector was insulated economically against the full consequences of the drought. The early part of the summer saw production surpluses, but at no time did agricultural production for the domestic market decrease to unbearable levels. In fact, many farmers responded to the drought in its early stages and increased acreages to make up for decreased yields.

The government's response to the drought was minimal. It began to act only at the beginning of August 1976: looking in detail at the plight of farmers and giving water authorities additional powers through the *Drought Act*. By October, there was considerable debate in parliament about subsidies and food pricing, but in fact the drought was over. The drought of 1976 cannot be dismissed as inconsequential; it did disrupt people's lives at all levels of society. The consequences were inevitable given the changes in water usage in the United Kingdom, especially the dependence of agriculture upon mains supplies and the increases in domestic water consumption. The national government was poorly prepared for the drought and still has not taken steps to rationalize water management in the United Kingdom to prepare for inevitable future droughts. This case illustrates two poorly realized facts about drought. Firstly, drought is ubiquitous: all countries can be affected by exceptional aridity. Drought is as likely in the United Kingdom (or Indonesia or Canada) as it is in the Sahel, given the right conditions. Secondly, the national response to drought in westernized countries is often similar to that of many Third World countries. The main difference is that in the Third World, governments are often inactive, usually because they are not economically able to develop mechanisms of drought mitigation and alleviation, while governments in westernized countries such as the United Kingdom (or other western European countries, which also experienced the 1976 drought) are not prepared to accept the fact that droughts can happen.

Those that lose

(Glantz, 1977; Bryson & Murray, 1977; Whittow, 1980; Gribbin, 1983; Hidore, 1983)

Africa's Sahel has succumbed to drought since cooling sea surface temperatures in the adjacent Atlantic became pronounced in the early 1960s. The seasonal switch of the intertropical convergence over Africa failed as a result. The consequences first affected, between 1968 and 1974, the region between Mauritania and Ethiopia. Human-induced feedback mechanisms, as described above, exacerbated drought conditions. The coping responses of people in the Sahel soon became inadequate to ensure the survival of populations. One of the reasons that Sahelian countries have suffered so dramatically is the relative inaction of national governments in responding to drought. In many of these countries, local groups have traditionally played a crucial role in minimizing the effects of drought. Farmers have always been able to adjust their mode of farming to increase the probability of some crop production, to diversify, to resort to drought-tolerant crops, or to respond instantly to changing soil and weather conditions. Nomads may resort to cropping under favorable conditions, follow regional migratory patterns to take advantage of spatially variable rainfall or, under extreme conditions, move south to wetter climates. All groups have strong kinship ties, which permit a destitute family to fall back on friends or relatives in times of difficulty. There is very little national direction or control of this type of activity, especially where national governments are weak and ineffective. In many of these countries, colonialism has weakened kin group ties and forced reliance upon cash cropping. Westernization, especially since the Second World War, has brought improved medical care that has drastically reduced infant mortality and extended lifespan. Unfortunately, birth rates have not decreased and all countries are burdened with some of the highest growth rates in the world: in the 1980s, 3 per cent and 2.9 per cent for Sudan and Niger, respectively. Technical advances in human medicine also applied to veterinary science, with the result that herd sizes grew to keep pace with population growth. In the Sudan between 1924 and 1973, livestock numbers increased fivefold, abetted by international aid development and the reduction of cattle diseases. The government responded to these increases during past droughts by providing money for construction of wells and dams. Otherwise, the national neglect was chronic. Subsistence farming is the primary economic activity of 80–90 per cent of people in the Sahel. Almost 90 per cent of the population is illiterate, and four of the countries making up the Sahel are among the twelve poorest countries in the world, and getting poorer.

During the first series of droughts, in the 1960s, the collapse of agriculture was so total that outward migration began immediately. By 1970, 3 million people in the west Sahel had been displaced and needed emergency food. The international response was minimal. When the drought hit Ethiopia, aid requests to the central government were ignored. As the drought continued into 1971, large dust storms spread southward. The United States was responsible for 80 per cent of aid and Canada, a further 9 per cent, even though many of the afflicted countries had once been French colonies. The fifth year of the drought saw not only community groups but also national governments overwhelmed by the effects. Outward migration of nomads led to conflict with pastoralists over wells, dams and other watering areas. Intertribal conflicts increased. Migration to the cities swelled urban populations, leading to high unemployment and large refugee camps, often resented by locals. Migrations occurred beyond political – but not necessarily traditional – borders. Up to a third of Chad's population emigrated from the country. Chronic malnutrition and the accompanying diseases and complaints – such as measles, cholera, smallpox, meningitis, dysentery, whooping cough, malaria, schistosomiasis, trypanosomiasis, poliomyelitis, hepatitis, pneumonia, tuberculosis, and intestinal worms – overwhelmed medical services. Chronic malnutrition usually results in death, but not necessarily by starvation. Lack of nutrition makes the body more prone to illness and disease, and less able to recover.

As far as aid efforts were concerned, 'the silence was deafening'. By 1972, every national government and international aid group knew of the magnitude of the disaster, but no plan for drought relief was instituted. The countries were either politically unimportant, or everyone believed the drought would end with the next rainy season. A survey of the natural disasters covered in the *New York Times* between 1968 and 1975 shows no references to the Sahelian drought. After 1972, national governments gave up and relied completely upon international aid to feed over 50 million people. Six west African countries were bankrupted. In the sixth year of the drought, the first medical survey was carried out. In Mali, 70 per cent of children, mostly nomads, were found to be dying from malnutrition. In some countries, food distribution had failed completely and people were eating any vegetation available. Until 1973, only the United States was contributing food aid. The Food and Agricultural Organization (FAO) of the United Nations did not begin relief efforts until May 1973, the sixth year of the drought, when it established the Sahelian Trust Fund. The difficulties of transporting food aid were insurmountable. Sahelian transport infrastructure was either non-existent or antiquated. Seaports could not handle the unloading of such large volumes of food. The railroad inland from Dakar, Senegal, was over-committed by a factor of 5–10. Where the railways ended, road transport was often lacking or broken down. Eventually, camel caravans were pressed into service to transport grain to the most destitute areas. In some cases, expensive airlifts were utilized. In 1973, only 50 per cent of the required grain got through; in 1974, the seventh year of the drought, this proportion rose to 75 per cent. In some cases, the aid was entirely ineffective. For instance, grain rotted in ports, found its way to the black market or was of low quality.

The environmental impact of the drought was dramatic. Grasslands, overgrazed by nomads moving out of the worst hit areas, subsequently fell victim to wind deflation. The process of desertification accelerated along all margins of the Sahara. Between 1964 and 1974, the desert encroached upon grazing land 150 km to the south. Yet the lessons of this drought were ignored. By the early 1980s, and despite the death toll of the early 1970s, the population of the Sahel grew by 30–40 per cent. At this time, five of the world's nine poorest countries made up the Sahel. Land degradation continued, with more wells being drilled to support the rebuilding and expansion of livestock herds. Governments could not control the population expansion, and relied heavily upon food imports to feed people – even under favorable conditions. The 1983 drought was a repeat performance. The same script was dusted off; the same national and international responses were replayed. This time, however, rather than beginning in the west and moving east, the drought began in Ethiopia and moved westward into the Sudan. If it had not been for a chance filming by a British television crew, the drought would again have gone unnoticed. This chance discovery evoked two international responses never previously seen: the first attempt to implant drought aid in an area before a drought, and the efforts of Bob Geldof. Both of these responses will be discussed at length towards the end of this chapter.

Besides the obvious point about weak, impoverished national governments being incapable or unwilling to deal with drought, there are two points to be made about the continuing drought in the African Sahel. Firstly, Sahelian droughts have large temporal variability. It has already been emphasized, in Chapter 2, that most droughts worldwide appear to coincide with the 18.6-year lunar cycle. What has not been emphasized is the fact that this lunar cycle may be superimposed upon some longer term trend. The Sahelian droughts persisted for three decades. Secondly, some developed countries may not have experienced their worst drought. For instance, the 1982–1983 Australian drought was the worst in 200 years. The next worst drought in 200 years occurred in 1993–1995 and again in 2002–2003. There is evidence from tree-ring studies that some droughts in Australia in the last two to three centuries have lasted as long as the present Sahelian conditions. Western countries should not be so ready to chastise the governments of Sahelian countries about their inability to deal with drought. Australia, and many other countries, has yet to be tested under similar conditions over such a long period – conditions that should be viewed as likely under our present climatic regime.

Those that win

(Charnock, 1986)

Given the persistence of the African drought, and the fact that so many people face a losing battle, the question arises whether any African country can really cope with drought. There is one notable exception to the bleak news, namely what has been achieved in Botswana. Botswana, in 1986, was in its fifth consecutive year of drought – a record that easily matches that of any country in the Sahel. Yet no one appeared to be dying from starvation, although two-thirds of its inhabitants were dependent upon drought relief. Botswana occupies the southern hemisphere equivalent of the Sahel region of Africa, at the edge of the Kalahari Desert. Long-term droughts have continually afflicted the country, with average rainfall varying erratically from 300 to 700 mm yr^{-1}. The country has become newly independent in the last two decades; but, unlike other decolonized African countries, it has not been wracked by intertribal conflict and has an administration that is open to self-criticism. After a single year of drought in 1980, the government commissioned an independent study on drought-relief efforts, finding measures severely wanting. Following that report, the government overhauled the storage and distribution network, to the point that 90 per cent of the population requiring food aid during the 1986 drought year received uninterrupted supplies. The government also instituted a program of early warning that tracked the intensity of the drought in Botswana. Every month rainfall figures were collected and analyzed to produce a contour map that pinpointed those areas favorable for cropping and those areas in stress. In 1983, a computer model was initiated that predicted soil moisture and maximum crop yield. The model takes rainfall data and makes projections of crop yield until harvesting, based upon the premise of 'no stress' and 'continuing stress'. This model is updated with daily rainfall readings fed into a centralized collecting agency via two-way radio from all parts of the country. The model permits farmers to ascertain the risk to crops if they are planted now, the risk to crops if they are cultivated, and the risk to crops if harvesting is postponed. Because soil moisture is calculated in the model, it is possible to determine the reserves of soil moisture from the previous crop year.

Botswana monitors the state, not only of its agriculture, but also of its people. Each month, two-thirds of its children under the age of five are weighed at health clinics, to detect undernourished children with a body weight less than 80 per cent the expected weight for their age. Areas where drought effects are manifesting as undernourishment can be pinpointed and reported to a central drought committee. This committee, with access to government channels, relays information about supplies back to local committees. The government firmly believes that the loss of rural income during drought, and its attendant unemployment, exacerbates the shortage of food. In the 1986 drought, the government temporarily employed over 70 000 people in rural areas, replacing one-half of lost incomes. These programs were not in the form of food-for-work but operated strictly on a cash-for-work basis. Projects, selected by local committees, were usually drought-mitigating. In addition, the drought program offered free seed, grants for land cultivation, subsidies for oxen hire, and mechanisms for selling drought-weakened animals at above market prices. Government-trained agricultural officers made efforts to enthuse the rural community, to provide the expertise to maintain farmers on the land during the drought, and to make certain that maximizing the farmers' potential did not result in degradation of available arable land. The emphasis of this program is on keeping the rural population on the land, and keeping them self-reliant at existing living standards. In the 1980s, two factors jeopardized the program: growing national apathy and a 3.9 per cent annual growth rate in population. Unfortunately, all these efforts took place before the outbreak of AIDS in Africa. At present, the greatest risk to drought mitigation in Botswana is the breakdown in sophisticated government-sponsored programs due to the death of talented civil servants as the result of this twentieth century scourge.

Laissez-faire: the Australian policy

(Waring, 1976; Tai, 1983)

Australia is a country plagued by drought, where efficient grazing and cultivation systems have been developed to cope with a semi-arid environment. Droughts rarely occur at the same time across the continent. Only the 1895–1903 drought affected the whole country. Most Australian farmers are financially viable as long as they experience a minimum of three consecutive harvests in a decade. This expectation makes it difficult to separate dry years from drought years. The 1982–1983 drought, which was the worst in Australian history up to that point in time, was preceded by several dry years in the eastern states. Even during the severest times in 1982, the federal government was slow to instigate drought relief schemes. This inability to perceive the farmer's plight was one of the reasons Malcolm Fraser's Liberal–Country Party coalition lost power in the March 1983 federal election. Response to drought in Australia, thus, falls into two distinct categories. The first line of defence occurs with the individual farmer's response, as discussed above. Second, when the farmer fails to profitably manage a property, then the state and federal governments attempt to mitigate a drought's worst effects and ensure that quick economic recovery is possible when the drought breaks. Unfortunately, this response is too piecemeal and at times politically motivated.

One of the first responses to drought is the perception that areas are being affected by exceptional deficiencies in rain. The Bureau of Meteorology maintains an extensive network of rain gauges, and classifies an area as drought-affected when annual rainfall is below the tenth percentile. Many states use this information and data on dam levels, crop, pasture and range condition, and field reports through government agencies and politicians, to declare districts drought-affected. This status is usually maintained until sufficient rainfall has fallen to recharge groundwater or regenerate vegetation to sustain agriculture at an economic level. State relief to farmers can take many forms. Initially, attempts will be made to keep stock alive. Subsidies may be given to sink deeper bores, import fodder to an area, carry out agistment of livestock in unaffected areas, or import water using road or rail tankers.

The 1982–1983 drought exemplifies the range of measures that can be adopted to lessen the hardship of drought. Over 100 000 breeding sheep were shipped from the eastern states to Western Australia, a distance of 3000 km. The Australian National Railways arranged special freight concessions and resting facilities en route, while the federal government provided a 75 per cent freight subsidy. In addition, assistance was given to farmers paying interest rates over 12 per cent, and a 50 per cent subsidy was provided for imported fodder for sheep and cattle. Despite such measures, droughts can be so severe that it is only humane to destroy emaciated stock. In the 1982–1983 drought, local councils forced to open and operate slaughtering pits were reimbursed their costs, and graziers were given a sheep and cattle slaughter bounty. In South Australia, each head of sheep and cattle generated a subsidy of $A10 and $A11, respectively. In addition, the federal government matched this sheep subsidy up to a maximum of $A3 per head. However, the main aim of the federal government's drought aid package, which totalled approximately $A400 million, was to prevent the decimation of breeding stock so vital to rebuilding Australia's sheep and cattle export markets.

Attempts at conserving and supplying water were also energetically pursued. The Victorian government cancelled indefinitely all leave for drilling-rig operators working in the mining industry, and redirected them to bore-drilling operations. The number of water tankers proved insufficient to meet demand, and the state resorted to filling plastic bags with water and shipping them where needed. In New South Wales, water allocations from dams and irrigation projects were cut in half and then totally withdrawn as the capacity of many dams dwindled to less than 25 per cent. Broken Hill's water supply was threatened as the Darling River dried up. Water in Adelaide, which is drawn from the Murray River, reached unsafe drinking limits as salinity increased through evaporation. To save trees within downtown Adelaide, trenches were dug around the trees and roots were hand-watered. Many other country towns and cities suffered a severe curtailment of water usage except for immediate domestic requirements.

While the 1982–1983 drought caused farmers economic hardship, the financial response of state and federal governments was relatively minor. The drought cost farmers an estimated $A2500 million and resulted in a $A7500 million shortfall in national income. Government relief (both state and federal) probably did not exceed $A600 million, with most of this aid directed towards fodder subsidies. In Australia, in most cases, the individual farmer must bear the financial brunt of droughts and 1982–1983 drought was no exception. Many farmers were forced to increase their debt load, with personal interest rates as high as 20 per cent, to the point where their farming operations became economically unviable even when favorable times returned. The continued high interest rates following this drought and the over-capitalization of many Australian farmers have ensured a rural decline throughout the latter part of the twentieth century.

INTERNATIONAL RESPONSE

International relief organizations

By far the largest and most pervasive international relief organizations are the ones associated with the United Nations or the Red Cross. The United Nations Disaster Relief Office (UNDRO) was established by the General Assembly in 1972, and financed partly through the United Nations budget and the Voluntary Trust Fund for Disaster Relief Assistance. The latter channel allows countries, sensitive about contributing to the overall running cost of the United Nations, to participate in disaster relief funding. UNDRO gathers and disseminates information about impending disasters (in the form of situation reports) to governments and potential donors. It conducts assessment missions and, working through 100 developing countries, uses trained, experienced personnel who can open channels of communication with government officials in disaster-prone areas, irrespective of the political beliefs of that government or area. It has a responsibility for mobilizing relief contributions and ensuring the rapid transport of relief supplies. During times of disaster, it can call upon other agencies for personnel to implement programs. It also has the duty of assessing major relief programs, and advising governments on the mitigation of future natural disasters.

There are a number of other United Nations offices that are involved less directly in drought relief. The United Nations Children's Fund (UNICEF), Food and Agriculture Organization (FAO), World Health Organization (WHO), and World Food Programme (WFP), while mainly involved in long-term relief in underdeveloped countries, can provide emergency services and advice. In addition, the World Meteorological Organization (WMO) monitors climate, and the United Nations Educational, Scientific and Cultural Organization (UNESCO) is involved in the construction of dams for drought mitigation. The International Bank for Reconstruction and Development (World Bank) finances these mitigation projects, as well as lending money to Third World countries for reconstruction after major disasters.

Temporary infrastructure to handle specific disasters can also be established from time to time. In response to the Ethiopian–Sudanese drought of the early 1980s, the United Nations Office for Emergency Operations in Africa (UNOEOA) was established to coordinate relief activities. This organization channeled $US4.5 billion of aid from 35 countries, 47 non-government organizations and half-a-dozen other United Nations organizations, into northern Africa during this drought. Credited with saving

35 million lives, this office is one of the success stories of the United Nations. It was headed by Maurice Strong, a Canadian with a long-time involvement in the United Nations and a reputation for being able to 'pull strings' – especially with uncooperative governments with obstructive policies. He persuaded Sudan's Gaar Nimeiny to open more refugee camps to relieve the desperate and appalling conditions in existing camps, and coerced Ethiopia's Colonel Haile Mengistu to return trucks (commandeered by the army for its war effort against Eritrean rebels) to the task of clearing the backlog of food stranded on Ethiopian docks.

The United Nations is not always the altruistic organization implied by these descriptions. Many of its branches, most notably the Food and Agricultural Organization (FAO), are politicized and riddled with inefficiency, if not outright cronyism. FAO, which has a $US300 million annual budget to develop long-term agricultural relief programs, in the 1980s spent 40 per cent of its budget on paternalistic and corrupt administration. At the height of the Ethiopian famine in 1984, FAO administrators delayed for 20 days in signing over $US10 million in emergency food aid, because the Ethiopian ambassador to FAO had slighted the FAO. The most efficient FAO field worker in Ethiopia was recalled during the drought because he inadvertently referred to himself as a UN, instead of a FAO, worker in statements to the press. Much of the FAO administration operates as a Byzantine merry-go-round, where the mode of operation is more important than the alleviation of drought in member countries.

Second in effort to the United Nations is the International Red Cross (Red Crescent in Islamic countries), which constitutes the principal non-government network for mobilizing and distributing international assistance in times of disaster. Founded in 1861, it consists of three organizational elements: the International Committee of the Red Cross (ICRC), an independent body composed of Swiss citizens concerned mainly with the victims of armed conflict; the 162 national societies, which share the principles and values of the Red Cross and conduct programs and activities suited to the needs of their own country; and the League of Red Cross Societies (LORCS), which is the federating and coordinating body for the national societies and for international disaster relief. The International Red Cross, because of its strict non-aligned policy, plays a unique and crucial role in ensuring disaster relief can reach refugees in areas stricken by civil war, or controlled by insurgency groups with little international recognition.

The aims of the Red Cross are duplicated by a number of religious and non-affiliated organizations. One of the largest of these is Caritas (International Confederation of Catholic Charities), which was established for relief work during the Second World War, and which afterwards continued relief efforts during natural disasters. The Protestant equivalent is the World Church service. Other denominational groups maintaining their own charities include the Salvation Army, Lutheran Church, Adventist Development and Relief Agency (ADRA) and the Quakers (who in the United States sent aid to Cuba after it was devastated by hurricanes). Religious charities also exist in Islamic, Hindu and Buddhist countries. Finally, there are a number of non-profit organizations that can raise money for immediate disaster relief, long-term reconstruction and disaster mitigation. These organizations include AustCare, CARE, World Vision, Oxfam, and Save the Children. Some have internalized administrations that pass on as much as 95 per cent of all donations, while others utilize professional fundraising agencies that may soak up as much as 75 per cent of donations in administrative costs. These organizations periodically collect monies using door-knock appeals, telethons, or solicitations through regular television, radio, newspaper, and magazine advertisements.

International aid 'flops'

Whenever the international community is mobilized to respond to large droughts and impending famine for millions of people, the question arises 'Does the aid get through?'. There are three difficulties with supplying aid to Third World countries. Firstly, the sudden increase in the volume of goods may not be efficiently handled because existing transport infrastructure in those countries is inefficient or non-existent. If relief is sent by boat, it may sit for weeks in inadequate harbors awaiting unloading. Once unloaded, the aid may sit for months on docks awaiting transshipment. Such was the case with grain supplies sent to Ethiopia during the drought at the beginning of the 1980s. If a drought has occurred in an isolated area, there simply may be no transport network to get relief to the disaster site. This problem may be exacerbated by the destruction of transport networks because of civil war.

Secondly, Third World countries may have an element of corruption or inefficiency to their administration. The corruption, if it exists, may lead to relief aid being siphoned off for export, ransom or sale on the black market. Finally, inefficient administrations may slow the relief aid with paper work or be unaware that relief aid is even necessary.

By far one of the biggest 'flops' to occur for these reasons was the international aid effort for the Sudanese drought of 1984–1985. This drought was the first in Sudan to be predicted and prepared for before its occurrence. In the previous two years, Ethiopia had been wracked by drought that went virtually unnoticed for a year, and was then exposed on British television. The Ethiopian drought generated considerable attention about why it had occurred, and how relief was getting through to the people. It became a foregone conclusion that the drought was spreading westward and would eventually enter Sudan, a country that had magnanimously permitted over 1 million Ethiopian refugees to cross its borders for relief aid. The management of the Ethiopian famine was a disaster in itself. As it had in 1975, the Ethiopian government at first failed to admit that it had a famine problem, mainly because the drought was centered in rebellious northern provinces. It continued to sell garden produce to other countries instead of diverting food and produce to the relief camps that had been set up or had spontaneously appeared. Authorities deliberately strafed columns of refugees fleeing the drought-affected areas and allowed relief aid to pile up in its eastern ports. At the same time, an enormous sum of money was spent to dress up the capital for a meeting of African nations.

The United States government recognized these problems and decided that a similar situation would not hinder relief aid for the impending drought in Sudan. It pledged over $US400 million worth of food aid and set about overcoming the distribution problem, which consisted in getting the food from Port Sudan on the Red Sea to the western part of Sudan, which was isolated at the edge of the Sahara Desert. The United States hired consultants to evaluate the distribution problem, and it was decided to use the railway between Port Sudan and Kosti, or El-Obeid, to ship grain west before the onset of the rainy season. The grain would then be distributed westward using trucks. The United States went so far as to warn the residents of the western part of Sudan that a drought was coming, and that food aid would be brought to their doorstep. The inhabitants of that part of the country were not encouraged to migrate eastward. Thus, there was going to be no repeat of the large and deadly migrations that occurred in Ethiopia.

However, no one assessed the Sudanese railway system's ability to handle transportation of such a large volume of grain. In fact, scrapped trains in western countries were in better shape than those operating on the Sudanese railway system. A decision was made to repair the railway, but inefficiencies delayed this for months. Finally, it was decided to move the food for most of the distance by truck. No one had evaluated the trucking industry in the Sudan. The operation absorbed most of the trucks in the country. The local truck drivers, realizing they now had a monopoly, went on strike for higher wages. By the time the railway and trucking difficulties were overcome, the drought was well-established, and few if any people had migrated to the east. They were now starving. When the transport situation was sorted out, the rains set in, and many trucks became bogged on mud tracks or while attempting to ford impassable rivers, or hopelessly broken down. Meanwhile, grain supplies were rotting in the open staging areas in the east, because no one had considered the possibility that the grain would still be undistributed when the rainy season broke the drought.

The decision was finally made to airlift food supplies into remote areas using Hercules aircraft, a decision that increased transport costs fourfold. But this was not the end of the story. The rains made it impossible for the planes to land, and there was no provision to parachute supplies to the ground. Like sheep in an Australian flood awaiting airdropped bales of hay, the starving inhabitants of west Sudan waited for bags of grain to be airdropped. The Hercules would fly as low and as slowly as possible, as bags of grain were shoved out without any parachutes to slow them down. The bags tumbled to the ground, bounced and broke, scattering their contents. The Sudanese then raced on foot to the scene to salvage whatever grain they could before it was completely spoilt in the mud.

PRIVATE RESPONSES: BOB GELDOF

(Geldof, 1986)

Without Bob Geldof there would have been no Band Aid or Live Aid: his name is now synonymous with spontaneous, altruistic, international disaster relief efforts – although he was not the first celebrity to take on massive famine relief. In 1891, Leo Tolstoy and his family organized soup kitchens, feeding 13 000 people daily in the Volga region of Russia following destruction of crops by inclement weather.

Bob Geldof was a musician from Ireland who worked with a band called the Boomtown Rats that had one international hit single in 1979, 'I don't like Mondays'. They were the number one band in England that year. Because he was a musician, Geldof had contacts with most of the rock stars at the time and was also aware of the sporadic efforts by musicians to help the underprivileged in the Third World. For example, following the 1970 storm surge in East Pakistan (now Bangladesh) and that country's independence, musicians organized a concert, the 'Concert for Bangladesh', to raise money to help overcome the disaster. It was staged free by several famous rock groups, and raised several million dollars through ticket sales and royalties on records.

Bob Geldof, like many other people in 1984, was appalled by the plight of the Ethiopians – and the fact that the famine remained virtually unknown to the outside world until a British television crew, led by reporter Michael Buerk, chanced upon it. Buerk's film showed the enormity of the disaster, and painted a dismal, hell-like picture. Children were being selectively fed. Only those who had a hope of surviving were permitted to join food queues. People were dying throughout the film. In the midst of this anarchy, the Ethiopian people stoically went on trying to live in the camps, which were over-crowded and lacking medical aid. Their honesty formed a striking contrast to the greed of most people viewing the show. Starving, Ethiopian refugees would not touch food in open stores unless it was given to them. No one appeared to be doing anything to help the refugees. The Ethiopian government had denied the existence of the drought, and the international monitoring agencies had failed to bring it to world attention. UNDRO in particular appeared hopelessly inefficient.

The emotional impact of the BBC film stunned all viewers, including Bob Geldof; however, a rock musician was hardly the most obvious candidate to organize and lead the largest disaster relief effort in history. The day after the screening of the documentary, Geldof began to organize, coerce, and lie to put together a group of musicians to produce a record for Christmas 1984, the profits of which were to be donated to Ethiopian relief. The group was made up of the most popular rock musicians in Britain at the time: Sting, Duran Duran, U2, Style Council, Kool and the Gang, Boy George and Culture Club, Spandau Ballet, Bananarama, Boomtown Rats, Wham, Ultravox, Heaven 17, Status Quo, David Bowie, Holly, and Paul McCartney. The band was called Band Aid, an obvious play on words, but chosen in reality because the name would make people aware of the futility of this one act of charity in overcoming the famine. In the end, Bob Geldof managed to convince all people associated with the record to donate their services free. The British government, after protracted lobbying, waived the value added tax. The proceeds of the record were to be put into a special fund to be distributed for relief aid as a committee saw fit. The record contained the song 'Do They Know It's Christmas?' on side one, and some messages from rock stars on side two. It cost £1 30p and reached number one the day it was released, six weeks before Christmas. Within weeks it was selling 320 000 copies per day and utilizing the pressing facilities of every record factory in Britain, Ireland and Europe to meet demand. People bought 50 copies, kept one, and gave the rest back for resale; they were used as Christmas cards, sold in restaurants, and replaced meat displays in butchery windows. Musicians in the United States (USA for Africa) and Canada (Northern Lights) subsequently got together to produce a similar type of album, which had the same success in North America. Over twenty-five Band Aid groups ('Austria für Afrika', 'Chanteurs Sans Frontiers', and three German groups amongst others) were formed worldwide, raising $US15 million for famine relief.

Following the success of the Christmas album, in 1985 Bob Geldof organized a live concert. It was to become the first global concert and would run for 17 hours on television, using satellite hook-ups. It would start in Wembley Stadium in London, and then end in Philadelphia with five hours of overlap. Famous bands would alternate on a rotating stage. The concert would also draw in television coverage of smaller concerts held in Australia and Yugoslavia. Countries would pay for the satellite link-up, and at the same time be responsible for organizing telethons to raise money. Paraphernalia would also be sold, with all profits going to the Band Aid fund. As much as possible, overheads were to be cost-free.

Critics said that it could not be done, that the organization would be too horrendous, the musicians too egotistical, that no one would watch it or give donations. A group of middle-aged consultants hired to evaluate the income potential of such a concert, said it would raise less than half a million US dollars. The concert was organized over three months and it was not known until the day of the show, 13 July 1985, exactly what groups would be participating. Many countries only telecast the concert at the last minute. The chaos, especially in Philadelphia, spelt disaster; but through the efforts of talented organizational people in television and concert promotion, the show was viewed by 85 per cent of the world's television audience.

The show blatantly, but sincerely, played upon the television audience's emotions. At one point, when telephone donations were lagging, a special video clip of a dying Ethiopian child was played to the words of the song 'Drive' by the group The Cars. As the child tried repeatedly to stand up, only to fall down each time, the song asked who was going to pick him up when he fell down. The clip brought a response that jammed the telephone lines worldwide. Donations were received from both communist and non-communist countries. One private donation from Kuwait totaled $US1.7 million. Ireland gave the most money per capita of any country, in total $US10 million. People there even cashed in their wedding rings and houses. In twenty-four hours Live Aid, as the concert became known, raised $US83 million.

Live Aid was to ad hoc disaster relief what Woodstock was to rock concerts. It became just one of many unique fundraising efforts. Geldof convinced people in many industries and recreational pursuits to run their own types of money-raising programs, spawning an alphabet soup of organizations including Actor Aid, Air Aid, Art Aid, Asian Live Aid, Bear Aid, Bush Aid, Cardiff Valley Aid, Fashion Aid, Food Aid, and so on. In France, which had not successfully linked in with the fundraising activity for Live Aid, students raised money in School Aid, whose success was ensured by the refusal of teachers to participate. In May 1986, Sport Aid saw joggers worldwide paying for the privilege of participating in a 10 kilometre run. Over 20 million joggers participated worldwide. Sport Aid raised nearly $US25 million, with donations being split between the Band Aid organization and UNICEF.

Bob Geldof personally oversaw the distribution of relief aid in Ethiopia and Sudan, and was responsible for ensuring, in a last desperate measure, that Hercules planes were used to salvage the Sudan drought relief operation. A committee, Live Aid Foundation, was established to distribute monies for relief efforts. It was run by volunteers, including professional business people, lawyers, and accountants. Bureaucratic administration was avoided as much as possible. A board of trustees (Band Aid Trust) was established to oversee the relief distribution. This board acted on the advice of a team of eminent academics who had regular working experience in Africa, from various institutions including the universities of Sussex and Reading; the School of Hygiene and Tropical Medicine, and the School of Oriental and African Studies in London; and Georgetown University in Washington. Of the money raised, 20 per cent went for immediate relief, 25 per cent for shipping and transport, and the remainder for long-term development projects vetted by these experts. In addition, the Band Aid Trust approached countries to undertake joint aid. For instance, the United States government was ideologically opposed to the Ethiopian government, but not to the Sudanese government. Band Aid teamed up with United States aid agencies to ensure that relief efforts were directed equally to these two countries. Band Aid handled the Ethiopian relief, while the United States tackled Sudan. Other countries were approached for transport and funding. The Australian government was asked to refurbish ten Hercules transport planes for food drops; but in the end donated only one plane without support funding. Other countries, faced with similar requests, donated all the support needed to keep their planes in the air. President Mitterrand of France, in one afternoon of talks with Geldof, pledged $US7 million in emergency relief aid for areas and projects recommended by Band Aid field officials.

The Band Aid organization also decided that money should support drought reconstruction and mitigation. Proposals for financing were requested, and then checked for need by a four-person field staff paid for by private sponsorship. Over 700 proposals were evaluated, with many being rejected because of their lack of relevance or grassroots participation. By the end of 1986, $US12 million had been allocated for immediate relief, $US20 million for long-term projects, and $US19 million for freight. Project funding went mainly to Burkina Faso, Ethiopia and the province of Eritrea, and Sudan. The largest sum was for $US2.25 million for a UNICEF program to immunize children, and the smallest was for $US3000 to maintain water pumps in refugee camps in eastern Sudan.

All these efforts had their critics. Many argued that Geldof's fundraising efforts were siphoning money that would have gone to legitimate aid organizations. Geldof always contended that the efforts were raising the level of consciousness amongst ordinary people. In fact, following the Live Aid concert, donations to Oxfam and Save the Children increased 200 per cent and 300 per cent, respectively. Geldof foresaw criticism that he might personally profit from his fundraising activities and refused to use one cent of the money for administration or personal gain. All work had to be voluntary, using equipment, materials, labor and fares that had been donated free and with no strings attached.

Of all the international relief programs, Bob Geldof's has been the most successful at pricking the consciences of individuals worldwide – regardless of their political ideology. Initially he tapped the music industry for support, but has shown that the methods for raising aid, with involvement of people from all walks of life, are almost endless. Geldof's program of raising money and distributing funds without any significant administration costs is one of the most successful and efficient ever, putting to shame the efforts of individual countries and the United Nations. The euphoria of Band Aid and Live Aid has not worn off; however, Geldof saw them as temporary measures. At the end of 1986, the Band Aid umbrella organization was wound up and converted to a standing committee. Donations are now being channeled, through this committee, to permanently established aid organizations. Since Ethiopia, there has been no other calamitous drought, no need for an international response, no testing of the First World's capacity for altruism in the face of a Third World disaster. Since the Ethiopian drought, communism has fallen, sectarianism has formally fragmented nations, greed has overwhelmed capitalistic economies, and terrorism has permeated the voids opened by poverty, disenfranchisement and failed nationalism. The inevitable specter of calamitous drought will rise again to challenge the sincerity of the ideals that Bob Geldof – for one brief moment – established in response.

CONCLUDING COMMENTS

The effects of drought go far beyond the immediate crisis years when communities and nations are trying to survive physically, socially and economically. Of all hazards, drought has the greatest impact beyond the period of its occurrence. This does not necessarily affect population growth. For instance, while many Sahelian countries suffered appalling death tolls in the 1968–1974 drought, high birth rates of 3 per cent ensured that populations had grown to pre-drought levels within 10–13 years. However, the population was skewed, with the number of children under the age of 18 making up 50 per cent of the population in some countries. Such a young population taxes medical and school facilities and overwhelms economic development. Unless exploited, such numbers of young people do not contribute directly to the labor force or to a nation's economic growth. In addition, survivors of droughts can weigh down family units and community groups with an excessive number of invalids, especially amongst the young. Many of the diseases that afflict malnourished children can lead to permanent intellectual and physical impairment. Malnourished children who survive a drought may never overcome their malnutrition, especially if they are orphaned, have many siblings, exist within a family unit that cannot recover economically from the drought, or are displaced from home areas where kinship ties can traditionally provide support.

Additionally, droughts pose difficult problems in recovery. If seed stock is consumed, it must be imported quickly to ensure that a country can re-establish the ability to feed itself. While many people migrate out of drought areas to towns where food can be more easily obtained, the transport infrastructure is not adequate to ship seed into isolated rural areas when these people return home. A similar situation applies to livestock. During major droughts, even in developed countries such as Australia, breeding stock may either perish or be used for food. Livestock herds require several years to be rebuilt, a process that can put a heavy financial strain upon a farm's resources because this period of rebuilding generates little income. In the longer term, droughts are often accompanied by human-induced or natural land degradation. Rangeland may be overgrazed during a drought to keep livestock herds alive, and in some cases foliage stripped from shrubs and trees to feed stock. Landscapes can be denuded of any protective vegetation, resulting in wholesale loss of topsoil during dust storms. In some places in New South Wales, Australia, up to 12 centimetres of topsoil blew away during the 1982–1983 drought. This soil can never be replaced; while growth of vegetation, required to protect the remaining topsoil, can be a slow process taking years to decades. This problem is especially acute in semi-arid areas. Even in the United States, where a major federal government program was initiated on the Great Plains following the mid-1930s drought, land rehabilitation was continuing when the next major drought struck in the mid-1950s. For developing countries strapped for foreign aid and burdened by debt, such long-term recovery and mitigation programs are receiving such a low priority that complete land degradation, caused by successive droughts, may become a permanent feature of the landscape.

REFERENCES AND FURTHER READING

Allison, G.B. and Peck, A.J. 1987. Man-induced hydrologic change in the Australian environment. *Bureau of Mineral Resources Geology and Geophysics* Report 282: 29–34.

Bryson, R. and Murray, T. 1977. *Climates of Hunger*. Australian National University Press, Canberra.

Charnock, A. 1986. An African survivor. *New Scientist* 1515: 41–43.

French, R.J. 1983. Managing environmental aspects of agricultural droughts: Australian experience. In Yevjevich, V., da Cunha, L., and Vlachos, E. (eds) *Coping with Droughts*. Water Resources Publications, Littleton, pp. 170–187.

Garcia, R. 1972. *Drought and Man*, Vol. 1 Pergamon, Oxford.

Geldof, B. 1986. *Is That It?* Penguin, Harmondsworth.

Glantz, M. H. 1977. *Desertification: Environmental Degradation in and Around Arid Lands*. Westview, Boulder, Colorado.

Gribbin, J. 1983. *Future Weather: The Causes and Effects of Climatic Change*. Penguin, Harmondsworth.

Hidore, J.J. 1983. Coping with drought. In Yevjevich, V., da Cunha, L., and Vlachos, E. (eds) *Coping with Droughts*. Water Resources Publications, Littleton, pp. 290–309.

Lockwood, J.G. 1986. The causes of drought with particular reference to the Sahel. *Progress in Physical Geography* 10(1): 111–119.

Morren, G.E.B. 1983. The bushmen and the British: problems of the identification of drought and responses to drought. In Hewitt, K. (ed.) *Interpretations of Calamity*. Allen and Unwin, Sydney, pp. 44–65.

Rosenberg, N. and Wilhite, D.A. 1983. Drought in the US Great Plains. In Yevjevich, V., da Cunha, L., and Vlachos, E. (eds) *Coping with Droughts*. Water Resources Publications, Littleton, pp. 355–368.

Scott, E.P. (ed.). 1984. *Life Before the Drought*. Allen and Unwin, London.

Tai, K.C. 1983. Coping with the devastating drought in Australia. In Yevjevich, V., da Cunha, L. and Vlachos, E. (eds) *Coping with Droughts*. Water Resources Publications, Littleton, pp. 346–354.

Waring, E.J. 1976. Local and regional effects of drought in Australia. In Chapman, T.G. (ed.) *Drought*. AGPS, Canberra, pp. 243–246.

Warrick, R.A. 1983. Drought in the U.S. Great Plains: shifting social consequences?'. In Hewitt, K. (ed.) *Interpretations of Calamity*. Allen and Unwin, Sydney, pp. 67–82.

Watts, M. 1983. On the poverty of theory: natural hazards research in context. In Hewitt, K. (ed.) *Interpretations of Calamity*. Allen and Unwin, Sydney, pp. 231–262.

Whittow, J. 1980. *Disasters: The Anatomy of Environmental Hazards*. Pelican, Harmondsworth.

Young, J.A. 1983. Drought control practices in England and Wales in 1975–1976. In Yevjevich, V., da Cunha, L. and Vlachos, E. (eds) *Coping with Droughts*. Water Resources Publications, Littleton, pp. 310–325.

Flooding as a Hazard

CHAPTER 6

INTRODUCTION

The earlier chapter on large-scale storms did not consider in detail the effects of flooding associated with tropical cyclones, extra-tropical storms or east-coast lows. Nor was it appropriate to discuss large-scale flooding in that chapter because some of the worst regional flooding in the northern hemisphere has occurred in spring in association with snowmelt. In addition, rainfall associated with thunderstorms is a major cause of flash flooding. This chapter examines flash flooding events and regional floods. Flash flooding refers to intense falls of rain in a relatively short period of time. Usually the spatial effect is localized. Modification of the overall landscape during a single event is minor in most vegetated landscapes; however, in arid, semi-arid, cultivated (where a high portion of the land is fallow) or urban areas, such events can be a major cause of erosion and damage. Steep drainage basins are also particularly prone to modification by flash floods because of their potential to generate the highest maximum probable rainfalls. Large-scale regional flooding represents the response of a major continental drainage basin to high-magnitude, low-frequency events. This regional flooding has been responsible historically for some of the largest death tolls attributable to any hazard. When river systems alter course as a response to flooding, then disruption to transport and agriculture can occur over a large area.

FLASH FLOODS

Magnitude and frequency of heavy rainfall

(Griffiths, 1976)

Short periods of heavy rainfall occur as a result of very unstable air with a high humidity. Such conditions usually occur near warm oceans, near steep, high mountains in the path of moist winds, or in areas susceptible to thunderstorms. A selection of the heaviest rainfalls for given periods of time is compiled in Figure 6.1. Generally, extreme rainfall within a few minutes or hours has occurred during thunderstorms. Events with extreme rainfall over several hours, but less than a day, represent a transition from thunderstorm-derived rain to conditions of extreme atmospheric instability. The latter may involve several thunderstorm events. In the United States, the Balcones Escarpment region of Texas (see Figure 6.2 for the location of major placenames mentioned in this chapter) has experienced some of the heaviest short-duration rainfalls in this category. Extreme rainfall over several days is usually associated with tropical cyclones, while extreme rainfall lasting several weeks to months usually occurs in areas subject to seasonal monsoonal rainfall, or conditions where orographic uplift persists. In the latter category, Cherrapunji in India, at the base of the Himalayas, dominates the data (9.3 m rainfall in

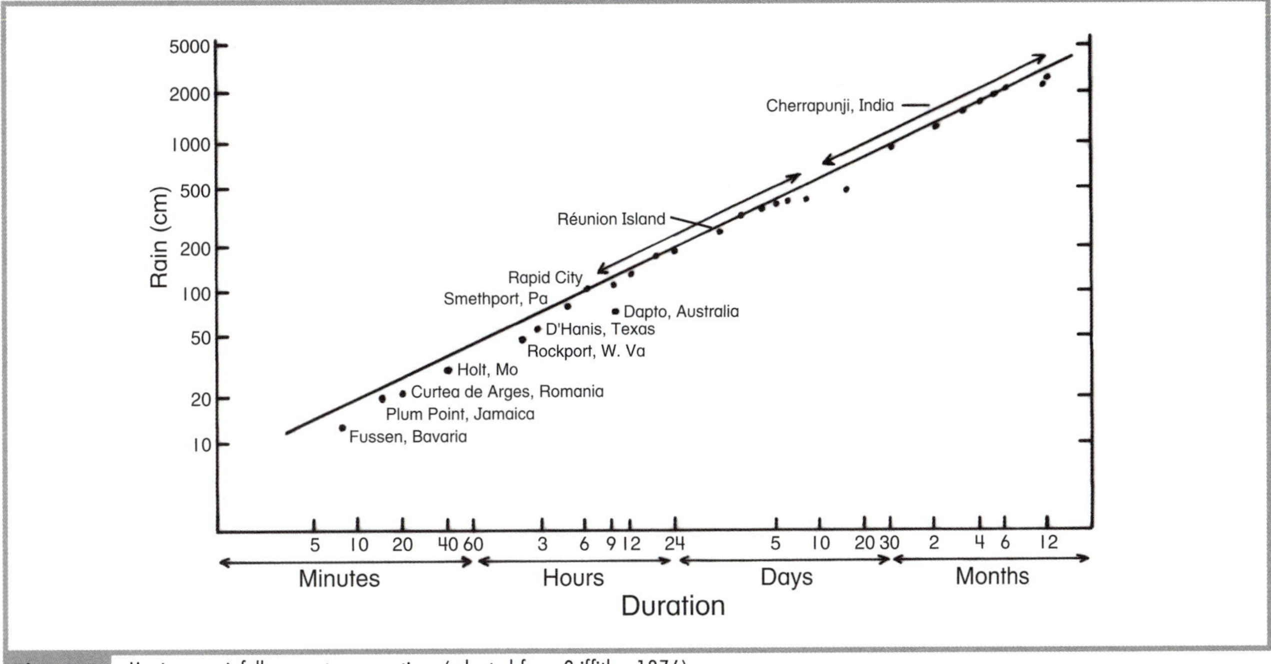

Fig. 6.1 Maximum rainfall amounts versus time (adapted from Griffiths, 1976).

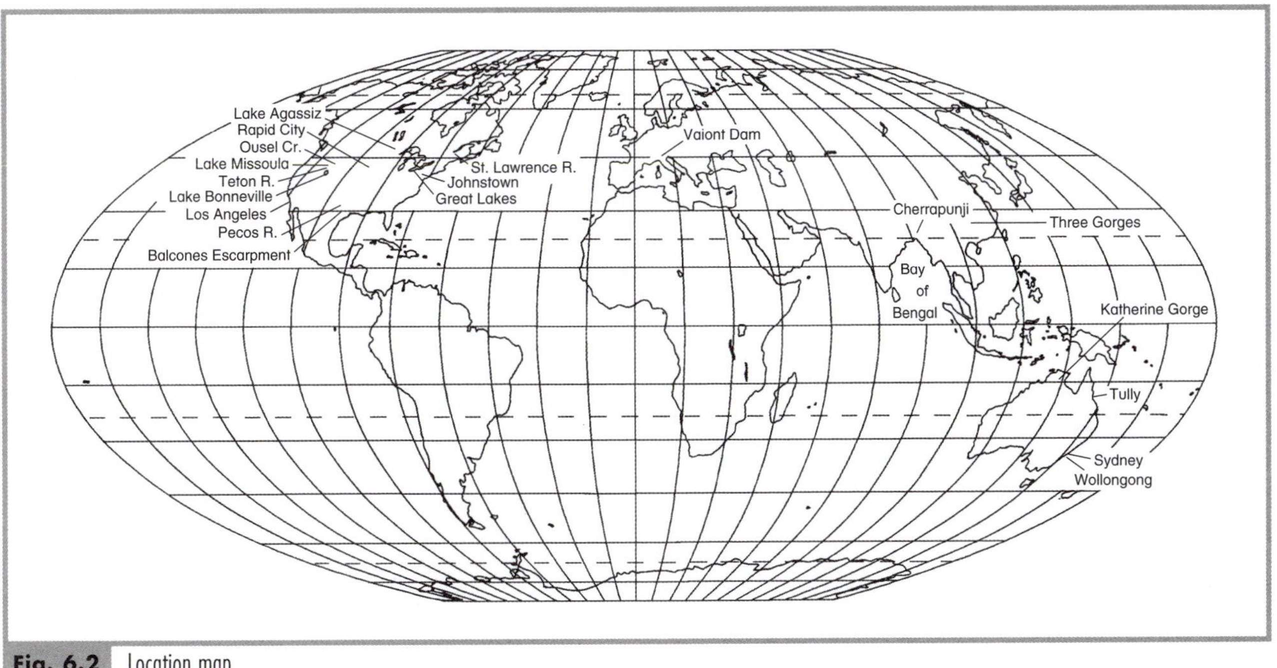

Fig. 6.2 Location map.

July 1861 and 22.99 m for all of 1861). In this region, monsoon winds sweep very unstable and moist air from the Bay of Bengal up over some of the highest mountains in the world. In Australia, the highest rainfall in one year (11.3 m) occurred near Tully in Queensland, a location where orographic uplift of moist air from the Coral Sea is a common phenomenon in summer.

The following equation (6.1) represents the line of best fit to the maximum rainfall data in Figure 6.1:

$$R = 0.42\, D^{0.475} \tag{6.1}$$

where R = rainfall in meters
D = duration in hours

While this equation fits the data well for periods above one hour, the line overestimates the amount of rain that can fall in shorter periods. The latter may merely represent the undersampling – by the existing network of *pluviometers* – of spatially small events dropping these large rainfall amounts over this shorter time span. The line of best fit probably represents a physical limit to the rate of condensation and droplet formation possible under the Earth's present temperature and pressure regime. The amounts of rainfall possible in less than one day are staggering. Within an hour, 0.42 m of rain is possible, and within a day, 1.91 m of rain can fall. The latter value represents more than the average yearly rainfall of any Australian capital city.

The large amounts of rainfall at Cherrapunji, India (or, for that matter, at Tully, Queensland), while exceptional, probably do not disturb the environment nearly as much as these intense but short falls. Over long time periods, locations normally receiving heavy rainfall have evolved dense vegetation that can absorb the impact of falling rain, and have developed drainage patterns that can handle the expected runoff. Most extreme short-term rainfalls occur in places where vegetation is sparser, and where drainage systems may not be best adjusted to contain large volumes of runoff. Here, dislodgement of topsoil by raindrop impact can also be high. Sheet flow occurs within very short distances of drainage divides and overland channel flow will develop within several meters downslope. As a result, sediment erosion and transport is high and rapid. For these reasons, flash flooding in arid and semi-arid regions can become especially severe.

In urban areas, where much of the ground is made impervious by roads or buildings and where drainage channels are fixed in location, flash flooding becomes more likely with much lower amounts of rainfall than indicated in the above graph. Since 1970, flash flooding in semi-arid or urban catchments appears to be increasing worldwide, including in the United States and Australia.

Flood power

(Baker & Costa, 1987)

The amount of work that a flood can perform, and the destruction caused by a flash flood, are not necessarily due to the high amounts of rainfall described above. Nor are high flood discharges a prerequisite for erosion. Rather, it is the amount of *shear stress*, and the stream power – the amount of energy developed per unit time along the boundary of a channel – that is more significant. It is because of stream power that floods in small drainage basins, as small as 10–50 km^2, can be more destructive than major floods on the Mississippi or Amazon rivers, with discharges 10–1000 times larger. This is particularly so where channels are narrow, deep and steep. Stream power per unit area of a channel is defined by the following equation:

$$\omega = \tau v \qquad (6.2)$$

where ω = power per unit boundary area
v = velocity
τ = boundary shear stress
$= \gamma R\, S$
where γ = specific weight of the fluid (9800 N m^{-3})
S = the energy slope of the flow
R = the hydraulic radius of the channel
$= A\,(2d + w)^{-1}$
where A = the cross-sectional area of the wetted channel
d = the mean water depth
w = the width of the channel

The parameters in Equation 6.2 are not that difficult to measure, because many rivers and streams have gauging stations that measure velocity and water depth during floods. Once the height of a flood has been determined, it is relatively simple to calculate the hydraulic radius. The slope of a channel is normally used in place of the energy slope; however, this may increase errors in discharge by 100 per cent during catastrophic flash floods. Problems may also arise in determining the specific weight of the fluid. The value quoted above is standard for clear water. During floods, however, waters can contain high concentrations of suspended material that can double the specific weight of floodwater.

Figure 6.3 plots stream power against drainage basin area for small flash floods, mainly in the United States, and for the largest floods measured in recent times. Some of these floods, such as the Teton River flood in Idaho on 5 June 1976, can be associated with the collapse of dams following heavy rains. The collapse of a dam can greatly increase the magnitude of stream power, because of its effect on hydraulic radius and the energy slope of the flow. In fact, dam collapses have led to the largest flash flood death tolls. For example, the Johnstown, Pennsylvania, flood of 31 May 1889 killed over 2200 people. The failure of dams during heavy rains can be attributed to neglect, inadequate design for high-magnitude events, geological location (for

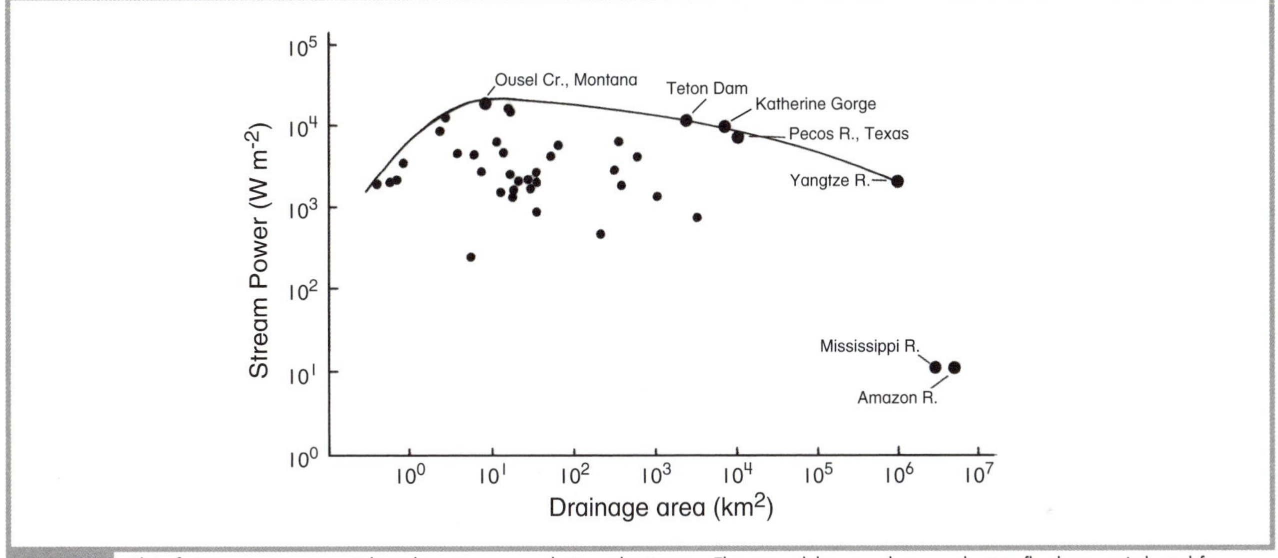

Fig. 6.3 Plot of stream power per unit boundary area versus drainage basin area. The curve delineates the upper limit in flood power (adapted from Baker & Costa, 1987).

example, built across an undetected fault line), or mischance. The greatest stream power calculated for any flood in the United States was for the 8 June 1964 Ousel Creek flood in Montana. During this flood, stream powers reached 18 600 W m^{-2}. Most of the powerful floods in the United States have occurred on quite small drainage basins because they have steeper channels. In comparison to the Ousel Creek flood, the large flood on the Mississippi River that occurred in 1973, had a stream power of 12 W m^{-2}.

For larger drainage basins, two conditions must be met to attain high stream powers. Firstly, the channel must be constrained and incised in bedrock. This ensures that high velocities are reached because water cannot spill out across a wide floodplain. Secondly, the upstream drainage basin must be subject either to high-magnitude rainfall events, or to high discharges. These two features severely limit the number of high stream power events occurring on large rivers. During the Teton Dam failure in Idaho in 1976, a stream power was reached in excess of 10 000 W m^{-2} because the river flowed through an incised channel. Similar magnitudes have occurred through the narrow Katherine River gorge in northern Australia. The Pecos River in Texas is also incised into a limestone escarpment subject to flash flooding, mainly during tropical cyclones such as Hurricane Alice, which penetrated inland in 1954. Finally, the Chang Jiang (Yangtze) River produces stream power values as high as flash floods on streams with drainage basins less than 1000 km^2. At the Three Gorges section of the Chang Jiang River, snowmelt and rainfall from the Tibetan Plateau concentrates in a narrow gorge. This river, in the 1870 flood, flowed at a depth of 85 m with velocities of 11.8 m s^{-1}.

Figure 6.3 does not plot the largest known floods. Stream power values for cataclysmic prehistoric floods associated with Lake Missoula at the margin of the Laurentian icesheet on the Great Plains, and with Lake Bonneville, were an order of magnitude larger. These resulted from the sudden bursting of *ice-dammed lakes* as meltwater built up in valleys temporarily blocked by glacier ice during the Wisconsin glaciation. The Missoula floods reached velocities of 30 m s^{-1} and depths of 175 m, producing stream powers in excess of 100 000 W m^{-2}, the largest yet calculated on Earth. Similar types of floods may have emptied from the Laurentian icesheet into the Mississippi River and from Lake Agassiz eastward into the Great Lakes–St Lawrence drainage system. Equivalent large floods also existed on Mars at some time in its geological history; however, if the high stream powers recorded on Earth were reached, the flows may have been as deep as 500 m – as gravity is lower on Mars.

The ability of these catastrophic floods to erode and transport sediment is enormous. Figure 6.4 plots mean water depth against the mean velocity for these flood events. Also plotted on this diagram is the point where flow becomes supercritical and begins to jet. *Supercritical flow* is highly erosive, and cannot be sustained in alluvial channels because the bed will be eroded so rapidly. It is also rare in bedrock channels because, once floodwaters deepen a small amount, the flow will then revert to being *subcritical*. Finally, Figure 6.4 plots the region where *cavitation* occurs. Cavitation is a process whereby water velocities over a rigid surface are so high that the vapor pressure of water is exceeded, and bubbles begin to form at the contact. A return to lower velocities will cause these bubbles to collapse with a force in excess of 30 000 times normal atmospheric pressure. Next to impact cratering, cavitating flow is the most erosive process known on Earth and, when sustained, is capable of eroding bedrock. Fortunately, such flows are also rare. They can be witnessed on dam spillways during large floods as the mass of white water that develops against the concrete surface towards the bottom of the spillway. Note that the phenomenon differs from the white water caused by air entrainment due to *turbulence* under normal flow conditions at the base of waterfalls or other drop structures. Cavitation is of major concern on spillways because it will shorten the life span of any dam, and may be a major process in the back-cutting of waterfalls into bedrock. Only a few isolated flash floods in channels have generated cavitation. However, some of the Missoula floods reached cavitation levels over long distances.

Large flood events have enormous potential to modify the landscape. The stream powers of cataclysmic floods, as well as many other common flash floods in the United States, are substantial enough to transport sand-sized material in wash load, suspended gravel 10–30 cm in size, and boulders several metres in diameter as bedload. This material can also be moved in copious quantities. Even stream powers of 1000 W m^{-2} are sufficient to move boulders 1.5 m in diameter. Many flash floods can reach this capacity. If cavitation levels are reached, bedrock channels can be quickly excavated. The Missoula floods, in the western United States, were able to pluck *basaltic* columns of bedrock, each weighing several tonnes, from streambeds, and carve canyons through bedrock in a matter of days. More importantly, as will be discussed below, present-day flood events are capable of completely eroding, in a matter of days, floodplains that may have taken centuries to form.

Synoptic patterns favoring flash flooding

(Hirschboeck, 1987)

The severity of catastrophic flooding is ultimately determined by basin physiography, local thunderstorm movement, past soil moisture conditions, and degree of vegetation clearing. However, most floods originate in anomalous large-scale atmospheric circulations that can be grouped into four categories as follows: (i) aseasonal occurrence or anomalous location of a definable weather pattern; (ii) a rare concurrence of several commonly experienced meteorological processes; (iii) a rare upper atmospheric pattern; or (iv) prolonged persistence in space and time of a general meteorological pattern. The degree of abnormality of weather patterns is very much a function of three types of pressure and wind patterns that are often part of the general atmospheric circulation. These patterns are shown in Figure 6.5. Meridional air flow represents north–south air movement and is usually associated with Rossby wave formation linked to the meandering

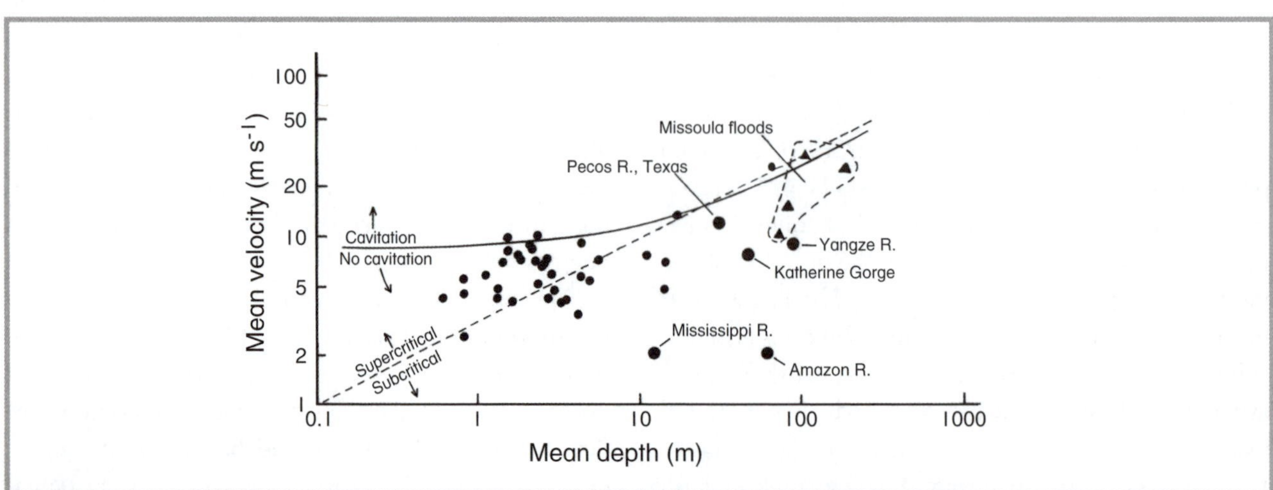

Fig. 6.4 Channel depth versus current velocity for various floods plotted together with boundaries for cavitation and supercritical flow (adapted from Baker & Costa, 1987).

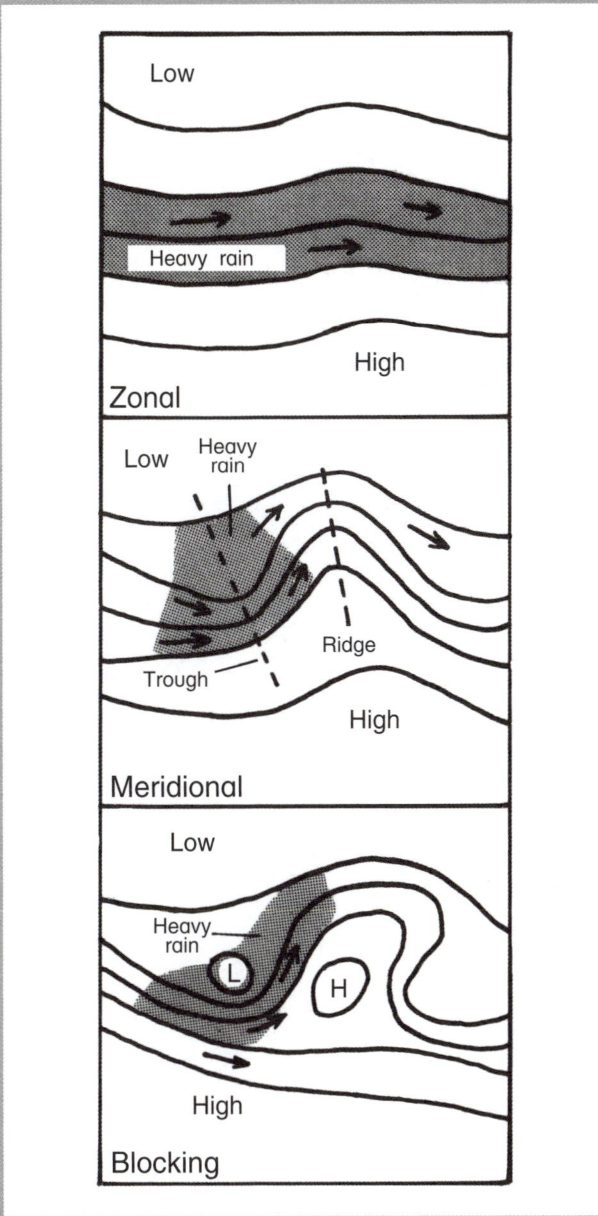

Fig. 6.5 Abnormal patterns of airflow, within general air circulation, leading to flash flooding (adapted from Hirschboeck, 1987).

path of the polar jet stream, especially over continents in the northern hemisphere. As stated in Chapter 2, the jet stream path can be perturbed by El Niño–Southern Oscillation events, and can be associated with persistent weather patterns giving rise to droughts. Exaggerated meridional flow also appears to be one of the most frequent atmospheric patterns generating flash flooding in the United States. Frequently, it is associated with blocking of a low- or high-pressure cell. The former occurs over warmer water while the latter is associated with the stalling or agglutination of mobile polar highs. Blocking in North America commonly occurs over the north Atlantic and Pacific Oceans. In the north Atlantic, blocking increases extra-tropical storm activity and enhances persistence of patterns.

Within this general picture of anomalous circulation are smaller features that are responsible for localized flash floods. These features can be classified into four categories: synoptic, frontal, mesohigh, and western events. Synoptic events occur when an intense mid-latitude cyclone and a semi-stationary front are linked to an intense, low-pressure trough overhead. Rainfall is widespread, persistent and, in local cases, heavy as thunderstorms develop repeatedly in the same general area. Such storms occur in the seasonal transitions between winter and summer pressure patterns over North America. Thus, they are common from spring to early summer, and in autumn. Frontal events are generated by a stationary or very slow-moving front with zonal air circulation. Upper air stability may exist, and rainfall is usually heavy on the cool side of any warm front, as warm air is lifted aloft. Such events occur most frequently in July–August in the United States. Mesohigh floods are caused by instability and convection following the outbreak of cold air that may or may not be associated with a front. For instance, a stationary front may develop supercell thunderstorms that force out, in a bubble fashion, a high-pressure cell. These storms occur in the late afternoon or evening and are a localized summer feature. Western-type events refer to a range of regional circulation patterns over the Rocky Mountains. These events tend to peak in late summer and are associated with either extremely meridional or zonal circulation.

Maximum probable rainfall

(Australian Bureau of Meteorology, 1984)

There are limitations to the concepts of recurrence interval and probability of exceedence presented in Chapter 3. In many cases, calculations of recurrence intervals for flash flooding events within the historical record are erroneous because they give no idea of the possible rainfalls within a given time interval. The use of probability of exceedence graphs as a technique to hypothesize rainfall amounts is not viable for four reasons. First, the occurrence of a large and rare event does not exclude one of the same magnitude or larger happening soon after. Second, large and rare events are clustered in time. Third, rare events may or may not appear in rainfall records because the rain gauge network may sample rainfalls at too gross a scale to detect flash flood events in small catchments. Finally, and more importantly, if the rainfall regime becomes wetter, then rare events of say 1:100 years may suddenly become common events of 1:20 years or less. For example, in eastern Australia, rainfall between 1948 and 1993 increased by 30 per cent over the previous thirty years. The period witnessed exceptional localized and regional rainfalls that could not possibly be accounted for using standard and accepted probability of exceedence diagrams. Some of these events will be described later in this chapter.

An alternative approach uses the concept of *maximum probable rainfall* that can fall in a catchment within a set period, given optimum conditions. The procedure was developed in the United States in the 1940s. Extreme rainfalls are generated by large, virtually stationary thunderstorms; or by mesoscale synoptic storm systems containing convective cells. The amount of precipitation that can fall over a set time in a given area is a function of the available moisture in a column of air, and the efficiency of any storm system in causing condensation and aggregation of water particles. Because intense cells have a limited size, the amount of rainfall falling over a catchment becomes a function of the catchment size. Small catchments approximating the area of a thunderstorm will receive more rainfall than large catchments in which a thunderstorm occupies only part of the area. The rate of condensation, while dependent upon climatic relationships, is also dependent upon topography. Rough terrain, in which elevation changes by 50 m or more within 400 m – especially in close proximity to the ocean – can trigger thunderstorms or trap convective cells for several hours. Many locations in the Wollongong area of Australia and along the Balcones Escarpment in Texas favor maximum probable rainfall because of rough topography. Figure 6.6 presents the maximum probable rainfalls that theoretically can fall within fixed periods for smooth or rough topography. For example, a small urban catchment of about 10 km^2 consisting of smooth topography can have maximum probable rainfalls of about 460 mm within an hour (point A in Figure 6.6). If it rains for three hours, this value can rise to 660 mm (point B). However, if the catchment is rough then the maximum probable rainfall for the three-hour period can amount to 840 mm – a 27 per cent increase (Point C). Because tropical air holds more moisture per volume than air at the poles, the values derived from

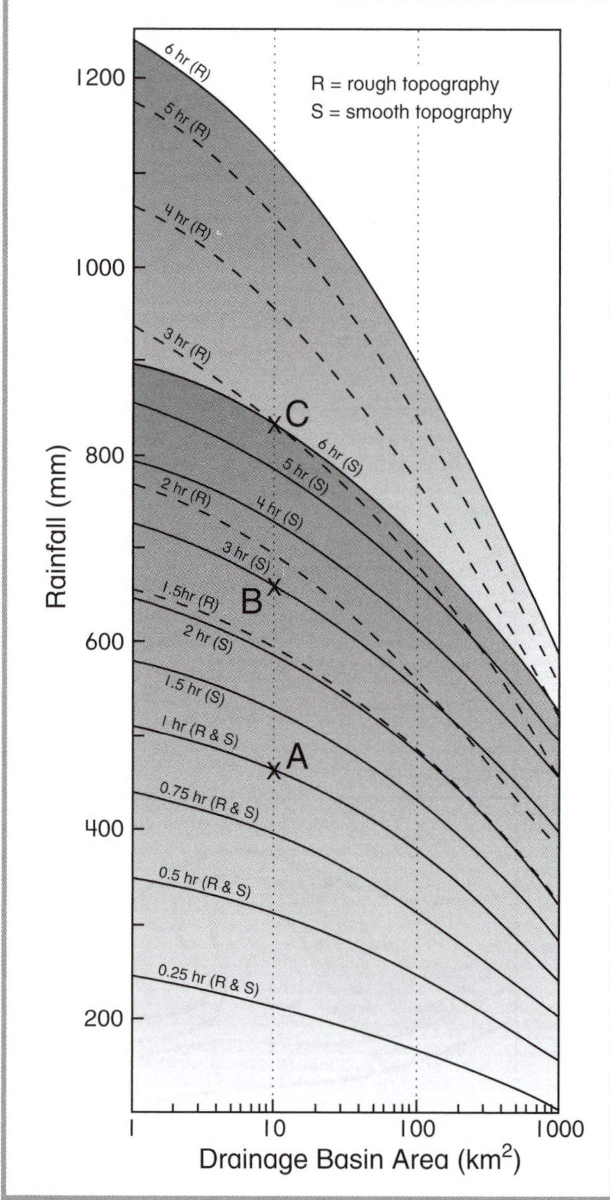

Fig. 6.6 Maximum probable rainfalls per unit time by drainage basin area (Australian Bureau of Meteorology, 1984).

Figure 6.6 must be corrected for latitude. Typically this correction is 0.65 at mid-latitude locations such as Sydney, Australia or Los Angeles, California. This latitudinal correction reduces the values mentioned above to approximately 300 mm, 430 mm, and 550 mm, respectively, at these two latter locations.

Figure 6.6 illustrates several factors about flash flooding. First, beware of small catchments. That babbling brook meandering intermittently through one's backyard can turn into a raging torrent under innocuous conditions. In Wollongong, Australia, flash flooding with the attendant movement of boulders has occurred in small creeks within 100 m of a similar-sized stream in which no rain has fallen. Second, after about one hour of continuously heavy rainfall, rough catchments receive 10–25 per cent more rainfall than smooth basins. That babbling brook running through a property with hilly topography is incredibly susceptible to flash floods. Finally, the volumes of rain that can fall within small catchments can be awesome. Figure 6.1 is a conservative depiction of the amounts of catastrophic rain that have been measured by our inadequate network of rain gauges. Within pristine catchments typical of the hilly topography of Sydney or Los Angeles, over 500 mm of rain can fall within the space of one hour. Thankfully rainfalls generating flash floods rarely continue for longer periods. When they do, the effects are catastrophic.

Flash flood events

(Bolt et al., 1975; Cornell, 1976; Maddox et al., 1980)

Flash flooding in the United States tends to be associated with summer storms and landfall of tropical storm systems. The Rapid City flood in the Black Hills of South Dakota, 9–10 June 1972, exemplifies the former aspect. One metre of rain fell in the space of six hours, an amount that had a recurrence interval of 1:2000 years. Flash floods swept down all major streams including Rapid Creek, where an old dam – built forty years earlier as a Depression relief project – collapsed, sending a wall of water through the downtown section of Rapid City. Over 230 people lost their lives and 2900 people were injured. Property damage exceeded $US90 million and included over 750 homes and 2000 cars. Texas is affected by the intrusion of tropical cyclones and easterly waves embedded in trade winds. Strong orographic uplift of these westerly moving systems can occur along the Balcones Escarpment. As a result, Texas has the reputation for being the most flood-prone region in the United States, and has recorded some of the highest rainfall intensities in the world (Figure 6.1). For example, during the Pecos River flood of June 1954, Hurricane Alice moved up the Rio Grande Valley and stalled over the Balcones Escarpment. The hurricane interacted with an upper air disturbance and turned into an intense, extra-tropical, cold-core cyclone. Over 1 m of rain was dumped in the lower drainage basin, resulting in flood-waters 20 m deep, with a recurrence interval of 1:2000 years. To date, this is the largest recorded flood event in Texas.

Urban flash floods

(Nanson & Hean, 1984; Australian Bureau of Meteorology, 1985; Riley et al., 1986a, b)

Urban areas exacerbate flash flooding. The breaking of the long drought in eastern Australia following the 1982–1983 ENSO event heralded the onset of extraordinary flash flooding in the Sydney–Wollongong region, as Walker circulation 'turned on' again over the following two years. The flooding was very localized and can be separated into two components: the Dapto flood of 17–18 February 1984 and the Sydney thunderstorms of 5–9 November 1984. In each case, slow moving or stationary convective cells developed in association with east-coast lows. Neither the localized convection nor the east-coast lows was unusual. Initially, the events represented the rare concurrence of two commonly experienced meteorological processes. Unfortunately, the pattern has occurred with alarming frequency throughout the remainder of the twentieth century.

The Dapto flood event began in the late evening of 17 February 1984 as a cold front pushed through the area in front of a high pressure cell centered south-east of Tasmania. As a result, moist onshore air flowed into the area at the same time as a small east-coast low developed south of Newcastle. This low tracked down the coast and then stabilized over the 500 m high Illawarra escarpment, north of the town of Dapto, at 7 am on 18 February. This was followed by development, during the remainder of the morning, of a complex, upper-level trough over the Wollongong region. These lows produced extreme instability and caused marked convergence of moist north-east and south-east airflow into the Wollongong area. Orographic uplift along the high escarpment dumped copious amounts of rain over the Lake Illawarra drainage basin over 24 hours (Figure 6.7). In a one-hour period on 18 February, 123 mm fell west of Dapto. Over a nine-hour period, 640 mm and 840 mm were recorded, respectively, at the base and crest of the escarpment. The heavier rainfall had a 48-hour recurrence interval of 1:250 years and is the greatest nine-hour rainfall recorded in Australia.

The resulting flooding of Lake Illawarra, however, had a 1:10-year frequency of occurrence. This case supports the United States evidence that flash flooding with severe consequences can occur in parts of a drainage basin with areas less than 50 km^2. Figure 6.7

shows the highly localized nature of the event. Within 5 km of the escarpment, rainfall amounts had dropped to less than 400 mm; within 10 km, amounts were less than 200 mm. Because any area on this map can be subject to this type of flooding – albeit rarely – it has been estimated that the Wollongong urban area could experience a Dapto-type event every 25 years. In the 20 years since 1984, five flash flood events of similar magnitude have occurred. Emergency organizations here should be prepared to handle high-magnitude, localized but rare floods in this region, with a much shorter expectancy rate than shown by individual events.

The Sydney metropolitan flooding that followed was not a single event, but rather a series of intense thunderstorms, which struck various parts of the metropolitan area on 5–9 November 1984. Accompanying the synoptic weather situation was a series of tornadoes and waterspouts that caused over $A1 million damage. The floods themselves cost the insurance industry $A40 million. One death resulted from lightning. This death toll is small considering the fact that most observers would class the lightning activity on the night of 8 November as the worst in living memory. Over the five-day period, rainfall totalled 550 mm south-south-east of the Sydney central business district. Most of the eastern and northern suburbs, as well as the Royal National Park south of Sydney, received totals in excess of 300 mm.

The floods originated from a low-pressure cell centered south-east of Brisbane on 5 November. This low moved slowly down the coast, rotating counter-clockwise around a strong, blocking high-pressure cell (maximum pressure 1032 hPa) that drifted from Adelaide to New Zealand over the period (Figure 6.8). The high-pressure cell directed consistent, moist, zonal flow onto the coastline. Concomitant with the movement of the high, a low-pressure trough developed parallel to the coastline. This trough spawned low-pressure cells that forced airflow over Sydney,

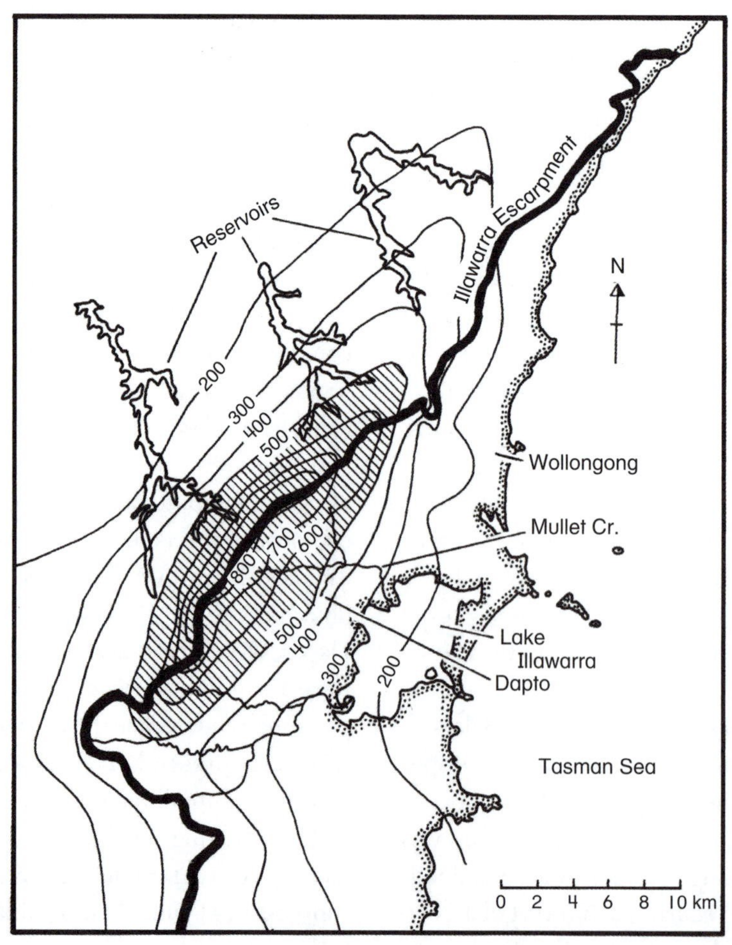

Fig. 6.7 Forty-eight-hour isohyets for the Dapto flood, Wollongong, Australia, 18 February 1984 (based on Nanson & Hean, 1984).

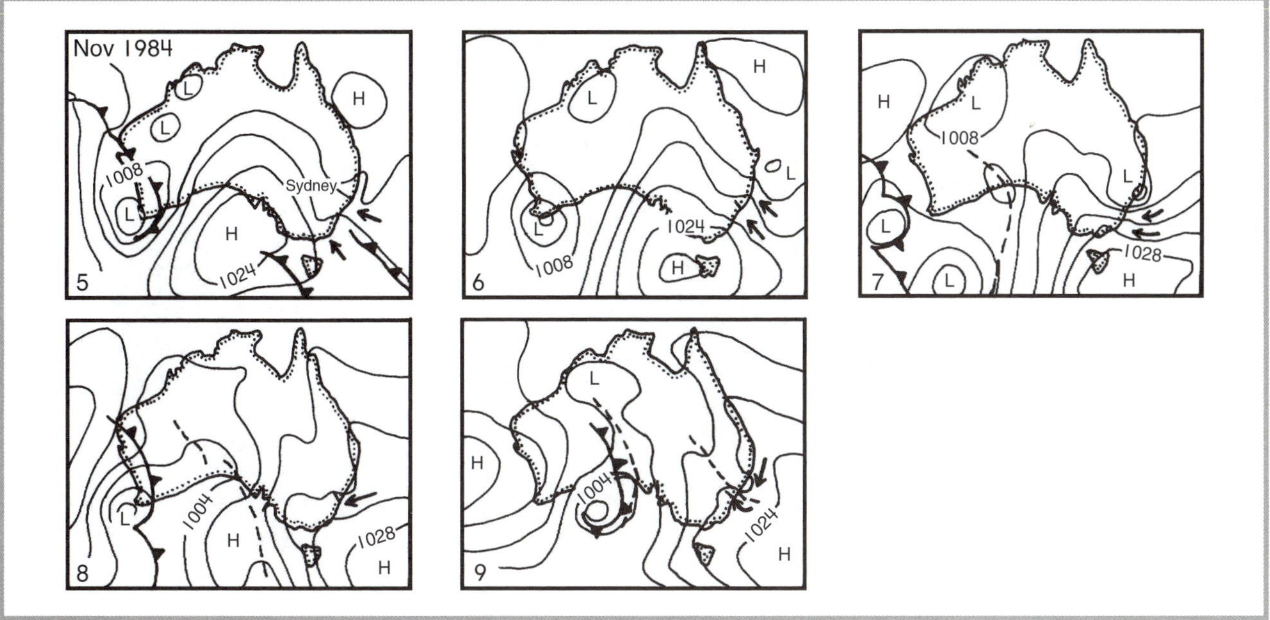

Fig. 6.8 Synoptic pattern for Sydney, Australia, flash floods, 5–9 November 1984.

resulting in intense atmospheric instability. The flooding occurred as three separate events during this period.

The first event on Monday 5 November began as thunderstorm activity at Cronulla (Figure 6.9a) and slowly moved northward over the city center. This thunderstorm activity originated as a series of intense cells associated with a low-pressure trough parallel to the coast and a separate low-pressure cell 50–100 km south of Sydney. A second trough existed in the interior of the continent (west of the Dividing Range), and drifted over Sydney in the early evening. Both of these troughs were responsible for separate rainfall events. Over 220 mm of rainfall fell within 24 hours over Randwick. Rainfall intensities for time spans under nine hours exceeded the 1:100-year event. The most serious flooding was caused by a storm drain at Randwick racecourse that was unable to cope with the runoff. Streets in adjacent Kensington bore the brunt of substantial flooding, which entered houses to depths of 0.5–1.0 m. Evening rush hour was thrown into chaos.

In the second event, on 8 November, a low-pressure trough was again present, this time directly over Sydney. Airflow converged into this trough over the northern suburbs. Seventeen convective cells were responsible for rainfall during the morning with the first two producing the heaviest falls. Over 230 mm of rain fell at Turramurra, in the northern suburbs, with falls of 125 mm between 7:15 and 8:15 am (Figure 6.9b). This latter intensity exceeded the 1:100-year event. The 20-minute rainfall intensity record at Ryde had a recurrence interval of between 50 and 500 years. Flooding during the second event began within 20 minutes of the commencement of heavy rainfall. Peak discharges occurred within five minutes of flooding. Many of the catchments that flooded were no more than 1 km long; however, flood depths up to 1.5 m became common, both in channels and in shallow depressions forming part of the drainage network. Many residents were unaware that these depressions were part of the flood stream network. Flooding occurred in the middle reaches of catchments as stormwater drains reached capacity and surcharged through access-hole covers. This *surcharging* capacity is deliberately built into Sydney's stormwater system, as it is prohibitively expensive to build an underground drainage network large enough to cope with rare events. Some of the flooding was due to storm inlet blockage by trees, loose debris or, in some cases, cars transported in floodwaters. It was even found that backyard paling fences could significantly block, divert, or concentrate flows, increasing flood levels locally by as much as 40 cm.

The third event occurred on the night of 8 November. Synoptically, it was very similar to the second event, with a trough cutting through Sydney and converging air into the city region. Rainfalls of 249 mm were measured at Sydney's Botanic Gardens, and values near 300 mm in and around the Royal National Park to the south (Figure 6.9c). Again, several cells were involved, and again peak intensities for time

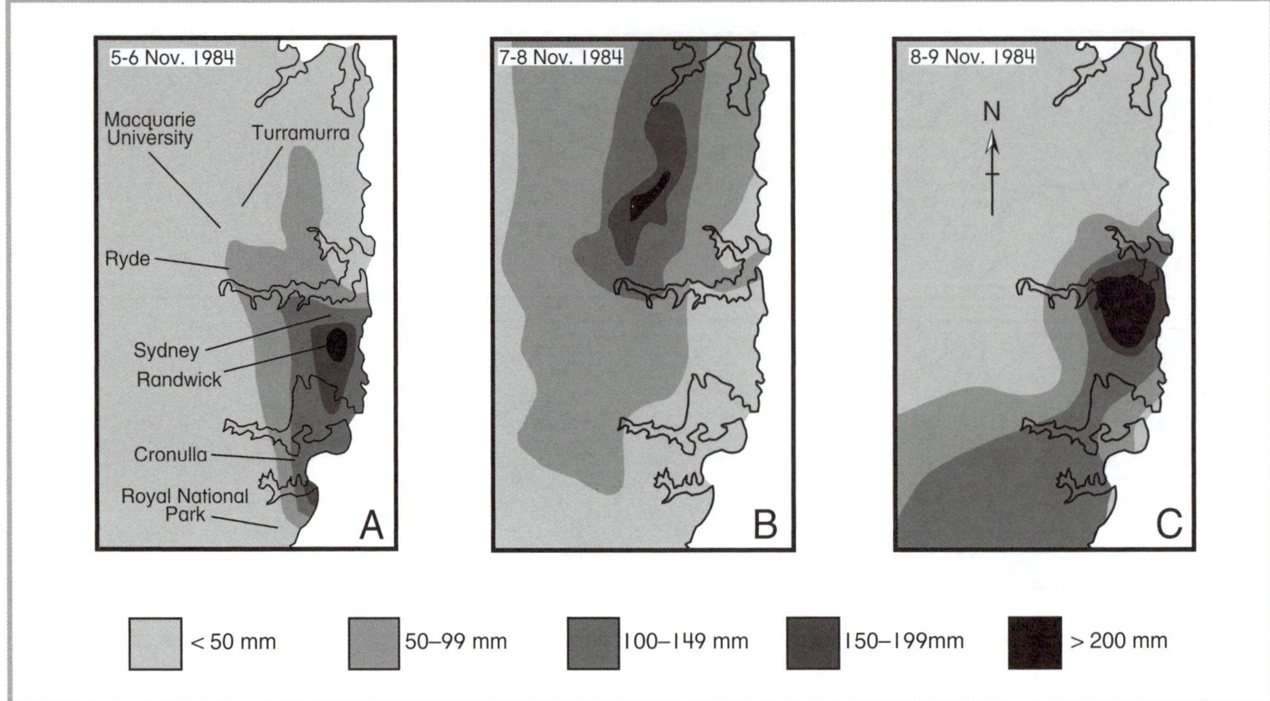

Fig. 6.9 Isohyets for three flash flood events in Sydney, Australia, 5–9 November 1984 (modified from Australian Bureau of Meteorology, 1985).

spans under nine hours exceeded the 1:100-year event. Many of the suburbs affected by this event had experienced their previous 1:100 year rainfall only four days previously. During this event, the road network supplanted natural drainage and took floodwaters away from storm drains normally expected to carry runoff. Because roadways have reduced frictional coefficients, flood peaks were almost instantaneous throughout the drainage system. Most of the damage during the third flash flood was caused by floodwaters reaching shopping centers at the base of steep slopes, sending walls of water and debris – in some instances including expensive cars – cascading through shops. In addition, the Tank Stream, which runs beneath the lower central business district of Sydney, was reactivated, flooding the Sydney Stock Exchange and causing several million dollars worth of damage. The State Library was also flooded, damaging rare exhibits.

There are three points about flash flooding in urban areas to be drawn from the events described above. Firstly, the probability of occurrence of a high-magnitude, localized rainfall event in a region can be an order of magnitude greater than the probability of the event itself. For example, the Dapto Flood was a 1:250-year event. The fact that any part of the Illawarra can experience such a localized storm indicates that the probability of a similar flood occurrence in the Wollongong region – along a 40 km stretch of coast – is about 1:25 years. Sydney experienced at least five 1:100-year events in the space of five days. In some cases, emergency services had to respond to all five events. Since then, Sydney has received the heaviest 48-hour rainfall on record. On 5–6 August 1986, an east-coast low dropped over 430 mm of rain in three days, causing rivers to flood from Bathurst, in the Blue Mountains, to the Georges River, in south Sydney. At least six people died, from drowning or being electrocuted by fallen power lines. Damage exceeded $A100 million, including over 3000 motor vehicles swamped by floodwaters. In this case, rainfall with a recurrence frequency of 1:100 to 1:200 years was experienced over a large area, much of which had been affected by sporadic 1:100-year events two years previously.

Secondly, urban flooding in most cases peaked within half-an-hour of the onset of intense rain. Much of the flash flooding was exacerbated by the structure of the urban drainage system. In the less densely built north-western suburbs of Sydney, surcharging of the storm drains along existing natural drainage routes caused flooding to homes and shops built in these areas. In the more densely built up eastern suburbs, the road network often took the place of the natural drainage system, causing damage to areas that would not normally receive this type of flooding.

Finally, it is very difficult to comprehend what the increased frequency in flooding means. While some might see it as a signal of Greenhouse warming, it appears to represent a shift towards extreme rainfall events in this region. Not only did the climatic patterns giving rise to one catastrophic flood reappear days later in the same location, but the patterns also tended to repeat themselves at other locations along the coast, and recur several times over the next few years. Since 1984, the 80 km stretch of coast encompassing Wollongong and Sydney area has experienced at least twelve similar events. A local coroner's inquiry into one recent event has shown that local authorities are almost unaware of this change in the rainfall regime. In New South Wales, the state government has realized that urban flooding has become more prevalent, and has added a $A25 surcharge to ratepayers' assessments to cover clean-up expenses and recover relief costs. Even this is a misconception of the nature of the climatic shift, because the surcharge is being applied to the western suburbs of Sydney, which in fact have not experienced the greatest number of flash floods. The greatest rainfall has occurred during thunderstorms over the central business district, which has a higher surface roughness coefficient because of tall buildings and parallel streets.

These events represent a significant increase in the occurrence of a natural hazard that has economic and personal ramifications for most of the four million residents in the region. Neither the significance of present flooding, nor its consequences, has been fully appreciated.

HIGH-MAGNITUDE, REGIONAL FLOODS

High-magnitude, regional floods usually present a disaster of national or international importance. Usually, large-scale drainage basins or many smaller rivers in the same region are flooded. Flooding appears to have increased in intensity and extent in recent years. As higher precipitation is one of the consequences of global warming, flooding has recently received the attention of media and relief organizations. Figure 6.10 plots the locations of major floods for fifteen years between 1985 and 2003. Flooding has occupied an extraordinary proportion of the world's land surface concentrated in two main regions: the tropics and the mid-latitude storm belts. This section describes historical flooding for three areas – the Mississippi River, eastern Australia, and China – to illustrate flooding's pervasiveness and high economic cost.

Mississippi River floods

(Bolt et al., 1975; Lott, 1993; Larson, 1996; Trotter et al., 1998; Public Broadcasting Service, 2000)

The Mississippi River drainage basin is the third largest in the world, and the largest in North America. The basin covers 3 224 000 km^2 or 41 per cent of the conterminous United States (Figure 6.11). Its headwaters can be divided into two areas, one originating at the continental divide in the Rocky Mountains, and the

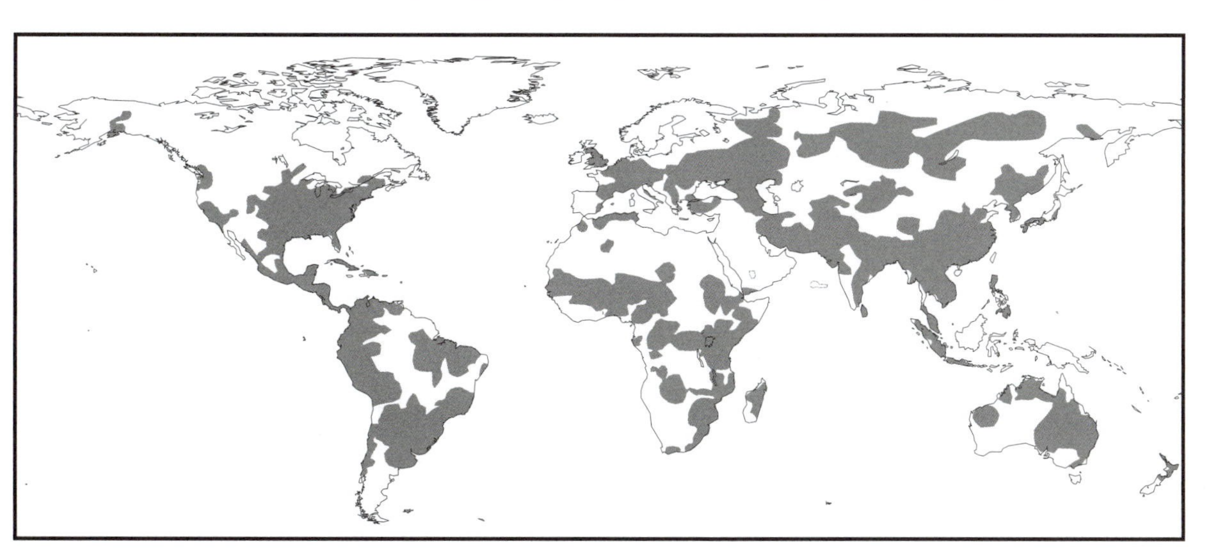

Fig. 6.10 Flood locations 1985–2003, excluding 1989, 1992, 1996, and 1997 (based on Dartmouth Flood Observatory, 2003).

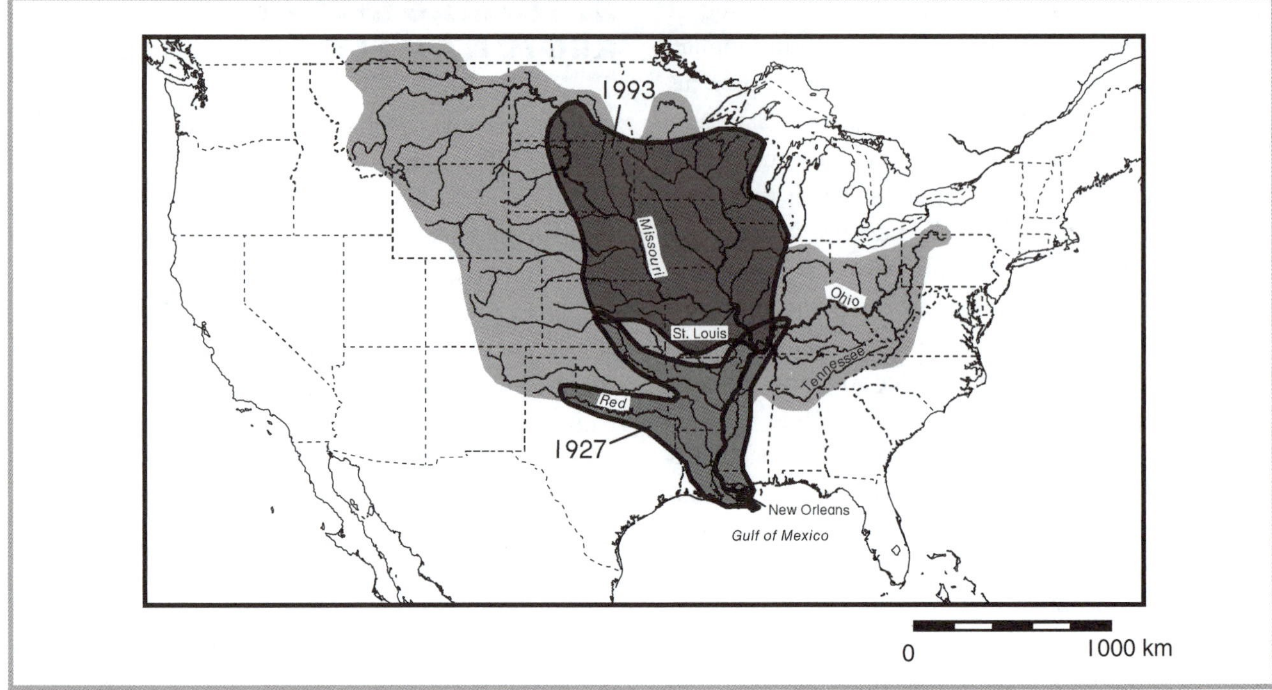

Fig. 6.11 The Mississippi River drainage basin and the extent of the 1927 and 1993 floods (based on Public Broadcasting Service, 2000).

other in the Great Lakes lowlands. The latter network has been developed only since the last glaciation, and includes an area that can accumulate large, winter snowfalls. Floods have been a constant feature of the Mississippi River since historical records began (Figure 6.12). Flooding can be divided into three periods: before 1927, between 1927 and 1972, and after 1972. Before 1927, flood control was considered a local responsibility. The flood of 1927 brought the realization that the Mississippi, which crosses state boundaries, could only be controlled at the national level. There were 246 fatalities, 137 000 buildings flooded, and 700 000 people made homeless as a result of this flood. Subsequently, the *Flood Control Act* of

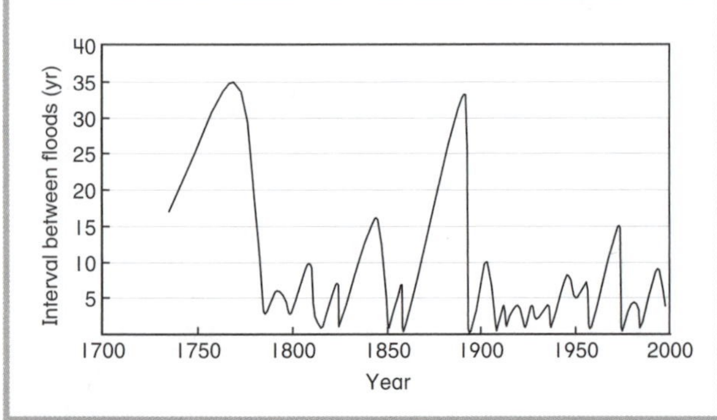

Fig. 6.12 Occurrence of major floods on the Mississippi River since 1700 versus interval between floods (based on Trotter et al., 1998).

1928 was passed and *levee* and reservoir maintenance and management were put under the auspices of state and federal authorities. Over 3000 km of *levees* and floodwalls, averaging 9 m in height in the lower part of the basin, were built to control flooding. Four floodways were also built to divert excess floodwaters into large storage areas, or into the Gulf of Mexico. Some parts of the channel susceptible to erosion were also stabilized with concrete mats, which cover the whole channel boundary. Finally, reservoirs were constructed on many of the tributaries to delay passage of floodwaters into the Mississippi. The period between 1927 and 1972 represents one where floods on the Mississippi were less severe and the mitigation works appeared to be successful.

Since 1972, the Mississippi Basin has been subjected to two periods of flooding that have tested these engineering works to the limit. The 1973 flood was the greatest to hit the basin to that point in time – 200 years since records began. Total rainfall for the six-month period exceeded the norm by 45 per cent in all parts of the Basin. On 28 April, the river peaked at a 200-year historical high at St Louis. Damage exceeded $US750 million, and 69 000 people were made homeless for periods up to three months. Without the flood mitigation works built since 1927, damage would have been as high as $US10 000 million, and 12.5 million hectares of land would have been flooded.

Because of localized decreases in river depth and higher flood heights, over 1300 km of main levees had to be raised after the 1973 flood.

As of 1993, the total damage caused by Mississippi flooding was equivalent to $US4.4 billion. The Great Flood of 1993 was worse (Figure 6.11). Mobile polar highs were stronger than normal in the spring of 1993, dragging cold air south into contact with warm air from the Gulf of Mexico. By April, the biggest floods on record were sequentially swamping the Missouri–Mississippi floodplain from north to south. The pressure patterns continued into summer with rainfalls exceeding 200–350 per cent of the norm. Flood levels reached 75–300-year recurrence intervals along major portions of the upper Mississippi and lower Missouri Rivers. On 1 August, the Mississippi crested at 15.1 m at St. Louis, 2.0 m above the previous record. Transport across the Mississippi for a distance of 1000 km was interrupted, 1000 levees collapsed, 55 people lost their lives, 48 000 homes in 75 towns were flooded, and 74 000 people were left homeless. The total damage bill across 12 states was $US18 billion. Fortunately, modelling forecast the disaster four months in advance, giving plenty of time for preparations and evacuation.

Great Australian floods

(McKay, 1979; Shields, 1979; Holthouse, 1986; Yeo, 2002; Australian Bureau of Meteorology 2003a, b)

The major rivers of eastern Australia originate within the Great Dividing Range, which parallels the east coast (Figure 6.13). Two systems flow westward. The Murray–Darling system and its tributaries – the Warrego, Condamine, Macintyre, Macquarie, Lachlan and Murrumbidgee – forms the largest river system in Australia, entering the southern ocean south of Adelaide. The other system consists of the Diamantina–Coopers Creek rivers, which flow into landlocked Lake Eyre. On the eastern side of the Dividing Range are a number of much shorter, but just as impressive, systems. These rivers consist of the Hunter, Macleay, Clarence, Brisbane, Burnett, Fitzroy, and Burdekin. Successive tropical cyclones or monsoonal troughs have in the past severely flooded all of these rivers, but never at the same time. Aboriginal legends from the Dreamtime imply that many river valleys have been flooded over widths of 50 km or more. Three flood events: the wet of 1973–1974, the Charleville–Nyngan floods of 1990 and the Katherine floods of 1998 – all

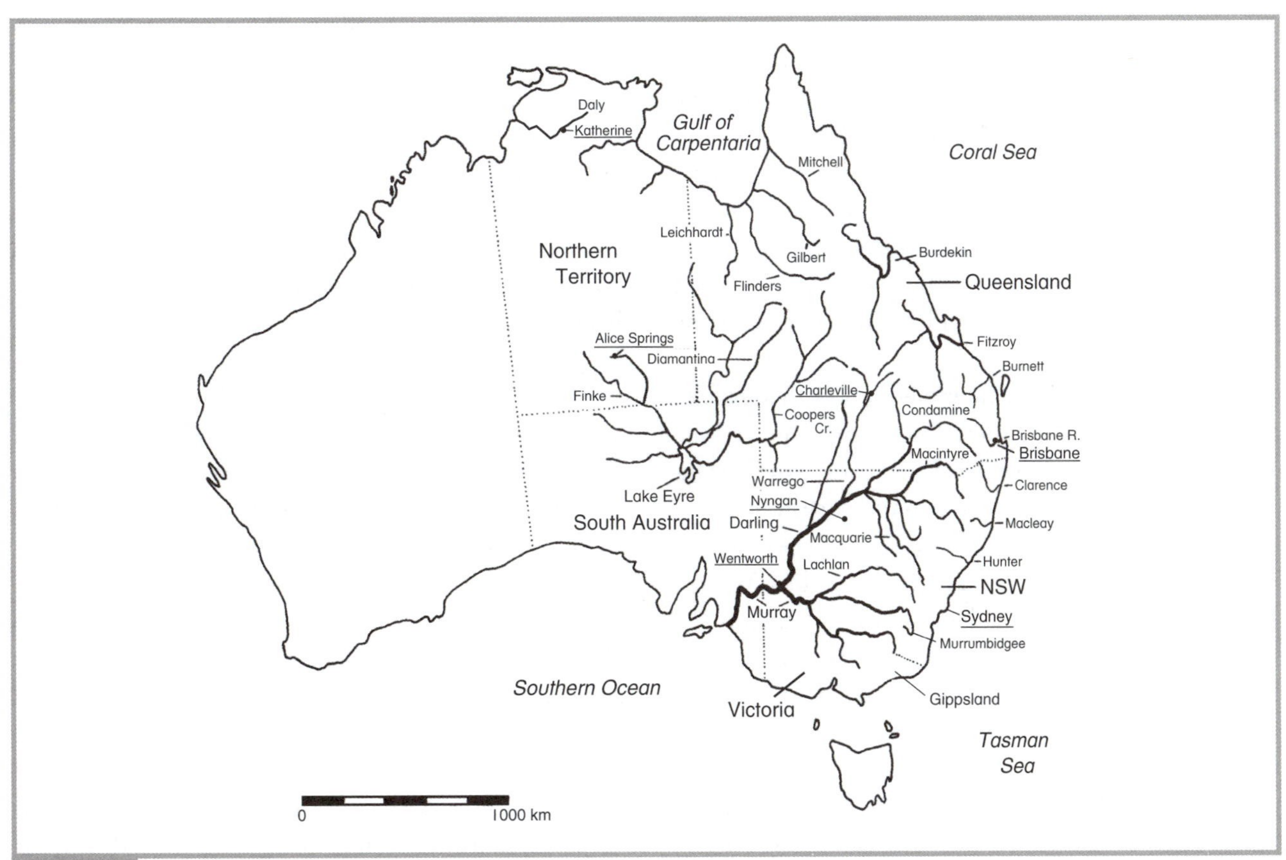

Fig. 6.13 Major rivers of eastern Australia.

related to ENSO – illustrate the broad catastrophic nature of flooding in eastern Australia that substantiate those Aboriginal legends.

Flooding in the summer of 1973–1974 coincided with one of Australia's wettest La Niña events in the twentieth century. Rainfall was torrential and continuous throughout most of January 1974, as the intertropical convergence settled over tropical Australia. On 25 January, Cyclone Wanda moved into the interior of Queensland and New South Wales, dumping in excess of 300 mm of rain in 24 hours over a wide area, and triggering massive flooding of all river systems. Rain in the catchment of the Brisbane River produced one of the worst urban floods in Australian history. These January floods represented the largest natural disaster to occur in Australia to that point in time. Flooding covered an area of 3 800 000 km^2, larger than the drainage basin of the Mississippi River. From Alice Springs to the Pacific Ocean, and from the Gulf of Carpentaria to the Murray River, military airlifts had to be arranged to supply isolated towns, cut off by floodwaters, with emergency food and stock fodder. Around the Gulf of Carpentaria, the tributaries of the Flinders River amalgamated to form a river 150 km wide. In northern New South Wales, torrential rainfall continued week after week, raising river levels in excess of 20 m. Floodwaters slowly flowed down the Diamantina and Coopers Creek drainage into the interior, filling Lake Eyre for only the fourth time this century. At Wentworth in New South Wales, farmland remained flooded for two years as successive floods came down the Darling and Murray in 1974–1975. Stock losses to sheep alone in New South Wales totalled 500 000. The scale and magnitude of flooding was unprecedented.

The Brisbane flood hit an expanding city that had not witnessed a major flood since 1893. Few suspected the flooding potential of the Brisbane River. Synoptically, the flooding of the city was due to the persistent recurrence of cyclones tracking over eastern Queensland in the space of four weeks, culminating in Cyclone Wanda. However, many of the flood's effects were exacerbated by human factors. Somerset Dam, built after the 1893 flood to contain similar events, was totally inadequate for this purpose; yet, many people believed that the dam protected them from flooding. The sprawling nature of growth in the Brisbane River Basin had also led to large-scale clearing of forest in rugged terrain, which enhanced runoff during the high-intensity rainfall periods of the 1974 flood. The character of the drainage basin was also altered. Urban development had seen some creeks filled in, while others had been lined with concrete. These latter channels were designed to contain only the 1:10-year flood. Roads, driveways, and houses had sealed large sections of the landscape. These modifications resulted in calculated flood discharges being twice that of the 1:100-year flood for adjacent vegetated catchments. Some urban catchments experienced the 1:1000-year flood event. Additionally, there was very little delay between time of peak rainfall intensity and peak flooding. Flash flooding occurred in urban catchments with flood peaks cresting within one hour of peak rainfall intensities. Many areas also had inadequate rain gauging stations, so that the exact amount of rain falling in some parts of the catchment was unknown, and the forecasting of flood levels impossible to determine.

The Brisbane flood also saw the collapse of organized disaster warnings. There was no central data processing authority, so that local flooding in key areas went unnoticed. Over 70 per cent of residents questioned afterwards about the flood had received no official warning. The media, in their efforts to report a major story from the field, sparked rumors and clouded the true picture of flooding. The rapid series of flash floods, in isolated parts of the catchment 24 hours prior to the main flood, led to public confusion and committed the Disaster Relief Organisation to what afterwards were evaluated as only minor disasters. The major flood peak was disastrous. All bridges across the Brisbane River were damaged or destroyed and 35 people drowned. At its height, the river broke its bank and ran through the central business district of Brisbane. In Ipswich, 1200 homes were destroyed. Overall, 20 000 people were made homeless. Only 11 per cent of residents who experienced the main flood received evacuation assistance from emergency organizations, and only 30 per cent acknowledged the clean-up help from any relief organization. Most relief came from friends and community groups. Over 40 per cent of victims received help from church contacts, and over 30 per cent of volunteers surveyed stated that they belonged to no organized group. Generally, most people applied themselves to evacuation, alternative accommodation, clean-up and rehabilitation with little reliance on government or social organizations for help. This raised serious questions about the efficiency of such organizations in disaster relief in Australia, and led to significant changes in the response to disasters by both private relief organizations and federal government agencies. Eleven months later these revisions were tested to the utmost when Cyclone Tracy destroyed the city of Darwin.

The floods of 1990 hit a region from central Queensland in the north to Gippsland in the south. On 18 April 1990, a strong upper-level low developed over the interior as a high-pressure cell stalled over the east of the continent. Up to 350 mm fell initially over central Queensland, triggering major floods of every creek, stream, and river. On 21 April, floodwaters converged suddenly on the Warrego River and swept through Charleville. Residents had to be rescued from rooftops by helicopter. Over the next few days both the rains and the floods slowly tracked southward, inundating one riverside town after the other. At Nyngan on the Bogan River, residents decided to ‘fight it out’ and began sandbagging the levees around the town. Television crews moved in to report nightly on their heroic but futile efforts. Over 200 000 sandbags were laid in four days.

In Australia, a national emergency can be declared only if someone reports it to the state government who then requests federal assistance. In anticipation of that call, Emergency Management Australia called in 500 troops who stationed themselves in the nearest town waiting to help. In the confusion and excitement of the flood, no one called for help. Finally, as the flood broke through the levee, television crews filmed people flinging themselves into the gaps in a last-ditch effort to stay the inevitable. Within an hour the town was flooded and helicopters evacuated the 2500 residents to Dubbo where the army had been waiting. The disaster continued southward as rivers sequentially broke previous flood heights by a metre or more. The low, which was blocked eastward by the Great Dividing Range, finally exited the continent through Gippsland, Victoria. Here, up to 350 mm of rain fell within 48 hours, causing major floods on all rivers flowing from the Snowy Mountains. In the space of four days, this single low-pressure cell had flooded a landmass the size of western Europe in one of the worst mass floods in Australian history.

The floods of 1998 illustrate the extent of flooding that can occur in Australia during a La Niña event. Five flood events occurred from Katherine in the north to the Gippsland in the south. The interior of the Australian continent receives most of its rainfall from tropical cyclones that can move inland long distances without the impediment of mountains. In January 1998, tropical Cyclone Les, which had developed over the Gulf of Carpentaria, moved westward and weakened into a rain depression. However, it stalled over Katherine in the Northern Territory and between 25 and 27 January, it dropped 400–500 mm of rain on the Katherine, Roper, and Daly River catchments. Floodwaters on the Katherine River rose to a record level 20 metres above normal. This was sufficient to send two metres of muddy water through the town of Katherine. The town’s 2000 residents left evacuation too late, fearing looters or believing that the flood threat was highly exaggerated. Over 1100 people had to be treated for injuries sustained while coping with the disaster.

The shift to a La Niña event is often marked by an increase in the frequency and intensity of east-coast lows. One such low caused heavy rains that flooded Gippsland on 23–24 June, while another caused flooding at Bathurst, west of Sydney, on 7–8 August. Finally, on 17 August, a small low developed along the 500 m escarpment backing Wollongong on the New South Wales coast. In the early evening, extreme rainfalls caused flash flooding that destroyed 700 cars and caused over $A125 million damage. In all areas, the flooding was not unique. Historic and palaeoflood evidence indicates that floods with up to three times the volume of the 1998 floods have occurred in all three regions in the last 7000 years. In Wollongong, the 1998 floods were simply part of a spectrum consisting of 80 or more floods that have occurred since the early 1800s. Much of the perception of increased flooding in Australia reflects media hype, increased urban growth and a stubborn resistance to learn from the lessons of previous floods.

Flooding in China

(Bolt et al., 1975; Czaya, 1981; Milne, 1986)

Flooding in China is mainly related to the vagaries of the Hwang Ho (Yellow) River channel. The Hwang Ho River – China’s River of Sorrow – drains 1 250 000 km^2 and flows over a distance of 4200 km (Figure 6.14). Throughout much of its upper course, it erodes yellow *loess*, and for 800 km of its lower length, the Hwang Ho flows without any tributaries on a raised bed of silt. Much of this material has entered the river because of poor land-use practices in the upper drainage basin in the last 2000 years. Today, the suspended load reaches 40 per cent by weight, making the Hwang Ho one of the muddiest rivers in the world. Where the river breaks onto the flatter coastal plain, *vertical aggradation* has built up a large alluvial fan-like delta from Beijing in the north, to Shanghai in the south. The river thus flows on the crest of a cone-shaped delta and builds up its bed very rapidly during flood events. As early as 2356 BC, the river channel was dredged to prevent silting. In 602 BC, the first series of levees was built to contain its channel. Despite these efforts, in the last 2500 years the Hwang Ho has broken its banks ten times, often switching its exit to the Yellow Sea over a distance of 1100 km on each side of the Shandong Peninsula (Figure 6.14). Associated with these breakouts has been massive flooding – with the result that China historically has suffered the greatest loss of life in the world from flood hazards.

Between 2300 BC and 602 BC the Hwang Ho flowed through the extreme northern part of its delta in the vicinity of Beijing. In 602 BC, the river moved slightly south to the region of Tianjin, and stayed there until 361 BC, whereupon it underwent a catastrophic shift south of the Shandong Peninsula. For the next 150 years, the Chinese expended considerable effort shifting it north again. In 132 BC, the Hwang Ho switched to a course near the present entrance, where dike building tended to keep it stabilized, with minor channel switching, on the northern side of the Shandong Peninsula until 1289 AD. For this period, the northern part of the alluvial plain aggraded – a process that steepened the topographic gradient southward. With the fall of the Sung dynasty, dike maintenance lapsed, and in 1289 AD the Hwang Ho again switched south of the peninsula, where it was allowed to remain for the next 660 years.

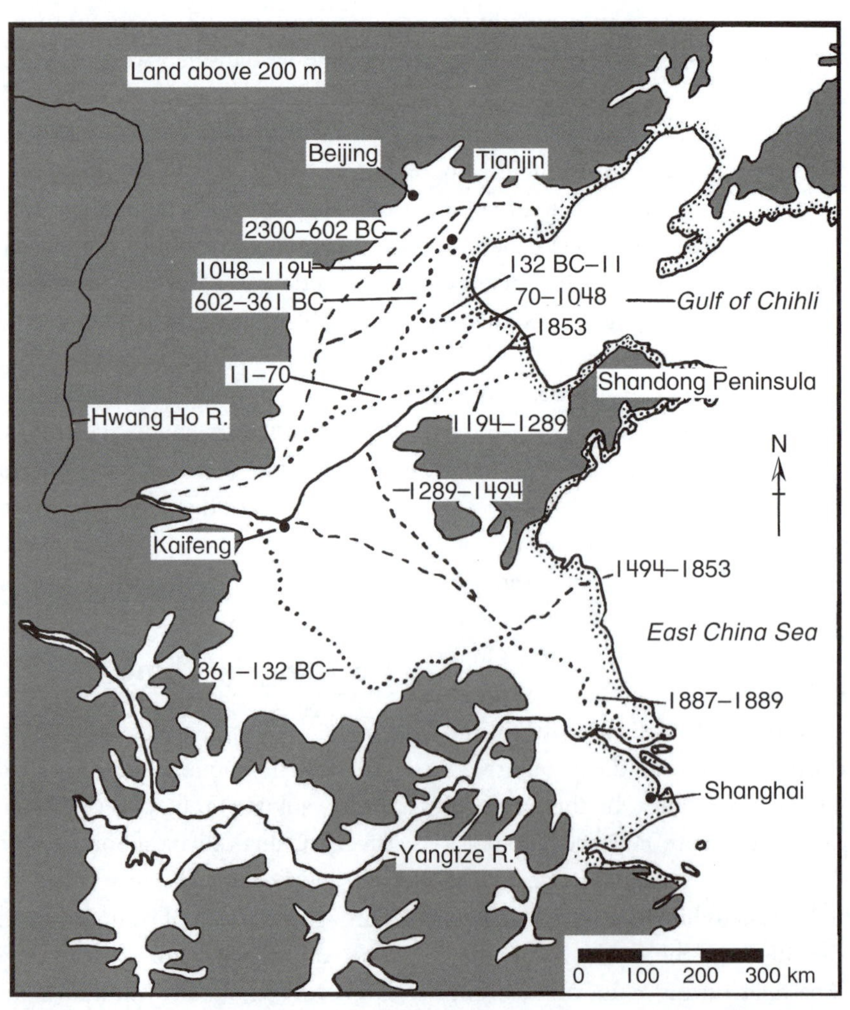

Fig. 6.14 Location of the Hwang Ho River channel: 2300 BC to the present (after Czaya, 1983). Dates shown are AD unless otherwise indicated.

The most destructive flooding occurred during this period, in 1332 AD, with seven million people drowning and over ten million people dying from subsequent famine and disease. By 1851, the Hwang Ho had aggraded its bed in the south to such an extent that it threatened to engulf the town of Kaifeng. Measured rates of aggradation exceeded 2 m yr^{-1}, and dykes more than 7 m high were built to contain its flow. Between 1851 and 1853, the Hwang Ho broke out northward and took on its present course. In 1887, a temporary breach saw the river move the furthest south it has ever been. It temporarily linked up with the Chang (Yangtze) River and, for a brief time, both rivers flowed out to sea through the same channel. The 1887 flood breached 22 m high embankments and flooded 22 000 km^2 of land to depths of 8 m. Over a million people lost their lives. Not only was the death toll severe, but the river also buried flooded land in meters of silt. Farmers relying upon the river in the northern part of the delta suddenly found themselves without irrigation water and facing starvation. Twentieth century floods have been comparatively less severe; however 200 000 and 300 000 people were killed in the floods of 1911 and 1931, respectively.

The natural tendency of the Hwang Ho to shift course has been greatly exacerbated during wartime. In 1938, General Chiang Kai-shek, in an attempt to prevent advancing Japanese armies taking the city of Chengchow, ordered the levee dikes to be dynamited. The diversion failed to stop the Japanese advance, but the ensuing floods killed a million unsuspecting Chinese, and possibly caused the death of 11 million people through ensuing famines and disease. It was not until 1947 that extensive engineering works forced the Hwang Ho permanently back into its prewar channel. Today, the Hwang Ho is stabilized for much of its length, and is kept to the north of the Shandong Peninsula. Because the river transports such large quantities of silt, its bed is continuing to aggrade and now sits 20 meters above its floodplain, the latter protected by an inner and outer set of dikes spaced 10 km apart. The Chinese government has embarked on a program of dam and silt basin construction to minimize the flooding effects of the Hwang Ho; however, the river is still not controlled, and will probably breach its dikes in the future during a large flood.

CONCLUDING COMMENTS

(Tropeano et al., 2000; Kunkel, 2003; Sheffer et al., 2003)

The wide scale nature of recent flood events is not unique in time. Rather it represents the continuation of a longstanding global hazard that has taken millions of lives. For example, in Canada, there has been no change in the frequency and intensity of extreme precipitation events during the twentieth century. Similarly, in the United States (where there has been an increase of 20–40 per cent in extreme precipitation over the long-term average of the last two decades of the twentieth century), flooding today is just as prominent as it was at the end of the nineteenth century. These changes are due to warmer sea surface temperatures in the tropical Pacific Ocean causing subsequent flooding in the United States. Likewise, what might appear to be the largest floods on record can be shown to be part of a continuum. For example, on 8–9 September 2002, the Gardon River in southern France reached a flood crest 14 m above normal following 680 mm of rain over 20 hours. However, this flood did not reach a cave lying 3 m higher that contained evidence of five separate flood events dating between 1400 and 1800. Similarly, on 14–15 October 2000, major flooding accompanied by landslides affected the Italian Alps. The floods were triggered by 400–600 mm of rain that fell within two days. However, many of the landslides represented the partial reactivation of much larger historical slides. In Australia, at least eleven comparable floods of similar magnitude have occurred at 20–40 year intervals since 1810. There is thus substantial evidence that flooding is ubiquitous. Recent floods across the globe have not been abnormal, but reflect the occurrence of events as part of an ongoing continuum.

REFERENCES AND FURTHER READING

Australian Bureau of Meteorology, Australia 1984. The Estimation of Probable Maximum Precipitation in Australia for Short Durations and Small Areas. *Australian Bureau of Meteorology Bulletin* No. 51.

Australian Bureau of Meteorology, Australia 1985. A Report on the Flash Floods in the Sydney Metropolitan Area Over the Period

5 to 9 November 1984. Special Report, Australian Bureau of Meteorology, Melbourne.

Australian Bureau of Meteorology, Australia 2003a. *Nyngan and Charleville, April 1990.* <http://www.bom.gov.au/lam/climate/levelthree/c20thc/flood8.htm>

Australian Bureau of Meteorology, Australia 2003b. *Katherine floods, January 1998.* <http://www.bom.gov.au/lam/climate/levelthree/c20thc/flood6.htm>

Baker, V.R. and Costa, J.E. 1987. Flood power. In Mayer, L. and Nash, D. (eds) *Catastrophic Flooding*. Allen and Unwin, London, pp. 1–21.

Bolt, B.A., Horn, W.L., MacDonald, G.A., and Scott, R.F. 1975. *Geological Hazards*. Springer-Verlag, Berlin.

Cornell, J. 1976. *The Great International Disaster Book*. Scribner's, NY.

Czaya, E. 1983. *Rivers of the World*. Cambridge University Press, Cambridge.

Dartmouth Flood Observatory 2003. *Global archive map of extreme flood events since 1985.* <http://www.dartmouth.edu/~floods/Archives/GlobalArchiveMap.html>

Griffiths, J.F. 1976. *Climate and the Environment: The Atmospheric Impact on Man*. Paul Elek, London.

Hirschboeck, K.K. 1987. Catastrophic flooding and atmospheric circulation anomalies. In Mayer, L. and Nash, D. (eds) *Catastrophic Flooding*. Allen and Unwin, London, pp. 23–56.

Holthouse, H. 1986. *Cyclone: A Century of Cyclonic Destruction*. Angus and Robertson, Sydney.

Kunkel, K. 2003. North American trends in extreme precipitation. *Natural Hazards* 29: 291–305.

Larson, L.W. 1996. The great USA flood of 1993. IAHS Conference on Destructive Water: Water-Caused Natural Disasters – Their Abatement and Control, Anaheim, California, June 24–28, <http://www.nwrfc.noaa.gov/floods/papers/oh_2/great.htm>

Lott, N. 1993. The summer of 1993: Flooding in the Midwest and drought in the southeast. *National Climatic Data Center Technical Report 93-04.*

Maddox, R.A., Canova, F., and Hoxit, L.R. 1980. Meteorological characteristics of flash flood events over the western United States. *Monthly Weather Review* 108: 1866–1877.

McKay, G.R. 1979. Brisbane floods: the paradox of urban drainage. In Heathcote, R.L. and Thom, B.G. (eds) *Natural Hazards in Australia*. Australian Academy of Science, Canberra, pp. 460–470.

Milne, A. 1986. *Floodshock: The Drowning of Planet Earth*. Sutton, Gloucester.

Nanson, G.C. and Hean D.S. 1984. The West Dapto Flood of February 1984: rainfall characteristics and channel changes. Department of Geography, University of Wollongong, Occasional Paper No. 3.

Public Broadcasting Service 2000. *Fatal flood*. <http://www.pbs.org/wgbh/amex/flood/maps/>

Riley, S.J., Luscombe, G.B., and Williams, A.J. (eds) 1986a. Proceedings Urban Flooding Conference: a conference on the storms of 8 November 1984. *Geographical Society N.S.W. Conference Papers* No. 5.

Riley, S.J., Luscombe, G.B., and Williams, A.J. 1986b. Urban stormwater design: lessons from the 8 November 1984 Sydney storm. *Australian Geographer* 17: 40–50.

Sheffer, N.A., Enzel, Y., Waldmann, N., Grodek, T., and Benito, G. 2003. Claim of largest flood on record proves false. *EOS, Transactions American Geophysical Union* 84(12): 109.

Shields, A.J. 1979. The Brisbane floods of January 1974, In Heathcote, R.L. and Thom, B.G. (eds) *Natural Hazards in Australia*. Australian Academy of Science, Canberra, pp. 439–447.

Tropeano, D., Fabio Luino, F., and Turconi, L. 2000. *Flooding in western Italian Alps, 14–15 October 2000.* <http://www.irpi.to.cnr.it/English/Events%2014-15%20october/Events%20.htm>

Trotter, P.S., Johnson, G.A., Ricks, R., Smith, D.R., and Woods, D. 1998. *Floods on the lower Mississippi: An historical economic overview.* <http://www.srh.noaa.gov/topics/attach/html/ssd98-9.htm>

Yeo, S. 2002. Flooding in Australia: A review of events in 1998. *Natural Hazards* 25: 177–191.

Fires in Nature

CHAPTER 7

INTRODUCTION

Of all natural hazards, the most insidious is drought. However, for some countries such as Australia and United States, droughts have not led to starvation, but to spectacular fires as tinder-dry forests ignite, grasslands burn and eucalyptus bushland erupts in flame. Of all single natural hazard events in Australia, bushfires are the most feared. Here, the litany of disasters over the past 150 years reads like the membership list of some satanic cult: 'Black Thursday', 'Black Friday', and 'Ash Wednesday' (Figure 7.1). For firefighters, state emergency personnel and victims, each name will invoke stories of an inferno unlike any other. North America has suffered just as badly in the past from fires, as have Europe and the former Soviet Union. Approximately 143×10^6 km^2 of the Earth's surface is covered by vegetation, of which 0.17 per cent burns on average each year. Even tropical rainforests can dry out and burn. For example, the U Minh forest in Vietnam (see Figure 7.2 for the location of major placenames mentioned in this chapter) ignited naturally in 1968. Large fires have also burnt through tropical vegetation in the Brazilian highlands and the Amazon Basin, especially in association with land clearing. Many urban dwellers would consider that, following wide-scale deforestation, there is no forest fire threat near cities. Fires are perceived to be a hazard only in virgin forest or, perhaps in

Fig. 7.1 William Strutt's painting *Black Thursday* (LaTrobe Collection, © and with permission of the State Library of Victoria). The title refers to the fires of 6 February 1851, which covered a quarter of the colony of Victoria, Australia.

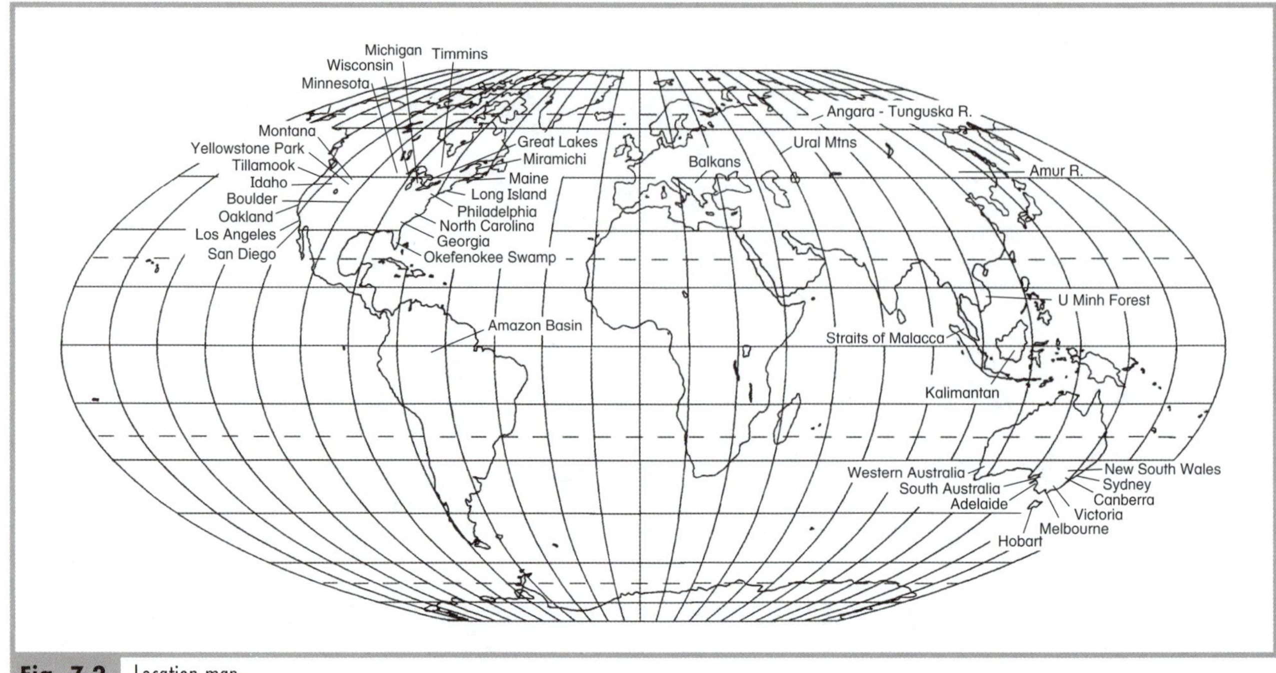

Fig. 7.2 Location map.

southern California or southern France, a hazard where urban expansion has encroached upon vegetated mountains. In fact, any country where climate is influenced by aridity or by periodic drought is susceptible to natural fires. This is particularly true for countries experiencing a *Mediterranean climate*, especially those where highly flammable eucalypts have supplanted more fire-resistant vegetation. More importantly, large areas of the northern hemisphere have undergone reforestation this century, with the abandonment of economically marginal farmland. Here, forest fires may again pose the severe hazard they were under initial settlement more than a century ago.

This chapter will examine the hazard of drought-induced natural fires. Those weather patterns favoring intense fires will be described first, followed by a discussion of the main causes of fire ignition. Major bushfire disasters will then be described, with particular emphasis upon the American and Australian environments, where some of the most intense conflagrations have been witnessed.

CONDITIONS FAVORING INTENSE BUSHFIRES

(Vines, 1974; Luke & McArthur, 1978; Powell, 1983; Voice & Gauntlett, 1984; Yool et al., 1985)

The potential for a major fire hazard depends upon the vegetation type, fuel characteristics, climate, and fire behavior. *Eucalyptus* is one of the most fire-prone vegetation types, having leaves with a high oil content that fosters burning. Such trees can rapidly renew foliage through epicormic shoots on the main trunk, or through shoots on buried and protected roots. After only a few years post-fire, they can become sites of conflagrations as severe as the original fire. *Eucalyptus* has been exported successfully from Australia and now covers significant areas of southern California, northern Africa, the Middle East, and India. Other fire-prone vegetation dominated by scrub oak, chamise, and manzanita exists in Mediterranean climates. Such vegetation communities are labelled '*chaparral*' in California, 'garrigue' or 'maquis' in France, 'phrygana' in the Balkans, 'matarral' in South America, 'macchin fynbosch' in South Africa, and 'mallee' or 'mulga' scrub in Australia. In regions affected by severe drought, fuel characteristics are very important to the spread and control of fires. Fuel characteristics can be separated into two categories: grass, and bush (mainly *Eucalyptus*). Grasslands produce a fine fuel both on and above the ground surface. Such material easily ignites and burns. In forests, the fine fuel consists of leaves, twigs, bark, and stems less than 6 mm in diameter. Apart from *Eucalyptus* leaf litter with its high oil content, most forest litter is difficult to ignite.

The arrangement of fuel is also important. Compacted litter is more difficult to burn and to extinguish than litter with a lot of air mass. Forests with dense undergrowth (especially pine plantations) will burn more readily than forests where the undergrowth has been removed. The moisture content of living and dead material also dictates the possibility of ignition and fire spread. Dead grasses readily absorb moisture, their moisture content fluctuating with the diurnal humidity cycle. Humidity increases at night, and grasses have their highest moisture content in the early morning. Monitoring of the proportion of cured grass and its moisture content can give a good index of the susceptibility of that fuel to ignition and burning. In *Eucalyptus* forest, the types of tree species condition fire behavior. Stringy-barks permit fire to reach treetops and throw off flaming bark, thus favoring *spot fires*, whereas smooth barks do not generate flying embers.

The climate before a fire season also conditions the fire threat. If fuel has built up, then climatic conditioning is less important; however, it still takes dry weather to cure and dehydrate grass and bush litter. This desiccation process is exacerbated by prolonged drought in the summer season. Typically, El Niño–Southern Oscillation (ENSO) events occurring at Christmas coincide with Australia's summer season. In California, La Niña events produce fire weather. During these types of events, dry conditions are guaranteed for several months, ensuring that fuels dry out more completely. Because such events can now be predicted months in advance, it is possible in many countries to warn of severe fire seasons before summer arrives.

The prediction of bush or forest fires may even extend beyond this time span. As pointed out in Chapter 2, the 11-year sunspot cycle appears in rainfall records for many countries, while the 18.6-year lunar cycle dominates the coincidence of worldwide drought or rainfall. Because natural conflagrations are most likely during major droughts, it is to be expected that forest fires evince one of these astronomical cycles. Vines (1974) has performed detailed work on the cyclicity of bushfires in Australia and Canada, and believes that cyclicities of 6–7 years and 10–11 years appearing in these countries directly reflect sunspot cycles. The 10–11-year periodicity also appears in the fire records of the southern United States since 1930, although at sunspot minima. In the Great Lakes region, major conflagrations between 1870 and 1920 coincided very well with peaks in the 11-year sunspot cycle. Logically, major fire seasons should also correlate with the 18.6-year lunar cycle; however, little research has been carried out to verify this conjecture.

Figure 7.3 illustrates the typical daily weather pattern leading to extreme bushfires in south-eastern Australia. A high-pressure system moves slowly south-east of Australia and ridges back over the continent. This directs strong, desiccating, north-west winds from the interior of the continent across its south-east corner. These winds are followed by a strong southerly change that may drop humidity but maintain high wind

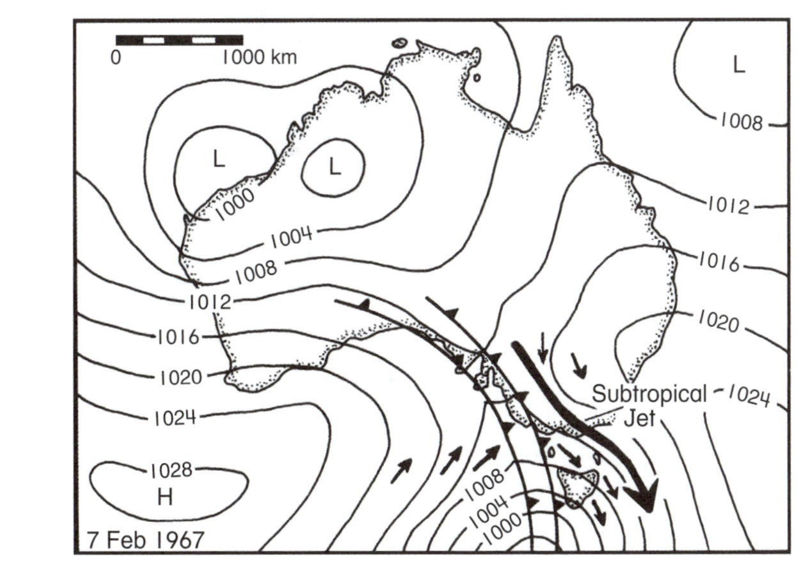

Fig. 7.3 Typical synoptic weather pattern conducive to bushfire conflagrations in south-eastern Australia.

speeds. Winds blowing from the interior of the continent warm adiabatically as they drop in elevation from the highland areas of Victoria to the coast, producing what is commonly known as a *föhn* wind. This adiabatic warming can push temperatures to 40°C or above at the coast. The process also reduces relative humidity to levels below 20 per cent. Similar winds exist elsewhere, especially in mountainous Mediterranean climates dominated by Hadley cells. Local variants of föhn winds are termed 'mono' in central California, 'Diablo' around San Francisco, 'Santa Ana' in southern California, 'sirocco' in southern Europe, and 'bora' in the Balkans. In Australia, strengthening of föhn winds is aided by convergence of isobars, and reinforced by a fast-moving subtropical jet stream in the upper troposphere on the western side of the high-pressure cell. Because this sequence of winds moves eastward across the continent over several days, there is a gradual shift in the bushfire threat from South Australia to Victoria, and then to New South Wales. In Western Australia, fire weather is characterized by positioning of high-pressure cells with north-easterly winds out of the interior.

Fire behavior and local topography ultimately determine the severity of the fire hazard. Often fires accelerate as the day progresses, because fuel moisture content drops and wind speed and atmospheric instability increase owing to diurnal heating. Additionally, as the fire intensifies during the day, convection is generated by the extra heat supplied to the atmosphere. Radiation from the fire dries out fuel in front of the fire, and spot fires can prepare forward areas for the passage of the main fire. Spotting in front of fires is quite common in Australia because of the presence of stringy-barks. Spotting ensures that convection occurs in front of the main fire, thus drawing the main fire-front towards the spot fires.

As wind speed increases, the rate of fire movement increases exponentially. A wind speed of 10 km hr^{-1} will move a fire – in Australian bushland of average fuel content – at the rate of 0.5 km hr^{-1}. At 20 km hr^{-1}, this rate increases to 0.8 km hr^{-1} and, at 40 km hr^{-1}, to 1.8 km hr^{-1}. If the fire actually reaches treetops and burns through the crowns of trees (Figure 7.4), it can accelerate and travel at speeds of 20 km hr^{-1}. With spotting, fires can move forward or 'propagate' faster than this, creating a firestorm that can travel at 60 km hr^{-1} or more. The rate at which a bushfire moves does not taper off with increased wind velocities.

Fig. 7.4 Crown fire in A) *Eucalyptus* forest, Australia (photograph by N.P. Cheney, National Bushfire Research Unit, CSIRO, Canberra)

Fig. 7.4 and B) a boreal forest in northern Ontario, Canada (photograph by Jim Bryant, Ontario Department of the Environment). Note the similarities between the fires. *Eucalyptus* is perceived as more flammable, but boreal forests produce three to four times more energy per unit time because of their greater biomass.

However, in grassland, fire movement decelerates at wind speeds above 50 km hr^{-1} because of fragmentation of the fire head. This is not significant when one considers that, under the same wind speed, grass fires can move about eight times faster than a bushfire. If fires move upslope, their rate of advance will increase because rising heat from combustion dries out fuel. Fires will also move faster upslope because the wind velocity profile steepens without the influence of ground friction. In rugged terrain, spot fires are more likely because embers can be caught by winds blowing at higher speeds at the tops of hills or ridges.

Atmospheric stability also controls fire behavior. If the atmosphere is unstable, then convective instability can be established by the fire's heating of the atmosphere. Once initiated, convective instability will continue even after the heat source is removed. Under conditions of extreme instability, fire whirlwinds are generated and can reach speeds exceeding 250 km hr^{-1}. These whirlwinds are of three types. The first is related to high combustion rates and generates a rotational updraft, which can move embers a couple of hundred meters beyond the fire-front. The second type represents a thermally induced tornado. This type originates in the atmosphere on the downwind side of the convective column or the lee side of topographic relief. The winds are capable of lifting large logs and producing gaseous explosions in the atmosphere. These vortices are particularly dangerous because wind is sucked towards the fire tornado from all directions. The third type of whirlwind is the post-burn vortex (Figure 7.5). It forms because of the heat emanating from a burnt area up to a day after the fire. These vortices are usually less than 20 m in diameter, but they can carry embers from burnt to unburnt areas.

Fig. 7.5 Whirlwind associated with a clearing after burning (photograph by A.G. Edward, National Bushfire Research Unit, CSIRO, Canberra).

CAUSES OF FIRES

(Luke & McArthur, 1978; Pyne, 1982; van Nao, 1982)

No matter how dry a forest or grassland becomes, there can be no fire unless it is ignited. The recorded cause of a fire depends upon how rigorously fire statistics are collected in individual countries. For instance, in European countries, 45 per cent of all fires are of unknown cause; however, in Canada, only 5 per cent of fires have unknown causes. In Europe, only 2 per cent of the known causes of bushfires can be attributed to natural causes such as lightning. Over half are the result of arson and 40 per cent, of human carelessness. The latter group includes burning that gets out of control, smoking, children playing, fires due to trains, and camp fires. In Canada, lightning is responsible for 32 per cent of all forest fires, whereas in the United States it accounts for fewer than 8 per cent of fires. The latter figure is misleading because lightning has played a major role in igniting fires in isolated areas of the United States. Between 1940 and 1975, over 200 000 fires were started by lightning in the western United States alone. This represents an average of 15.6 fires per day. Single storm events or pressure patterns can also produce a large number of lightning fires in a short space of time. Between 1960 and 1971, the western United States experienced six events that produced 500–800 lightning fires each. The largest such event, in June 1940, started 1488 fires.

Humans are the greatest cause of fires. Arson by mentally disturbed people accounts for 32 per cent of all fires in the United States and 7 per cent in Canada.

In southern California, nearly a third of all fires are started by children, most of whom are boys playing with matches. Throughout the twentieth century in the United States, economic downturns have brought increased incendiarism, because firefighting affords temporary job opportunities for the unemployed. The major difference between arson and lightning as a cause of fire is the fact that lightning-induced fires tend to become conflagrations because they occur in isolated areas, while arson-induced fires tend to occur in accessible areas that are relatively easy to monitor and control. In the United States 2000 fire season, lightning caused 7659 fires burning out 4930 km^2 of vegetation. In contrast, humans caused 96 390 fires that burnt across 10 050 km^2. Of this number, 30 per cent were due to arson and consumed 34 per cent of the area burnt.

In Australia, the causes of fire are more numerous because of the variety in vegetation, climate, and land-use. Additionally, prescribed burning as a fire prevention procedure is practiced more in this country than any other part of the world. This is despite the fact that prescribed burning was used in the south-east United States for clearing undergrowth and minimizing the outbreak of major fires. Between 1966 and 1971 in Western Australia, 36 per cent of all fires resulted from burning-off; only 8 per cent were due to lightning. Vehicles, farm machinery, and trains ignited 15 per cent of all fires, with carelessness accounting for 21 per cent. In the Northern Territory, most fires are caused by either lightning or burning-off. Here, Aborigines traditionally light fires to clear the landscape so they can catch game. In eastern Australia, where the population is greater, arson has become a major cause of fires in recent years. Single individuals have been found to light more than one fire on a high fire-risk day. Because the spring burning-off season can encroach upon early warm weather, notably during ENSO years, the incidence of fires caused by burning-off has also increased in recent years.

BUSHFIRE DISASTERS: WORLD PERSPECTIVE

(Cornell, 1976; van Nao, 1982; Seitz, 1986; Couper-Johnston, 2000)

Bushfires as a hazard are usually evaluated in terms of loss of life, property damage (burning of buildings or entire settlements), and the loss of timber. While loss of life can still occur in northern hemisphere temperate forests, most destruction is evaluated in terms of the loss of marketable timber. This statement needs qualification because timber has no value unless it can be accessed, and there is no imperative to access it unless it is close to centers of consumption. In North America, it is mainly the forests lying within 1000 km of dense urban development that are used to supply building timber, unless there is a need for timber with specific qualities – such as the Californian Redwood. Beyond this distance, forests are mainly used to supply pulp and paper in the form of newsprint. There are large areas in the Canadian and Alaskan boreal forest where no harvesting of trees takes place because it is uneconomic. These areas generally are uninhabited and no attempt is made to suppress forest fires. Fires are allowed to burn uncontrolled, and may take weeks or the complete summer season to burn out. In northern Canada, it is not uncommon for such fires to cover thousands of square kilometers.

In the Middle Ages, large areas of Europe were subjected to forest fires. Damage and loss of life has gone virtually undocumented, although it is estimated that today's conflagrations cover only 20 per cent of the area they did centuries ago. Today, large tracts of deciduous, mixed and chaparral forests remain in which fires pose a serious threat. In Europe as recently as 1978, over 43 000 fires consumed 440 000 hectares of forest and as much again of grasslands and crops. These figures are not remarkable. No European country has had more than 2 per cent of its total forest cover burnt in a single year. Individual fires on a much larger scale have occurred throughout the northern hemisphere. In October 1825, over 1 200 000 hectares of North American forest burned in the Miramichi region of New Brunswick, together with 320 000 hectares in Piscataquis County, Maine. The Wisconsin and Michigan forest fires of 1871 consumed 1 700 000 hectares and killed 2200 people. Areas of similar magnitude were burnt in Wisconsin in 1894 and in Idaho and north-western Montana in 1910. The Porcupine Fire of 1911 raced unchecked from Timmins, Canada, north through virgin forest for over 200 km, killing scores of settlers on isolated farms. During the 1980 Canadian fire season, virtually the whole mixed forest belt from the Rocky Mountains to Lake Superior was threatened. Over 48 000 km^2 of forest were destroyed, with a potential loss equivalent to the timber required to supply 340 newsprint mills for one year. The destruction only got worse – with 73 880 km^2 and 64 040 km^2 of forest going up in smoke in 1989 and 1994, respectively. The magnitude of this burning was matched along the Amur River (separating China and Siberia) by the Great Black Dragon Fire of 1987, which burnt through 73 000 km^2, killed 220 people and left 34 000 homeless. The areas of these last three fires are equivalent to the area of Scotland, and are 13 times larger than the Yellowstone Fire of 1988.

All of these fires pale in significance when compared to the Great Siberian fires of July–August 1915. In that year, up to 1 million km^2 of Siberia, from the Ural Mountains to the Central Siberian Highlands, burned (an area 2–20 times greater than the extent of the 1980 fires in Canada). An area the size of Germany, covering 250 000 km^2, was completely devastated between the Angara and Lower Tunguska Rivers. These gigantic fires were brought about by one of the worst droughts ever recorded in Siberia. Most of the *taiga* and larch forests were desiccated to a combustible state, producing crown fires. Over 500 000 km^2 of peat dried out and burnt to a depth of 2 m. The amount of smoke generated was 20–180 $\times$ 10^6 tonnes, a figure equivalent to the amount of smoke estimated to be produced in a limited-to-extreme nuclear war leading to a nuclear winter. Thick smoke was lifted more than 12 km into the sky. In the immediate area, visibility fell to 4–20 m and to less than 100 m as far as 1500 km away. As a result, the solar radiation flux was significantly attenuated at the ground, dropping average temperatures by 10°C. At the same time, long wave emission at night was suppressed such that the diurnal temperature range varied by less than 2°C over a large area. These below-average temperatures persisted for several weeks; however, harvests in the area were delayed by no more than two weeks. Little evidence of the fires was detectable outside Siberia. Because of the isolation of the area and the low population density, reported loss of life was minimal. The above documentation of large fires implies that conflagrations of this magnitude are frequent occurrences in the boreal forests of the northern hemisphere. Forest fires presently pose the most serious natural hazard in the two largest countries of the world, Canada and Russia.

Nor are fires restricted to temperate latitudes. Since 1980, large fires have swept through the tropics of Indonesia and Amazonia, exacerbated by ENSO events and human malpractice. During the 1982–1988, 1994, and 1997–1998 ENSO events, 32 000 km^2, 51 000 km^2, and 50 000 km^2, respectively, of rainforest burnt across Indonesia, mainly in Kalimantan. During the 1997–1998 ENSO event, smoke due to slash-and-burn agriculture, forest clearing, and firing of peatlands enveloped over 100 million km^2 of South-East Asia in an acrid, pea-soup haze that reduced visibilities to 2–100 m. At one point the pollution reading at Pontianak in west Kalimantan reached 1890 mg m^{-3}, 40 times the level considered safe by the World Health Organization. In Indonesia, about one million people suffered health problems and 40 000 had to be hospitalized. At one school, visibility was so low that school children had to be roped together to keep them from getting lost. About 2.6 $\times$ 10^9 tonnes of carbon – 1000 times more than produced by the Siberian fires of 1915 – were released into the atmosphere, an amount equivalent to 40 per cent of global emissions that year from fossil fuels. Because of reduced visibility, two ships collided in the Straits of Malacca, killing twenty-nine people. The haze was also implicated in the crash of a Garuda Airbus into a mountain at Medan with the loss of 234 passengers and crew. Ultimately, 7000 professionals and 31 000 volunteers fought the fires. As Indonesia's resources became overwhelmed, the international community responded with cash donations, medical assistance, and firefighting expertise at a level usually reserved for a major earthquake or drought. Indonesia's gross national product dropped 2.5 per cent due to the decline in its tourist and forest industries.

United States

(Pyne, 1982; Romme & Despain, 1989; Firewise, 1992; ThinkQuest Team, 2001)

Fire history

In North America at least 13 fires have burned more than 400 000 hectares. In the twentieth century, almost all such mega-fires have begun as controlled burns, or as wildfires that at some point could be considered controllable. And almost all severe fire seasons were preceded by drought. The intensity of any fire is not necessarily dependent upon the growth of biomass during wet seasons, but on the availability of dried litter or debris generated by disease or insect infestation, on windstorms, previous fires, or land clearing. This point is well-recognized. For example, the 1938 New England hurricane mentioned in Chapter 2 left a trail of devastation in the form of uprooted trees and felled branches. Considerable effort was spent cleaning up this debris before it could contribute to major forest fires. Fire-scarred areas take decades to regenerate, and regeneration in these areas is particularly susceptible to future burning. In North America, large burnouts such as the Tillamook in Oregon, burnt originally in 1933, spawned numerous fires for the next 30 years. Indians who cleared or fired forests for agriculture, firewood, hunting, and defense conditioned much of the early fire history of North America. Thus, the forests that met white settlers were park-like and open. While European settlement adopted many native American fire practices, the demise of such practices made possible the reforestation of much of the north-east and began the period of large-scale conflagrations. A similar process followed the decline of prescribed burning – first by the turpentine operators and then by the loggers in the pine forests of the south-east. The most dramatic effects are seen on the Prairies, where forests have reclaimed a grassland habitat as land was converted to cultivation or rangeland. In southern California, urban development has seen grasslands treed simply for aesthetic reasons. In each case, Europeans have attempted to supplant natural fire cycles with total suppression to protect habitation.

The onset of forest fires in reforested regions of the United States was dramatic. Disastrous fires swept through the north-east in 1880, and it became a policy to restrict settlement in forest reserves. That policy led to reduced burning-off and a build-up of forest fuel that resulted in 400 000 hectares burning from New York to Maine. The fires were perceived as a threat to timber reserves and watershed flood mitigation. They resulted in a concerted effort to set up fire lookouts in the north-east, and prompted the passage of the *Weeks Act* in 1911 to protect the watersheds of navigable streams and to subsidize state fire protection efforts. These efforts – along with the formation of the Civilian Conservation Corps (CCC) in the 1930s – saw the demise of large wildland fires; however, urbanization had expanded into forest areas. The nation's first urban forest fire occurred in October 1947 at Bar Harbor, Maine. Over 200 structures were burnt and 16 lives lost. This scene was repeated over a wide area in 1963 when fires threatened suburbs in New York and Philadelphia, and destroyed over 600 structures on Long Island and in New Jersey. These fires led to incorporation of municipal fire units into the state protection system, and the signing of interstate agreements for fire suppression assistance. By the 1970s, the latter had spread nationally and internationally. Today bushfires also plague southern California's urban areas, which since the 1950s have encroached upon fire-prone scrubland. Little thought has been given to building design and layout, with a jumble of organizations responsible at various levels for fire suppression. The costs of firefighting operations in southern California as a result have reached high levels. In 1979, the Hollywood Hills and Santa Monica mountain fires involved 7000 firefighters at a cost of a million dollars per day for a month.

In the south-east of the United States, reforestation and regeneration of undergrowth presented, by 1930, large reserves of biomass that fuelled major fires during droughts. In 1930–1932, Kentucky, Virginia, Georgia, and Florida suffered devastating fires. Further fires between 1941 and 1943 saw the reintroduction of prescribed burning. In 1952, 800 000 hectares of forest burned across Kentucky and West Virginia; in 1954–1955, 200 000 hectares burned in the Okefenokee Swamp and, in 1955, 240 000 hectares burned in North Carolina. These occurrences led to the establishment, in North Carolina, of one of the most efficient fire suppression organizations in the country and, in Georgia, of the first fire research station. This station produced the nation's leading methods in prescribed burning.

In the virgin forests of the Lake states, similar-sized fires have occurred historically. However, the death tolls have been horrendous. Here, disastrous fires occurred as settlers first entered the region, together with logging companies and railways. The logging produced enormous quantities of fuel in the form of slash on the forest floor; the railways provided sources of ignition from smokestacks; and the settlers provided bodies for the resulting disaster. Between 1870 and 1930, one large fire after another swept through the region in an identical pattern. Most of the conflagrations occurred in autumn following a summer drought. The worst fires occurred in 1871, 1881, 1894, 1908, 1910, 1911 and 1918. The Peshtigo and Humboldt fires of 1871 took, respectively, 1500 and 750 lives in Wisconsin; the Michigan fires of 1881 killed several hundred; the Hinckley, Minnesota, fire in 1894 took 418 lives; and finally the Cloquet fire in Minnesota in 1918 killed 551 people. The fires were so common, that most residents treated fire warnings nonchalantly, to their own detriment.

The great fires of 1871 were the first, and by far the worst, to sweep the Lake states. Ironically they occurred in October on the same day as the Great Chicago fire. In the weeks preceding the fires, much of the countryside was continually ablaze. Almost all able-bodied men were employed in fire suppression, and many inhabitants had inaugurated procedures to protect buildings. On 8–9 October the fires broke into firestorms only ever witnessed on a few other occasions. Flames leapt 60 m into the air and were driven ahead at a steady rate of 10–16 km hr^{-1} by spot fires, intense radiation, and updrafts being sucked into the maelstrom. The approach of the flames was heralded by dense smoke, a rain of firebrands and ash, and by combusting fireballs of exploding gases. Winds of 100–130 km hr^{-1} picked up trees, wagons and flaming corpses. The fire marched with an overpowering roar like continual thunder or artillery barrages. Hundreds were asphyxiated hiding in cellars, burned to death as they sheltered in former lakebeds converted to flammable marshes by the drought, or trampled by livestock competing for the same refuges. The Lake fires were certainly the worst in documented history and show how vulnerable any forest can be under the right fuel and climatic conditions.

Recent disasters

Despite these efforts to identify the fire regime and come to terms with urban encroachment on forest, fires in North America continue to be a major hazard. Three events in recent years illustrate the range of problems: the Yellowstone fire of 1988 because it pitted advocates for natural burns against those for total fire suppression; the Oakland fire of 20 October 1991 because it characterizes the urban threat; and the 2000 fire season because it shows how widespread the problem can be during regional drought.

The Yellowstone fire was impressive for its intensity. More area was burnt in the region on 20 August 1988 than during any decade since 1872. The fires were ignited in June by lightning following six years of above-normal rainfall. For the first month, the fires were allowed to burn unchecked; but in mid-July, after 35 km^2 had been burnt, a decision was made to contain them. A week later this area had doubled and a month later 1600 km^2 had been incinerated despite massive efforts at suppression. The dramatic increase in burning was due to drought conditions that were the worst since the 1930s. Winds as high as 160 km hr^{-1} stoked the fires despite the 9000 firefighters brought in to contain them. By mid-August, at least eight separate fires were burning at rates as high as 20 km day^{-1} with temperatures exceeding 32°C. By the time the fires were finally extinguished, 5666 km^2 of forest had been burnt. Some of the best natural scenery in the United States had been scarred for decades and a debate was ignited pitting those who favored a laissez faire attitude to fires against those who favored complete suppression. Supporting the former view was the fact that in the previous twelve years, 235 fires had been ignited by lightning, allowed to blaze, and had died out after burning an average of 0.4 km^2 of forest. The largest fire destroyed only 30 km^2 and had threatened no lives. Slowly people came to realize that the 1988 fires were an indication that a 'natural order' was being re-established – one that had been upset for a century. Suppression of fires for over a century had allowed an ecosystem to develop that would otherwise have been returned, by fires every ten to 20 years, to a more natural state. The forests in Yellowstone generally become more flammable as they age. This is because old-growth forests accumulate greater amounts of dead biomass at ground level and are susceptible to crown fires because the trees are taller with unpruned lower limbs. The Yellowstone fires had reached this limit in the 1930s, but benign climate and the introduction of fire suppression had extended the fire-prone state. The suppression consisted of 'Smokey the Bear' technology and heroics. Aircraft and helicopters scoured the landscape for the first sign of fire, water bombers attacked even the most remote flame, and smoke jumpers were parachuted into sites within hours of ignition. All fires were to be suppressed by 10 o'clock in the morning. The fires of 1988 were part of a natural 200–300-year cycle of fire that was delayed by 30–40 years of efficient fire suppression.

Opposing fire suppression was the view that nature should be aided by prescribed burning. Had the Yellowstone forests been culled artificially by sanctioned arson, under weather conditions disadvantageous to fire, then the smoke that hid the scenery, affected health and deterred the tourists would not have been so extreme. The Yellowstone fires fuelled the view that forested landscapes, burnt or unburnt, have different values to diverse interest groups.

The Oakland, California, fire of October 1991 was not the first fire to sweep the area. From 1920 to 1995 over 3500 structures were lost to fire in the East Bay

Hills – the worst event being the loss of 584 homes in 1923. This urban–wildfire conflict was exacerbated by the dominance of fire-prone vegetation such as *Eucalyptus*. Monterey pine, chaparral, grass, and ornamental species such as junipers and cedars were also present. All are highly flammable, generate intense radiation, and produce easily blown embers. The fires were aided by the weather. The region had experienced five years of drought while frosts the previous winter had increased the amount of dead vegetation in the forest. On the day the firestorm occurred, relative humidity dropped to 16 per cent and temperatures rose to a record 33°C. The local föhn winds enhanced the firestorm. These winds became turbulent over the local terrain and reached gusts of 90 km hr^{-1}. Finally, an inversion layer developed below 1000 m, containing within the lower atmosphere the heat generated by the fires and preheating vegetation in front of the flames.

The disaster began as a small brush fire on 19 October 1991 in the hills above Oakland. This fire was easily controlled, but it continued to smolder beneath a 20 cm thick carpet of pine needles. At 10:45 am on 20 October, flames reappeared and developed into a wildfire that lasted three days and became the worst urban wildfire disaster in United States history. Within 15 minutes of reignition, the fire became a firestorm that generated its own weather and proceeded along several fire fronts. Within 30 minutes, eight pumping stations and ten water reservoirs were lost due to power failures. Within an hour, the fire – spreading at a rate of 1.67 m s^{-1} – destroyed 790 buildings. Local firefighting resources were overwhelmed and a Bay-wide call for assistance was issued. Eighty-eight fire trucks, six airplanes, 16 helicopters and over 1400 service personnel responded to the fires. However, fire units from outside the area found that their hose connections were incompatible with Oakland's hydrant system. Communications amongst emergency services failed within the first twelve hours as call lines became congested. The rugged terrain interfered with radio and cellular (mobile) phone transmissions. Abandoned vehicles, narrow roads, downed power lines and the thick smoke trapped in the inversion layer hampered firefighting efforts. The same restrictions hindered evacuation of residents. Nineteen people were killed trying to flee the fires; eleven of these died trapped in a traffic jam. By the end of the first day, ten key reservoirs were drained and firefighters had to rely upon tankers for water supplies. The fires were extinguished only because they began to burn more slowly as they reached flatter terrain, and because the winds diminished. After burning for three days, the firestorms had burnt through 6.4 km^2 of affluent neighborhoods, destroyed 3469 structures worth an estimated $US1.6 billion, injured 150 people and taken 25 lives.

Following the 1998–1999 La Niña event in the United States, the 2000 fire season was conditioned by widespread drought from Florida in the south-east to Washington in the north-west. Across the country, 122 827 fires burnt across 34 000 km^2 – more than double the annual average – destroying 861 buildings, killing 11 people, and costing $US1.6 billion to suppress. At the peak of the fire season in Florida, more than 500 wildfires broke out each day. Eight of the ten worst fires were concentrated in the north-west. This situation stretched the country's firefighting capacity. At the season's peak on 29 August, 1249 fire engines, 226 helicopters, 42 air tankers and 28 462 people were fighting the fires. Eventually personnel and equipment had to be brought in from Canada, Mexico, Australia, and New Zealand. And still the carnage intensified. In October 2003, fifteen fires in southern California burnt out 3000 km^2, destroying 3640 homes and 1174 other structures (Figure 7.6).

Fig. 7.6 An aerial view of total devastation in a housing estate following the southern Californian fires on 29 October 2003. Note the lone house that survived unscathed (photo 031029-N-4441P-013 taken by Photographer's Mate 2nd Class Michael J. Pusnik, Jr., of the U.S. Navy assigned to the 'Golden Gators' of Reserve Helicopter Combat Support Squadron Eighty Five (HC-85). Used with permission of Director, Naval Visual News Service, U.S. Navy Office of Information. <http://www.news.navy.mil/list_single.asp?id=10314>).

These recent fires have led to a shift in the way fires are dealt with in the United States. Prior to the Yellowstone fire of 1988, fire suppression took precedence. This led to the accumulation of abundant fuel stores and more intense firestorms when fire inevitably broke out. Since then, protection of human life and of property have become the first and second priorities, respectively, in fire prevention and management. An emphasis has also been placed upon urban wildfires, including fuel reduction programs and education about hazard minimization. While government bodies and agencies are mainly responsible for fire management, private property owners are also being encouraged and educated in fire prevention and protection in urban vegetated areas. Importantly, wildfires are now recognized as a critical natural process for the health of ecosystems. The Forest Service has instituted a new fire management policy called Fire 21. Unless lives and property are threatened, fires will be allowed to burn as part of their natural ecological role.

Australia

(Luke & McArthur, 1978; Powell, 1983; Oliver et al., 1984; Voice & Gauntlett, 1984; Webster, 1986; Pyne, 1991)

Conditions

Australian bushfires are unique, because they represent the present-day occurrence of deadly fires previously witnessed only in the United States and Canada before the 1940s. Australian bushfires are also unique because of the highly flammable nature of Australian forest vegetation. Globally, Australian bushfires compare in size to the largest recorded in North America. While loss of life is now minimal in North America because of efficient evacuation procedures and the nature of settlement, in Australia it appears to be increasing because of expansion of urban populations into rugged bush country. Mention has been made of the free-burning nature of forest fires in northern Canada. While attempts are made in Australia to put out fires, many in inaccessible hill country are left to burn. These fires pose a threat directly if they change direction under strong winds, and indirectly because they provide a source for fire spotting beyond the limits of the main fire-front. Australian bushfires are also impossible to control once they become wildfires, because of the high flammability of the bush. Some of the most intense fires were the conflagrations in Hobart in 1967 and south-eastern Australia in 1983.

The map in Figure 7.7 shows that the spatial occurrence of fires in Australia is seasonally controlled. The tropical zone has a winter–spring fire season, because the summer is usually too wet and grasses will not cure and dry out before spring. The fire season progresses southward and across the continent along the coasts through the spring. By late spring–summer,

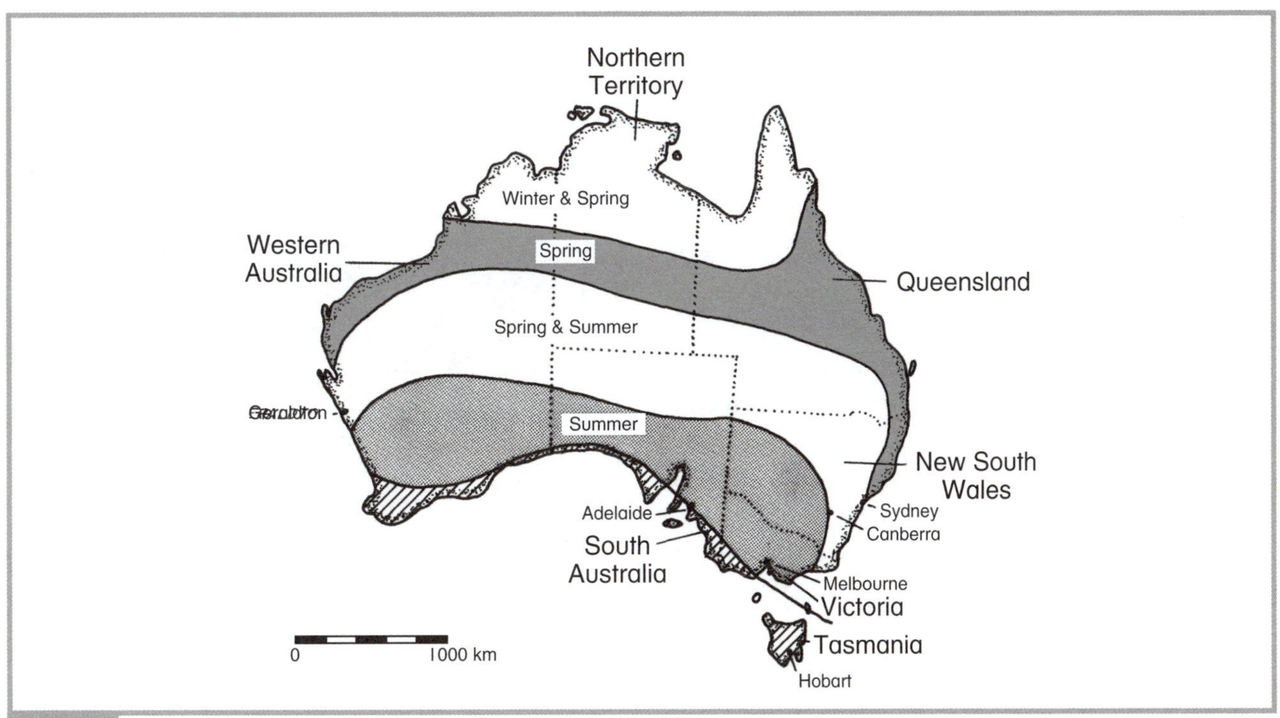

Fig. 7.7 Pattern of seasonal bushfires in Australia (after Luke & McArthur, 1978).

the fire season has peaked in a line stretching from Geraldton to Canberra and Sydney. By mid-summer, the zone has moved to the southern part of the continent. At this time, hot, desiccating winds from the interior can blow towards the coast as mobile polar highs stall over the Tasman Sea (Figure 7.3). Historically, this is the time of severest bushfires and greatest loss of life in Australia. Finally, the southern extremities of the continent experience bushfires as vegetation dries out at the end of summer.

Bushfires are not necessarily restricted to the seasons shown in Figure 7.7. In the central spring and summer zone, severe bushfires have been known to occur in the autumn. For example, in 1986 the worst bushfires of the season broke out in early April following a dry summer and record-breaking temperatures for the month.

Historic disasters

The worst fires in Australian history are well-documented and have occurred in the fire-prone belt of southern South Australia and Victoria. On 14 February 1926, fires ignited under temperatures in excess of 38°C and relative humidity of less than 15 per cent. Sixty lives were lost and an untold number of farms, houses and sawmills burnt out. In the 1931–1932 fire season, 20 lives were lost. The 13 January 1939 fires occurred under record-breaking weather conditions. Fires raced through extremely dry bush as temperatures rose to 46°C and humidity dropped as low as 8 per cent. Over 70 lives were lost in numerous fires in the south-eastern part of the country. The 1943–1944 season was a repeat. Over 49 people died in grass and bushfires under oppressively hot and dry weather. Fires in the Dandenongs in 1962 killed 14 people and burnt 450 houses. On 8 January 1969, 23 people died in bushfires as temperatures exceeded 40°C and winds topped 100 km hr^{-1}.

Since the Second World War, there have been four fire events that stand out as exceptional: the Hobart fires of 7 February 1967, the Ash Wednesday fires of 1983, the Sydney fires of January 1994, and the Canberra maelstrom of 18 January 2003. The last three have occurred at decadal intervals, reflecting one of the major climate cycles outlined in Chapter 2.

The Hobart, Tasmania, fires of 1967 were preceded by a season of above-average rainfall and heavy growth of grasses. As drought set in and this vegetation dried out, numerous controlled burns were lit to diminish the fire risk; however, by 7 February, 81 fires had been left burning around Hobart under the misconception that rain would extinguish them (Figure 7.8). On that day, temperatures reached 39°C with humidity as low as 18 per cent, conditions not witnessed in a century. The fires were driven by north-west winds in excess of 100 km hr^{-1} generated by the pressure pattern shown in Figure 7.3. By noon, fires had raced through the outer districts of Hobart and were entering the suburbs. By the time the fires were controlled, 67 people were dead, 1400 homes and 1000 farm buildings destroyed, and 50 000 sheep killed. Insurance claims totalled $A15 million. The fires in Hobart clearly represent a natural hazard exacerbated by human development and attitudes. They also signify a major shift – towards urban conflagrations – in the nature of bushfires in this country. Urban settlements in Hobart had expanded into hilly, picturesque bushland under the misconception that metropolitan firefighting and water services would be available to counteract any increased fire risk. The unattended spot fires burning around Hobart before 7 February stretched these resources to the limit, scattering equipment outside the city and

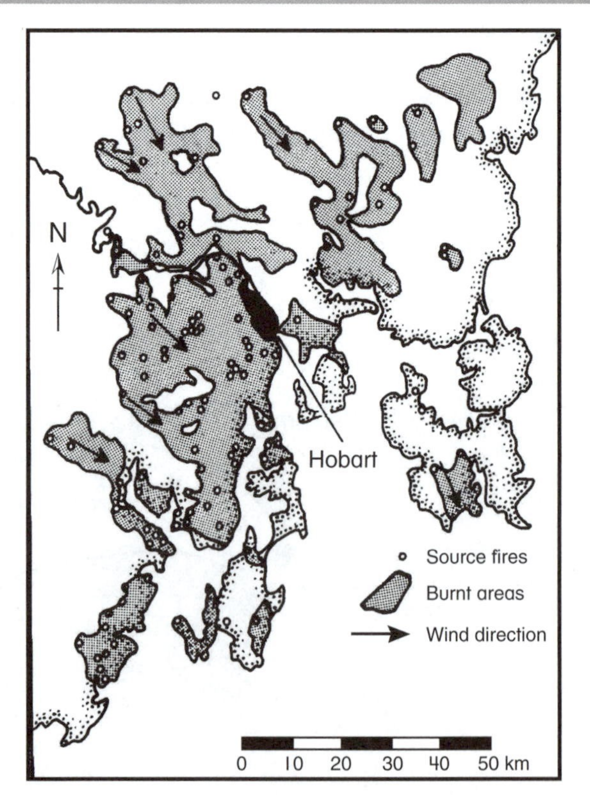

Fig. 7.8 Extent of bushfires and location of spot fires prior to Hobart, Tasmania, bushfires, 7 February 1967 (adapted from Cheney, 1979).

reducing the efficacy of communication. On the day of the disaster, many communities had to rely upon their own resources to cope with the blazes because professional assistance was unavailable or non-contactable.

The Ash Wednesday fires of 16 February 1983 caused substantial death and injury. The 1982–1983 El Niño–Southern Oscillation event produced the worst drought in Australia to that point in time. Bushland and grasslands were desiccated. While the weather conditions of 16 February had occurred often throughout the summer, temperature on the day rose above 40° C accompanied by winds gusting over 60 km hr^{-1}. Weather forecasts accurately predicted one of the hottest and driest days of the summer. Hot, dry air flowed strongly from the interior of the continent, reinforced by the subtropical jet stream in the upper atmosphere. A strong cool change was forecast to cross the south of the continent in the afternoon and early evening, bringing some relief to the desiccating conditions. For weeks, total fire bans had been enforced and broadcast hourly over much of south-eastern Australia. Bushfire organizations and volunteer firefighters in South Australia, Victoria, and New South Wales were put on alert. Every human resource was available and ready.

There was little anyone could do to stop the fires. By noon, much of Adelaide was ringed by fires and at 3:30 pm the cold change swept through. The change dropped temperatures 10°C, but the winds increased and drove the fires to the north-east. In Victoria, fires broke out around Mt Macedon to the north of Melbourne and in the Dandenong Ranges to the east, with hot, dry north-westerly winds gusting up to 70 km hr^{-1}. Other fires developed in rural scrubland in western Victoria and in the coastal resort strip west of Geelong. Firefighters speedily brought under control all but eight of the 93 fires that broke out. The cool change, rather than bringing relief as expected, turned the fires into raging infernos. Fire fronts were shifted 60–90 degrees and pushed forward by winds averaging 70 km hr^{-1} and gusting at times to 100–170 km hr^{-1}. The conflagrations that swept the Adelaide and Melbourne areas were unprecedented. Only three to four times per century could such a combination of conditions be expected anywhere in Australia. Even under ideal fuel conditions, it is physically impossible to generate fire intensities in the Australian bush greater than 60 000–100 000 kW m^{-1} (because of their greater biomass, boreal forests can have values up to 2 500 000 kW m^{-1}). On Ash Wednesday, thousands of houses, farms, and towns were subjected to the maximum limit of these intensities as fires bore down on them at velocities close to 20 km hr^{-1}. Aluminum tyre rims and house sidings melted into puddles as air temperatures reached 660°C. The intense heat generated thermal tornadoes that snapped, at ground level, the trunks of whole stands of trees. Trees in front of the fire exploded first into flames because of the intense radiation and were then consumed by the maelstrom. Such conditions had not been witnessed since the firestorms that swept through the cities of Hamburg and Tokyo as the result of fire bombing in the Second World War. People fleeing in cars were outrun and incinerated. Hundreds of people survived only by quick thinking and luck. At McMahons Creek east of Melbourne, 85 people escaped the inferno by sheltering 30 m underground in the 1 m spacing between water pipes servicing a dam. At the town of Cockatoo, in the Dandenongs, over 150 townspeople survived in the local kindergarten, in the middle of a playground as their town burnt down around them. Along the Otway coast, people fled to the safety of the local beaches, and then into the ocean as the sand became too hot to bear or wind-blasted them.

Seventy-six people died; 3500 were injured, many with burns that would cripple them for the rest of their lives; over 300 000 sheep and 18 000 cattle were killed; over 500 000 hectares of urban forest and pasture burnt in a single day; 300 000 km of fencing was destroyed; and 1700 homes and buildings were consumed (Figure 7.9). In South Australia alone, 40 per cent of the commercial pine forest was burnt. Insurance claims reached $A200 million, with total property losses exceeding $A500 million. Channel 7's national television news simply ended its evening program by silently listing the 20 or more towns that had ceased to exist that day.

The Ash Wednesday bushfires were a rare occurrence in Australia. No modern technology, no prescribed burning, no preparation by home-owners or municipal councils, no improvement in prediction, no number of firefighters or amount of equipment could have stopped those fires from occurring nor negated their effects. The weather conditions leading up to Ash Wednesday were matched only by those preceding the 1939 bushfires. However, the fires exemplify the fact that there is a weather pattern that tends to occur prior

Fig. 7.9 Devastation to the community of Fairhaven along the Otway coastline of Victoria following the Ash Wednesday bushfires, 16 February 1983. Note similarity of devastation to that in Figure 7.6 (photograph © and reproduced courtesy of the *The Age* newspaper, Melbourne).

to extreme bushfires in south-eastern Australia – and this pattern should be researched and predicted. Warnings of this particular hazard and research into the design of buildings and the structuring of communities to prevent fire intrusion must be undertaken to minimize the threat. Issues relating to prescribed burning practices, and pitting conservationists against fire authorities in attempts to prevent another Ash Wednesday, must be resolved. The Ash Wednesday bushfires of 1983 were the worst natural disaster in Australia up to that time. Research, planning, and preparation for major bushfires must be seen in this country as the prime challenge for natural hazard mitigation. These activities, coupled with an effective education program on bushfire survival techniques (staying inside a house or car until the fire passes over is the safest behavior), offer hope of reducing the death toll due to this hazard in Australia.

The Sydney 1994 and Canberra 2003 fires were mainly urban events. The lessons learnt during the Hobart fires of 1967 were practiced during the Sydney fires but ignored in the Canberra fire of 18 January 2003. The Sydney fires were part of a larger fire disaster affecting 600 km of the eastern seaboard of New South Wales. Drought was a major factor as eastern Australia slipped into the 1990–1995 El Niño-Southern Oscillation event. With the advent of fires in early summer, fire authorities knew that climate conditions were deteriorating rapidly and that worse was to come. For the first time ever, the fires would not stop at the edge of Sydney. Rather, they would penetrate to the heart of the city along green belts and through beautified suburbs planted during the 1960s and '70s with highly flammable native trees and shrubs. Following Bureau of Meteorology forecasts for desiccating conditions in early January, and as the number of fires around the city grew from 40 to 148 between 3 and 8 January, authorities planned accordingly, setting in place a military-style operation to save lives and property. Convoys of urban and rural firefighting equipment totaling 1700 units were summoned from as far away as Melbourne and Adelaide, distances of over 1000 km. Firefighters were flown in from across the continent and from New Zealand. At the peak of the threat, on 'Hell Fire' Saturday, 8 January, 70 000 firefighters – three-quarters of whom were volunteers – were battling blazes that would consume 800 000 hectares of bush. On that day, air temperatures climbed to 40°C and humidity dropped to 4 per cent accompanied by winds of 70 km hr^{-1} blowing out of the continent's hot interior. The firefighters were supported by 26 000 people providing food, accommodation and logistical support; and assisted by 16 fixed-wing flying water bombers and 80 helicopters including three Erickson aircranes – *Elvis*, *The Incredible Hulk* and *Georgia Peach*. Sewage tankers, petrol tankers, and concrete mixers assisted in carting water. Taxis, ambulances, chartered buses, and ferries were used to evacuate some of the 20 000 people fleeing the fires. The scale of the response surpassed anything witnessed during urban wildfires in the United States – including the 1991 Oakland fires. The effort paid off. Over a three-week period, only four people died and just 170 homes, valued at $US50 million, were lost. A grateful Sydney held a ticker tape parade in honor of everyone who had prevented a much larger disaster – that at the beginning of the fires was predicted to destroy 5000 homes.

In contrast to the Sydney fires, the Canberra fire of 18 January 2003 was not well-managed. Again drought was a prominent factor, but this time its severity was the worst in recorded history. Fires in the previous year had burnt more area in the surrounding state of New South Wales – 6000 km^2 – than in any other year including the 1993–1994 fire season. Lightning in the week preceding 18 January had sparked numerous fires in the rugged Brindabella Ranges 50 km west of the city. As the high-pressure pattern responsible for most catastrophic fires in eastern Australia developed (Figure 7.3), hot air from the arid interior of the continent was drawn over the city and humidity plunged due to adiabatic heating. By 2:00 pm temperatures had reached 40°C, humidity had dropped to less than 4 per cent – reaching 0 per cent at one station – and, ominously, wind speeds began to increase to 130 km hr^{-1}. Within half-an-hour, fires had raced out of the mountains, through plantations of *Pinus radiata* (these trees have higher biomass than eucalypts) and into the outer suburbs of Canberra. As in Hobart 35 years before, firefighters had been distracted by fires burning on the outskirts of the city. When the fires reached Canberra, 1000 firefighters, 14 aircraft, 119 fire trucks, and 54 water tankers were fighting three blazes outside the city. Only 12 pieces of equipment were available within the urban area to cope with the fires. The unprecedented firestorm that developed snapped off, like a lawnmower cutting grass, trees the size of telephone poles 1 m above the ground, flung vehicles through the air, ripped doors off fire trucks, deroofed houses, hurled bits of roofing tile through car windows like bullets, and carried pieces of wood as large as compact discs more than 12 km away. The ground trembled and the sky was filled with a sound similar to a fleet of 747s taking off. Fireballs slammed into houses, palls of dense smoke spontaneously burst into flame and, as if by magic, houses suddenly ignited one by one like matches being lit against a reddened sky. Daylight was supplanted by darkness, eerily lit by the feeble headlights of fleeing cars and sprays of racing embers. The helicopters, including an Erickson aircrane, which had so successfully contained urban fires in eastern Australia over the previous decade, were helplessly grounded by the lack of visibility. Communications between the fire command center and field units broke down in chaos. Fire hydrants failed, gas meters ignited torching adjacent houses, and exasperated firefighters broke down in tears. The fires raged for 12 hours, killing four people, destroying 431 homes, and causing $250 million damage including destruction of the internationally renowned Mt Stromlo astronomical observatory. It was an urban fire that should never have been allowed to happen and shattered any confidence that Australia had learned from the lessons of previous fires.

CONCLUDING COMMENTS

Of all natural hazards occurring in Australia, bushfires now have the potential to be the major cause of property damage and loss of life. The Ash Wednesday fires of 1983 took more lives and destroyed more property in scattered semi-rural communities than Cyclone Tracy did in moving through the center of Darwin. Dense settlement, which is now occurring in rugged bushland on the outskirts of major Australian cities, is increasing the probability of another major bushfire disaster like Ash Wednesday or the 2003 Canberra fire. All too often, images of horrendous conflagration fade within a few years from people's minds. A drive through the outskirts of Melbourne or the Adelaide Hills will show that houses, destroyed in 1983 because they were not fire-resistant, have been reconstructed exactly as they were before the fire. In New South Wales, at least 10–20 homes each year over the last 20 years have been destroyed by bushfires around urban centers. In the northern suburbs of Sydney, homes with wooden decks and exteriors have been built up the inaccessible slopes of steep gullies.

Throughout this book, the 1:100-year event is emphasized as an event that has not yet occurred in most of Australia. Overseas, in countries settled for longer, structures designed to withstand that same event have been shown to be inadequate over periods longer than 100 years. If 1:100-year bushfire maps were constructed for settled areas in southern Australia, many newer suburbs, as well as some of the older inner suburbs, would obviously lie within bushland that has undergone recurrent burning.

Oddly, just when America is abandoning prescribed burning in national parks, Australia is adopting it. This is the reverse of the pattern twenty years ago. The debate whether to burn or not will, in the end, be resolved by defining which policy best mitigates loss of life. However, prescribed burning can also kill. In the middle of winter on 8 June 2000, four volunteer firefighters died in a Sydney controlled burn that got out of control. Similarly, another controlled burn too close to the summer fire season, in the Goobang National Park, burnt 14 000 hectares of farmland killing 5000 sheep and destroying 140 km of fencing. One of the reasons that prescribed burning has been reintroduced in Australia is that, despite the world's best-developed volunteer firefighting organization, urban conflagrations are increasing. This volunteer service will be severely tested and ultimately fail within the next 20 years as the urban bushfire threat continues to escalate in south-eastern Australia.

Nor is this threatening situation confined to Australia. The same is occurring in southern California, which has similar fire weather and vegetation but more rapid urban expansion. What is not easily recognized is the degree to which urban settlement has extended into forest areas in all parts of the United States. Since the first major urban forest fire in 1947, urbanization has continued at a rapid pace, and forests have completely reclaimed large tracts of abandoned farmland near settlements and cities. Each urban wildfire takes on a distinct identity – for example the Black Tiger Fire of July 1989 near Boulder, Colorado, or the Stephan Bridge Road Fire of May 1990 in Crawford County, Michigan. Since the mid-1950s, this urbanization has taken on a different aspect as millions of urban dwellers have chosen to leave major cities during summer for cottages in nearby woodlands. This occurs in the north-eastern Appalachians, northern Michigan, and Wisconsin. In Canada, up to 50 per cent of the population of Toronto and Montreal escape the summer's heat each weekend by traveling into the Canadian Shield forest. Historically, these forests have been subjected to infernos. Given the propensity of humans to play with fire, the regeneration of forests and the tendency for warmer summers, these woodlands will undoubtedly witness the recurrence of large catastrophic fires similar to those that affected the Great Lakes region in the late nineteenth century.

REFERENCES AND FURTHER READING

Cheney, N.P. 1979. Bushfire disasters in Australia 1945–1975. In Heathcote, R.L. and Thom, B.G. (eds) *Natural Hazards in Australia*. Australian Academy of Science, Canberra, pp. 72–93.

Cornell, J. 1976. *The Great International Disaster Book*. Scribner's, NY.

Couper-Johnston, R. 2000. *El Nino: The Weather Phenomenon That Changed the World*. Hodder and Stoughton, London.

Firewise 1992. *The Oakland/Berkeley Hills fire*. <http://www.firewise.org/pubs/theOaklandBerkeleyHillsFire/>

Luke, R.H. and McArthur, A.G. 1978. *Bushfires in Australia*. AGPS, Canberra.

Oliver, J., Britton, N.R. and James, M.K. 1984. The Ash Wednesday bushfires in Victoria, 16 February 1983. *James Cook University of North Queensland Centre for Disaster Studies, Disaster Investigation Report* No.7.

Powell, F.A. 1983. Bushfire weather. *Weatherwise* 36(3): 126.

Pyne, S.J. 1982. *Fire in America: A Cultural History of Wildland and Rural Fire*. Princeton University Press, Princeton.

Pyne, S.J. 1991. *Burning Bush: A Fire History of Australia*. Allen & Unwin, Sydney.

Romme, W.H. and Despain, D.G. 1989. The Yellowstone fires. *Scientific American* 261(5): 21–29.

Seitz, R. 1986. Siberian fire as 'nuclear winter' guide. *Nature* 323: 116–117.

ThinkQuest Team 2001. *2000 fire season*. <http://library.thinkquest.org/C0119184/english_text/historical_fires_2000_fire_season.shtml>

van Nao, T. (ed.) 1982. *Forest Fire Prevention and Control*. Martinus Nijhoff, The Hague.

Vines, R.G. 1974. Weather Patterns and Bushfire Cycles in Southern Australia. *CSIRO Division of Chemical Technology, Technical Paper* No. 2.

Voice, M.E. and Gauntlett, F.J. 1984. The 1983 Ash Wednesday fires in Australia. *Monthly Weather Review* 112(3): 584–590.

Webster, J.K. 1986. *The Complete Australian Bushfire Book*. Nelson, Melbourne.

Yool, S.R., Eckhardt, D.W., Estes, J.E. and Cosentino, M.J. 1985. Describing the bushfire hazard in southern California. *Annals of the Association of American Geographers* 75(3): 417–430.

Oceanic Hazards

CHAPTER 8

INTRODUCTION

In Chapter 3, waves were often mentioned as one of the main agents of erosion and destruction by storms. In fact, destructive waves can occur without storm events, and in association with other climatic and oceanographic factors such as heavy rainfall and high sea levels, to produce coastal erosion. In this sense, waves form part of a group of interlinked hazards associated with oceans. This chapter examines these and other oceanographic hazards. Wave mechanics and the process of wave generation will be outlined first followed by a description of wave height distribution worldwide. This section will conclude with an appraisal of the hazards posed by waves in the open ocean.

Wind is the prime mechanism for generating the destructive energy in waves. In cold oceans, seas or lakes, strong winds can produce another hazard – the beaching of drifting sea-ice. Sea-ice is generally advantageous along a coastline that experiences winter storms, because broken sea-ice can completely dampen the height of all but the highest waves. When frozen to the shoreline as shorefast ice, sea-ice can completely protect the shoreline from any storm erosion. However, floating ice is easily moved by winds of very low velocity. These winds might not generate appreciable waves, but they can drive sea-ice ashore and hundreds of metres inland. The force exerted by this ice can destroy almost all structures typically found adjacent to a shoreline. For this reason, sea-ice must be considered a significant hazard in any water body undergoing periodic freezing and melting. This chapter will describe the problem of drifting sea-ice, its worldwide distribution, and the measures taken to negate its effects.

It was pointed out in Chapter 3 that waves might not be important as a hazard if the full intensity of a storm reaches landfall during low tide. Such was the case during Cyclone Tracy in Darwin, Australia, in 1974 (see Figure 8.1 for the location of all major placenames mentioned in this chapter). Even with a 4 m storm surge, waves did not reach the high tide mark, because the storm made landfall during low tide in an environment where the tidal range is about 7 m. More insidiously, low waves can cause considerable damage along a shoreline if sea levels are either briefly elevated above normal levels, or rising in the long term. It is generally believed that sea level is presently rising worldwide (*eustatic* rise) at a rate of 1.0–1.5 mm yr^{-1}. The most likely cause of this rise is the *Greenhouse effect*. Human-produced increases in CH_4, CO_2 and other gases are responsible for a general warming of the Earth's atmosphere. This causes sea level to rise because of thermal expansion of ocean waters. Note that this cause is more significant than melting of the world's icecaps. Even if recent warming is not due to the Greenhouse effect, the implications are the same. The recent and threatened rise in sea level poses a

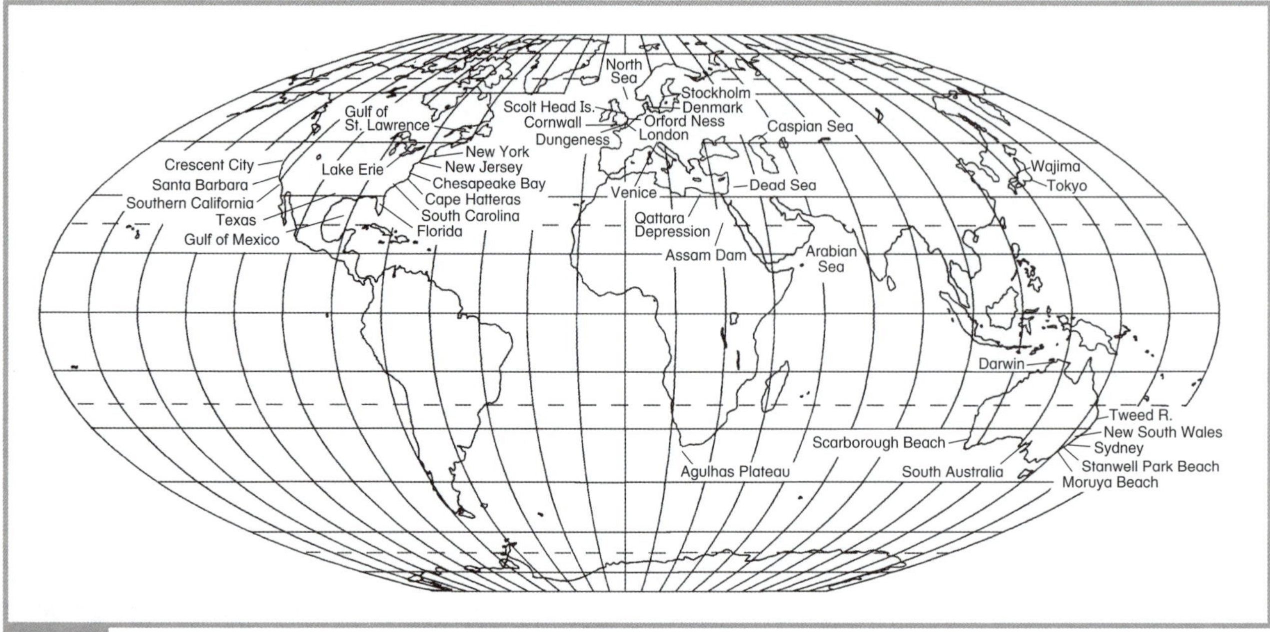

Fig. 8.1 Location map.

natural hazard to many coastal areas, especially where major world cities exist close to the limits of present sea level. The rise in sea level is also believed to be responsible for the accelerating trend in sandy beach erosion worldwide. Sea level behaviour and beach erosion are more complex than outlined in the above scenarios. This chapter will describe worldwide sea level trends, and examine the causes of sea level fluctuations. The chapter will conclude with an evaluation of the factors responsible for beach or general coastal erosion. The processes whereby rising sea level can induce beach erosion will be described, followed by an examination of other causes, such as increasing rainfall and storminess on the east coasts of North America and Australia. The relative importance of these causes will be evaluated using as an example part of a data set compiled for Stanwell Park Beach, Australia.

WAVES AS A HAZARD

Theory

(Wiegel, 1964; Beer, 1983; Dalrymple, 2000)

The basic form and terminology of a linear wave is described schematically in Figure 8.2. This diagram has been used to characterize simply general wave phenomena, including ocean waves, tides, storm surges, seiching, and tsunami. The waveform in Figure 8.2 has the following simple relationships in deep water:

$$L = CT \quad (8.1)$$

$$C = gT(2\Pi)^{-1} \quad (8.2)$$

then $$L = 1.56T^2 \text{ in meters} \quad (8.3)$$

where L = wavelength
C = wave velocity
T = wave period
g = acceleration due to gravity
$\Pi = 3.1416$

Wave motion is transmitted through the water column with water particles circumscribing elliptical paths that decrease exponentially in height and width with depth. If there is no measurable oscillation of water particles before the seabed is reached (point A in Figure 8.2), then the wave is considered to be in deep water. If oscillations are still present when the seabed is reached (point B in Figure 8.2), then the wave is in shallow water and affected by the seabed. The height and energy of a shallow-water wave are determined, not solely by waveform, but also by the nature of inshore topography. Under these conditions wave height can be increased by shoaling, by *wave refraction* – where offshore *bathymetry* has focused wave energy onto a particular section of coastline such as a headland – or by *wave diffraction* around headlands, *tidal inlets*, and harbor entrances. Figure 8.3 exemplifies each of these conditions, which are controlled in magnitude by wave height, period, and water depth. Generally, the rate at which energy moves with the wave – the energy flux – remains virtually constant as a wave travels through

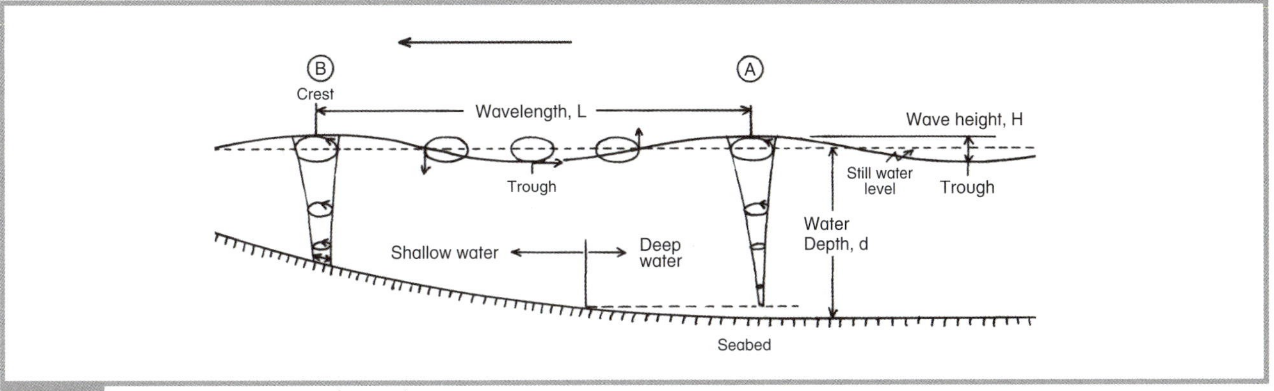

Fig. 8.2 Schematic representation of wave characteristics.

increasingly shallow water towards shore. However, the speed at which the waveform is traveling decreases with decreasing water depth. To maintain a constant energy flux as a wave shoals into shallower water, wave height must increase. The fact that wave velocity slows as water depth decreases also determines wave refraction. If one part of a wave crest reaches shallower water before another part, then the wave crest bends towards the shallower water segment. The process is analogous to driving one wheel of a vehicle off a paved or sealed road onto a gravel verge or shoulder. The wheel that hits the gravel first slows down dramatically, while the wheel still on the road continues to travel at the original road speed. The vehicle as a result veers towards the slower rotating wheel and careers off the road. For waves, the effect produces a bending of the wave crest towards shallow water, an effect that is particularly prominent around headlands or reefs that protrude out into the ocean.

Diffraction is the process whereby energy moves laterally along a wave crest. The effect is noticeable when a wave enters a harbor between narrow breakwalls. Once the wave is inside the wider harbor, the crest width is no longer restricted by the distance between the breakwalls, but gradually increases as energy freely moves sideways along the wave crest. In some circumstances, wave height can increase sideways at the expense of the original wave height. Such an effect often accounts for the fact that beaches tucked inside harbor entrances and unaffected by wave refraction experience considerable wave heights and destruction during some storms.

The destructiveness of a wave is a function of the wave height and period. Height determines wave energy, the temporary elevation of sea level, the degree of overwashing of structures, the height of water oscillations in the surf zone, and the distance of wave *run-up* on a beach. Because most waves affecting a coastline vary in period and height over a short period of time, oceanographers talk about a range of wave heights included in wave spectra. For engineering purposes, the highest one-third or one-tenth of waves, called the 'significant wave height', is used. Wave period determines wave energy, *power*, and shape in shallow water. The latter aspect determines whether wave-generated currents at the bed move seaward or shoreward. If seaward, then the beach erodes. The following equations define wave energy and power for a single wave in deep ocean water:

$$E = 0.125\rho g H^2 L \tag{8.4}$$

$$P = C_g E\,(CT)^{-1} \tag{8.5}$$

where
E = wave energy
P = wave power
H = wave height
L = wavelength
C_g = the group velocity
ρ = density of water

Group velocity is the speed at which a group of waves moves forward. In deep water, it is exactly 0.5 times the wave velocity such that individual waves tend to move forward through a set of waves and disappear at the front. In very shallow water, the group velocity is equal to the wave velocity. There is a significant difference between wave energy and power. Wave energy is the capacity of a wave to do work. Power is the amount of work per unit time. The difference can be explained simply by comparing a greyhound to an elephant. A greyhound is light and when it rubs up against you, you hardly feel it. An elephant is bulky and when it rubs up against you, you are pushed aside. The elephant,

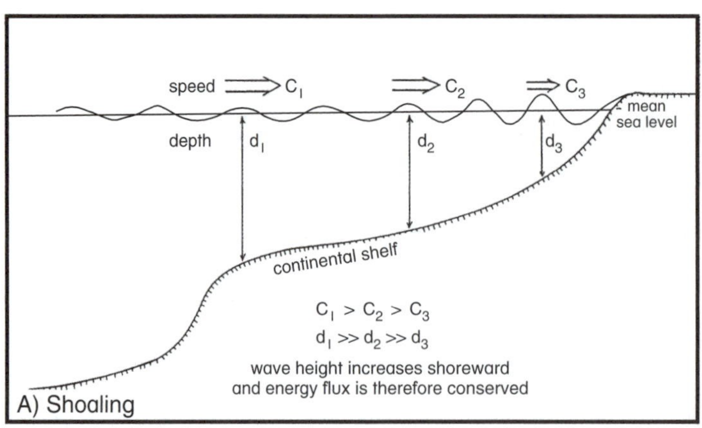

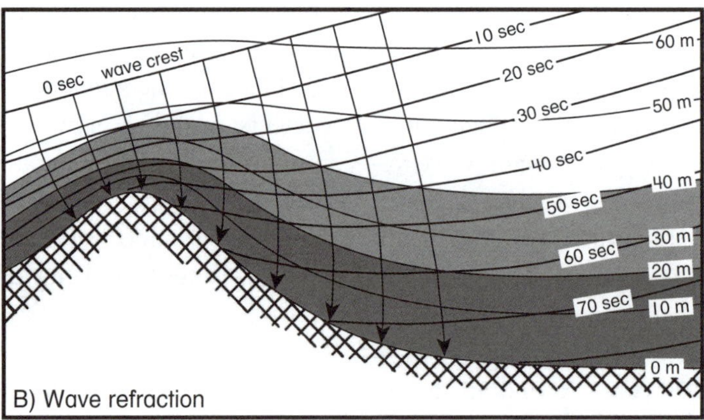

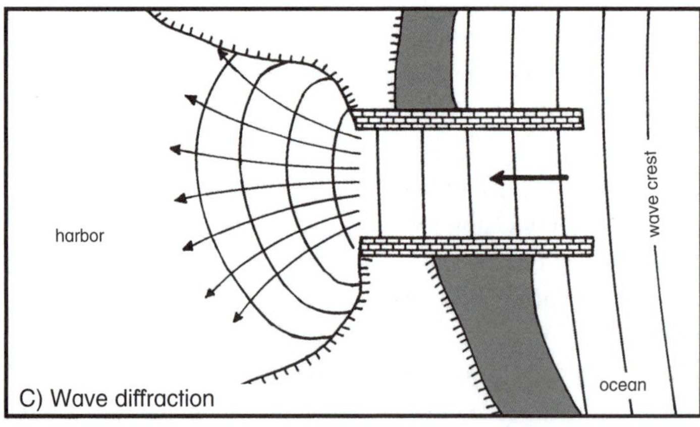

Fig. 8.3 Schematic diagrams of wave behavior in shallow water, A) shoaling, B) refraction, and C) diffraction.

because of its larger mass, exerts more energy against you than the greyhound. A greyhound can move fast and, when it hits you, you are bowled over. An elephant moves slowly and, when it hits you, you move easily out of its way. Here, the greyhound has more power than an elephant (unless of course the elephant decides to charge). Power as defined here applies to an infinite train of waves.

Wavelength is much greater than wave height, so that most of a wave's energy and power is controlled by wavelength, which is related to the wave period. However, a change in wave height has a much more dramatic affect on changing wave energy and power than a change in wavelength. Table 8.1 illustrates this effect for the most common wave periods in swell environments. (Note that these values would be

Table 8.1 Wave energy and power for typical ocean waves (based on Wiegel, 1964; Dalrymple, 2000).

Wave period (s)	Length (m)	Height (m)	Energy m^{-1} of wave crest in one wave (103 joules)	Power m^{-1} of wave crest (103 watts)	Power relative to 1 m high wave
10.0	156	1.0	0.196	9.79	1.0
		1.4	0.384	19.20	2.0
		2.0	0.784	39.18	4.0
		3.0	1.763	88.15	9.0
		7.0	9.598	479.92	49.0
14.0	306	1.0	0.384	13.71	1.0
		2.0	1.536	54.85	4.0
		3.0	3.455	123.41	9.0

reduced 2.4 per cent in fresh water because of its lower density.) A doubling in wavelength leads to a doubling in wave energy and power, whereas a doubling in wave height leads to a quadrupling in wave energy and power. In the coastal zone under strong winds, wave height can change more rapidly than wavelength. The amount of wave energy in a storm wave can become abnormally large (Figure 8.4). In swell environments, wave heights of 1.0 m and periods of 10 seconds are common; however, if a storm wave increases in height to 7 m – a value typical of the storm waves off the eastern coasts of the United States or Australia – then wave energy will increase almost fiftyfold. This is the reason why storm waves can be so damaging in such a short period of time.

World distribution of high waves

(PO.DAAC, 2003)

The generation of waves represents transference of energy from wind to the water body. The height to which a wave will grow depends upon four factors: wind speed, the duration of the wind, the length of water over which wind blows – termed fetch – and the processes leading to decay. Generally, wave height is dependent first upon fetch, second upon wind speed and last upon duration. Before the advent of *wave-rider buoys* to measure waves along a coastline, or altimeters in satellites to measure wave period and heights over large areas, data on wave characteristics depended upon ship observations and hindcasting procedures. Hindcasting theoretically estimates wave height and period based on wind duration, speed, and fetch derived from synoptic pressure maps. Except on small bodies of water or near shore where the resolution of imaging is too low, satellites have effectively eliminated the necessity for hindcasting to determine wave height. Altimeters in satellites now measure wave heights with an accuracy of 2.5–4.0 cm. Instruments were first used on SEASAT in 1978, then on GEOSAT from 1985 to 1988, and ERS-1 and 2 from 1991, TOPEX/POSEIDON from 1992, and Jason-1 from 2001. CNES, the French space agency, and NASA, the United States space agency, jointly operate the latter two satellites. Waves measured using satellite altimeters range from 1 to 5 m at daily and monthly levels. Figure 8.4 illustrates these patterns for January and July 1995, which are typical of these months. Maximum wave heights correspond to belts of maximum winds generated over oceans by mid-latitude low-pressure cells (Figure 2.3). Wave climate in the extra-tropical north Pacific and north Atlantic is especially seasonal, with much rougher conditions in the northern (*boreal*) winter than in the summer. The Southern Ocean has high waves throughout the year, but is roughest in the southern (austral) winter. Near the equator, larger waves are produced either by swell propagating from higher latitudes, or by seasonal winds (for example, monsoons and tropical storms). The Arabian Sea is particularly rough from June to August, coinciding with the south-west monsoon.

Waves as a hazard at sea

(Beer, 1983; Bruun, 1985; Kushnir et al., 1997; Zebrowski, 1997; Lawson, 2001)

The effect of waves as a hazard really depends upon where the wave is experienced. In the open ocean, freak, giant, or rogue waves pose a hazard to navigation. This does not involve a single wave. Instead, it

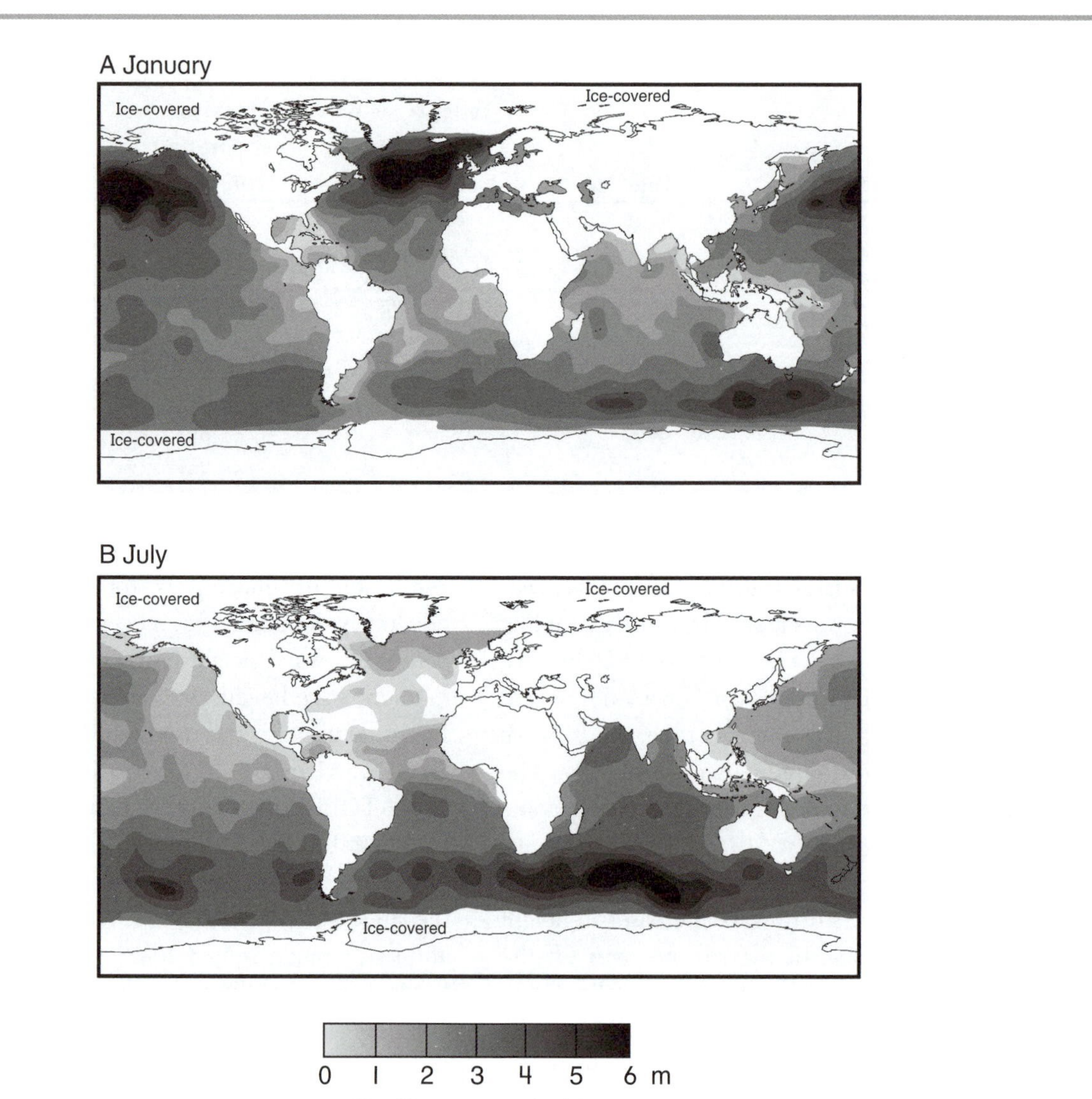

Fig. 8.4 Worldwide distribution of altimeter wave heights measured by the TOPEX/POSEIDON satellite A) January 1995, B) July 1995 (Jet Propulsion Laboratory, 1995a, b).

involves the meeting of several waves of different height, period, and direction that exist in the spectrum of waves generated by winds over the ocean. For example, the energy per meter of wave crest in a 4 m wave with a period of 10 seconds in deep water is 31.4×10^5 joules (Equation 8.4). The energy of a 3 m wave of the same period is 17.6×10^5 joules. If the two waves intersect then the summation of energy is 49×10^5 joules. Using Equation 8.4, the resulting wave height is 5 m. This wave height is unique in time and space. If a ship happens to be at the point of confluence at that time, then it is under threat of capsizing. Such waves have always plagued shipping, but today they also pose a potentially serious hazard to oil platforms, with potential ecological damage and investment loss.

Sometimes waves will increase in height when they meet an opposing current. The region along Africa's south-eastern coast is notorious for high waves due to current interaction. Here, waves generated in the roaring forties travel against the Agulhas current moving south, trapped between the submerged Agulhas Plateau and the continental shelf. Waves steepen in this region as a result. One confluence of wave crests steepened by this current swamped and sank a supertanker, the *Gigantic*, in the late 1970s. There are other regions of the world – including the east coast of New South Wales where storm waves often move into the East Australian Current – that are as dangerous.

Algebraic summation of interacting waves does not explain the incidence of rogue waves currently being measured. In the North Sea, the overall height of waves is increasing, with freak waves in excess of 17 m being recorded around many platforms. In the Gorm field, five waves in excess of 8 m were recorded in a 13-hour period on 6 September 1983. All of these waves are the result of chaos described best by the non-linear Schrödinger equation. This formula can predict waves four times the average wave height. Non-linear interaction can also produce waves of the same size, often in sets of three to five monstrous waves. Perhaps the largest recorded wave occurred on 7 February 1933 when the US *Ramapo* struck a storm with winds as high as 11 on the Beaufort scale. After a week of constant storm conditions, the ship leaned into the trough of a wave and was met by a wall of water 34.1 m high. Other waves of significance were the 26 m wave that hit Statoil's Draupner gas platform in the North Sea on New Year's Day 1995, the 29 m wave that rocked the QE2 crossing the North Atlantic in 1995, and the 30 m wave that almost downed a helicopter during the Sydney Hobart Yacht Race on 27 December 1998. Such waves have even toppled oil platforms, as was the case with the Ocean Ranger drilling rig on the Grand Banks of Newfoundland on 15 February 1982. Since 1969, monstrous waves as high as 36 m have probably been responsible for the disappearance of more than 200 supertankers.

Regional wave heights can also increase substantially over time. For instance, waves in the north Atlantic Ocean have increased in height by 20 per cent since 1980. Largest wave heights offshore from Cornwall have risen from 11.9 m in the 1960s to 17.4 m in the late 1980s, while mean wave heights have increased from 2.2 m to 2.7 m over this period. These heights represent a 32 per cent increase in wave energy, an increase that may not have been accounted for in the engineering design of shoreline or offshore structures. Intensification of the north Atlantic wind field has amplified these wave heights.

Waves as a hazard on rocky coasts

Simple wave theory can also explain how rough seas can sweep people unexpectedly off rocks. Presently one person dies each week in Australia as the result of being swept off rock platforms. This is currently the most common natural cause of death in Australia. While tourists – especially those with little knowledge of a coastline – appear at risk, the majority of deaths involve anglers. Fishing is the most popular recreational activity in Australia and many anglers view rock ledges and platforms as the best spots to fish.

Four factors exacerbate the hazard. First, if people have walked along a beach to get to a headland, they can be lulled into a false sense of security because, due to wave refraction, wave heights can be up to 50 per cent less along the beach than off the headland (Figure 8.3). Second, wave heights are diminished 5–20 per cent at the coast because of frictional interaction with the seabed. Along rocky coasts, especially in New South Wales where the continental shelf has a width as narrow as 12 km, the steep nature of the offshore contours minimizes this frictional loss to less than 5 per cent. Third, the term 'freak wave' typically relates to the interaction of wave spectra rather than to a single large wave. Waves rarely arrive along a coastline with a constant wave period, height, or direction. It is quite common, for example, for wave sets along the New South Wales coast to come from two different directions. In embayments, any directional variation in wave approach to the shoreline tends to be minimized because of wave refraction. On headlands, wave refraction is not efficient because headlands protrude seaward from the coast and offshore topography is steep. Here, the directional variation in wave approach is greatest and the potential for interaction between wave crests is at a maximum. Thus, wave crests from two different directions can often intersect, resulting in exaggerated wave heights due to the summation of the energy in the two waves. This combined wave will have a much larger run-up height along rocks than that produced by waves individually. Finally, waves along a beach will often break at some distance from shore and dissipate their energy across a surf zone. While set-up in the surf zone and run-up onto the beach can significantly and temporarily raise water levels at shore, most of the energy has been lost in the surf zone. Additionally, wave reflection back from the shoreline is minimal because beaches tend to be flat. Waves on headlands, however, break by surging at the shoreline. Energy loss occurs over the rocks rather than offshore in a wide surf zone. As well, the steep nature of headlands results in considerable reflection of energy seaward, a process that can temporarily raise sea levels several metres from shore. It is possible for the next incoming wave to be superimposed upon this supra-elevated water. This process can generate freak waves that can override rock platforms even when the tide is low (Figure 8.5).

Fig. 8.5 Deep-water swell waves of 4 m breaking over the breakwall to Wollongong harbor, Australia (photograph courtesy John Telford, Fairymeadow, New South Wales).

The crucial parameters in the generation of freak waves are high seas and variable wave conditions. These conditions may be associated with periods of stormier or windier weather. If such conditions have not occurred for some period, then people's awareness of wave behavior – especially along rocky coasts under high seas – may be poor. This ignorance is the main reason for a significant loss of life due to people being swept off rocks by waves in Australia.

SEA-ICE AS A HAZARD

Ice in the ocean

(Defant, 1961; Gross, 1972; Vinje, 2001; Anisimov & Fitzharris, 2001; Long et al., 2002)

Sea-ice has two origins: through the freezing of seawater and by the discharge of ice from glaciers into the ocean. The former mechanism is the more common method of forming sea-ice. Because the ocean has a salinity of 31–35 ‰ (parts per thousand), it freezes at temperatures around –1.9°C. New sea-ice does not contain pure water but has a salinity of 5–15 ‰. The faster ice forms, the more saline it is likely to become. As ice ages, the brine drains away, and the ice becomes salt-free and clear. These characteristics can be used to identify the age and origin of sea-ice. At –30°C, sea-ice can form at the rate of 10 cm day^{-1}. Typically in the Canadian Arctic, a winter's freezing cycle will produce ice 2–3 m thick. At this point, ice insulates the underlying water and freezing is substantially reduced. The process is similar for the freezing of ice in lakes and rivers; but because of low salinities, ice forms faster in these latter environments. The density of seawater is 1027 kg m^{-3}. With an air content of 1 per cent, 89 per cent of floating ice is submerged. If the air content rises to 10 per cent, as it can do with some icebergs and shelf-ice, then the submerged part is reduced to 81 per cent by volume.

Sea-ice can be divided into four categories. The first consists of the polar shelf- and sea-ice forming a permanent cover, 3–4 m thick, over 70 per cent of the Arctic Ocean and surrounding the Antarctic continent (Figure 8.6). This ice is very hummocky but may crack open in summer to form leads. In the following winter, leads will be squeezed shut by the shifting ice forming pressure ridges tens of meters high. Sea-ice may also melt patchily, especially in the Antarctic, forming open water areas called polynyas. Since the mid-1970s, the melting of Arctic sea-ice has increased. The area covered by sea ice has diminished 21 per cent between 1955 and 2000. This varies from a high of 6 per cent per decade since 1970 in the Atlantic, to 1.5 per cent per decade in the Canadian Arctic and Bering Sea (see Figure 8.6 for placenames mentioned in this section). While the trend has been attributed to global warming, melting has been a longer term process. Figure 8.7 illustrates this trend in the Arctic over the past two centuries using data from Davis Strait between Canada and Greenland, the Greenland Sea east of Greenland, and the Nordic Sea in the north Atlantic. In the latter region, sea-ice extent in April has decreased by 33 per cent (0.79×10^6 km^2) with a substantial proportion of this reduction occurring before industrial activity accelerated in the middle of the twentieth century. The dotted lines in Figure 8.7 indicate that changes in sea-ice extent are synchronous across the Arctic. The limit of sea-ice in the north Atlantic retreats when the North Atlantic Oscillation (NAO) is positive, due to an increased northward flux of warm air. However, off Labrador, the opposite effect is evident. This correlation fits the pattern, associated with the NAO, of warmer temperatures in the north-east Atlantic and cooler ones in eastern North America. Concomitant with this retreat, sea-ice thinned by 0.13 m between 1970 and 1992. In the eastern Siberian Sea, ice thickness decreased from 3.1 m in the 1960s to 1.8 m in the 1990s. In some locations, especially between Greenland and Norway, ice cover has thinned by 40 per cent over the past three decades. More dramatic has been the disappearance of iceshelves around the Antarctic Peninsula. Five iceshelves, amounting to 5000 km^2 of iceshelf, have collapsed due to regional warming of 2.5°C in the last

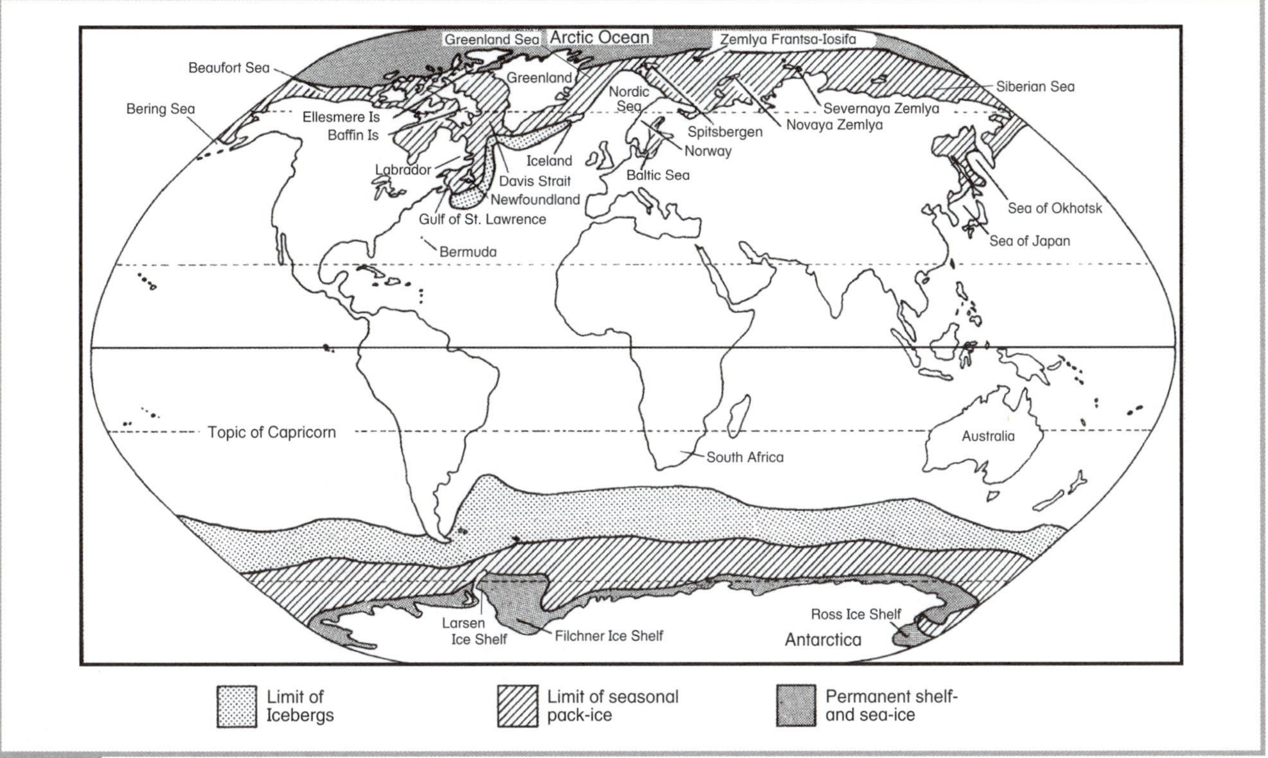

Fig. 8.6 Worldwide distribution of permanent polar shelf- and sea-ice, seasonal pack-ice, and icebergs (after Gross, 1972).

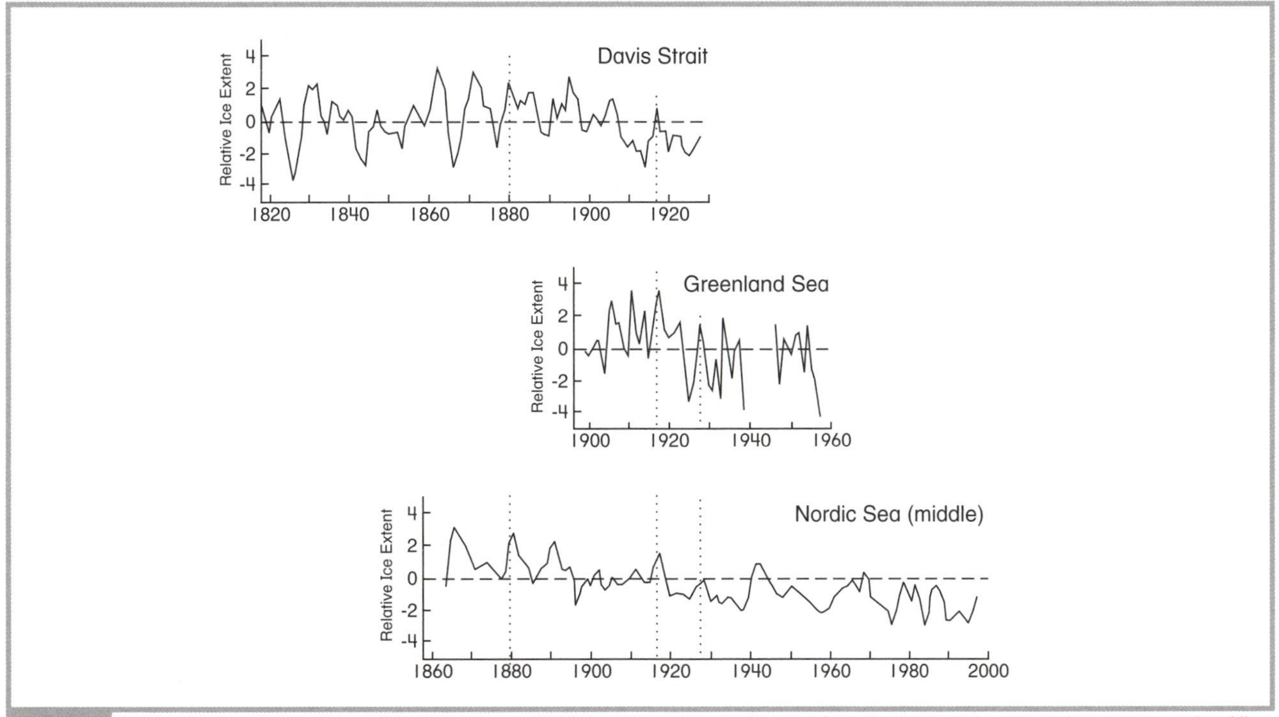

Fig. 8.7 Composite April sea-ice extent, 1818–1998, compiled from records in Davis Strait (Defant, 1961), Greenland Sea (Skov, 1970), and middle Nordic Sea (Vinje, 2001).

50 years. The largest collapse was the Larsen A shelf in 1995 (3000 km^2). This was matched in areal extent again in 1999. Because this ice is already floating, it has had no impact on sea level.

The second type of sea-ice is the annual pack-ice that mainly forms within the seas of Japan and Okhotsk, the Bering and Baltic Seas, the Gulf of St Lawrence and along the south-east coast of Canada.

Pack-ice also forms in inland lakes where winter freezing takes place. Pack-ice generally extends more southward in the northern hemisphere along the western sides of oceans because of prevailing southward-moving cold currents at these locations. In particularly severe winters, pack-ice can extend towards the equator beyond 50° latitude in both the southern and northern hemispheres. If the following summers are cold, or the ice build-up has been too severe, then it is possible for large areas of pack-ice not to melt seasonally.

Excessive pack-ice between Greenland and Iceland terminated the Greenland Viking colony because it prevented contact with Europe after the middle of the fourteenth century. Except for the early 1970s, pack-ice limits in the Arctic have retreated poleward in the twentieth century, concomitant with global warming. Pack-ice becomes a hazard only when it drifts into shore, or when it becomes so thick that navigable leads through it have a serious chance of closing and forming pressure ridges. Whaling and exploration ships have been crushed in this way. In the summer of 1985–1986 and 2001–2002, the Australian Antarctic research program was severely hampered by thick pack-ice that closed in, stranding supply vessels and crushing and sinking a ship. In the Canadian Arctic, thick pack-ice has historically posed a threat to shipping and increased the cost of re-supplying northern settlements. Pack-ice terminated numerous journeys to the Canadian Arctic in the nineteenth century, the most famous of which was the loss of the Franklin expedition.

The third type of sea-ice consists of shore-fast ice. This ice grows from shore and anchors to the seabed. It effectively protects a shoreline from wave attack during winter storms but, upon melting, can remove significant beach sediment if it floats out to sea. This type of ice poses a unique hazard at the shoreline and will be discussed later.

The fourth category of sea-ice consists of icebergs. Icebergs form a menace to shipping as exemplified by that most notorious disaster, the sinking in April 1912 of the *Titanic* with the loss of 1503 passengers and crew. Icebergs in the northern hemisphere originate in the north Atlantic, as ice calves from glaciers entering the ocean along the west coast of Greenland and the east coast of Baffin Island. Icebergs coming from the east coast of Greenland usually drift south, round the southern tip of Greenland, and join this concentration of ice in the Davis Strait. Along the southern half of the west coast of Greenland, icebergs collect in fjords and discharge into the Strait at fortnightly intervals on high tides. Along the northern half of the Greenland coast, glaciers protrude for several kilometers into the sea and icebergs calve directly into Davis Strait. About 90 per cent of all icebergs in the northern hemisphere collect as swarms in Davis Strait during severe periods of pack-ice, and move south into the Labrador current as the pack-ice breaks up in spring. These icebergs then pass into the Atlantic Ocean, off Newfoundland, with widely varying numbers from year to year. Fewer than 10 per cent of icebergs originate in the Barents Sea from glaciers on the islands of Spitsbergen, Zemlya Frantsa-Iosifa (Franz Josef Land), Novaya Zemlya, and Severnaya Zemlya.

Whereas winds determine the direction that pack-ice moves, ocean currents control the movement of icebergs. Because of Coriolis force, ice drifts to the right of the wind in the northern hemisphere and to the left in the southern hemisphere. Icebergs in the north Atlantic usually have heights of tens of meters, although some are up to 80 m in height and over 500 m in length. The most southerly recorded sighting in the northern hemisphere was near Bermuda at 30.33°N. In the Antarctic, icebergs usually originate as calved ice from the Ross and Filchner iceshelves. Here, glaciers feed into a shallow sea forming an ice bank 35 and 90 m above and below sea level, respectively. The production of icebergs in the Antarctic is prolific, probably numbering 50 000 at any one time. Icebergs breaking off the Ross Ice Shelf may reach lengths of 100 km. The largest yet measured was 334 km by 96 km with a height above water of 30 m. This one tabular ice block had a volume that exceeded twice the annual ice discharge from the entire Antarctic continent. In 2000, an iceberg – named B15 and measuring 295 km by 37 km – broke off the Ross Ice Shelf. The number of observed icebergs has increased fivefold; however, technological advances in iceberg observation account for most of this increase. In addition, the shelves are undergoing the end of a major calving cycle that recurs every 50–100 years. Tabular icebergs have been found as far north as 40°S, with the most northerly recorded sighting being 26.05°S, 350 km from the Tropic of Capricorn. This is unusual because the Antarctic circumpolar current between 50° and 65°S forms a warm barrier to northward ice movement. About 100 smaller tabular icebergs, up to 30 km^2 in size, also exist in the Arctic Ocean, drifting mainly in a clockwise direction with the currents in the Beaufort Sea. These ice rafts have calved from five small iceshelves on Ellesmere Island. Because they have little chance of escaping south to warmer waters, individual blocks can persist for 50 years or more.

In the southern hemisphere, icebergs pose little threat to navigation because they do not frequently enter major shipping lanes. However, it is not unknown for shipping between South America and Australia, and between Australia and South Africa, to be diverted north because of numerous sightings of icebergs around 40–50°S. North Atlantic icebergs do enter the major shipping lanes between North America and Europe, especially off the Grand Banks, east of the Newfoundland coast. In severe years, between 30 and 50 large icebergs may drift into this region. Historically, they have resulted in an appallingly high death toll. Between 1882 and 1890, icebergs around the Grand Banks sank 14 passenger liners and damaged 40 others. The iceberg hazard climaxed on 14 April 1912 – a year of a strong El Niño – when the 'indestructible', 46 000 tonne *Titanic* scraped an iceberg five times its size. The encounter sheared the rivets off plates below the waterline over two-fifths the length of the ship. So apparently innocuous was the scrape that most passengers and crew members did not realize what had happened. Within three hours, the *Titanic* sank dramatically, bow first. Only a third of the passengers survived in lifeboats, while over 1500 people died as the ship sank. Immediately, the United States Navy began patrols for icebergs in the area. In the following year, 13 nations met in London to establish and fund the International Ice Patrol, administered by the United States Coast Guard. The Coast Guard issues advisories twice daily and all ships in the north Atlantic are obliged to notify authorities of the location of any icebergs sighted. Since the Second World War, regular air patrols have also been flown. With the advent of satellites, large icebergs can be spotted on satellite images, while the more threatening ones can be monitored using radio transmitters implanted on their surface. Individual ships can also spot icebergs using very sophisticated radar systems.

While the iceberg hazard to shipping has all but disappeared, the hazard to oil exploration and drilling facilities has increased in recent years, especially in the Beaufort Sea and along the Labrador coastline. Because over 80 per cent of their mass lies below sea level, large icebergs can gouge the seabed to a depth of 0.5–1.0 m, forming long, linear grooves on the continental shelf. The icebergs could therefore gouge open a pipeline carrying oil or gas across the seabed. Drifting icebergs also have enough mass and momentum to crumple an oil rig. Not only could there be substantial financial loss with the possibility of deaths, but the resulting oil spill could also pose a significant ecological disaster. Attempts have been made to divert icebergs by towing them away using tugboats; however, the large mass of icebergs gives them enormous inertia, and the fact that they are mostly submerged means that they are easily moved by any ocean current. At present, waves pose the most serious threat to oil and gas activities in ice-prone waters, but it is only a matter of time before an iceberg collides with a drilling-rig or platform. Present disaster mitigation procedures for this hazard consist simply of evacuating personnel from threatened facilities. Such procedures have been used on several occasions over the last few years.

Ice at shore

(Taylor, 1977; Kovacs & Sodhi, 1978; Bruun, 1985)

Nowhere is the destructive effect of ice displayed more than at the shoreline. During winter, the icefoot, which is frozen to shore at the waterline, protects the beach from storm wave attack. Pack-ice, which develops in the open ocean or lake, also attenuates the fetch length for wave development. The middle of winter sees the whole shoreline protected by ice. In spring, the pack-ice melts and breaks up, whereupon it may drift offshore under the effect of wind. At this time of year, shorelines are most vulnerable to ice damage if the wind veers onshore and drives the pack-ice shoreward. At shore, pack-ice has little problem overriding the coastline, especially where the latter is flat. This stranding ice can pile up to heights of 15 m at shore, overriding wharfs, harbor buildings, and communication links. Ice can actively drive inland distances up to 250 m, moving sediment as big as boulders 3–4 m in diameter. On low-lying islands in the Canadian Arctic, shallow pack-ice, 1–2 m thick, can push over the shoreline along tens of kilometers of coastline (Figure 8.8). Large ridges of gravel or boulders – between 0.5 m and 7 m in height, 1–30 m in width and 15–200 m in length – can be plowed up at the shoreline. The grounding of ice and subsequent stacking does not require abnormally strong winds. Wind velocities measured during ice-push events have rarely exceeded 15 m s^{-1} (54 km hr^{-1}). Nor is time important: ice-push events

Fig. 8.8 Pack-ice 1 m thick pushing up gravel and muds along the south shore of Cornwall Island, Canadian Arctic. Scars in upper part of the photograph represent ice push at slightly higher sea levels.

have frequently occurred in as little as 15 minutes. All that is required is for loosely packed floating ice to be present near shore, and for a slight breeze to drive it ashore.

Ice pushing can also occur because of thermal expansion, especially in lakes. With a rapid fall in temperature, lake ice can contract and crack dramatically. If lake water enters the cracks, or subsequent thawing infills the crack, warming of the whole ice surface can begin to expand the ice and push it slowly shoreward. If intense freezing alternates with thawing, then overriding of the shore can be significant for lakes as small as 5 km^2. Larger lakes tend not to suffer from ice expansion because ice will simply override itself, closing cracks that have opened up in the lake.

The main effect of grounded sea-ice or ice push is the destruction of shore structures. In lake resorts in the north-eastern United States and southern Canada, docks, breakwalls, and shoreline buildings are almost impossible to maintain on small lakes because of thermal ice expansion. On larger lakes, and along the Arctic Ocean coastline in Canada, northern Europe, and Russia, wind-driven ice push has destroyed coastal structures. In the Beaufort Sea area of North America, oil exploration and drilling programs have been severely curtailed by drifting pack-ice. Sites have had to be surrounded by steep breakwalls to prevent ice overriding. In the Canadian High Arctic, oil removal was delayed for years while the government tried to determine the cheapest and safest means of exporting oil. Pipelines laid across the seabed between islands are exposed to drifting icebergs, or to ice-push damage near the shoreline. Alternatively, harbor facilities for oil tankers are threatened by ice push. Even where ice push is not a major problem, drifting sea-ice can push into a harbor and prevent its being used for most of the short (two to three months), ice-free navigation season.

SEA LEVEL RISE AS A HAZARD

Introduction

Sea level rise also poses a significant long-term hazard. Over the past century, there has been a measurable rise in global (eustatic) sea level of 15.1 ± 1.5 cm. In the early 1980s, this rate was viewed as accelerating to 3 mm yr^{-1}, fuelling speculation that the Earth was about to enter a period whereby sea levels could rise 50–100 cm eustatically in the next 100 years. A sea level rise of this magnitude has destructive implications for the world's coastline. Beach erosion would accelerate, low-lying areas would be permanently flooded or subject to more frequent inundation during storms, and the baseline for watertables raised. Much of the rise in sea level in the twentieth century has been attributed to the melting of icecaps, concomitant with increasing temperatures because of global warming. Whether global warming is occurring and whether it is natural or human-induced are moot points. Global warming and thermal expansion of surface ocean waters are forecast to cause sea level rises of 20–40 cm by 2100. This section will examine the evidence for sea level rise, its causes, and the reasons for sea level fluctuations.

Current rates of change in sea level worldwide

(Cazenave et al., 2002; Nerem et al., 2002; Colorado Center for Astrodynamics Research, 2003)

Figure 8.9 presents sea level curves measured on tide gauges, 1930–1980, for five major cities generally perceived as being threatened by a rise in ocean levels: London, Venice, New York, Tokyo and Sydney. The cities are not representative necessarily of eustatic change in sea level because local (*isostatic*) effects operate in some areas. London is in an area where the North Sea Basin is tectonically sinking, Venice suffers from subsidence due to groundwater extraction, and Tokyo has had tectonic subsidence due to earthquake activity. However, the diagrams illustrate that some of the world's major cities would require extensive engineering works to prevent flooding given predicted rates of sea level rise. In addition, Figure 8.9 presents the changes in sea level for three cities: Stockholm; Wajima, Japan; and Crescent City on the west coast of

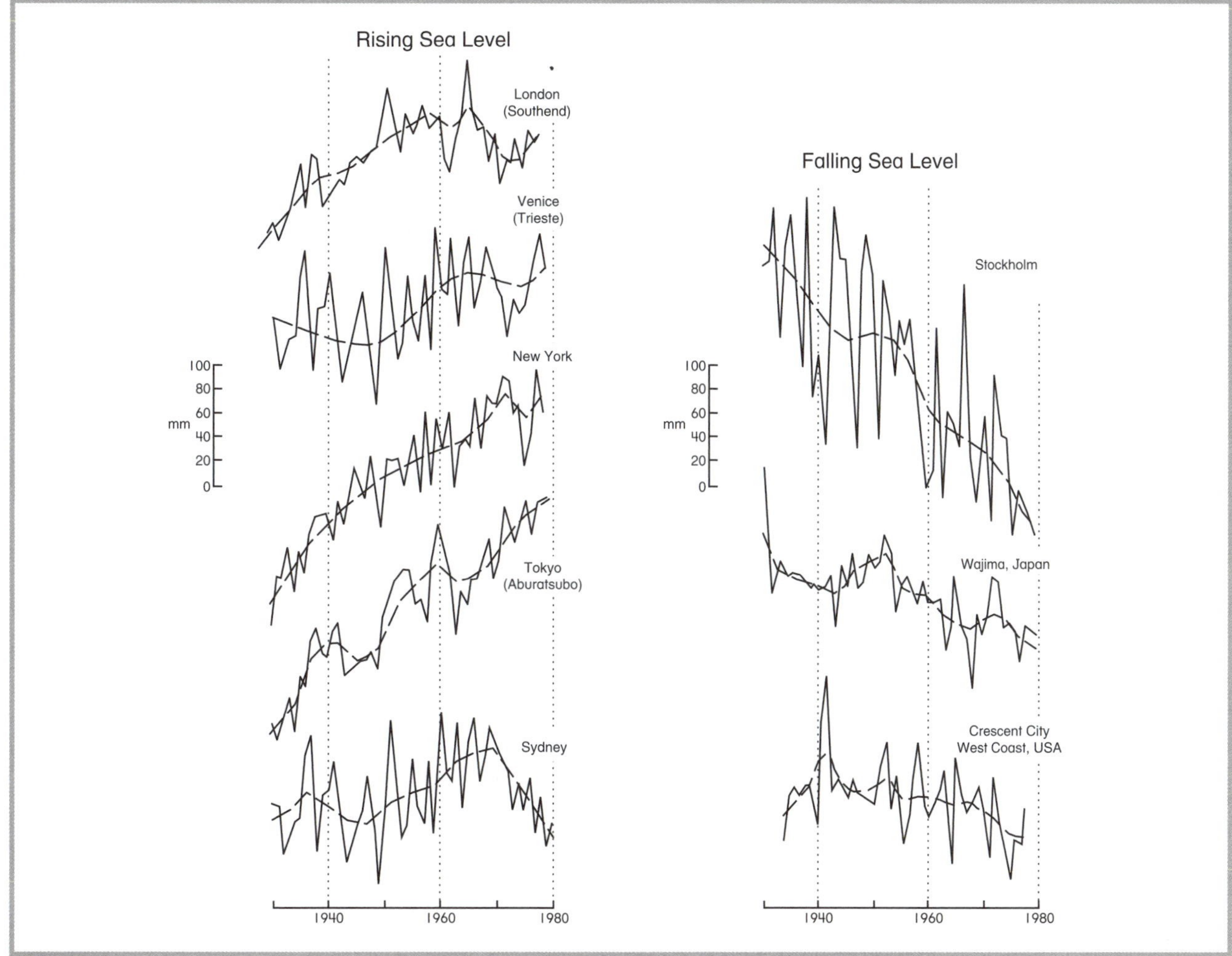

Fig. 8.9 Annual changes in sea level, 1930–1980, at select tide gauges where sea level is either rising or falling.

the United States where uplift very much affects sea level. In the case of Stockholm, land is still rising as the result of glacial deloading over 6000 years ago. Tectonic uplift affects the other two cities. All records indicate that sea level can at times fall even at places such as London and Venice where it is rising in the long term. More importantly, Figure 8.9 indicates that sea level records are 'noisy'. Globally, the average annual variability in sea level – as measured by tide gauges – is 35 mm. This is 20–35 times the hypothesized annual rate of global sea level rise.

Figure 8.10 presents the trends in sea level between 1960 and 1979 at shoreline tide gauges having at least ten years of data. Note that tide gauges are not the optimum means for measuring sea level changes. Their primary purpose is to measure tides in harbors for navigation – not sea level due to global warming. They are also unevenly spread across the globe. For these reasons, only the overall trend in sea level along a shoreline is plotted in Figure 8.10. Large areas of consistent sea level change are difficult to delineate because rates can change from positive to negative over distances of a few hundred kilometers or less. This is the case in northern Europe and eastern Asia. For example, rates range from 24 mm yr^{-1} of submergence along the south-east coast of Japan, to 6.8 mm yr^{-1} of emergence along the north-west coast. Sea level has fallen since 1960 along significant stretches of coastline, mainly in western Europe, western North America, and eastern Asia. Increasing sea levels occur in the Gulf of Mexico; along the eastern coastlines of North America, South America, and Japan; and the shorelines of southern Europe.

Satellite altimeters described above that measure wave heights provide better information about sea level than tide gauges when results are averaged over time. The patterns of sea level change measured by these altimeters over the open ocean between January 1993 and June 2002 are also plotted in Figure 8.10. There is very little correspondence between the

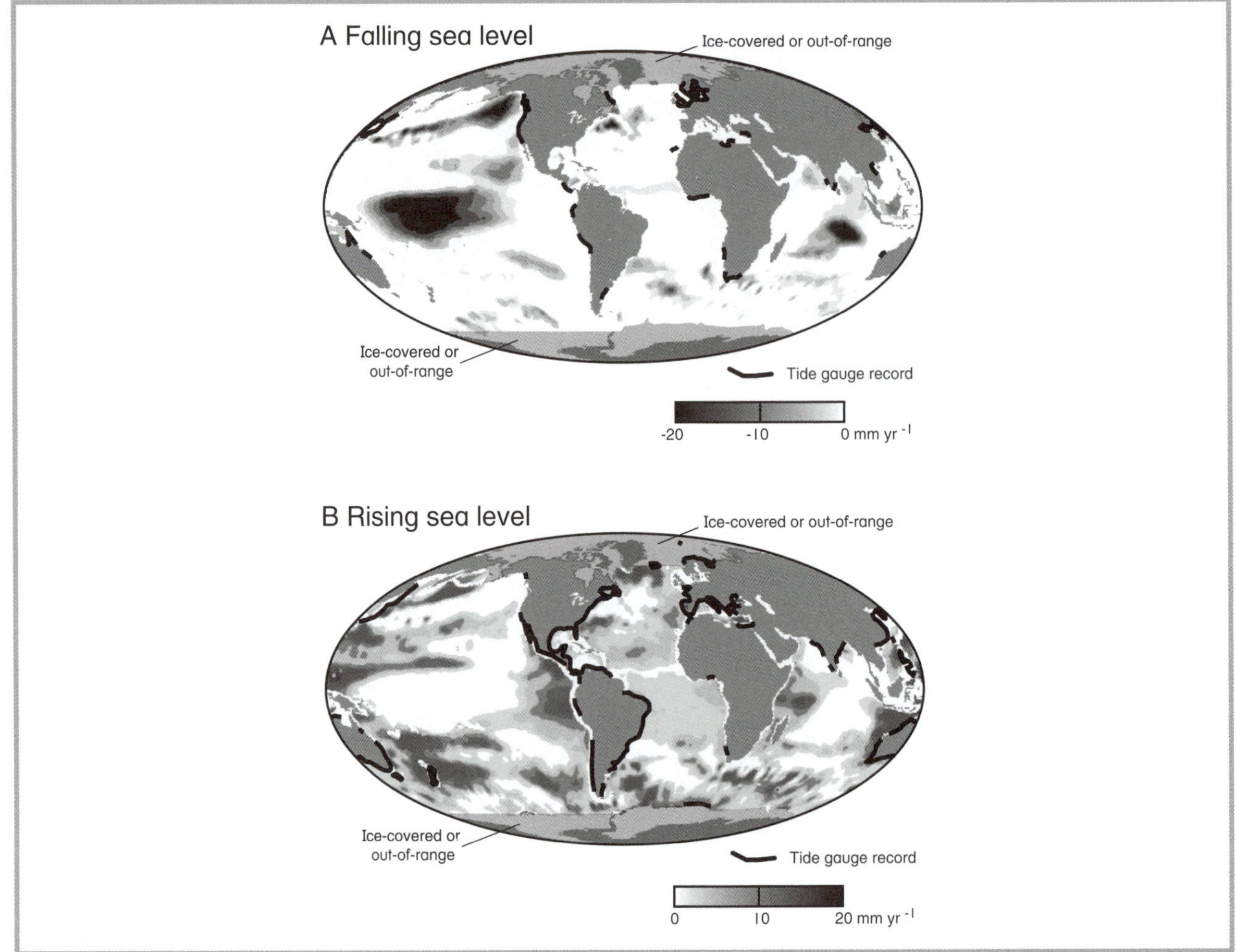

Fig. 8.10 Changes in global sea level measured at coastal tide gauges between 1960 and 1979 (from Bryant, 1988) and from altimeters on the TOPEX/POSEIDON satellite between January 1993 and December 1999: A) falls, B) rises (from Cazenave et al., 2002).

measurement of sea level in the open ocean and that measured at shore using tide gauges. More than likely, the two types of instruments are measuring separate phenomena. Tide gauges are measuring ocean basin edge effects along irregular coastlines. These levels are affected by a myriad of processes that can be grouped under four headings: 1) changes due to daily weather fluctuations affecting temperature, wind stress, and atmospheric pressure; 2) seasonal variation due to heating of water or runoff from rivers; 3) inter-annual effects due to major oscillations in pressure and heating in individual oceans; and 4) tectonics. Satellites measure changes in the overall volume of water in the oceans. Their results indicate that daily, monthly, and inter-annual changes are dominated by *steric* factors related to heating of the ocean surface and attributable primarily to changes in the Southern, North Atlantic, and North Pacific Oscillations. For example, sea level is falling in the central Pacific at a rate that exceeds 15 mm yr^{-1}; however, it is rising at similar rates along the adjacent coastline. This is the signature of repetitively strong ENSO and La Niña events. There are also positive and negative changes at the boundary of the Antarctic circumpolar current in the south Pacific Ocean. Overall, global sea level has been rising recently (Figure 8.10b). This is borne out in Figure 8.11, which plots the average of all altimeter readings between 1992 and 2002. Sea level is increasing at the rate of 1.85 ± 0.42 mm yr^{-1}, a value that is remarkably similar to that determined using tide gauge data.

Factors causing sea level rise

(Bryant, 1987; Sahagian et al., 1994; Gornitz, 2000)

Thermal expansion of oceans is not the sole cause of rises in sea level during the twentieth century. Other factors may be responsible, including deceleration of large-scale ocean *gyres* or melting of alpine glaciers. For instance, the rise in sea level during the twentieth

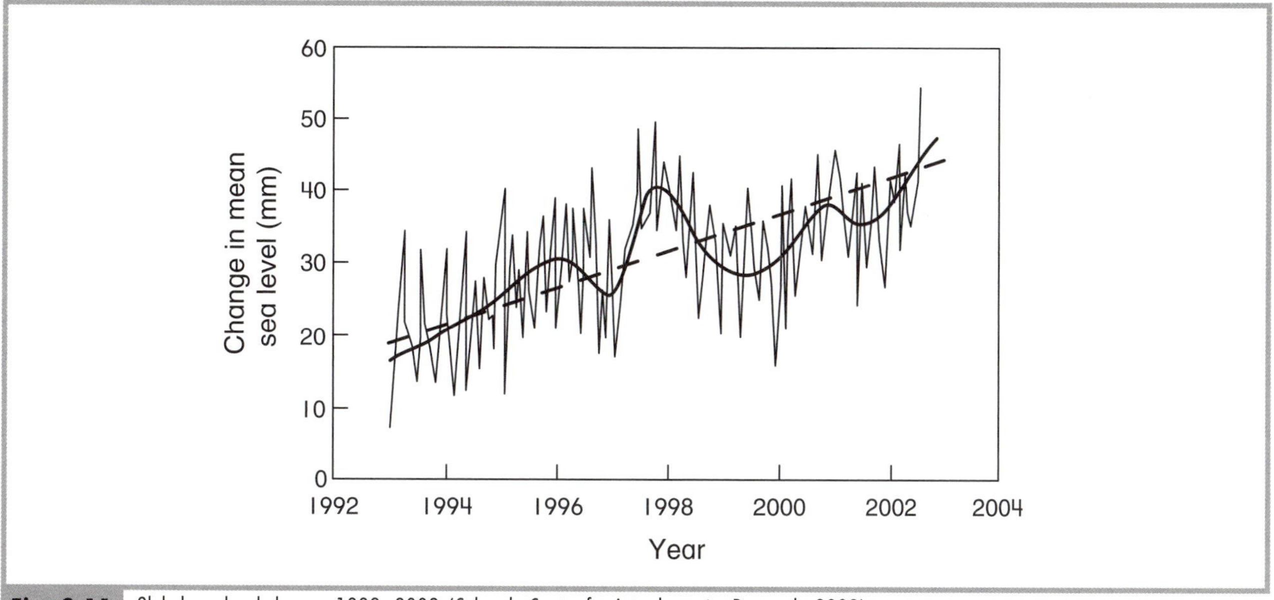

Fig. 8.11 Global sea level change, 1992–2002 (Colorado Center for Astrodynamics Research, 2003).

century required only a 20 per cent reduction in alpine ice volume. Anthropogenic effects in fact dominate global sea level changes. The two main mechanisms are through damming and groundwater extraction. These effects work in opposition to each other. Dams, built mainly since 1950, presently impound between 530 km^3 and 740 km^3 of water. This is equivalent to a decrease in sea level of 1.3–1.8 mm yr^{-1}. On the other hand, groundwater extraction is raising sea level. The United States obtains 20 per cent of its annual water consumption from groundwater. This amounts to 123 km^3, equivalent to a rise in sea level of 0.3 mm yr^{-1}. If the source of the whole world's annual water consumption of 2405 km^3 were proportional to that in the United States, then it would equate to a sea level rise of 1.3 mm yr^{-1}. Other activities that alter the global hydrological balance include deforestation, draining of wetlands, lake reclamation, and irrigation. In total, human activity could be reducing sea level by 0.9 mm yr^{-1}. This means that the natural rate of sea level rise could be as high as 2.75 mm yr^{-1}. Continued dam construction will maintain the human buffer on this natural rise, but there is a finite limit to the amount of water that can be impounded. By the year 2020, this limit should be reached and the rate of measured sea level rise could double. Fortunately, there is an economical solution for negating future sea level rises. Seawater could be diverted, with minimal effort, into depressions such as the Dead Sea, the Qattara Depression in Egypt, Lake Eyre in Australia and the Caspian Sea. These basins have very large areas below sea level that could easily hold increased ocean volumes.

Global sea level and the hydrological cycle

(Bryant, 1993)

The cyclicity present in Figure 8.11 is problematic. Certainly, a high frequency component of about four months exists with amplitudes of 15–20 mm. A longer term fluctuation of about two years' duration also exists. The latter is probably the quasi-biennial oscillation, which has a periodicity of 2.2 years. This oscillation is a stratospheric phenomenon that appears in various countries' rainfall records. This implies that the fluctuations are climatically induced, reflecting seasonal and inter-annual variations in the global hydrological cycle. The tide gauge records presented in Figure 8.9 demonstrate this aspect, despite the dominance of local factors. Changes in sea level recorded on tide gauges are remarkably coherent across the globe. For example, sea levels in southern California rise at the same time as those in southern Australia decrease; sea level at Sydney, Australia, tends to rise one year before it does at London and Venice; and sea level at London or Venice tends to rise one year before it does at New York or Tokyo. This coincidence accounts for about 62 per cent of the annual variation in sea level fluctuations amongst these five cities. As Figure 8.9 illustrates, sea levels oscillate 40–150 mm every 3–5 years. This periodicity is characteristic of the Southern Oscillation, which has the greatest effect on rainfalls, especially in the southern hemisphere. During ENSO events, rainfall decreases over Australia, South-East Asia, India, and Africa. At these times, water drains from continents, accumulates in the oceans, and causes global sea level to rise. There is a lag in time as water spreads out from these source regions across the globe. During La Niña events, increased monsoonal rain falls over Australia, India, South-East Asia, and Africa and accumulates in rivers, lakes, and the ground. As a result, global sea level falls. The land area involved in these alternating drought and flood cycles is approximately 16×10^6 km^2, equivalent to 10–11 per cent of the ocean's surface area. The rainfall difference between dry ENSO and wet La Niña phases is typically 200–400 mm spread over a period of 2–3 years. The volume of rainfall involved, when averaged over the oceans, equates with a change in global sea level of 22–45 mm, a value that approximates the average annual variability of 35 mm present in tide gauge records and the satellite altimeter data. As will be shown subsequently, these fluctuations have an impact on coastal erosion that may be just as important as any long-term rise in sea level.

BEACH EROSION HAZARD

Introduction

There is little argument that 70 per cent of the world's sandy beaches are eroding at average rates of 0.5–1.0 m yr^{-1}. Some of the reasons for this erosion include increased storminess, coastal submergence, decreased sediment movement shoreward from the continental shelf (associated with leakage out of beach compartments), shifts in global pressure belts and, hence, changes in the directional components of wave climates. As well, humans this century have had a large impact on coastal erosion. No single explanation has worldwide applicability because all factors vary in regional importance. A lack of accurate, continuous, long-term erosion data complicates evaluation of the causes. Historical map evidence spanning 100–1000 years has been used in a few isolated areas, for instance on Scolt Head Island in the United Kingdom, Chesapeake Bay in the United States, and on the Danish coast; however, temporal resolution has not been sufficient to evaluate the effect of climatic variables. Aerial photographs exist only in number since the 1930s and often suffer from insufficient ground control for accurate mapping over time. Ground surveying of beaches was rarely carried out before 1960 and is often discontinuous in time and space. Examples of long time series beach change data include: the 16-year record for Scarborough Beach, Western Australia (this also goes back discontinuously into the 1930s); the fortnightly record for Moruya Beach, New South Wales, from 1973 to the present; and the periodic profiling of the Florida coastline at 100 m intervals since 1972. These examples are isolated occurrences. Coastal geomorphologists are still not exactly certain why so much of the world's coastline is eroding. This section will examine some of the more popular means of explaining the causes of this erosion.

Sea level rise and the Bruun Rule

(Schwartz, 1967; Bryant, 1985)

The general emphasis on rising sea level – an assumption made without firm evidence – has led to this factor being perceived as the main mechanism causing eroding beaches. Beach retreat operates via the *Bruun Rule*, first conceived by Per Bruun, a Danish researcher, and formulated by Maurice Schwartz of the United States. The Bruun Rule states that, on beaches where offshore profiles are in equilibrium and net longshore transport is minimal, a rise in sea level must lead to retreat of the shoreline. The amount of retreat is not simply a geometrical effect. Bruun postulated that excess sediment must move offshore from the beach, to be spread over a long distance to raise the seabed and maintain the same depth of water as before (Figure 8.12). The Bruun Rule implies that beaches do not steepen over time as they erode, but maintain their configuration, be it barred, planar or whatever. Generally, a 1 cm rise in sea level will cause the shoreline to retreat 0.5 m. Numerous studies – ranging from field measurements at various coastal locations such as high-energy exposed tidal environments, open ocean beaches and confined lakes, to experiments in wave tanks – have substantiated this rate. The Bruun Rule operates by elevating the zone of wave attack higher up the beach. This is exacerbated by the fact that the baseline of the beach watertable is also raised. Water draining from the lower beach-face decreases sediment shear resistance, and enhances the possibility of sediment liquefaction, causing seaward erosion of sand and shoreline retreat. The liquefaction process is described in more detail in the chapters on earthquakes and land instability.

Unfortunately, while the Bruun Rule has been proven universally, there are limitations to its use. The

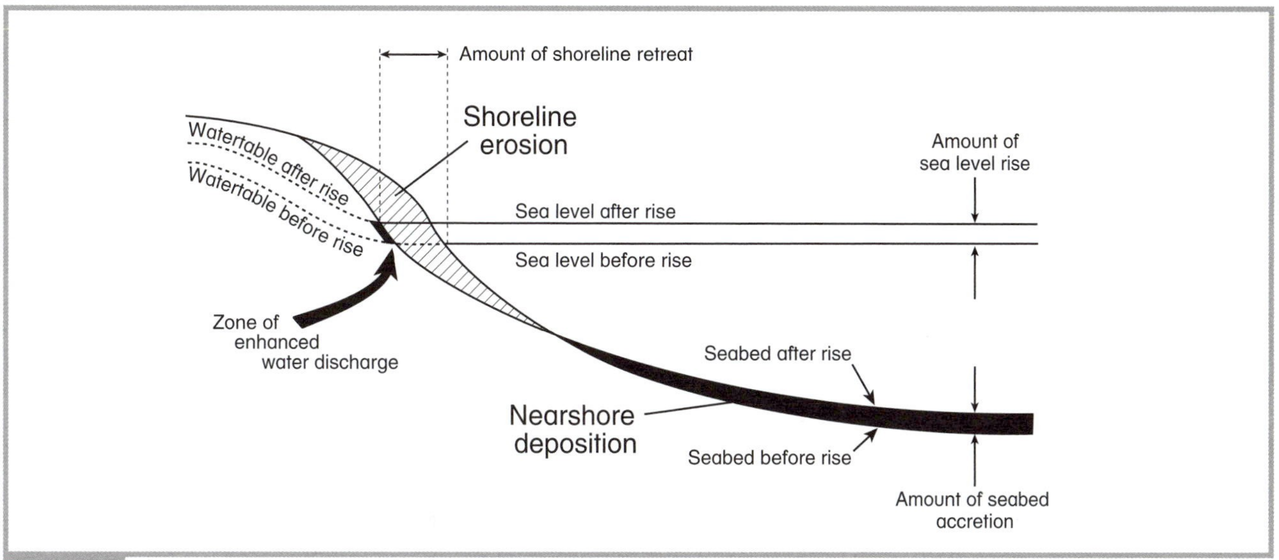

Fig. 8.12 The Bruun Rule for the retreat of a shoreline with sea level rise.

actual volume of sediment needed to raise the seabed as sea level rises depends upon the distance from shore and the depth from which waves can move that material. The Bruun Rule was formulated when it was believed that this depth was less than 20 m. Individual waves can move sediment offshore to depths exceeding 40 m, and well-defined groups or sets can initiate sediment movement out to the edge of the continental shelf. Coastlines, where Bruun's hypothesis was most successfully applied, include the east coasts of the United States and Japan, where sea level is rising rapidly. There are many areas of the world where sea level is rising at slower rates, or declining, and coastal erosion is still occurring. It would appear that sea level change as a major, long-term cause of beach erosion has evolved from evidence collected on, and applied with certainty to, only a small proportion of the world's coastline. A dichotomy also exists between measured rates of beach retreat and the present postulated eustatic rise in sea level. Common magnitudes of beach retreat in the United States, and elsewhere in the world, range from 0.5–1.0 m yr^{-1}. According to the Bruun Rule, shoreline retreat at this rate would require sea level rises three to seven times larger than the maximum rise of 3 mm yr^{-1} that may presently be occurring globally. Obviously, factors other than sea level rise must be involved in beach erosion.

The actual rise in sea level may be irrelevant to the application of the Bruun Rule. As shown above, sea level oscillations at all timescales are one of the most widely occurring aspects of sea level behavior. More sections of the world's coastline have significant oscillations than simply have rising sea level. The emphasis upon the Bruun Rule has been upon rising levels, but the rule can also predict accretion where sea level is falling. If the Bruun Rule is to account for long-term beach erosion in these locations, then what has to be taken into account is the imbalance between the amount of erosion during high sea level fluctuations and the amount of accretion when sea level drops. It takes a shorter time to remove material seaward during erosional, rising sea level phases – storm events – than it does to return sediment landward during accretional, falling sea level periods. If eroded sediment that has been carried seaward cannot return to the beach before the next higher sea level phase, then the beach will undergo permanent retreat. Bruun's hypothesis only provides the mechanism by which these shifts in sea level operate. The more frequent the change in sea level, the greater the rate of erosion.

Other causes of erosion

(Dolan & Hayden, 1983; Bryant, 1985, 1988; Dolan et al., 1987; Bird, 1996; Gibeaut et al., 2002)

Clearly, more than one factor can account for erosion of the world's coastlines. Table 8.2 lists these additional factors. Human impact on beach erosion is becoming more important, especially with this century's technological development in coastal engineering. In some countries such as Japan, breakwalls have been used as a ubiquitous solution to coastal erosion, to the extent that their use has become a major reason for continued erosion. Wave energy reflects off breakwalls, leading to offshore or *longshore drift* of sediment away from

Table 8.2 Factors recognized as controlling beach erosion (based on Bird, 1988).

Exclusively Human-induced

- Reduction in longshore sand supply because of construction of breakwalls.
- Increased longshore drift because of wave reflection off breakwalls behind, or updrift of, the beach.
- Quarrying of beach sediment.
- Offshore mining.
- Intense recreational use.
- Revegetation of dunes.

Exclusively Natural

- Sea level rise.
- Increased storminess.
- Reduction in sand moving shoreward from the shelf because it has been exhausted or because the beach profile is too steep.
- Increased loss of sand shoreward from the beach by increased wind drifting.
- Shift in the angle of wave incidence due to shifts in the location of average pressure cells.
- Reduction in sediment volume because of sediment attrition, weathering, or solution.
- Longshore migration of large beach lobes or forelands.
- Climatic warming in cold climates, leading to melting of permanently frozen ground and ice lens at the coastline.
- Reduction in sea-ice season leading to increased exposure to wave attack.

Either Human-induced or Natural

- Decrease in sand supply from rivers because of reduced runoff from decreased rainfall or dam construction.
- Reduction in sand supply from eroding cliffs, either naturally, or because they are protected by seawalls.
- Reduction in sand supply from dunes because of migration inland or stabilization.
- Increase in the beach watertable because of increased precipitation, use of septic tanks or lawn watering.

Source: Based on Bird, 1988

structures. Classic examples of breakwall obstruction or longshore sediment drift, leading to erosion downdrift, include the Santa Barbara harbor breakwall in California, and the Tweed River retaining walls south of the Queensland Gold Coast in eastern Australia. Offshore mining may affect refraction patterns at shore or provide a sediment sink to which inshore sediment drifts over time. Erosion of Point Pelée, Ontario, in Lake Erie, can be traced to gravel extraction some distance offshore. Many cases also exist, along the south coast of England, of initiation of beach erosion after offshore mining. Dam construction is very prevalent on rivers supplying sediment to the coastline. Californian beaches have eroded in many locations because dams, built to maintain constant water supplies, have also trapped sediment that maintains a stable, coastal sand budget. The best example is the case of the Aswan Dam in Egypt. Since its construction, the Nile Delta shoreline has retreated by 1 km or more. Even the simple act of revegetating dunes can cause shoreline retreat if the sand budget of a beach is fixed. Sand blown into vegetated dunes cannot return to the beach on offshore winds, so the beach slowly narrows over time.

Two little-known factors may be more important than any of the above. Each time someone walks across a beach, that person becomes a mini-excavator. A person who does not disrobe and endeavor to shake all clothing free of sand after swimming can remove 5–10 g of sand on each visit to the beach. On popular bathing beaches, this can amount to several tonnes per kilometer of beach per year. Secondly, since the Second World War rapid urban expansion near the coastline has led to modification of urban drainage channels and watertables. Water pumped from coastal aquifers to supply urban water may not only cause local subsidence – as is the case for most cities along the United States east coast – but may also lead to accelerated beach erosion because that water may raise local watertables if it is passed through septic systems or used to water lawns.

Natural factors are perhaps still the major causes of beach erosion. Long-term changes have been best documented on beach forelands along the Channel and east coast of England at Dungeness, Orford Ness, and Scolt Head Island. These forelands are tens of kilometers wide and appear to have developed since the Holocene rise in sea level, following the last glaciation. However, some are slowly migrating along shore. Material is eroded from the updrift ends and deposited as a beach ridge on the downdrift end. In addition, shorter term beach erosion has manifested itself in the twentieth century, mainly because of the change in global climate that has occurred since 1948. This change has not been one of magnitude as much as one of variation: the storms follow longer periods of quiescence, record rain and floods break severe droughts, and the coldest winter on record is followed by the warmest summer. Unfortunately, investigations of this climate shift have focused on temperature change, whereas the shift also involves dramatic changes in coastal rainfall regimes, and changes in pressure cell location leading to an upsurge in storminess.

While sea level rise enhances beach erosion by elevating beach watertables, logically any factor affecting watertables can also cause beach erosion. For instance, a sandy backshore absorbing rainfall acts as a time-delaying filter for subsequent discharge of this water to the sea through the lower foreshore. The process can raise the watertable substantially for periods up to two months after heavy rainfalls. Many coastlines, especially those where seasonal orographic rainfall is pronounced, can receive over 1000 mm of precipitation within a month. Under these intense rainfalls, even sand suspension on the shore face by groundwater evulsion is possible. Rainfall may be better related to beach change than sea level. This aspect can be shown for Stanwell Park Beach (40 km south of Sydney, Australia): a compartmentalized, exposed, ocean beach with no permanent longshore leakage of sediment. The beach over time has oscillated between extreme erosion and accretion, but with no long-term trend in either state (Figures 2.11 and 8.13). Figure 8.14 plots the relationship defined between average deviations from the mean high tide position and (a) monthly sea level and (b) monthly rainfall for this beach between 1895 and 1980. Rainfall explains more of the change in shoreline position over time than does sea level (8.7 per cent versus 6.6 per cent). Despite all of the above factors, the most logical cause of beach erosion is increased storminess. Figure 8.14(c) plots, between 1943 and 1980, the relationship between the high tide position and accumulated heights of storm waves for the previous year on Stanwell Park Beach. Each 1 m increase in storm wave height led to 0.47 m of beach retreat – a value equivalent to the effect generated by a 1 cm rise in mean sea level. The relationship is twice as strong as that produced by rainfall and/or a rise in sea level. In fact, storms are often accompanied by higher sea levels and heavy rainfall. Storms concurrent with these phenomena can explain 40 per cent of the variation in shoreline position at Stanwell Park Beach. In contrast, periods without storms favor beach accretion.

Fig. 8.13 Difference in beach volume, Stanwell Park, Australia. Above, the beach after sustained accretion in February 1982. Below, the same beach when eroded severely by a storm in August 1986 (photograph courtesy of Dr Ann Young, Department of Geography, University of Wollongong).

This effect is also recognizable elsewhere. Evidence from the barrier island coastline of eastern North America, between New Jersey and South Carolina, not only indicates large spatial variation in long-term rates of erosion of barrier islands induced by storms, but also shows that particularly large storms tend to erode or overwash the shoreline more in the places having the

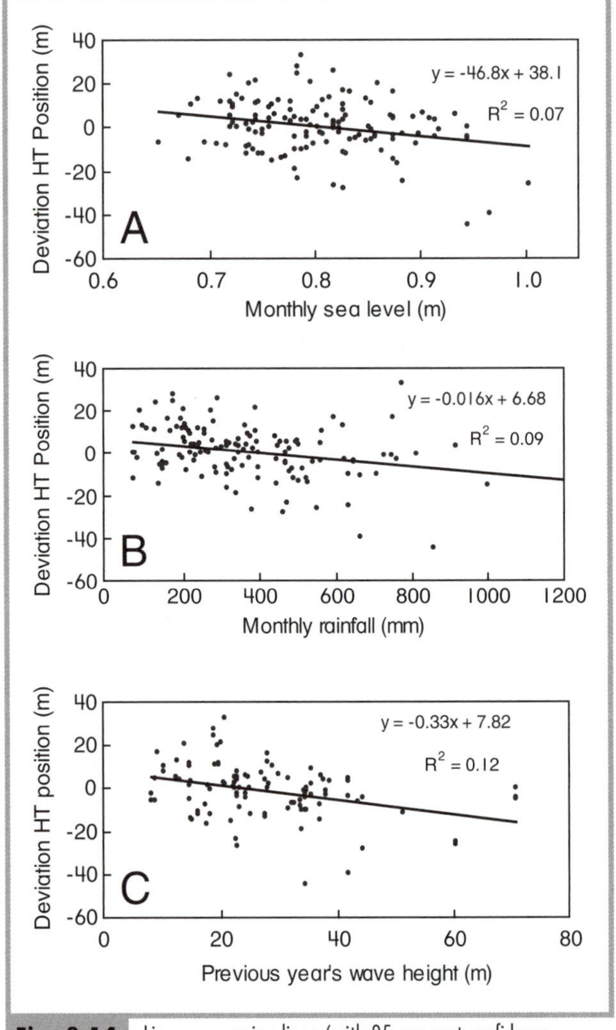

Fig. 8.14 Linear regression lines (with 95 per cent confidence limits) between average deviation from the mean high tide position on Stanwell Park Beach, Australia and A) monthly sea level, B) monthly rainfall, and C) height of previous year's storm waves (Bryant 1985, 1988). Graphs A) and B) are for the period 1895–1980, while graph C) is for the period 1943–1980.

highest historical rates of erosion produced by lesser events. Beaches along this section of coastline are eroding on average 1.5 m yr^{-1}. Islands with more southerly exposure have slower rates of erosion, a fact implying that the direction of storm tracking and wave approach are important. Along this coastline between 1945 and 1985, waves exceeded 3.4 m in height an average of 64 hours annually. East-coast lows and strong onshore winds generated by anticyclones accounted for 13 and 24 per cent, respectively, of this total. Most storm waves were produced by storms tracking along the polar front eastward across the continent, especially south of Cape Hatteras. The Great Ash Wednesday Storm of 1962 had little effect upon long-term erosion over this 42-year period, with its impact embedded in the overall storm wave climatology.

In contrast, normally low-energy coastlines affected periodically by big storms respond differently. Here, rare but big storms generate waves with enough energy to exceed a threshold for complete beach erosion. Beaches undergoing this type of change occur in the Gulf of St Lawrence and the Gulf of Mexico where tropical storms are the main cause of shoreline retreat of barrier islands. Along the Texas coast of the Gulf of Mexico, tropical storms frequently affect the coastline. For example, between 1994 and 1998, eight tropical storms struck the north-western Gulf. However, only tropical storm Josephine in October 1996 generated waves that exceeded the threshold for significant erosion by overwashing vegetated dunes. In this low-energy environment, this threshold is exceeded by any storm generating a storm surge of 0.9 m and waves in excess of 3 m lasting more than 12 hours. Smaller storms, even though they can persist five times longer, do not cross this threshold. In the Gulf of St Lawrence, *episodic* events that change the barrier coastline are even rarer because this region lies at the northern boundary of hurricanes, which here occur as infrequently as once every twenty-five years. Barrier shorelines remain spatially stable for years only to undergo landward retreat by tens of meters during these events. More dramatic is the puncturing of barriers to form temporary tidal inlets that then infill slowly over the following decade.

CONCLUDING COMMENTS

While sea level rise may be responsible for recent sandy beach erosion (the Bruun Rule), this factor may be subservient to the role played by rainfall and storms. Rainfall controls beach position through its influence on the watertable – as does rising sea level. Storms are a continuous process operating over the time span of decades or longer. Individual storms are not necessarily responsible for beach erosion. However, if storms occur in clusters, then beaches erode because there is less time for beach recovery before the next storm. Three variables: rainfall, sea level and storms, form a suite of interrelated variables. Analysis of data for Stanwell Park Beach illustrates this aspect. Groups of environmental parameters that individually can be linked to erosion, collectively and logically, operate together. Furthermore, changes in these environmental parameters reflect the large-scale climatic change that the Earth is presently undergoing. The more dramatic and measurable increases in temperature have unfortunately detracted from the more significant nature of that change, namely the increasing variability of climate since 1948. This increase accelerated in the 1970s and shows no sign of abating. The world's beaches will continue to respond to this change; unfortunately, it will be in an erosive manner.

Finally, a note of caution: this chapter has cast doubt on the relevance of emphasizing a uniform, worldwide rising sea level, and shown that this factor is not solely responsible for eroding beaches. Erosion of the world's beaches may not be so common, or so permanent, as generally believed. For example, when averaged over the long term, the high tide record for Stanwell Park Beach (Figure 2.11) is stable even though sea level off this coast is currently thought to be rising by 1.6 mm yr^{-1}; rainfall has increased about 15 per cent since 1948; and there are no presently active sediment sources. In Florida, detailed surveys at 100 m intervals around the coast have shown that beaches did not erode between 1972 and 1986, even though sea level rose and rainfall increased and sand supplies remained meager. Here, the absence of storm activity over this time span may be the reason for stability.

REFERENCES AND FURTHER READING

Anisimov, O. and Fitzharris, B. 2001. Polar regions (Arctic and Antarctic). In McCarthy, J.J., Canziani, O.F., Leary, N.A., Dokken, D.J. and White, K.S. (eds.) *Climate Change 2001: Impacts, Adaptation and Vulnerability – Contribution of Working Group II to the Third Assessment Report of IPCC*. Cambridge University Press, Cambridge, pp. 801–842.

Beer, T. 1983. *Environmental Oceanography*. Pergamon, Oxford.

Bird, E. C. F. 1996. *Beach Management (Coastal Morphology and Research)*. Wiley, Chichester.

Bruun, P. (ed.) 1985. *Design and Construction of Mounds for Breakwalls and Coastal Protection*. Elsevier, Amsterdam.

Bryant, E.A. 1985. Rainfall and beach erosion relationships, Stanwell Park, Australia, 1895–1980: worldwide implications for coastal erosion. *Zeitschrift für Geomorphologie Supplementband* 57: 51–66.

Bryant, E.A. 1987. CO_2-warming, rising sea level and retreating coasts: review and critique. *Australian Geographer* 18(2): 101–113.

Bryant, E.A. 1988. The effect of storms on Stanwell Park, NSW beach position, 1943–1980. *Marine Geology* 79: 171–187.

Bryant, E.A. 1993. The magnitude and nature of 'noise' in world sea level records. In Chowdhury, R.N. and Sivakumar, M. (eds) *Environmental Management, Geo-Water & Engineering Aspects*. Balkema, Rotterdam, pp. 747–751.

Cazenave, A., Do Minh, D., Cretaux, J.F., Cabanes, C., and Mangiarotti S. 2002. *Interannual sea level change at global and regional scales using Jason-1 altimetry*. <http://topex-www.jpl.nasa.gov/science/invest-cazenave.html>

Colorado Center for Astrodynamics Research 2003. *Long term sea level change: global*. <http://ccar.colorado.edu/research/gmsl/PDF/msl_global_noib.pdf> (URL defunct as of 2004)

Dalrymple, R. 2000. University of Delaware Wave Calculator. <http://www.coastal.udel.edu/faculty/rad/wavetheory.html>

Defant, A. 1961. *Physical Oceanography* v. 1. Pergamon, New York.

Dolan, R. and Hayden, B. 1983. Patterns and prediction of shoreline change. In Komar, P.D. (ed.) *CRC Handbook of Coastal Processes and Erosion*. CRC Press, Boca Raton, pp. 123–150.

Dolan, R., Hayden, B., Bosserman, K. and Isle, L. 1987. Frequency and magnitude data on coastal storms. *Journal of Coastal Research* 3(2): 245–247.

Gibeaut, J.C., Gutierrez, R., and Hepner, T.L. 2002. Threshold conditions for episodic beach erosion along the Southeast Texas Coast. *Transactions of the Gulf Coast Association of Geological Societies* 52: 323–335.

Gornitz, V. 2000. Impoundment, groundwater mining, and other hydrologic transformations: Impacts on global sea level rise. In Douglas, B.C., Kearney, M.S., and Leatherman, S.P. (eds) *Sea Level Rise: History and Consequences*. Academic Press, San Diego, pp. 97–119.

Gross, M.C. 1972. *Oceanography: A View of the Earth*. Prentice-Hall, Englewood Cliffs, New Jersey.

Jet Propulsion Laboratory 1995a. *'TOPEX/POSEDON: Wave height, January'*. <http://education.gsfc.nasa.gov/experimental/all98invProject.Site/Pages/trl/inv47WAVE_HEIGHT_JAN.html>

Jet Propulsion Laboratory 1995b. *'TOPEX/POSEDON: Wave height, July'*. <http://education.gsfc.nasa.gov/experimental/all98invProject.Site/Pages/trl/inv4-7WAVE_HEIGHT_JUL.html>

Kovacs, A. and Sodhi, D.S. 1978. Shore Ice Pile-up and Tide-up: Field Observations, Models, Theoretical Analyses. United States Army Corps Eng. Cold Regions Research and Engineering Laboratory Report.

Kushnir, Y., Cardone, V.J., Greenwood, J.G., and Cane, M. 1997. On the recent increase in North Atlantic wave heights. *Journal of Climate* 10: 2107–2113.

Lawson, G. 2001. Monsters of the deep. *New Scientist* 30 June: 28–32.

Long, D.G., Ballantyne, J., and Bertoia, C. 2002. Is the number of Antarctic icebergs really increasing?. *EOS, Transactions – American Geophysical Union* 83: 469, 474.

Nerem, R. S., Mitchum, G.T., Giese, B.S., Leuliette, E.W., and Chambers, P. 2002. *An investigation of very low frequency sea level change using satellite altimeter data*. <http://topex-www.jpl.nasa.gov/science/invest-nerem.html> (URL defunct as of 2004)

PO.DAAC (Physical Oceanography Distributed Active Archive Center) 2003. *NEREIDS ocean topography overview*. Jet Propulsion Laboratory, Californian Institute of Technology <http://nereids.jpl.nasa.gov/cgi-bin/ssh.cgi?show=overview>

Sahagian, D.L., Schwartz, F.W., and Jacobs, D.K. 1994. Direct anthropogenic contributions to sea level rise in the twentieth century. *Nature* 367: 54–57.

Schwartz, M.L. 1967. The Bruun theory of sea-level rise as a cause of shore erosion. *Journal of Geology* 75: 76–92.

Skov, N.A. 1970. The ice cover of the Greenland Sea. *Meddelelser om Grønland* 188(2).

Taylor, R.B. 1977. The occurrence of grounded ice ridges and shore ice piling along the northern coast of Somerset Island, N.W.T. *Arctic* 31(2): 133–149.

Vinje, T. 2001. Anomalies and trends of sea ice extent and atmosphere circulation in the Nordic Seas during the period 1864–1998. *Journal of Climate* 14: 255–267.

Wiegel, R.L. 1964 *Oceanographic Engineering*. Prentice-Hall, Englewood Cliffs, New Jersey.

Zebrowski, E. 1997. *Perils of a Restless Planet: Scientific Perspectives on Natural Disasters*. Cambridge University Press, Cambridge.

PART 2

GEOLOGICAL HAZARDS

Causes and Prediction of Earthquakes and Volcanoes

CHAPTER 9

INTRODUCTION

Of all natural hazards, earthquakes and volcanoes release the most energy in the shortest time. In the past 40 years, scientists have realized that the distribution of earthquakes and volcanoes is not random across the Earth's surface, but tends to follow crustal plate boundaries. In the past 20 years, research has been dedicated to monitoring these regions of crustal activity with the intention of predicting – several days or months in advance – major and possibly destructive events. At the same time, planetary studies have led to speculation that the clustering of earthquake or volcanic events over time is not random, but tends to be cyclic. This knowledge could lead to prediction of these hazards decades in advance. Before examining these aspects, it is essential to define how earthquake intensity is measured, because earthquakes are always characterized by their magnitude. This aspect will be examined first, followed by a description of the distribution of earthquakes and volcanoes over the Earth's surface and some of the common causes of these natural disasters. The chapter concludes with a discussion on the long- and short-term methods for forecasting earthquake and volcano occurrence.

SCALES FOR MEASURING EARTHQUAKE INTENSITY

(Holmes, 1965; Wood, 1986; Bolt, 1993; National Earthquake Information Center, 2002)

Seismic studies were first undertaken as early as 132 AD in China, where crude instruments were made to detect the occurrence and location of earthquakes. It was not until the end of the nineteenth century that these instruments became accurate enough to measure the passage of individual *seismic* waves as they traveled through, and along, the surface of the Earth. The characteristics of these waves will be described in more detail in the following chapter. The magnitude or intensity of these waves is commonly measured using scales based on either measurements or qualitative assessment. The first measured seismic scale was the M_s scale established by Charles Richter in 1935. Richter defined the magnitude of a local earthquake, M_L, as the logarithm to base ten of the maximum seismic wave amplitude (in thousandths of a millimeter) recorded on a standard seismograph at a distance of 100 km from the earthquake epicenter. The scale applied to Californian earthquakes and to the maximum amplitude of any type of seismic wave. Each unit increase on this scale represented a tenfold increase in the amplitude of ground motion. The scale was later standardized for any earthquake using seismometers at any distance from the epicenter and surface (Rayleigh) waves at a period of 18–22 seconds and a wavelength of 60 km. The characteristics of seismic waves will be described in more detail in the following chapter. This magnitude scale is called surface magnitude or M_s. Alternatively, the amplitude of the initial seismic wave (P-wave), m_b, having a period of 1–10 seconds can be used, because it is not affected by the focal depth of the source of the earthquake. This measure is called the body-wave magnitude. The body-wave magnitude can be related to the M_s scale as follows:

$$m_b = 2.9 + 0.56\, M_s \tag{9.1}$$

Overall the m_b value is less representative than the M_s one of earthquake magnitude. The energy released by an earthquake can be related to these scales as follows:

$$\log_{10} E = 4.8 + 1.5M_s \tag{9.2}$$

$$\log_{10} E = 5.0 + 2.6M_b \tag{9.3}$$

The energy values are defined in ergs. For each unit increase in surface magnitude, M_s, energy increases by 31.6 times. None of these scales is appropriate for large seismic events. Here, the seismic moment, M_w, is now used. This scale is based upon the surface area of the fault displacement, the average length of movement and the rigidity of the rocks fractured.

Table 9.1 lists the annual frequency of earthquakes over the twentieth century on the M_s scale, while Table 9.2 lists the largest events over the same period. No historical earthquake has exceeded a magnitude of 9.0 on the M_s scale. An earthquake larger than M_s= 8.0 occurs about once a year. Its impact is catastrophic. About 18 earthquakes occur each year in the 7.0–7.9 range. These earthquakes are deadly near population centers. About 120 and 800 earthquakes occur annually in the 6.0–6.9 and 5.0–5.9 ranges, respectively. The former earthquakes are deadly near population centers with inferior construction standards, while the latter can still cause significant property losses. The largest earthquakes appear to occur along the western edges of the North and South American Plates, and in China (Figure 9.1).

The M_s scale requires instrumentation. It cannot be used to assess the magnitude of prehistoric earthquakes or ones occurring in countries without a network of seismic stations. Instead, the Mercalli scale (or its modified form) is used. This scale, summarized in Table 9.3, can be related approximately to the M_s scale. The Mercalli scale qualitatively estimates the strength of an earthquake using descriptions of the type of damage that has occurred close to the origin or *epicenter* of the earthquake in built-up areas. Determining

Table 9.1 Annual frequency of earthquakes by magnitude on the M_s scale.

Magnitude	Frequency
>8.0	1
7.0–7.9	18
6.0–6.9	120
5.0–5.9	800
4.0–4.9	6200
3.0–3.9	49000
2.0–2.9	365000

Table 9.2 Some of the largest earthquakes, by magnitude, on the M_s scale.

Date	Location	Magnitude, M_s	Death-toll
1 November 1755	Lisbon, Portugal	8.7	70 000
31 January 1906	Andes (Columbia)	8.6	500
16 August 1906	Valparaiso, Chile	8.6	20 000
18 April 1906	San Francisco, US	8.3	3 000
3 January 1911	Tienshan, Kazakhstan	8.4	452
16 December 1920	Kansu, China	8.6	200 000
1 September 1923	Tokyo, Japan	8.3	143 000
3 March 1933	Japanese Trench	8.1	3 064
15 August 1950	Assam, India/China	8.6	1 526
22 May 1960	Chile	8.5	2 231
27 March 1964	Alaska	8.6	100
26 January 2001	Gujarat, India	7.6	20 000

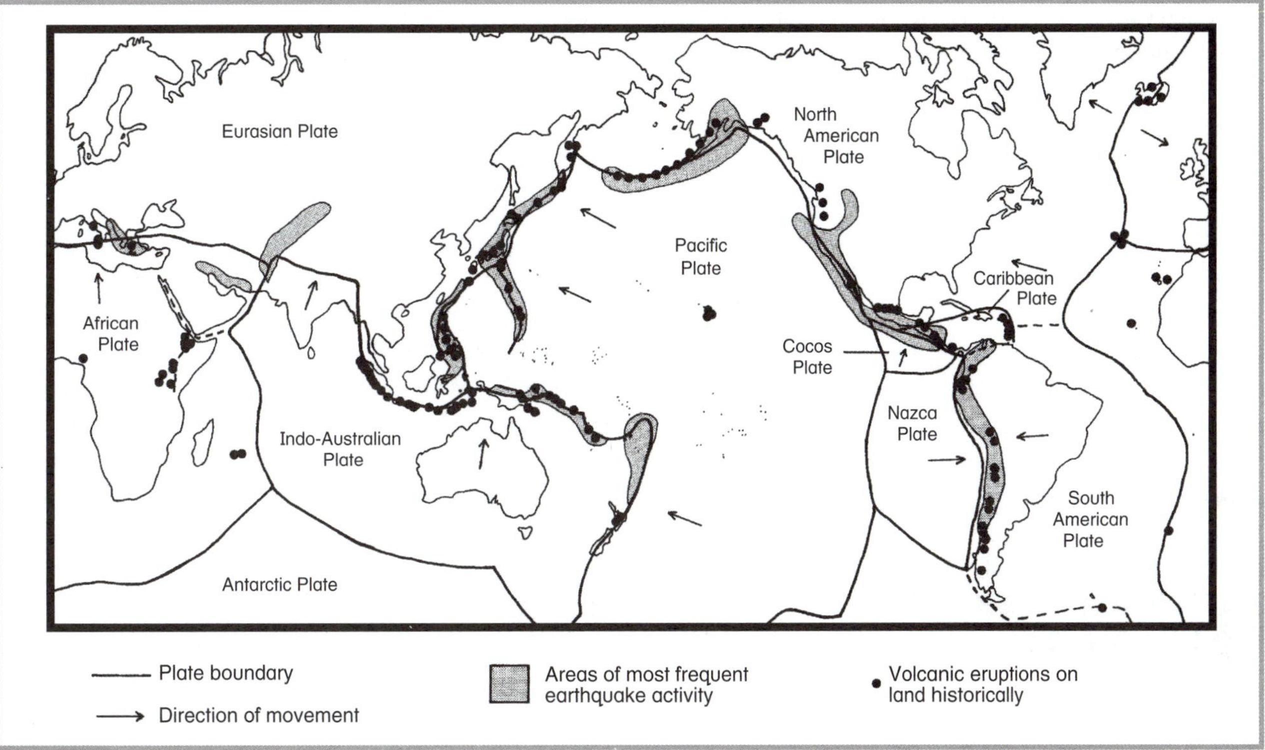

Fig. 9.1 Distribution of plate boundaries, intense earthquakes and historical land-based volcanic eruptions (based on Press & Siever, 1986; Bolt, 1993).

Table 9.3 The Mercalli scale of earthquake intensity.

Scale	Intensity	Description of effect	Maximum Acceleration in mm s^2	Corresponding Richter Scale
I	Instrumental	detected only on seismographs	<10	
II	Feeble	some people feel it	<25	
III	Slight	felt by people resting; like a large truck rumbling by.	<50	<4.2
IV	Moderate	felt by people walking; loose objects rattle on shelves.	<100	
V	Slightly Strong	sleepers awake; church bells ring.	<250	<4.8
VI	Strong	trees sway; suspended objects swing; objects fall off shelves.	<500	<5.4
VII	Very Strong	mild alarm; walls crack; plaster falls.	<1000	<6.1
VIII	Destructive	moving cars uncontrollable; chimneys fall and masonry fractures; poorly constructed buildings damaged.	<2500	
IX	Ruinous	some houses collapse; ground cracks; pipes break open.	<5000	<6.9
X	Disastrous	ground cracks profusely; many buildings destroyed; liquefaction and landslides widespread.	<7500	<7.3
XI	Very Disastrous	most buildings and bridges collapse; roads, railways, pipes, and cables destroyed; general triggering of other hazards.	<9800	<8.1
XII	Catastrophic	total destruction; trees driven from ground; ground rises and falls in waves.	>9800	>8.1

the magnitude of an earthquake on this scale does not depend upon seismographs, but instead upon actual field reports and pictures of damage to cultural features, mainly buildings. The scale has some quantitative basis in that the categories on the scale increase with increasing acceleration of the shock wave through the crust. When the maximum acceleration exceeds 9.8 m s^{-2}, which is the acceleration due to gravity, the shock wave of the earthquake hits the Earth's surface so hard that objects are tossed up into the air and trees can be physically hammered from the ground. These accelerations can be used to construct maps of *seismic risks*, which will be described in more detail in the following chapter.

DISTRIBUTION OF EARTHQUAKES AND VOLCANOES

(Hodgson, 1964; Doyle et al., 1968; Bolt et al., 1975; Blong, 1984)

Figure 9.1 shows the global distribution of areas of most frequent earthquake activity together with historically active volcanoes on land. Superimposed on this map are the locations of active plate boundaries. The most striking feature is the correspondence between the occurrence of these two hazards and the boundaries of crustal plates. Active volcanoes occur in three locations: near convergent plate margins, at divergent plate margins and over *mantle* hot spots. Briefly, convergent plate boundaries are either *subduction zones*, where two sections of the Earth's crust are colliding with one plate being consumed or subducting beneath the other; or mountain-building zones, where two or more plates are colliding with one overriding the other. The ocean *trenches* and the Pacific *island arcs* occur above subduction zones, while the Himalayas and European Alps occur within mountain-building zones (see Figure 9.2 for the location of all major placenames mentioned in this chapter). Although both these regions give rise to earthquakes, subduction zones are far more important in accounting for the majority of earthquakes and almost all explosive volcanism. While volcanoes at these locations produce only 10–13 per cent of the *magma* reaching the Earth's surface, they are responsible for 84 per cent of known eruptions and 88 per cent of eruptions with fatalities. Most of these eruptions occur around the Pacific Ocean, along what is termed the Pacific 'ring of fire'. Plate boundaries through Indonesia, Italy, and New Zealand have the longest record of activity. Between 1600 and 1982, 67 per cent of all known deaths resulting from volcanoes occurred in Indonesia, on the convergent boundary of the Indo–Australian and Eurasian Plates. Mantle hot spots are the next most important region of volcanic activity. The most notable hot spot includes the Hawaiian Islands, which lie in the middle of the Pacific Plate. Most divergent plate margin volcanoes occur along mid-ocean ridges where the crust is separating. The latter is the cause of volcanism in Iceland. Volcanoes on

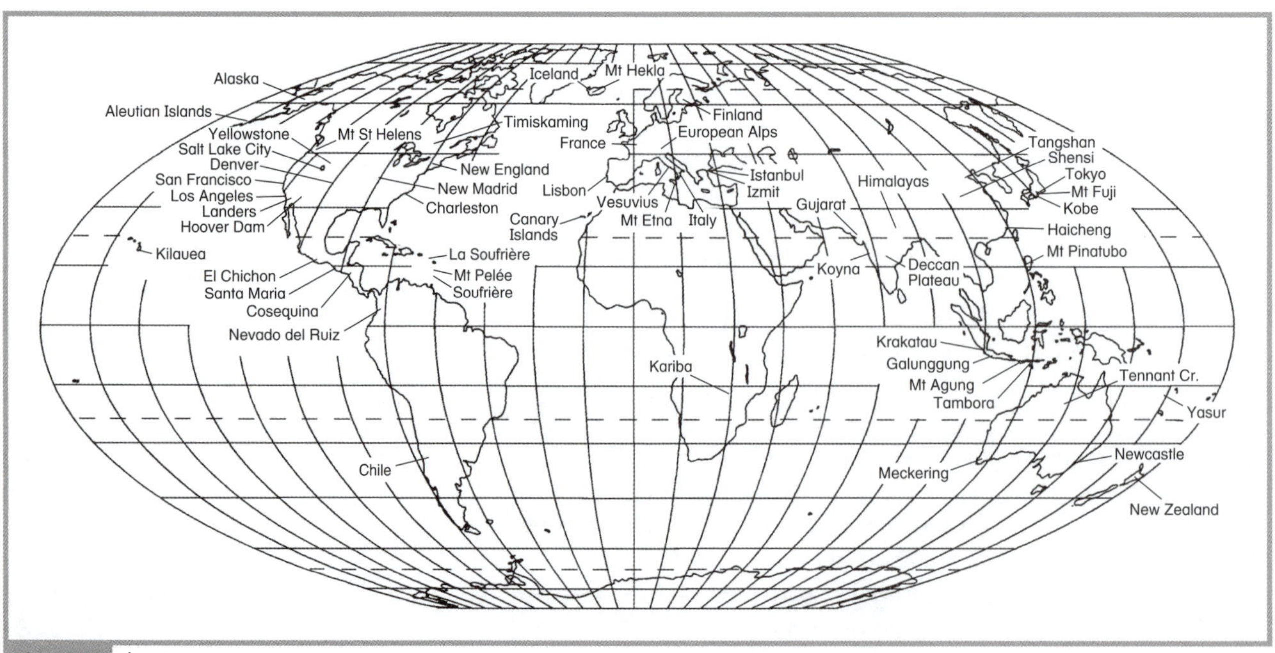

Fig. 9.2 Location map.

plate boundaries or over hot spots are usually associated with earthquakes.

About 75 per cent of earthquake energy is released in the upper 60 km of the crust along plate boundaries. These are termed *shallow-focus* events compared to earthquakes that occur up to 700 km below the crust beneath subduction zones. Little if any seismic activity originates at depths greater than 700 km. There are however a significant number of earthquakes that are not associated with either volcanic activity or plate boundaries, but in fact have an *intra-plate* origin. Many intra-plate earthquakes are not associated with any known faults or historical earthquake activity. For example, in China, the earthquake record goes back to 780 BC, and shows that some of the largest and most destructive earthquakes have occurred in a belt from Tibet to Korea, as much as 1000 km from the nearest plate boundary. The most destructive earthquake in terms of loss of life occurred on 23 January 1556 in Shensi province on the Hwang Ho River. These intra-plate earthquakes result from crustal stresses due to continued uplift in the Himalayas, to sinking in the Shensi *geosyncline* or movement along the north-western edge of the north China plain. The Tangshan earthquake of 28 July 1976, which took 250 000 lives, occurred in this latter region. In North America, such intra-plate earthquakes occured at Timiskaming in 1935 in the middle of the Canadian Shield, at New Madrid along the Mississippi River between 16 December 1811 and 7 February 1812, and at Charleston, South Carolina, on 31 August 1886. The most recent intra-plate earthquake occurred on 29 January 2001 at Gujarat, India. It killed 50 000 people.

Of all locations that should be *aseismic*, the Australian continent continues to register the most intra-plate earthquakes. The Tennant Creek, Northern Territory, earthquakes of 22–23 January 1988 are the largest recorded in Australia, registering slightly more than 7.2 on the Richter scale. Before these, the Meckering earthquake of 1968 in Western Australia was the largest, registering 6.8. This was one of the few earthquakes in Australia to be associated with observed faulting. Until December 1989, no one had been killed by these earthquakes in Australia and property damage was minimal. The Newcastle earthquake of 28 December 1989 changed this dramatically. In ten seconds, an earthquake of magnitude 5.5 on the Richter scale shook the center of Newcastle, New South Wales. It killed 12 people and left a damage bill of close to $A1000 million. Almost all of these earthquakes occurred within the upper crust. Shallow earthquakes undoubtedly represent strain release of energy by fracturing. The source of stress probably originates from the thermal imbalance in the upper mantle and may be related to convective currents under Australia.

Figure 9.3 compares the return interval in years for earthquakes of various magnitudes between California, a seismically active region, and the supposedly aseismic

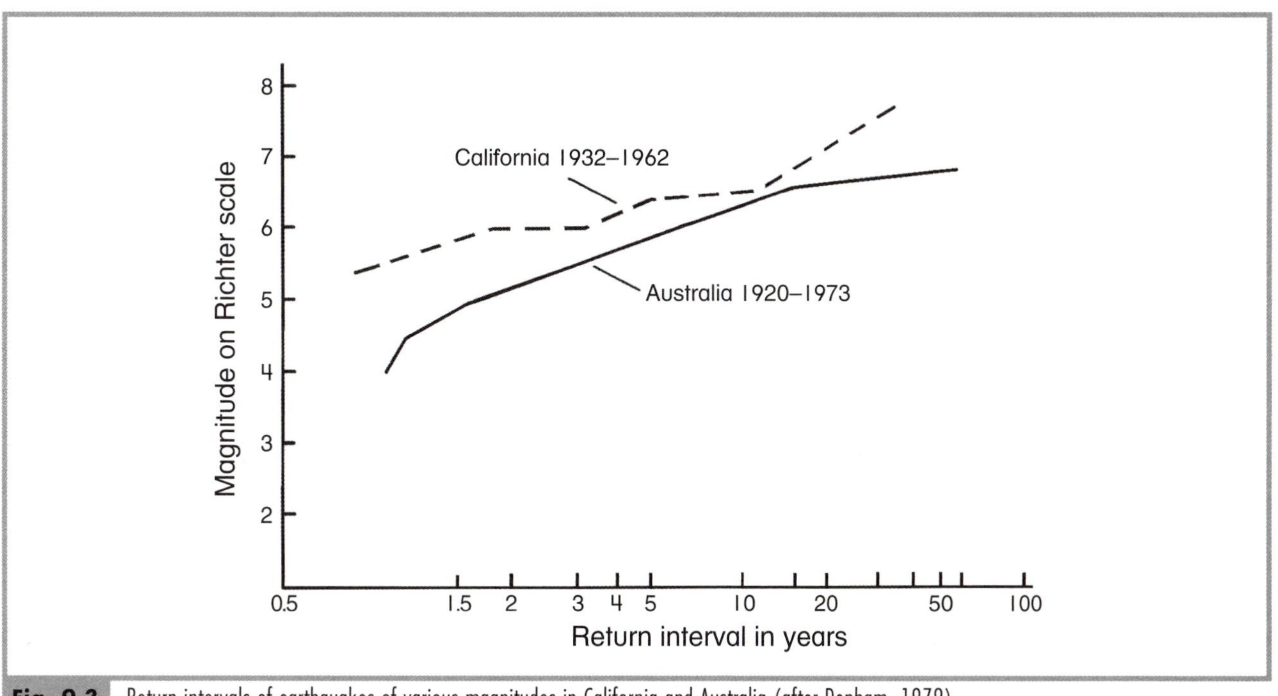

Fig. 9.3 Return intervals of earthquakes of various magnitudes in California and Australia (after Denham, 1979).

Australian continent. The data for Australia do not fit a straight line because the most seismic part of this continent is so isolated, with some events going unrecorded. In any 100-year period, Australia can expect an earthquake of magnitude 6.5 or larger. The Meckering earthquake of 1968 and the Tennant Creek earthquake of 1988 in fact exceeded this value. In the last 75 years, there have been 18 earthquakes that have exceeded 6 on the Richter scale. The 1 in 5-year event has a magnitude of 5.8 on this scale. In contrast, California can expect an earthquake in excess of 8 on the Richter scale at least once in 100 years. Several historical earthquakes, including the San Francisco earthquake of 1906, have exceeded this value. In California, the 1 in 5-year event exceeds 6.4 on the Richter scale. The mean annual maximum earthquake in California is about 0.5 orders of magnitude larger than that in Australia. This difference becomes more substantial when it is realized that California is about one-twentieth the size of Australia.

This review on the distribution of earthquakes emphasizes three facts. Firstly, while earthquakes are most likely to occur along plate boundaries, the largest earthquakes in terms of death toll and destruction have not been associated with plate margins. Secondly, no continent or region can be considered aseismic. Even those that are perceived as being aseismic have a remarkably high incidence of earthquakes above 6 on the Richter scale. And finally, while a major damaging earthquake is imminently expected along the San Andreas Fault, major earthquakes have occurred in the eastern part of North America, which if they occurred near populated centers today would be just as destructive as expected ones in California.

CAUSES OF EARTHQUAKES AND VOLCANOES

(Holmes, 1965; Bolt et al., 1975; Whittow, 1980; Wood, 1986; Ritchie & Gates, 2001; Matthews et al., 2002; Heki, 2003; Jones, 2003)

Plate boundaries

Most volcanism occurs along the plate boundaries in subduction zones, and along lines of crustal spreading such as the mid-Atlantic ridge. The theory of continental drift states that the crust in the center of oceans spreads because of diverging convection cells in the mantle. These cells rise towards the Earth's crust and spread out at the surface, resulting in a ridge with a rift in the middle (Figure 9.4). This rift is infilled with molten mantle material, which at weak points or fractures can form volcanoes. As the crust continues to spread apart, volcanoes migrate away from the center of the ridging. If conduits joining the volcanoes to the magma pool can remain open, then the volcanoes will remain active. However, if the crust spreads faster than *lava* can be fed to the volcano, the conduits block, the volcano becomes extinct, and newer ones develop closer to the rift.

Where plate boundaries are colliding, a different process occurs. Part of one plate may be driven beneath another (Figure 9.5). This plate drags down with it crustal material and forms a trench. The crustal

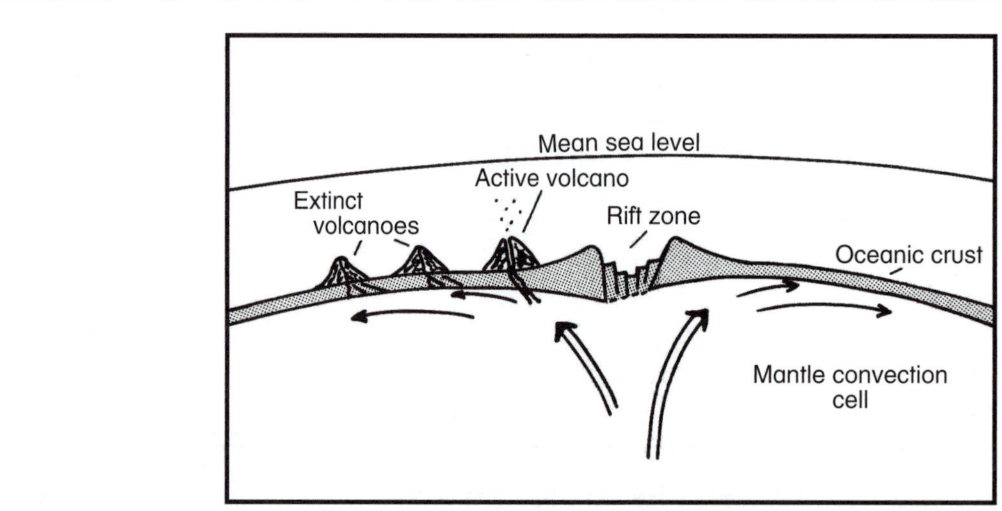

Fig. 9.4 Origin of volcanism associated with mid-ocean plate spreading.

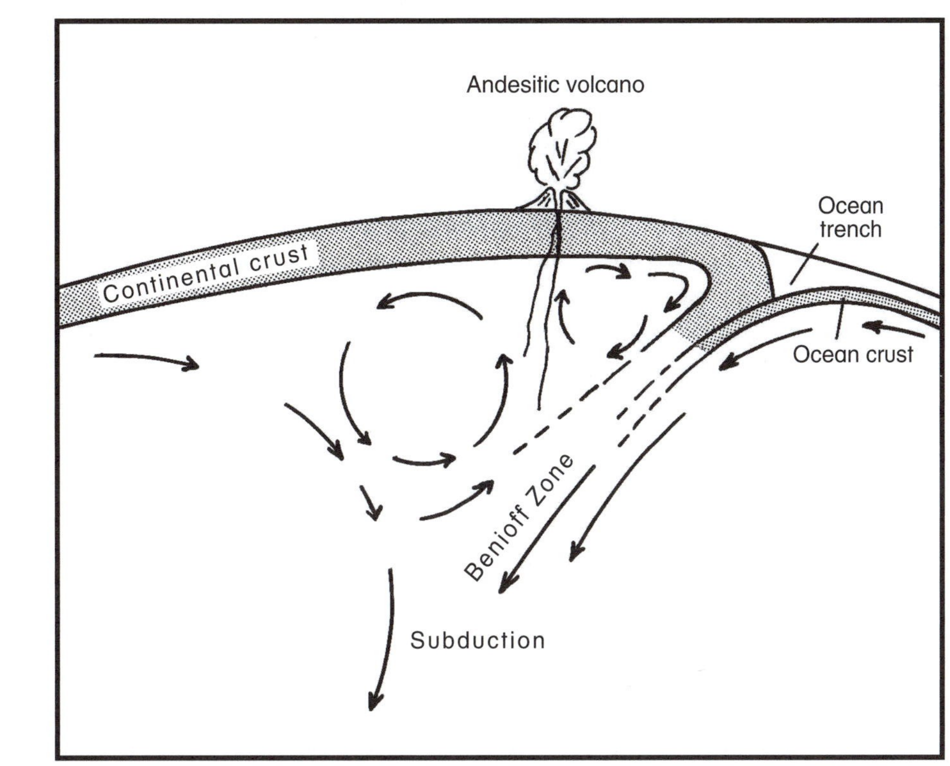

Fig. 9.5 Origin of volcanism over a subduction zone (after Holmes, 1965).

material is melted and incorporated into the mantle. A large zone of thrusting, termed the *Benioff zone*, forms to depths of 700 km with intense heat generation. This zone builds up heat because of magma extrusion, friction generated by earthquake activity and radioactive decay of crustal sediments. Beneath the overriding crust, less dense, heated magma (mainly *andesite* and *rhyolite*) rises from the Benioff zone into the upper layers of the mantle, before breaking sporadically through the crust to build volcanic cones. This magma contains gases derived from contact with groundwaters or from chemical reactions in dehydrated parts of the crust that are being subducted. The western Pacific Ocean ring of fire represents the consumption of the Pacific Plate by the Eurasian and Indo–Australian Plates. A chain of volcanoes – an island arc – forms parallel to the subduction zone. The Japanese, Indonesian, and Aleutian Islands are classic examples of volcanic island arcs built in this manner.

The Benioff zone is also a major zone of earthquake activity because it represents thrust faulting of two plates over a wide area. Deep-focus earthquakes at 70–700 km depth occur mainly beneath the continental side of the islands surrounding the Pacific and Indian Oceans. Over 75 per cent of all earthquakes occur in the plate rim of the Pacific Ocean. The Andes Mountains can be included in this category as they represent an uplifted version of an island arc. As the depth of focus increases in the Benioff zone, the epicenters of deep earthquakes move further inland towards the centers of volcanic activity. At times of volcanic eruptions, earthquakes may occur because of volcanic explosions, shallow magma movements within the crust, or sympathetic tectonic earth movements. The first two factors result from release of pressures in the crust, or changes due to expansion or deflation of the magma volume.

Plate collisions can also result in one continental crust overriding another, as is the case with the Indo–Australian Plate overriding the Eurasian Plate. Few if any volcanoes are generated in this situation because the crust of the Earth generally thickens at these locations, and fracturing down to the mantle is rare. Plates can also slide past each other as in the case of the Pacific and North American Plates along the San Andreas Fault. Not only earthquakes but also volcanoes can develop because fractures, called 'transcurrent faults', open up at right angles to the line of activity. Most of the volcanic activity in the Rocky Mountains and the Andes is of this nature.

Hot spots

A similar situation to volcano generation at separating plates can be produced if a hot spot in the mantle remains nearly stationary over time. A hot spot represents a long-lived plume of magma 10–100 km in diameter rising from the edge of the Earth's core 2900 km beneath the Earth's surface. Volcanoes will develop above the hot spot as crustal material is slowly melted from below. If a crustal plate drifts over the pseudo-stationary hot spot, then volcanoes drifting away from the spot become extinct, and newer volcanoes develop in their place. The Hawaiian Islands owe their origin to drifting of the Pacific Plate westward over a hot spot (Figure 9.6). This process has been occurring for over 40 million years. The oldest activity occurs at Midway Island and a series of *seamounts* northwest of the Hawaiian Islands. On the latter islands, the active eastern island is about 3–4 million years younger than the furthermost western and dormant Kauai Island. Rifting in the Hawaiian area ensures that fractures penetrate to the magma source below. The melting of crustal material above the hot spot, and subsequent fracturing and sinking of the weakened crust, produces this rifting.

The Hawaiian Island chain indicates that the Pacific Plate has drifted away from the East Pacific Rise north-west towards subduction zone trenches. Other island chains in the Pacific Ocean support this hypothesis. Figure 9.7 plots the location of main island strings in the Pacific Ocean associated with drifting of the Pacific Plate over different hot spots. These hot spots probably represent the location of upward flowing convection currents in the mantle that are responsible for the spreading of plates away from the East Pacific Rise. Note that not all volcanoes appear above the ocean surface. Many are submerged. The Caroline, Tuamotu, Society, and Austral volcanic island chains all parallel movement of the Hawaiian Islands, and thus indicate a north-west movement of the Pacific Plate away from the East Pacific Rise. The Sala-y-Gomez Island and Galapagos Island chains on the east side of the Rise indicate migration of plates eastward. At least 25 primary hot spots have been identified, two of which lie beneath continental crust under the eastern seaboard of Australia and Yellowstone in the United States

The theory of hot spots has fundamental flaws. For example, the crust under Iceland and the island of Hawaii, which are viewed as classic examples, does not have anomalous heat flow suggestive of rising magma. The site of volcanism in the Canary Islands on the east side of the mid-Atlantic ridge also does not appear to be moving with the plate. Instead, eruptions have occurred randomly beneath the islands. Finally, the composition of volcanic material discharged above hypothesized hot spots is not always indicative of deep mantle material. An alternative hypothesis suggests

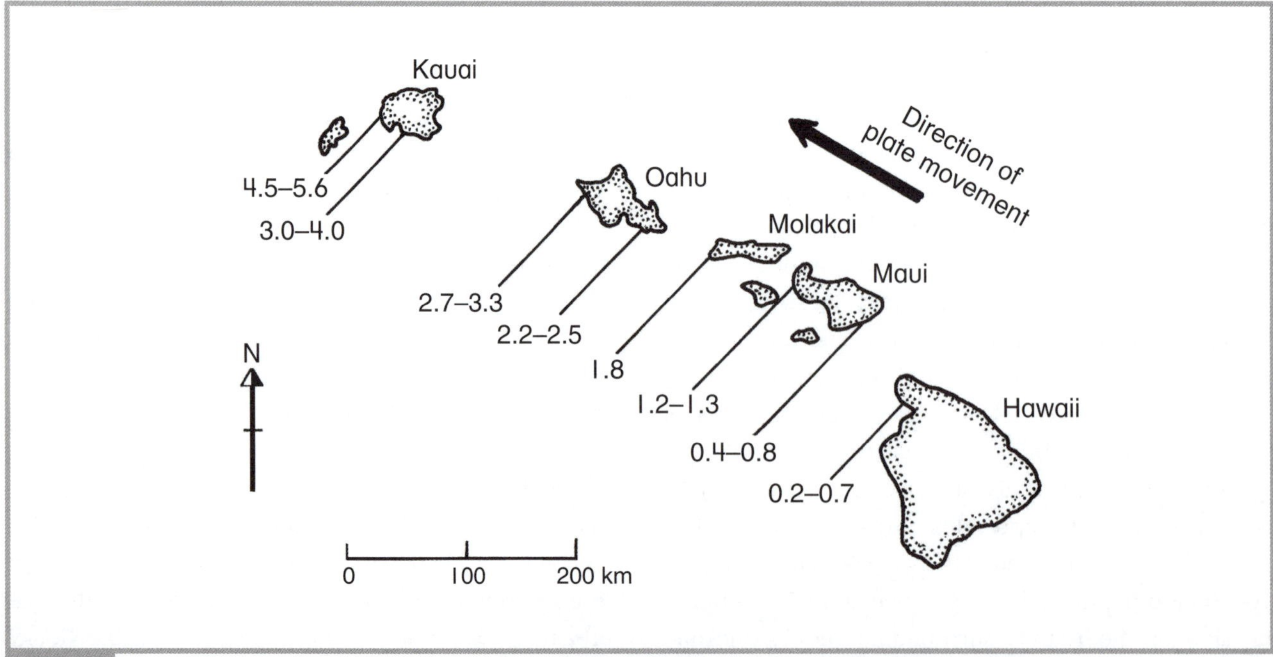

Fig. 9.6 Age and direction of movement of Hawaiian Islands drifting over a hot spot (after Muller & Oberlander, 1984).

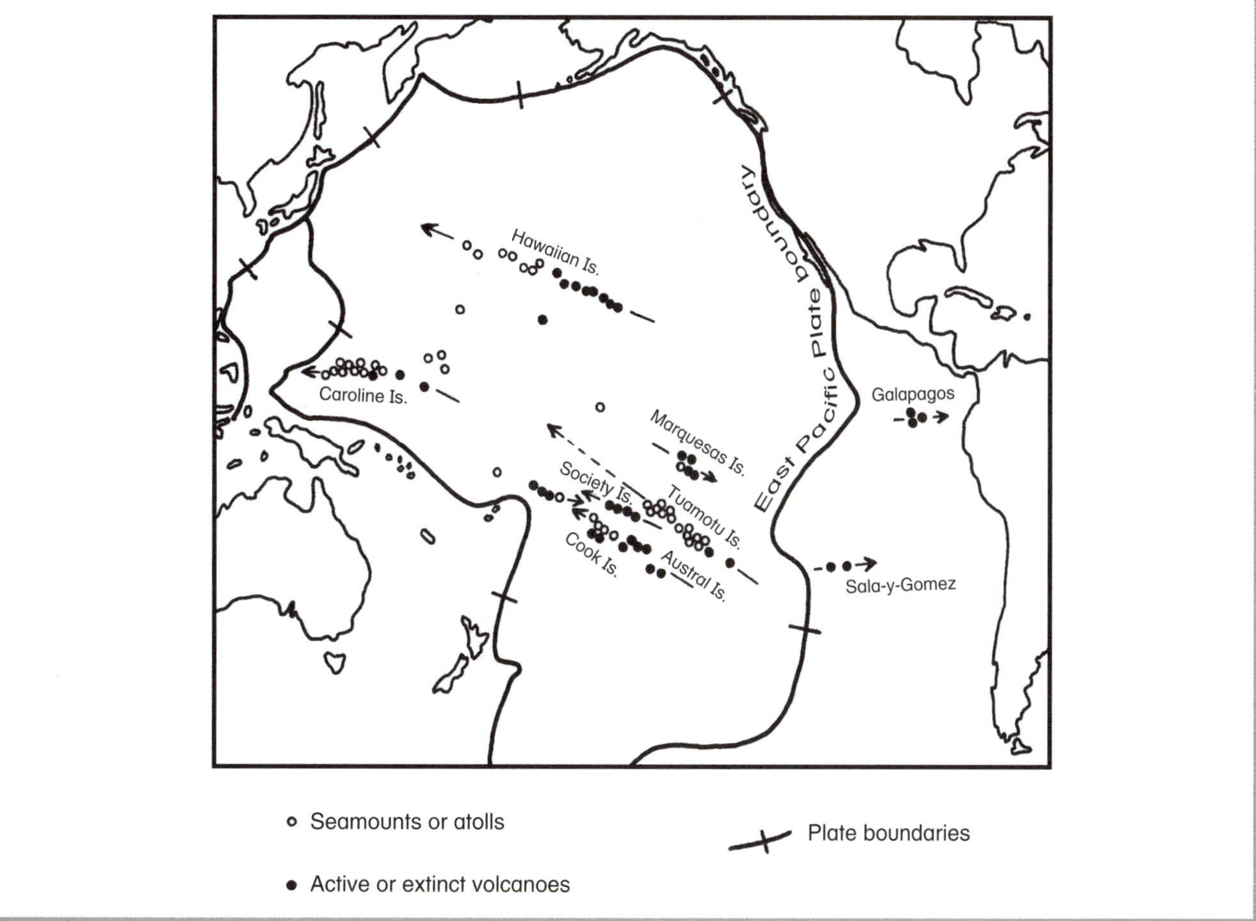

Fig. 9.7 Location of other hot spots and associated island strings on the Pacific Plate (based on Holmes, 1965).

that lines of volcanic activity represent tearing of the Earth's crust perpendicular to mid-ocean ridges. The Hawaiian and Tuamotu Island chains certainly fit the locations where stress fractures should form via this mechanism. It is now viewed that intra-plate volcanism is more complex than originally theorized in the 1970s using the concept of hot spots.

Other faulting and dilatancy

The above factors account for the location of most volcanic activity and a significant number of earthquakes. Most damaging earthquakes simply represent the rapid release of strain energy stored within elastic rocks. Such earthquakes are termed tectonic earthquakes. The Earth's crust is being continually stretched and pulled in different directions as plates move relative to each other, and as forces act within the plates themselves. Even minor crustal movement sets up elastic strain within the surface of the Earth, which may be greater than the internal strength of the layers of bedrock. This strain builds up a reservoir of energy much like a coiled spring in a child's toy. Where this strain is excessive, ruptures or faults occur to relieve the pressures that are being built up. Fracturing may occur gradually, or as a sudden series of shocks that radiate outwards from the strain zone in an uneven fashion, dependent upon the spatial variation in rock strength.

The type of faulting or rupturing that occurs depends upon the characteristics of the fault. Figure 9.8 shows the typical range of faults that occur with earthquakes. The zone of earthquake influence is narrowest for the strike-slip fault, whereas normal or thrust faulting produces a wider zone of influence. However, the difference is really dependent upon the angle of the fault. For example, as the dip of the San Andreas Fault is virtually vertical, earthquake damage rapidly diminishes within 20 km of the fault line. This is why moderate earthquakes this century in southern California have not caused widespread damage. On the other hand, the high magnitude earthquakes in Chile in 1960, and Alaska in 1964, occurred on low-angled reverse faults (20° or less in the case of Alaska) associated with subduction zones, a feature that extended

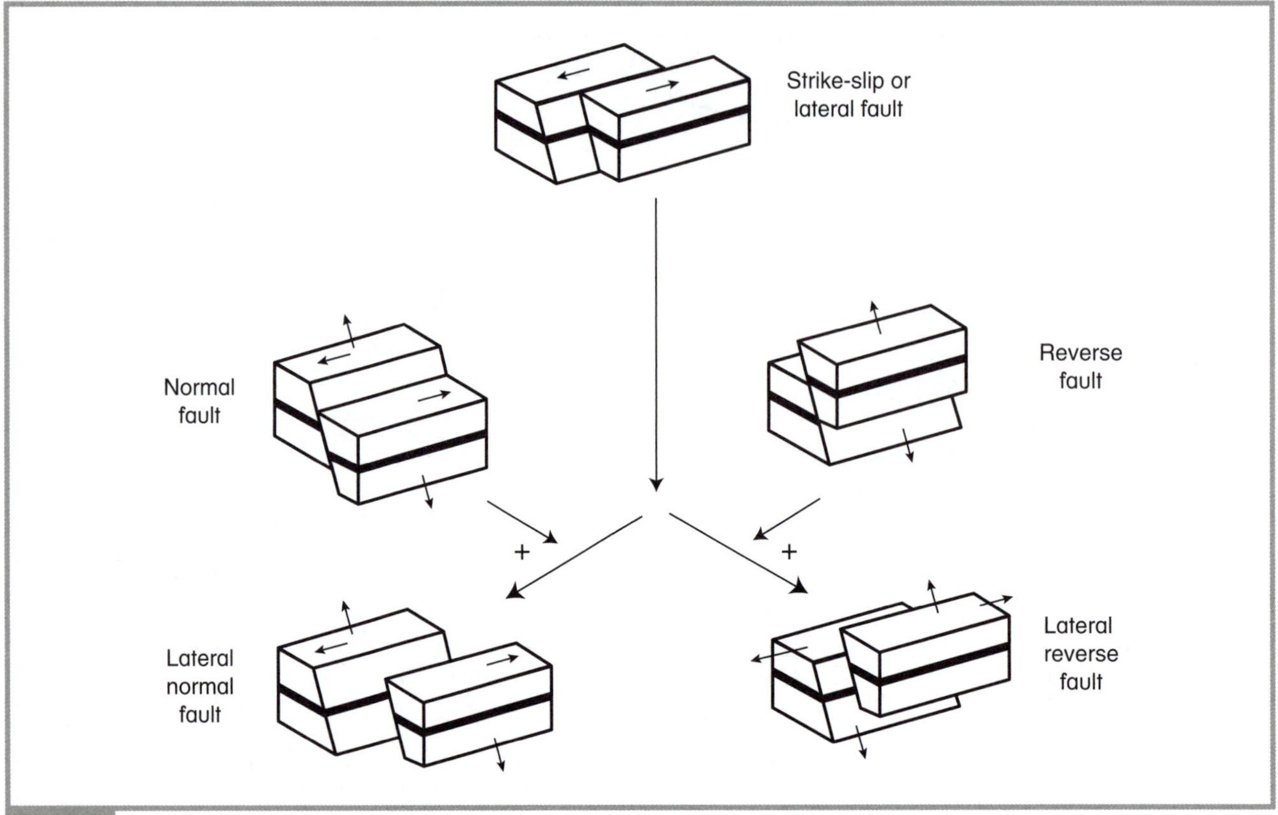

Fig. 9.8 Types of faults associated with earthquake activity.

damage over a much larger area. Weathered material above bedrock can absorb considerable amounts of the fault displacement, such that faulting occurring in underlying bedrock may not be present at the Earth's surface. Elastic strain can be relieved slowly over time as *tectonic creep*. Such creep can be monitored and used as a tool for earthquake prediction, a facet that will be discussed later.

Earthquakes can be triggered by sudden loading or deloading of the Earth's crust along most faults. The coincidence of tropical cyclones and earthquakes has already been noted in Chapter 3. Here the mechanism was linked to the rapid onset of storm surges on continental shelves. The loading threat may be even more subtle. In Japan, records over the last 1500 years indicate that intra-plate earthquakes with magnitudes greater than 7 are three times more likely to occur in spring. The cause is related to snow accumulation during winter that puts stress on reverse and strike-slip faults. Rapid melting in spring then unloads the crust suddenly, triggering earthquakes. As little as 1 m equivalent depth of water may be involved in the process.

Faulting and earthquakes can also operate via dilatancy in crustal rocks. At depths greater than 5 km, the pressure due to the weight of overlying rock is equal to the strength of unfractured rock. The shearing forces bringing about sudden brittle failure and frictional slip can never be obtained, because the rock deforms as a *plastic*. However, the presence of water provides a mechanism for sudden rupture by reduction of the effective friction along any crack boundary. If crustal rocks in the upper crust strain without undergoing plastic deformation, they may crack locally and expand in volume. The process is known as *dilation*. Cracking may occur too quickly for immediate groundwater penetration, but water will eventually penetrate the cracks, providing lubrication for any remaining stresses to be released. This model has prospects for predicting earthquakes because it invokes measurable changes in ground levels, electrical conductivity and other physical factors preceding tremors.

Added water (dams and rain)

The dilatancy theory can also explain the presence of earthquakes around reservoirs. Reservoir earthquakes occur in both seismic and non-seismic zones, and are unrelated to rock type or local faulting patterns. Interest in this phenomenon started in 1935 after the building of the Hoover Dam, which impounded Lake Mead on the Colorado River. The area was not noted for earthquake activity before the dam was constructed. Since that time, however, over 1000 earthquakes have been generated with enough strength to be felt by local inhabitants. Since then, this effect has been noted following the construction of 10–15 other reservoirs worldwide including the Kariba Dam on the Zambezi River in southern Africa and the Koyna Dam near Mumbai, India. It should be noted that not all artificial reservoirs have generated earthquakes, even when they have been located in active seismic zones. However, enough moderate earthquakes registering 5–6.5 on the Richter scale have been produced at Koyna, India, to cause the loss of 177 lives. It appears that reservoir water penetrates the underlying bedrock, reducing effective frictional resistance along fractures, and permitting slippage significant enough to generate earthquakes. Mining, water and oil extraction, and waste-fluid disposal underground have also resulted in local seismic activity. For example, wastewater from chemical warfare manufacturing was pumped underground at Denver, Colorado, between 1962 and 1965, resulting in measurable tremors. However, the effects of these latter operations appear minor compared to those produced so far by reservoir construction. Finally, there is a statistical correlation between the wettest time of year and eruptions of Yasur, Mt Etna, Mt St Helens, and Soufrière on the island of Montserrat. Rain immediately preceded recent eruptions of the latter volcano on three different occasions. Rain seeps into the volcano, and upon contact with the magma chamber, turns to steam causing a blast.

PREDICTION OF EARTHQUAKES AND VOLCANOES

Clustering of volcanic and seismic events

Volcanoes

(Gribbin, 1978; Lamb, 1972, 1982; Blong, 1984; Pandey & Negi, 1987)

The prediction of earthquakes and volcanoes can be broken down into two time spans: long-term prediction beyond a period of several years, and short-term prediction from a couple of years to several days or hours before the event. Long-term prediction of volcano and earthquake activity involves the delineation of cycles in activity over decades or centuries, and clustering of events.

Over geological time, periods of volcanism have not occurred as random events. Over the past 250 million years, volcanic activity appears to be enhanced every 33 million years with a minor peak occurring every 16.5 million years. The greatest peak in activity occurred 60–65 million years ago, corresponding to the formation of the Deccan Plateau in India, and the Cretaceous mass extinction that wiped out the dinosaurs. Periods of volcanic activity correlate with other mass extinctions, major reversals in the Earth's magnetic field, meteor or comet impact-cratering and other terrestrial processes. The cause of these catastrophic events is related to the passage of the solar system through the *galactic disc*. Our galaxy – the Milky Way – is a flattened spiral of stars that, side-on, looks like a plate. The solar system rotates relative to the center of the galaxy, passing from one side of the disc to the other. A complete circuit takes 33 million years, such that the solar system passes through the axis of the galaxy every 16.5 million years. At present, the solar system is passing through the galactic disc and volcanism has been increasing over the last 2 million years.

Extended periods of volcanism represent a major change in the *isothermal*, circulatory, and convective behavior of the mantle. As a result, major plate movements take place, leading to faster crustal spreading rates, to broadening of oceanic ridges and to sea level rises. More importantly, increased volcanic dust during major periods of volcanic activity screens out enough solar radiation, reducing *photosynthesis* and disturbing the food web sufficiently, to cause the disappearance of many land- and ocean-based species. Astronomical periodicities at the galactic level are sufficient to account for most periods of volcanism on the Earth, concomitant with increased comet impacts, *magnetic reversals*, climatic change and, ultimately, mass extinctions. Volcanism must be one of the most far-reaching and profound processes affecting the Earth geologically.

Blong (1984) has summarized some interesting data on the frequency of volcanic eruptions over the past 10 000 years. Over the last 500 years, individual volcanoes have erupted at the median rate of once every 220 years. The number of eruptions per century, when plotted on log probability paper (Figure 9.9), forms a straight-line relationship similar to the frequency of occurrence of most natural hazard events. About

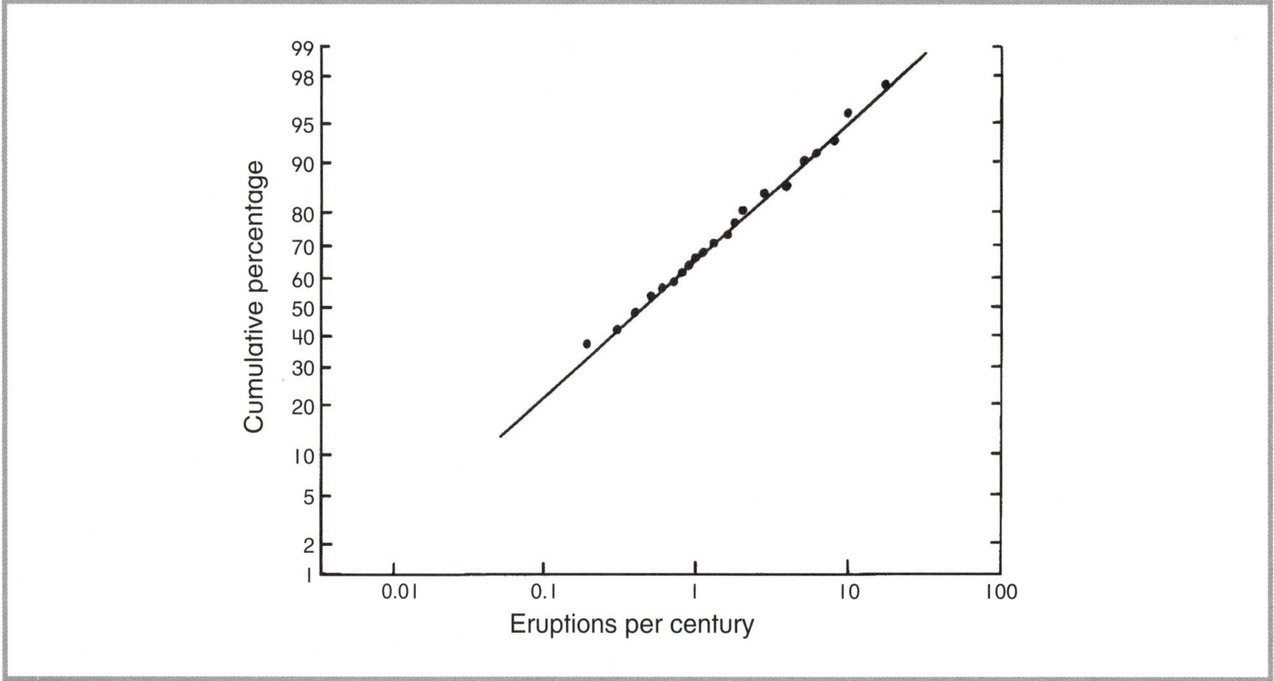

Fig. 9.9 Probability diagram of eruptions of individual volcanoes per century (Blong, ©1984, with permission Harcourt Brace Jovanovich Group, Australia).

20 per cent of volcanoes erupt less than once every 100 years, and 2 per cent less than once every 10 000 years. Both the El Chichon eruption in Mexico in March 1982, and the Pinatubo eruption in the Philippines in June 1991, exemplify this latter aspect. Neither had any historic record of eruption and both appeared extinct. The 25 most violent volcanic eruptions occurred with a median frequency of 865 years. Of the more than 5500 eruptions by 1340 volcanoes since the last glaciation, only 40 per cent are known to have erupted in the historic past. Thus, most of the world's extinct volcanoes could become active in the future. On average, one extinct volcano erupts every five years.

While the above analysis shows the randomness of eruptions, the historical record of volcanic activity based on dust that volcanoes have thrown into the atmosphere illustrates that eruptions are clustered over time. The amount of dust can be evaluated accurately by examining the attenuation of solar radiation by volcanic dust and measuring the widths of tree-rings (Figure 9.10). Professor Lamb in the United Kingdom compiled indices of volcanic activity based on solar radiation attenuation going back to the sixteenth century. His index is referenced to the amount of dust produced by the Krakatau eruption of 1883 (base value of 1000) and is known as the *Dust Veil Index*. The Santorini eruption in the Aegean in 1470 BC produced an index value of 3000–10 000, while Mt Vesuvius in 79 AD generated a value of 1000–2000. Solar radiation attenuation, lower temperatures, and increased frosts reduce tree-ring growth. These effects are enhanced following major volcanic eruptions, so that tree-rings should have increased density and diminished width. A record of the 20 lowest tree-ring density measurements for the last 400 years, from Europe and North America, is also plotted in Figure 9.10. These records show that the eruption of Huaynaputina, in Peru in 1601, had the greatest impact on tree-rings, while the eruption of Tambora in 1815 produced the greatest amount of dust in the atmosphere.

There have been some truly cataclysmic eruptions over the past 300 years. The Tambora eruption on the island of Sumbawa in Indonesia, during its main eruption in 1815, and the Cosequina eruption in Nicaragua in 1835, both released four times as much dust as Krakatau. These eruptions occurred simultaneously with many smaller ones, producing a period of sustained volcanic activity beginning in 1750 and terminating with the eruptions of Mt Pelée in Martinique, Soufrière on St Vincent Island, and Santa Maria in Guatemala in the Caribbean Sea area in 1902. The tree-ring record indicates that the period between 1600 and 1710 was also one of persistent volcanism. After the volcanic eruption of Katmai, Alaska, in 1912,

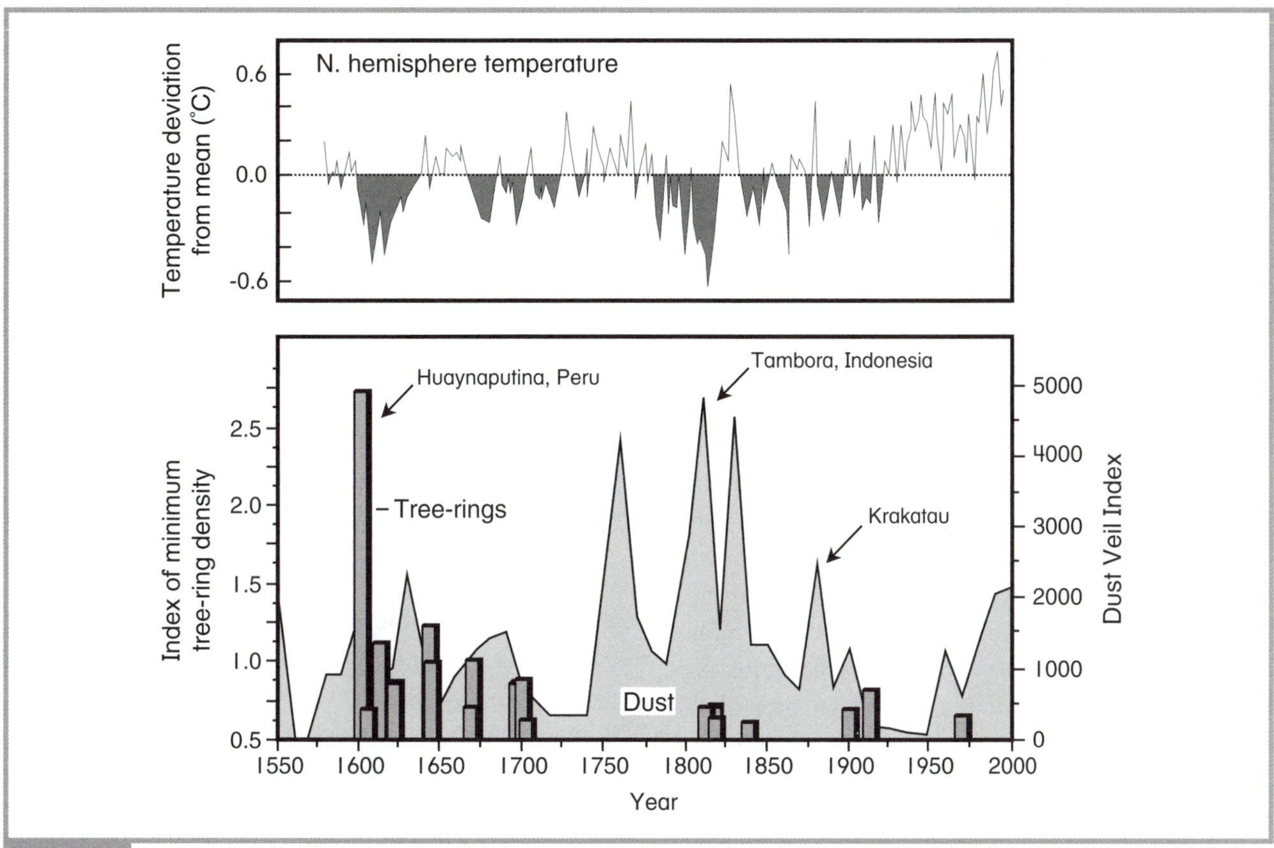

Fig. 9.10 Recent volcanism (1600–2000) as determined by Lamb's Dust Veil Index and tree-ring density measurements. The top panel plots northern hemisphere air temperature over the same time span (based on Gribbin, 1978; Lamb, 1985; and Jones et al., 1995).

there was not one significant dust-producing volcanic eruption that affected the northern hemisphere climate until Mt Agung in Bali in 1961. Mt Hekla, Iceland, in 1947 produced 100 000 $m^3 s^{-1}$ of ash, which reached as far as Finland; however, the eruption was short-lived and did not inject significant debris into the stratosphere. Mt Agung represented a mild reawakening of activity worldwide, which did not become intense until the Mt St Helens eruption. Besides Mt St Helens, there have been major eruptions of Galunggung, western Java, Indonesia, in April 1982; El Chichon in March 1982; Nevado del Ruiz, Colombia, in November 1985; Augustine, Alaska, in 1986; and Pinatubo in June 1991. Because of these eruptions, the Dust Veil Index climbed above 2000 during the last two decades of the twentieth century.

Also plotted in Figure 9.10 is the average temperature of the northern hemisphere, referenced to 1880. Dust emissions have a significant impact on global temperature 2–3 years after a major eruption. For example, dust from the Krakatau, Indonesia, eruption of 27 August 1883, reduced solar radiation in France by 10–20 per cent over a three-year period. The correspondence between the Dust Veil Index and global temperature is strong (*regression coefficient* of –0.65, which is significant at the 0.01 level). Periods of increased volcanism decrease surface air temperatures and have had a marked impact on society. For instance, volcanism in the 1780s produced extremely cold winters and drought throughout western Europe and Japan. In France, the adverse climatic events exacerbated the social conditions leading to the French Revolution in 1789. In 1816, the dust and sulfate emissions from the Tambora eruption, in the previous year, produced temperature drops of 1°C in New England, western South America, Europe, and the south-west Pacific. In New England, the event was known as the 'year without a summer'. Frost occurred in every month of the year in eastern North America, crops failed in Wales and central Europe leading to famine, and the South-East Asian monsoon was very intense. In contrast, the summer was hot in the Mississippi Valley, the Yukon, and north-western Russia. This eruption, and a series of others, was conducive to climatic conditions that favored the first global epidemics of typhus and cholera between 1816 and 1819.

The heightened volcanic activity of the last four centuries indicates that the Earth may be witnessing a period of increased volcanism not seen for 10 000 years.

Earthquakes (Whittow, 1980)

There is some evidence to suggest that regional earthquake activity is clustered. Figure 9.11 plots the magnitude of earthquakes based on the Richter scale in Japan since 684, the Mediterranean for the period since 1900, and the United States since 1857. Japan, which is seismically very active, has experienced a series of high-magnitude earthquakes centered on the eighth, sixteenth and twentieth centuries. In the nineteenth century, earthquakes appeared to cluster every 50 years. The most seismic period in the record occurred after the Second World War, but the most destructive earthquake in Japan in recent times occurred in 1923 at Tokyo. The Kobe earthquake of 17 January 1995 may simply have been part of this enhanced twentieth century activity. In the Mediterranean, earthquakes occur more regularly (Figure 9.11b). Since 1920, there have been four clusters of earthquakes, centered around 1930, 1941, 1953–1956, and 1977. Except for Turkey, the Mediterranean region has been quiescent in the latter part of the twentieth century. In North America, clustering of earthquakes has also occurred along the western continental margin. The most intense phase of seismic events has occurred since 1980. Other periods of clustering occurred around 1906 – just before the San Francisco earthquake – 1930, 1951, and 1964. However, the clustering is not nearly so pronounced as in the Mediterranean region. If clustering of earthquakes is real, then volcanic eruptions and earthquake activity should also be coherent because they have the same underlying *geophysical* causes.

Interaction between earthquakes and volcanoes (Hill et al., 2002)

There is a high occurrence of volcano activity immediately after earthquakes greater than 6.5 in magnitude. For example on 29 November 1975, Kilauea volcano erupted 30 minutes after a 7.5 magnitude earthquake on its southern flank. Cordon Caulle volcano in central Chile erupted two days after the great 9.5 M_w magnitude Chilean earthquake of 22 May 1960. While these eruptions occur within days of an earthquake, the effect may extend to hundreds of years afterwards up to 1000 km away from earthquake epicenters. The causes are due to static stress changes near the epicenter of an earthquake, stress changes associated with the slow relaxation of the lower crust, or dynamic stresses triggered by the earthquake. Static stress changes are caused by changes in pressure in a magma body lying close to an earthquake. The change may squeeze the magma upward like toothpaste from a toothpaste tube or enhance melting adjacent to the magma. More likely, the stress change may trigger the formation of bubbles in the magma and the freeing of conduits linking the magma chamber to the surface. These processes probably caused Mt Fuji in Japan to erupt in 1707, one month after an 8.2 magnitude earthquake; and Mt Pinatubo to erupt in 1991 following a 7.7 magnitude earthquake 100 km away. It is also possible that inflation of the magma chamber may trigger earthquakes that positively feed back on the inflation process. This pairing of events within ten years of each other seems to be characteristic of major eruptions of Vesuvius. Earthquake swarms over periods of 7–50 years are also related to viscous relaxation of the lower crust. After an earthquake, relaxation of shear stresses in the upper 15–20 km of crust triggers a concomitant relaxation in the underlying plastic crust. This effect induces volcanic eruptions either immediately or 30–35 years afterwards up to 1000 km from the earthquake epicenter.

The most likely process causing volcanic activity after earthquakes takes place in the magma. Seismic waves can dislodge bubbles held by surface tension on the floor and walls of a magma chamber. The same seismic waves can also increase bubble size. During the dilating phase of a seismic wave, bubbles grow as gas escapes under lower pressure from the magma. During compression, the gases dissolve back into the magma. However, periods of dilation are longer than those of compression. Hence, as the seismic waves pass through magma chambers, the bubbles grow in size. Within a few hours, the 28 June 1992 Landers, California, earthquake triggered 14 other earthquakes up to 1200 km away. Many of these earthquakes occurred in the southern part of the Long Valley Caldera where it is estimated that the Landers earthquake increased pressure by 7–10 MPa. Finally, earthquakes may dislodge crystals that have precipitated out of the magma body. Because these crystals are denser than the magma, they settle downward, forcing gas-rich magma upward. This latter magma is

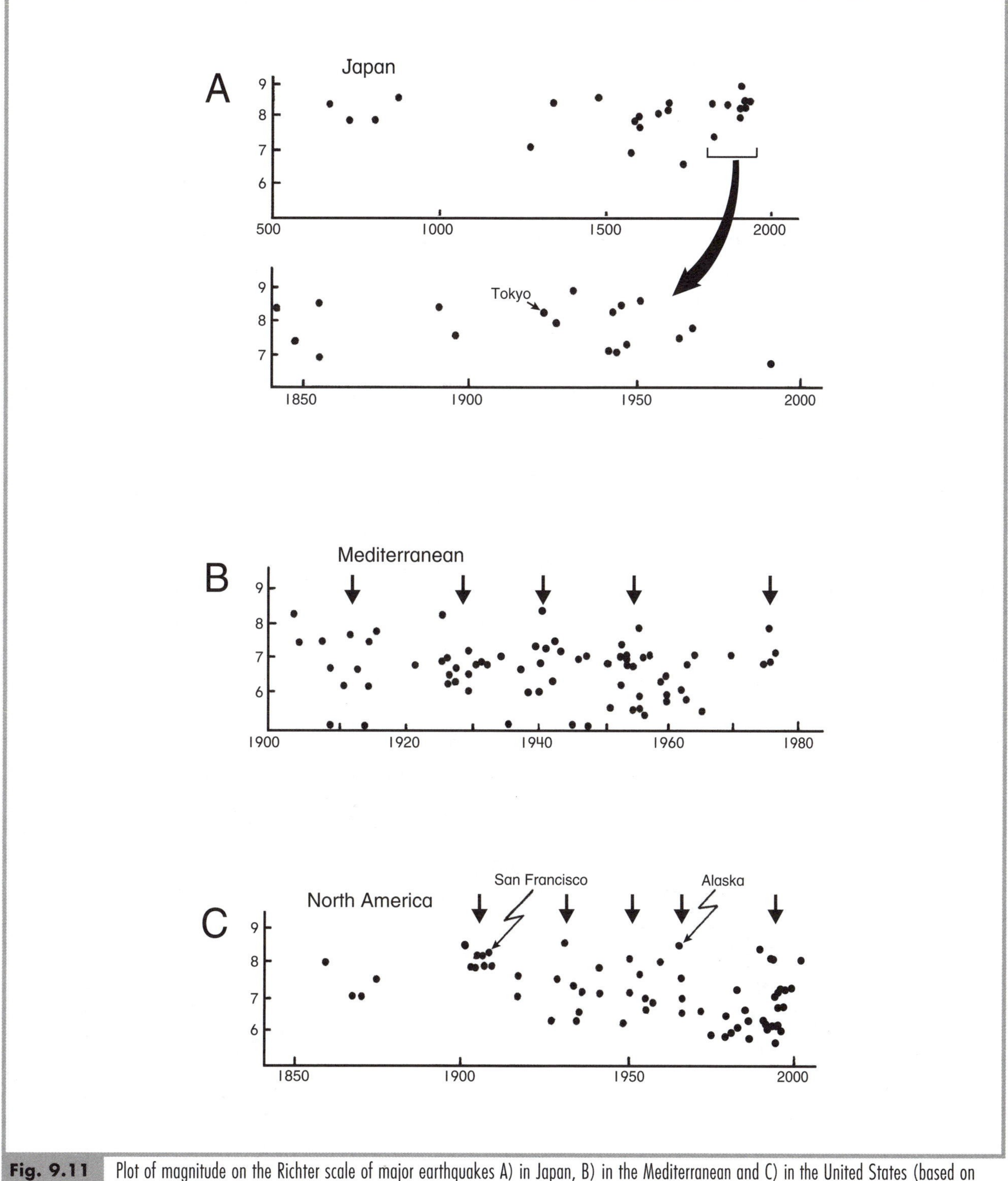

Fig. 9.11 Plot of magnitude on the Richter scale of major earthquakes A) in Japan, B) in the Mediterranean and C) in the United States (based on Whittow, 1980). Arrows indicate clustering. Updated from <http://earthquake.usgs.gov/activity/past.html>

subject to a pressure reduction, releasing gas that increases buoyancy and leads to further upward movement. This molten material can continue to rise upward in a runaway process at the rate of 10 cm s^{-1}, eventually rupturing the surface within a week of the triggering earthquake.

Prediction of seismic activity

Earthquake cycles (Wood, 1986)

In 1975, research into earthquake movements in swamp deposits in the Los Angeles area indicated that there had been eight major earthquakes since 565 AD, spaced at intervals of between 55 and 275 years. The average return period was 160 years, with the last major earthquake occurring in 1857. Based upon this periodicity, the next major earthquake above 8 on the Richter scale is imminent in the Los Angeles area within the next 20–30 years. This forecast is not illogical given the fact that most earthquakes occur along plate boundaries, which are slowly moving past each other at rates averaging 10 cm yr^{-1}. If the frictional drag between two plates moving relative to each other is not released continually, then forces will build up over time, culminating in a single major jump of the plates along a fault line. Because the plates are moving at a continual rate, the historic record of fault displacements should provide the return period or frequency of earthquakes in a region. For instance, if the average displacement along a fault is 5 m for each earthquake, then the fault should be active every 50 years, given the above rate of plate movement, 10 cm yr^{-1}. Around the Pacific Ocean, the return period for earthquakes is between 75 and 300 years. More significantly, once a plate has moved along a fault line in an area, the probability of a similar sized event occurring in the next few decades after the earthquake is perceived as small. In the area of the 1960 Chilean earthquake, the return period of large earthquakes for the past 300 years averages 80 years. The next major earthquake here is not expected until 2040 ± 20 years.

Around the Pacific region, plates are moving so consistently that anomalous regions that are not moving can be easily detected. For instance, in the Alaskan region, plate movements have generated continual earthquake activity over the past 150 years as stresses build up to crucial limits and are periodically released at various points along the plate margin. However, at some locations, the stresses may not be released easily; these points appear in the historical record as abnormally aseismic, while adjacent regions are seismically active. These locations, called *seismic gaps*, are prime sites for future earthquake activity, especially if earthquakes have not occurred in the region in the previous 30 years. The Alaskan earthquake of 1964 filled in one of these gaps as did the Mexican earthquake of 19 September 1985. Major seismic gaps as of 2003 are shown in Figure 9.12. The Los Angeles area of California is located at a seismic gap, as is much of the Caribbean Sea.

Away from plate boundaries, the overall rate of movement along faults is an order of magnitude slower

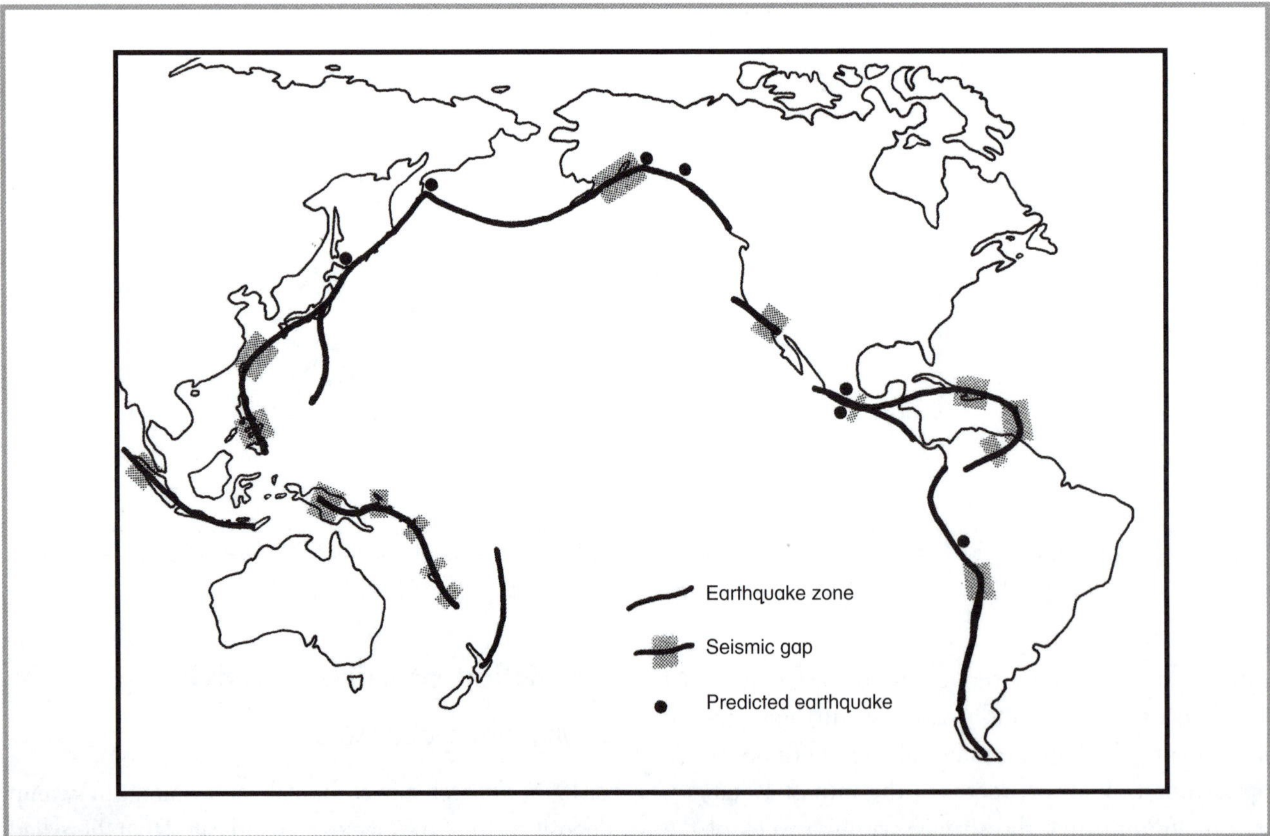

Fig. 9.12 Location of major seismic gaps in the Pacific 'ring of fire' (Coates, 1985).

than around the Pacific Rim. Here, the repeat time of an earthquake becomes much larger. This has significance for areas in the middle of plates that have experienced isolated earthquakes in historical times. For instance, the New Madrid earthquakes in the Mississippi Valley in 1811–1812, and the Charleston, South Carolina, earthquake in 1886 appear to be anomalous (Figure 9.2). The concept of earthquake repeat-cycles would indicate that these regions are highly unlikely to experience a similar intensity earthquake for hundreds, if not thousands, of years. There are also active fault lines where little or no major earthquake activity has been recorded historically. For instance, the Wasatch fault, which passes through Salt Lake City, appears to be more active than others in the region. Yet, historically it has never experienced a major earthquake. For these types of faults, major earthquakes may occur at 500–1000-year frequencies.

Short-term prediction of seismic activity

(Rikitake, 1976; Ward, 1976; Simpson & Richards, 1981; Kisslinger & Rikitake, 1985; Lighthill, 1996)

Of all natural hazards, earthquakes generate the largest and longest range of associated phenomena that can be used to foreshadow the timing of the impending event. Research into earthquake prediction has been carried out since the late nineteenth century. In 1891, the Japanese government set up an earthquake investigation committee, which collected historical records of earthquakes in Japan as far back as 416 AD. In 1925, the Earthquake Research Institute was established at Tokyo University to give earthquake research a scientific impetus. Following Japan's economic recovery in the 1950s, earthquake prediction was intensified. In China, where historical records go back 3000 years, a major effort at identifying earthquake precursors was initiated after 1950. On 5 February 1975, the city of Haicheng became the first major inhabited area to be evacuated before the occurrence of a significantly large earthquake. In the months leading up to the earthquake, ground tilting was noticed together with unusual animal behavior and water spurting from the ground. The city was evacuated 12 hours before the earthquake struck. The earthquake measured 7.2 on the Richter scale and, despite the destruction of 90 per cent of the city, only a few lives were lost.

Precursors of earthquakes group into the following five categories: land deformation, seismic activity, geomagnetic and *geo-electric activity*, ground water, and natural phenomena. Land deformation studies are based on the concept that the strain energy building up in the Earth's crust will evince itself in minor lateral or vertical distortions at the Earth's surface. Much of the Earth's surface has been surveyed with benchmarks or triangulation stations. In some countries, these stations have been accurately surveyed to form an extensive triangulation network. By resurveying stations within this network every decade, it is possible to detect gross movement, ranging from centimeters to meters, in the Earth's crust. It is also possible to relate these stations to tide gauges, to look for land movements relative to sea level in coastal areas. On a smaller scale, it is now possible with infra-red and laser survey equipment to mount survey stations on opposite sides of active fault zones and to measure earth movements as small as 1 mm over distances of several kilometers. This has been carried out along key sections of the San Andreas Fault. Points showing rapid changes in earth movement along the fault may signal future earthquake activity nearby. Crustal deformations can also be measured using tiltmeters and strain gauges. Increases in strain signal earthquake activity within months, while rapid increases in tilting of the Earth's crust (called *tilt steps*) give warnings of activity within hours.

On a smaller timescale, *foreshocks* or *microseisms* also indicate an impending earthquake. However, the nature of the warning is paradoxical and depends upon the area. Some earthquake zones give foreshocks days to hours before the main shock, while others evince a decrease in microseismic activity before the shock. Changes in seismic wave velocities through rock can also be used to foreshadow an earthquake. Strain changes the velocity at which a seismic wave will travel through the ground. By measuring the velocities of human-induced shocks across known earthquake zones, it is possible to forecast major earthquakes years in advance. Decreases in strain of up to 20 per cent have been measured for compression-type waves before earthquakes.

There are also geomagnetic and geo-electric precursors. It has already been pointed out that the Earth's geomagnetic activity varies with the sunspot cycle, a relationship that may be linked to earthquake activity. Geomagnetic activity also varies spatially and temporally over the surface of the Earth. Anomalous changes in geomagnetic activity in the order of 4–20 gammas have been measured up to ten years before local earthquake activity. The Earth's ability to transmit currents, or its *electric resistivity*, can foreshadow earthquake activity. Changes in resistivity are monitored by sending electrical pulses through the ground and measuring their strength some distance away. Decreases in resistivity of 10–15 per cent have been detected several months before major earthquakes, while stepped increases have been detected several hours in advance, even for earthquakes some distance from the monitoring site. Electrical currents, termed *telluric* currents (or Earth currents), are also continually flowing through the Earth's crust. Changes in the order of 2 mV have been observed hours before small earthquakes.

Many minerals such as quartz have *piezoelectric* properties. When stress is applied to rocks containing these minerals an electrical current is generated. Changes in the electrical potential between two rods bored into the crust can be used to measure the magnitude of stress building up along a fault line before an earthquake. The best example of this technique is the controversial VAN method – named after the initials of Professors P. Varotsos, K. Alexopoulos and K. Nomicos who developed it to forecast earthquakes in Greece. The method allegedly predicts the magnitude and epicenter of earthquakes, greater than M_S 5, within an error of 0.7 units and 100 km respectively, several hours to days in advance.

Groundwater fluctuations in wells can foreshadow earthquakes with magnitudes greater than 5 on the Richter scale, 0.5–10.0 days before the event. This technique is used extensively now as a forecasting tool in China. The longer the existence of anomalous water levels, and the wider the anomaly, the larger the impending earthquake. Changes in groundwater may be related to tectonic strain induced by crustal movements, or to the dilatancy phenomenon. A feature associated with changes in groundwater level is the change in the water's radon content. Radon, especially in mineral waters, will increase exponentially years before an earthquake, and then decrease rapidly to previous levels after the event. The increase in radon may reflect increased water movement through cracks opened by dilatancy, permitting more of this isotope to be incorporated into groundwaters.

There are also natural indicators of impending earthquakes. Catfish in Japan become very active before earthquakes and even jump out of the water. In addition, local fish catches around Japan increase significantly just before earthquake activity. Ground animals will leave tunnels, stabled animals will become restless, and dogs will start barking hours before an earthquake. Snakes, weasels, and worms deserted the ancient city of Helice, Greece, days before it was completely destroyed by an earthquake in 373 BC. Before the Lisbon earthquake of 1755, worms crawled from the ground and covered the surface. The evacuation of the city of Haicheng, China, in February 1975 – just before it was destroyed by an earthquake – followed observations of strange animal behavior: geese flew into trees, pigs became aggressive, chickens refused to enter coops, and rats appeared drunk. In the San Francisco zoo, animals have been found to group together by species about half-an-hour before an earthquake. Animal behavior is now being monitored daily in zoos and marine lands around San Francisco to incorporate animal behavior into an earthquake prediction system.

It is not exactly clear why animal behavior is prognostic of earthquakes. A response to long wave electromagnetic radiation has been given as one reason for animal panic. Animals may also be supersensitive to vibrations and *ultrasound* generated by small earthquakes preceding the main event. However, reptiles and birds that exhibit unusual behavior do not hear ultrasound. In addition, ultrasound from deep earthquakes is absorbed by rock, leaving only very low frequency sound to reach the surface. In this case, humans as well as animals should hear the sounds. Another reason for odd animal behavior may be an animal's extreme sensitivity to the smell of methane, known to leak from the ground during tremors. However, many burrowing animals that continually tolerate natural methane as they dig also panic before earthquakes. Finally, a barrage of *electrostatic* particles emanating from the ground may precede tremors. Animals with fur or feathers are very sensitive to electrostatic charges, and the sudden increase in quantity of such particles may simply irritate them to the point of panic or flight. Whatever the cause, scientists in China and other parts of the world now view the monitoring of animal behavior as the best means of forecasting the occurrence of major earthquakes.

Unusual weather changes may also be a harbinger of impending disaster. Mists close to the ground, as well as glowing skies, have been reported before earthquakes. In some cases, the light appears to emanate from the ground as a flash of flame. While this luminosity has often been dismissed, there is a new theory that suggests charge separation occurs deep in the Earth's crust under the pressures that build up before earthquakes. Igneous rocks contain minerals with paired oxygen atoms. Under stress, these break, with a negative ion remaining trapped in the crystal lattice of the mineral and the positive charge flowing to the Earth's surface. The positive charge can combine with an electron to give off infra-red light. Prior to the devastating 26 January 2001 earthquake in Gujarat, India, increased infra-red emissions were detected by satellite around the quake's epicenter. Otherwise, the charge builds up to 400 kV over a short distance at which point it ionizes air causing a luminous plasma – known as earthquake lights – to form. Alternatively, laboratory studies have shown that rock crushed under pressure gives off luminescence.

The prediction of earthquakes is still not an exact science. China, because it faces such large death tolls from earthquakes, has developed the most advanced predictive techniques. Five major earthquakes struck China in the early- to mid-1970s. All but one was predicted early enough to permit orderly evacuation of people from buildings with minimal death or injury. The exception, the Tangshan earthquake of 28 July 1976, took the lives of 250 000 people – the highest number of lives lost in an earthquake in two centuries. While the early warning signs of that earthquake were observed, they were too faint to arouse concern.

Randomness versus clustering

(Lighthill, 1996; Bak, 1997; Geller et al., 1997; Stein, 2003)

Foolproof prediction of large earthquakes is still not possible. The consensus today is that earthquakes occur randomly. If the cumulative frequencies of earthquakes are plotted against magnitude for various depths through the Earth's crust, the resulting curves have a similar shape following a power law. This is characteristic of systems that are self-organized and contain critical thresholds. For example, if you build up a pile of dry sand, it will reach a critical slope angle where failure will occur. The scale of this failure is random, from individual grains to massive sections of the sand pile. Massive failure occurs much less frequently than does the movement of individual grains. If the cumulative frequency of failure is plotted against volume, the resulting curve fits a power curve. More importantly, small events often trigger much larger ones, but there is no direct relationship between the two. The movement of a single grain may trigger another grain to move or the collapse of half the sand pile – or anything in between. This is characteristic of earthquakes, which are often preceded by swarms of smaller earthquakes, but you can never predict which one will trigger the 'big' one.

Most of the short-term precursors have only been identified as such following an earthquake. These alleged signals vary greatly from earthquake to earthquake and rarely can be detected as a regional phenomenon surrounding the epicenter of an earthquake. Crucially, the track record at predicting earthquakes has failed dismally. Even the success of predictions for the 1975 Haicheng earthquake can be challenged. Official records now indicate that 1328 were killed and 16 980 injured. For every precursor event linked to an earthquake, there are hundreds that have produced false alarms. For example, the crust around Palmdale in Los Angeles swelled rapidly by as much as 20 cm in 3 years in the 1970s. After the expenditure of millions of dollars on monitoring, the bulge was found to be an artefact of measurement error. The area is still awaiting an earthquake. The VAN method using geo-electric activity has also failed at predicting earthquakes in Greece, because the signals being measured are indiscernible from background noise. Each false prediction is just another statistical 'nail in the coffin' of earthquake prediction.

The seismic gap concept is flawed. Many earthquakes occur in swarms with the leading earthquake not necessarily being the largest one. Earthquakes should be viewed as chaotic *geophysical* phenomena, generated by a spectrum of seismic waves with varying amplitudes and periods. If this is the case, then earthquakes behave as white noise. One of the aspects of such systems is that earthquakes recur at the same location rather than in areas that are more quiescent. For example, earthquake activity along the western North American Plate boundary increased in the years leading up to the San Francisco earthquake of 1906. A predominance of magnitude 5 earthquakes in a region usually indicates, several months in advance, the imminent occurrence of a much larger earthquake. These smaller earthquakes often ring the future epicenter, forming what is known as a *Mogi doughnut*, named after the discoverer of the effect.

Some of these fine details about the spatial structure of earthquake swarms are being teased out. Earthquakes are produced by changes in stress at an epicenter. This stress is made up of two components, one parallel to the plane of a fault, and the other perpendicular to it. If shear stress along the fault exceeds frictional resistance, or the stress pressing against the fault from the sides is relaxed, the crust on both sides of the fault will slip, producing an earthquake. The total stress field is known as Coulomb stress. During a major earthquake, Coulomb stress can alter by less than 3 bars. The stress that appears at the epicenter of an earthquake on a fault line simply shifts elsewhere along the fault because stress must be conserved. The best example of this phenomenon occurred in Turkey. Here, the 17 August 1999 earthquake at Izmit that took 25 000 lives was the twelfth to occur along the North Anatolian fault since 1939. Each event simply shifted the stress along the fault, triggering an earthquake at another location a few years later. The Izmit earthquake reduced stress along 50 km of fault line, shifting it to Düzce 100 km to the east, and towards Istanbul in the west. A magnitude 7.1 earthquake subsequently affected Düzce in November 1999. Fortunately, this stress shift was monitored and publicized. Engineers pressured authorities in Düzce to close schools two months beforehand. Many of these schools were destroyed in the November earthquake. The stress build-up towards Istanbul remains and experts predict that the annual probability of an earthquake in the capital has now risen from 1.9 per cent to 4.2 per cent. The probability of an earthquake occurring in this city before 2030 has risen from 48 per cent to 62 per cent.

Stress mapping has also been used to explain the pattern of earthquakes that followed the 28 June 1992 Landers earthquake. Records show that there is a 67 per cent chance of another large earthquake along the San Andreas Fault within a day of a 7.3 magnitude earthquake. The Landers earthquake was of this magnitude, and within three hours, a 6.5 magnitude earthquake occurred at Big Bear, 40 km to the southwest. Coulomb stress had not only shifted along the San Andreas Fault but also laterally onto a parallel fault line. Seven years later, in 1999, a 7.1 magnitude earthquake occurred at Hector Mine, 40 km to the north of Landers. Swarms of minor earthquakes that followed the Landers earthquake also occupied areas of higher stress. Worldwide, 61 per cent of aftershocks greater than magnitude 5 – that have occurred within 250 km of the epicenter of a magnitude 7 or greater earthquake – correspond to areas where Coulomb stress has increased. Seismicity never completely shuts down around the center of an earthquake. Earthquake activity simply increases away from the epicenter. This may explain why the number of earthquakes in the San Francisco area decreased after 1906 relative to the preceding 60 years. Earthquake activity has simply shifted elsewhere along the San Andreas Fault. The Loma Prieta earthquake of 1989 may have been the beginning of a return to the previous rate of activity in the San Francisco Bay area.

Prediction of volcanoes

(Decker, 1976; Tazieff & Sabroux 1983; Smith, 1985)

While the first outburst of volcanic activity can now be predicted, it is at present almost impossible to predict the direction or intensity of activity that follows. To date, only a handful of eruptions have been forecast, the earliest being the renewed activity of Kilauea, Hawaii, in November 1959. Most of these predictions so far have been for eruptions that involved fluid magma and did not threaten life. Volcanic eruptions that consist of viscous magmas, or that become explosive, still cannot be predicted. For example, while the renewed activity of Nevado del Ruiz, Colombia, in November 1985, was noted, the timing of the final eruption could not be predicted, and 20 000 people lost their lives in the ensuing heated mudflows. There have also been some notable and costly false alarms for these types of volcanoes. On 12 April 1976, the residents of Guadeloupe in the West Indies were told to evacuate because of an imminent eruption of La Soufrière volcano. Over 75 000 people heeded the warning and were evacuated. They waited for 15 weeks until the volcano finally produced a small, very harmless eruption on 8 July. It cost $US500 million to maintain the evacuation and brought economic ruin to many. Such predictions, while soundly based, only tend to weaken the credibility of subsequent warnings.

The techniques for predicting volcanic eruptions or activity are just as varied as, but more technical than, those for predicting earthquakes. Precursors for volcanoes group into the following four categories: land deformation, seismic activity, geomagnetic and geoelectric effects, and gases. Ground deformations around volcanoes are due to the subterranean movements of molten magma. These movements can be vertical, lateral, or oblique. They manifest themselves at the surface by tilting of the Earth's surface, which can be measured using tiltmeters. The movements can be fast or slow, positive or negative. Differences in movement can occur over short distances. Large, sudden tilts generally herald a violent eruption while slight tilts of increasing frequency foreshadow the movement of magma closer to the Earth's surface. Tiltmeters on the northern side of Mt St Helens measured dramatic inflation at the rate of 0.5–1.5 m day^{-1} preceding the eventual eruption. Tiltmeters have also been used in Hawaii to accurately forecast eruptions of Kilauea.

Studies of seismology have dominated volcanology since the early 1900s. As magmas flow through subterranean channels, they apply stress to rocks, which can fracture and set off seismic activity. This applies mostly to volcanoes characterized by fluid magma. Earthquakes set off by volcanoes differ from tectonic ones in that they occur at depths of less than 10 km, and are low in magnitude. Volcanic tremors can be divided into two groups. The first group consists of prolonged, continuous volcanic vibrations due to resonance of fluid magma flowing through cracks. Some explosive volcanoes, notably Mt St Helens, were beset with this type of tremor. These long period vibrations have proved highly successful at forecasting impeding eruptions. The most notable success was the prediction twenty-four hours in advance of the eruption of Mt Redoubt in Alaska on 2 January 1990. The second group consists of spasmodic or regular vibrations, the frequency of which indicates the origin and nature of the magma. All forecast eruptions of Kilauea, Hawaii, have been based upon tilting and earthquake precursors. However, not all seismic activity associated with volcanism can be used with certainty to predict subsequent activity. While most eruptions are usually, but not always, preceded by swarms of earthquakes, the presence of tremors can also represent collapse of rock into emptying magma channels, a process that may indicate cessation of volcanic activity. In a few cases, the subsidence of earthquake activity may also signal an eruption.

Active volcanoes are dominated by temporally and spatially variable geomagnetic fields. Volcanoes contain a high content of ferromagnetic minerals, which can set up changes in the local magnetic field. Magnetization, however, is reduced by increasing temperature, vanishing completely above 600°C. The magnetic field of a volcano is thus reliant upon the proximity and temperature of molten magma near the surface. As hot magma between 200 and 600°C approaches the surface, the geomagnetic field should therefore decrease. Magnetization can also be enhanced by increasing pressure and stress exerted by flowing magma as it approaches the surface. This process is termed *piezomagnetism*, and at present is being researched extensively as a new technique for prediction. Geoelectric measurements involving the resistivity of the subsurface layers of a volcano and the change in the telluric currents can give an image of the behavior of magma at depth. Resistivity depends upon the nature of the rock, its water content, salinity, and temperature. Telluric currents depend upon geological structure, *lithology*, moisture content, and temperature. At present, these methods are used to define the structure of the volcano, including the presence of natural conduits, which may become the preferred pathways for continued magma movement.

The analysis of gaseous constituents exhaled from a volcano constitutes one of the best techniques for understanding and forecasting eruptive activity. Unfortunately, while the technique is so informative, it is restricted by the need to chemically analyze the gas constituents immediately they are vented from the volcano. Many gas studies are still in their infancy, mainly because techniques for sampling and analyzing gases from venting volcanoes are still being developed. The most common gases vented by a volcano are H_2O, CO_2, SO_2, H_2, CO, CH_4, COS, CS_2, HCl, H_2S, S_2, HF, N_2 and the rare gases helium, argon, xenon, neon and krypton. The chemical nature of these gases depends upon the maturity of the magma melt, because not all gas components have the same solubility at a given pressure. The type of gas also depends upon the amount of crustal material relative to original magma that is incorporated into the melt. If the magma is chemically stable, the *partial pressure* ratios between various gases can give an indication of the pressure and temperature of the magma en route to the surface. For example, the partial pressure ratio of $CO:CO_2$, HF:HCl and $H_2O:CO_2$ can all be used as qualitative geothermometers. The first two ratios increase, and the last one decreases, as temperature increases. The ratios of $CO:CO_2$, $H_2:H_2O$, and $H_2S:SO_2$ are sensitive to changes in the conditions that control the thermodynamic equilibrium of the gas phase, and can be used to distinguish between *hydrothermal* and magma flows. The ratios $SO_2:CO_2$ and S:Cl, plus the absolute amount of HCl, increase immediately prior to eruptions, while the ratio $He:CO_2$ decreases. These changes result from the different solubilities of individual gases and from the ascent of fresh magma to the surface. Isotope studies of various gases can also be used to delineate the different sources of material reaching the magma pool from below.

CONCLUDING COMMENTS

This chapter has shown that earthquakes and volcanoes are no longer poorly understood phenomena that invoke fear, superstition, and pagan attempts at control because of a lack of knowledge about their origin and behavior. While earthquakes and volcanoes can still kill a large number of people, and can occur with extreme suddenness, our knowledge is reaching the stage where a significant number of these events can be predicted in space and time to permit measures to be taken to minimize the loss of life. Since the widespread acceptance of continental drift theory, and the accurate measurement of thousands of earthquake epicenters, the most likely locations of future earthquake activity have been mapped for the globe. Sound scientific observations and instrumental measurements are defining the best precursors of imminent tremors or eruptions. The success of such techniques in China in the 1970s is exemplary; the failure of authorities in California to install sensitive strain gauges along the San Andreas Fault, which could have forecast the 1989 San Francisco earthquake, is alarming. Complacency or disbelief will always plague our attempts to predict both earthquakes and volcanic eruptions. Both traits can be justified because there is still a large random component to the occurrence of these hazards. The Tangshan earthquake of 1976 and the eruptions of Mt St Helens in 1980 and El Chichon in 1982 are only a few of the recent examples supporting this statement.

The next two chapters will describe in more detail the exact nature of, and phenomena associated with, earthquakes and volcanoes. As well, some of the more significant events in recorded history will be described to reinforce the view that these two hazards are still amongst the most destructive natural events to afflict humanity.

REFERENCES AND FURTHER READING

Bak, P. 1997. *How Nature Works: The Science of Self-Organized Criticality*. Oxford University Press, Oxford.

Blong, R.J. 1984. *Volcanic Hazards: A Sourcebook on the Effects of Eruptions*. Academic Press, Sydney.

Bolt, B.A. 1993. *Earthquakes*. W.H. Freeman and Co., New York.

Bolt, B.A., Horn, W.L., MacDonald, G.A., and Scott, R.F. 1975. *Geological Hazards*. Springer-Verlag, Berlin.

Coates, D.R. 1985. *Geology and Society*. Chapman and Hall, New York.

Decker, R.W. 1976. State of the art in volcano forecasting. In *Geophysical Predictions*. National Academy of Science, Washington, pp. 47–57.

Denham, D. 1979. Earthquake hazard in Australia. In Heathcote, R.L. and Thom, B.G. (eds) *Natural Hazards in Australia*. Australian Academy of Science, Canberra, pp. 94–118.

Doyle, H.A., Everingham, I.B., and Sutton, D.J. 1968. Seismicity of the Australian continent. *Journal of the Geological Society of Australia* 15(2): 295–312.

Geller, R.J., Jackson, D.D., Kagan, Y.Y., and Mulargia, F. 1997. Earthquakes cannot be predicted. *Science* 275: 1616–1617.

Gribbin, J. 1978. *The Climatic Threat*. Fontana, Glasgow.

Heki, K. 2003. Snow load and seasonal variation of earthquake occurrence in Japan. *Earth and Planetary Science Letters* 207: 159–164.

Hill, D.P., Pollitz, F., and Newhall, C. 2002. Earthquake-volcano interactions. *Physics Today* 55(11): 41–47.

Hodgson, J.H. 1964. *Earthquakes and Earth Structure*. Prentice-Hall, Englewood Cliffs, New Jersey.

Holmes, A. 1965. *Principles of Physical Geology*. Nelson, London.

Jones, N. 2003. Volcanic bombshell. *New Scientist*, 8 March: 32–37.

Jones, P.D., Briffa, K.R., and Schweingruber, F.H. 1995. Tree-ring evidence of the widespread effects of explosive volcanic eruptions. *Geophysical Research Letters* 22: 1333–1336.

Kisslinger, C. and Rikitake, T. (eds). 1985. *Practical Approaches to Earthquake Prediction and Warning*. Reidel, Berlin.

Lamb, H.H. 1972. *Climate: Present, Past and Future*. v. 1 & 2, Methuen, London.

Lamb, H.H. 1982. *Climate, History and the Modern World*. Methuen, London.

Lamb, H.H. 1985. Volcanic loading: The dust veil index. *Carbon Dioxide Information Analysis Center Numeric Data Package Collection*, Dataset No. NDP013.DAT, Oak Ridge National Laboratory, Oak Ridge, Tennessee, USA.

Lighthill, J. (ed.) 1996. *Critical Review of VAN: Earthquake Prediction from Seismic Electrical Signals*. World Scientific Publishing, London.

Matthews, A.J., Barclay, J., Carn, S., Thompson, G., Alexander, J., Herd, R., and Williams, C. 2002. Rainfall-induced volcanic activity on Montserrat. *Geophysical Research Letters* 29(13) doi:10.1029/2002GL014863.

Muller, R.A. and Oberlander, T.M. 1984. *Physical Geography Today* (3rd edn). Random House, New York.

National Earthquake Information Center 2001. *Frequency of occurrence of earthquakes based on observations since 1900*. United States Geological Survey, <http://neic.usgs.gov/neis/general/magnitude_intensity.html>

National Earthquake Information Center 2002. *Measuring the size of an earthquake*. <http://neic.usgs.gov/neis/general/measure.html>

Pandey, O.P. and Negi, J.G. 1987. Global volcanism, biological mass extinctions and the galactic vertical motion of the solar system. *Geophysical Journal* 89(3): 857–868.

Press, F. and Siever, R. 1986. *Earth* (4th edn). Freeman, New York.

Ritchie, D. and Gates, A. E. 2001. *Encyclopedia of Earthquakes and Volcanoes*. Facts on File, New York.

Rikitake, T. 1976. *Earthquake Prediction*. Elsevier, Amsterdam.

Simpson, D. W. and Richards, P. G. (eds) 1981. *Earthquake Prediction*. American Geophysical Union, Washington.

Smith, J. V. 1985. Protection of the human race against natural hazards (asteroids, comets, volcanoes, earthquakes). *Geology* 13: 675–678.

Stein, R.S. 2003. Earthquake conversations. *Scientific American* 288: 60–67.

Tazieff, H. and Sabroux, J.C. (eds) 1983. *Forecasting Volcanic Events*. Lange and Springer, Berlin.

Ward, P.L. 1976. Earthquake prediction. In *Geophysical Predictions*. National Academy of Science, Washington, pp. 37–46.

Whittow, J. 1980. *Disasters: The Anatomy of Environmental Hazards*. Pelican, Harmondsworth.

Wood, R. M. 1986. *Earthquakes and Volcanoes*. Mitchell Beazley, London.

Earthquakes and Tsunami as Hazards

CHAPTER 10

TYPES OF SHOCK WAVES

(Hodgson, 1964; Holmes, 1965; Whittow, 1980)

Earthquakes are shock waves that are transmitted from an epicenter, which can extend from the surface to 700 km beneath the Earth's crust. Earthquakes generate a number of types of waves, illustrated in Figure 10.1. A primary or P-wave is a compressional wave that spreads out from the center of the earthquake. It consists of alternating compression and dilation, similar to waves produced by sound traveling through air. These waves can pass through gases, liquids and solids, and undergo refraction effects at boundaries between fluids and solids. P-waves can thus travel through the center of the earth; however, at the core–mantle boundary they are refracted producing two shadow zones, each 3000 km wide, without any detectable P-waves on the opposite side of the globe.

The second type of wave is a shear or S-wave, which behaves very much like the propagation of a wave down a skipping rope that has been shaken up and down. These waves travel 0.6 times slower than primary waves. While the velocity of a primary wave through Earth depends upon rock density and compressibility, the rate of travel of a shear wave depends upon rock density and rigidity. Shear waves will travel through the mantle, but not through the Earth's rigid core. Thus, there is a shadow zone on the opposite side of the Earth that does not record S-waves. This zone overlaps completely the two shadow zones produced by refraction of P-waves at the core–mantle boundary. The spatial distribution and time separation between the arrivals of P- and S-waves at a seismograph station can be used to determine the location and intensity of an earthquake. Three stations receiving both P- and S-waves are necessary to determine the exact position of any epicenter.

Earthquakes also generate several different long or L-waves trapped between the surface of the Earth and the crustal layers lower down. These waves do not transmit through the mantle or core, but spread slowly outwards from the epicenter along the surface of the globe. Their energy is dissipated progressively from the focal point of the earthquake. One type of L-wave, the Rayleigh wave, behaves similarly to an ocean wave, while another type, the Love wave, literally slithers back and forth through the crust. This latter type of wave is responsible for much of the damage witnessed during earthquakes, and produces the shaking that makes it impossible to stand up. Long waves are the slowest of all seismic waves, taking about 20 minutes to travel a distance of 5000 km along the Earth's surface. All three types of waves, because they travel differently through dissimilar rock densities and states of matter, have been used to delineate the structure of the Earth's core, mantle, and crust.

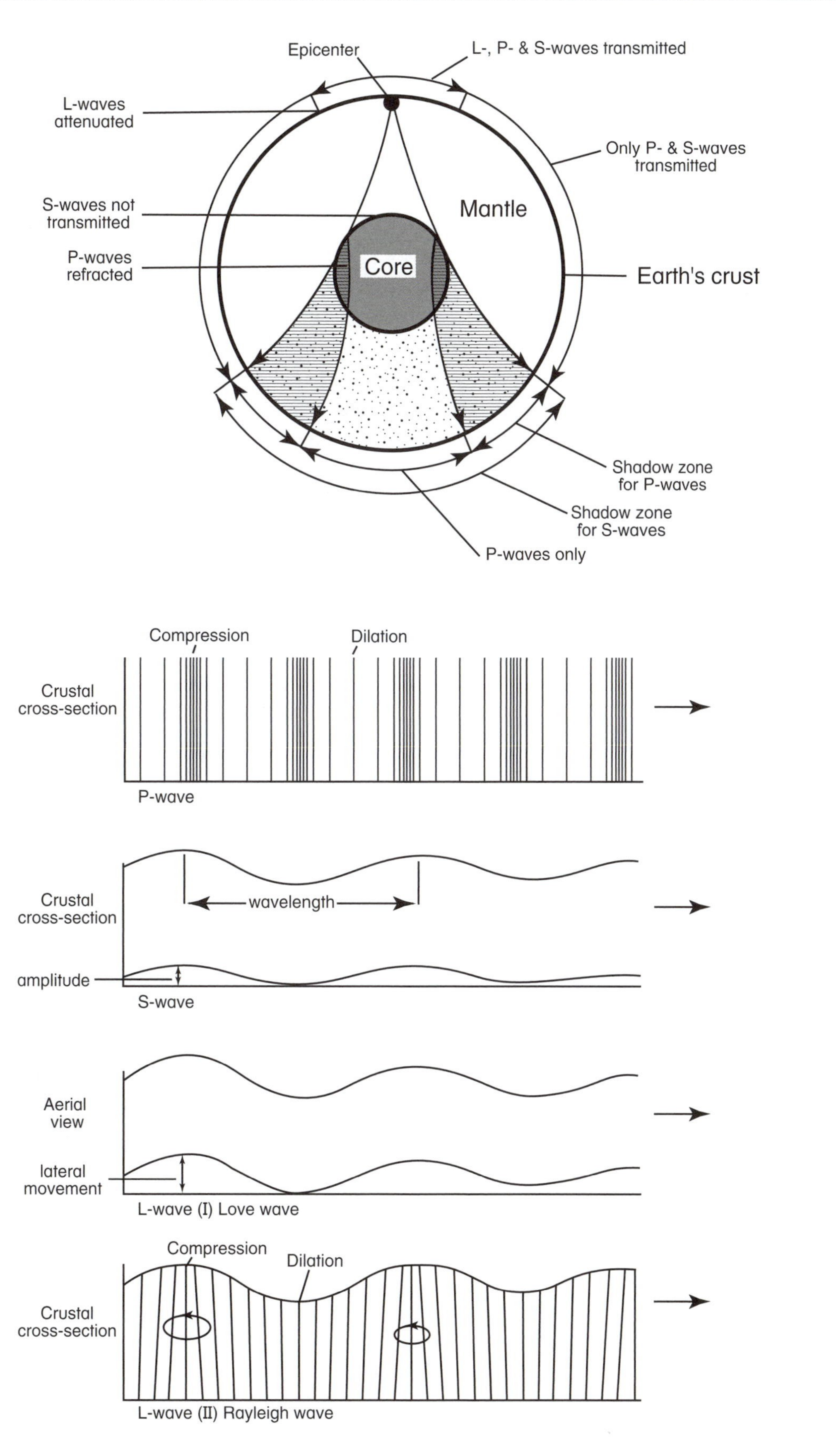

Fig. 10.1 Description of seismic wave types and their travel behavior through the Earth (adapted from Whittow, 1980).

SEISMIC RISK MAPS

(Keller, 1982; Giardini et al., 1999)

Maps can be constructed, based upon the historical record of seismic waves, to give probability estimates of the peak ground wave acceleration possible for any region over a fixed period. These accelerations are usually expressed as the 10 per cent probability over a fifty-year period. In terms of recurrence interval, they represent a 1:475 year event. Engineers can use such maps to assess the lateral forces acting on objects, such as rigid buildings, to determine the risk of their collapse during earthquakes. Figure 10.2 presents the seismic risk for the world's landmasses based upon historical data. Obviously, the estimated risk is a function of the historical record and spatial detection of earthquakes. For example, much of Australia (see Figure 10.3 for the location of all major placenames mentioned in this chapter), which lies in the middle of a continental plate, has a higher risk than Africa, which contains the Great *Rift Valley* and a historically seismic region in Algeria. In Australia, seismic detection is better developed than in most of Africa. In North America, publicity is overexaggerated regarding the seismic risk in southern California around the San Andreas Fault zone. The northern Rocky Mountains and the lower St Lawrence estuary are equally at risk. Large urban centers such as Salt Lake City, Seattle, Montreal, and Quebec City are all seismic. These latter regions have a 20 per cent probability in 100 years of a destructive earthquake corresponding to an intensity of 8 or greater on the Mercalli scale, and 6.5 or greater on the M_s scale. The seismic risk in Turkey, Iran, and the region bounding the Tibetan Plateau exceeds that in North America. Seismic risk, however, does not equate to damage or loss of life. These latter factors depend upon the establishment of building codes by planners and engineers, and their enforcement, especially in urban areas, by government. For example, the 17 August 1999 Izmit earthquake (M_s = 7.4) in Turkey, where codes are lax, killed 25 000 people compared to the similarly sized, 17 October 1989, earthquake (M_s = 7.1) in San Francisco – where codes are strictly enforced – which killed only 62 people.

EARTHQUAKE DISASTERS

General

(Holmes, 1965; Cornell 1976; Whittow, 1980; Ritchie & Gates, 2001)

Each year on average, earthquakes kill 10 000 people and cause $US400 million property damage. In the period 1964–1978, over 445 000 people lost their lives from earthquakes or earthquake-related hazards. Table 10.1 lists significant earthquakes in terms of loss of life since 800 AD. As with tropical cyclones, there have been some notable events. In 1290, in China, two earthquakes – in the Gulf of Chihli, and at Beijing – took

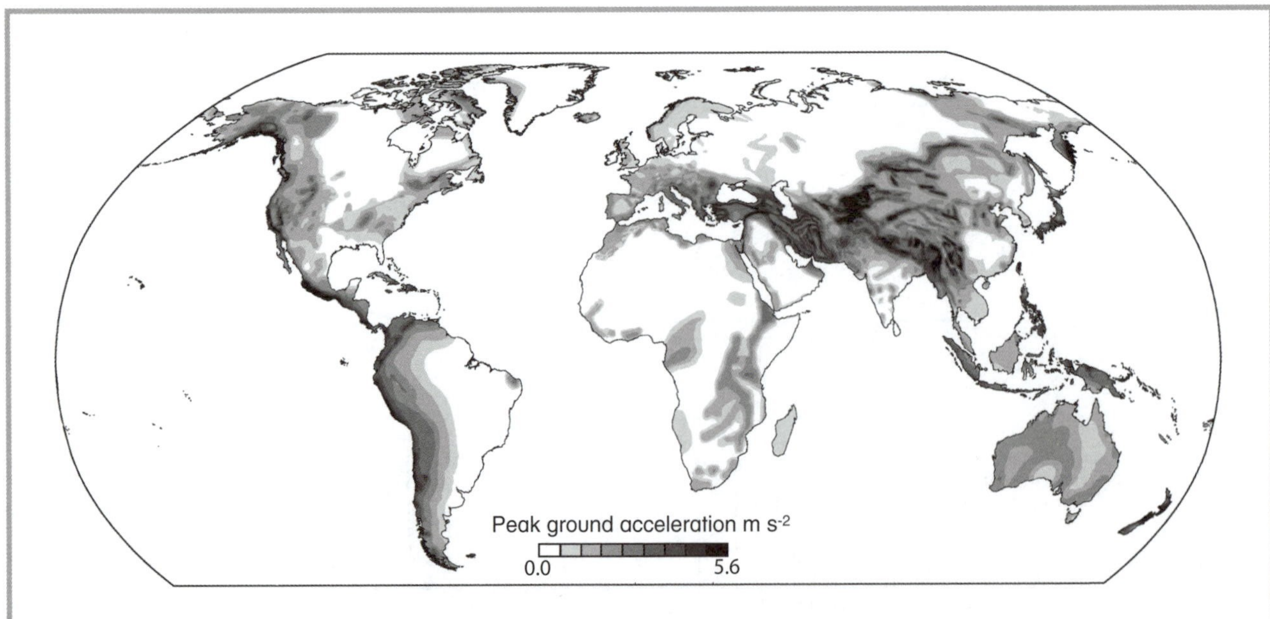

Fig. 10.2 Global seismic risk map. Shading defines peak horizontal accelerations with a probability of exceedence of 10 per cent in 50 years or a recurrence interval of 475 years (based on Giardini et al., 1999).

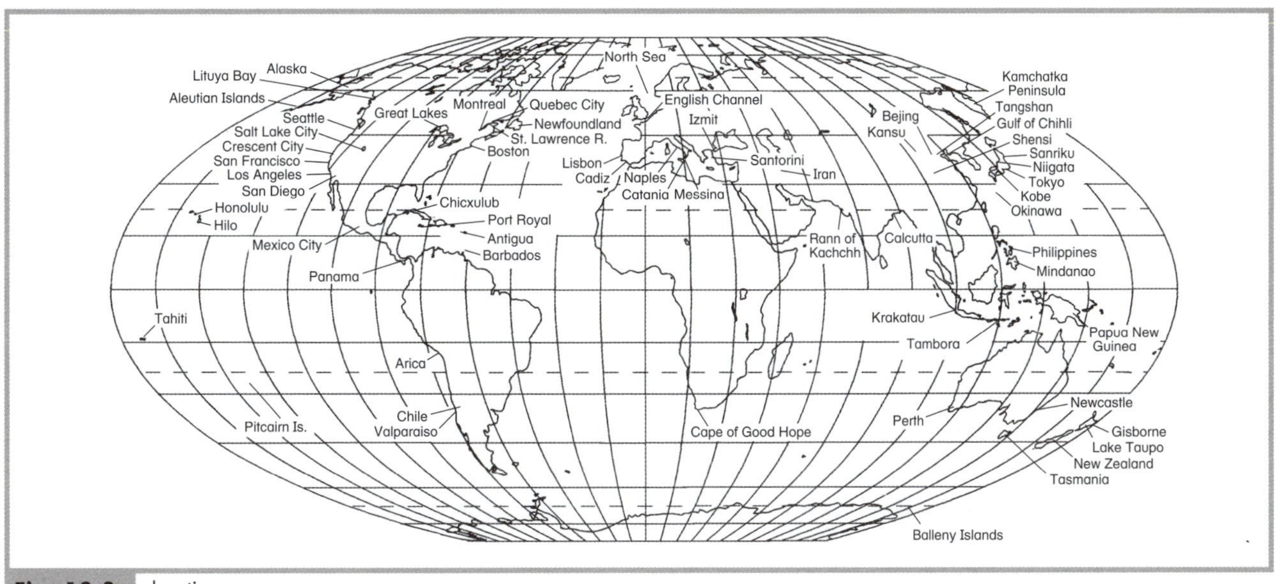

Fig. 10.3 Location map.

100 000 lives. The Shensi event in China on 23 January 1556 claimed 830 000 lives. That earthquake struck loess country, and caused the collapse of thousands of human-made caves dug into cliffs and used as homes. Another earthquake at Kansu, China, in 1920 killed 200 000 people. The Calcutta earthquake of 11 October 1737 wiped out half the city, killing 300 000 people. In 1923, the Tokyo earthquake – which is discussed below in more detail – killed 143 000 inhabitants. Italy has been struck many times

Table 10.1 Major historical earthquakes, by death toll (United States Geological Survey, 2002)

Date	Location	Death toll	M_s Magnitude (if known)
23 January 1556	Shensi, China	830 000	
11 October 1737	Calcutta, India	300 000	
27 July 1976	Tangshan, China	255 000	8
9 August 1138	Aleppo, Syria	230 000	
22 May 1927	Xining, China	200 000	8.3
22 December 856	Damghan, Iran	200 000	
16 December 1920	Kansu, China	200 000	8.6
1 September 1923	Tokyo, Japan	143 000	8.3
September 1290	Chihli, China	100 000	
November 1667	Shemakha, Caucasia	80 000	
18 November 1727	Tabriz, Iran	77 000	
28 December 1908	Messina, Italy	70 000	7.5
1 November 1755	Lisbon, Portugal	70 000	8.7
25 December 1932	Kansu, China	70 000	7.6
31 May 1970	Peru	66 000	7.8
21 June 1903	Silicia, Asia Minor	60 000	
11 January 1693	Sicily, Italy	60 000	
23 March 893	Ardabil, Iran	50 000	
4 February 1783	Calabria, Italy	50 000	
20 June 1990	Iran	50 000	7.7
30 May 1935	Quetta, Pakistan	30 000	7.5

by earthquakes, with ones near Naples and Catania, Sicily, in 1693, and at Messina on 29 December 1908, taking between 60 000 and 150 000 lives each. The San Francisco earthquake on 18 April 1906 was the worst American earthquake, killing 498 people, with much of the city being destroyed by ensuing fires. Most recently, the 1976 Tangshan earthquake claimed approximately 250 000 people, making it one of the worst earthquakes on record.

While many earthquakes stand out historically, very little research has been performed on the causes of some of the largest such as the Calcutta earthquake. In modern times, four to five events stand out because of their magnitude and the nature of the disaster. The 1906 San Francisco earthquake occurred in the vulnerable San Andreas Fault area of California and wiped out a major United States city in a country that believed itself to be immune from such disasters. The Tokyo earthquake of 1923 totally destroyed the largest city in Japan. The Chilean earthquake of 1960 marked a revival in large earthquake activity after a decade of small events, and produced one of the largest widespread *tsunami* events in the Pacific Ocean. The Alaskan earthquake of 1964 took few lives, but literally shook the globe like a bell. The 1976 Tangshan earthquake, notable for one of the most massive death tolls in the twentieth century, struck an area not renowned for seismic activity at a time when the Chinese believed that they could predict major earthquakes. The latter event is also notable for the cloak of secrecy surrounding its aftermath. Not only did China refuse international aid, but it also released very few details of the event until five years afterwards. Two of these events, the 1964 Alaskan earthquake and the 1923 Tokyo earthquake, are discussed in more detail below.

Alaskan earthquake of 27 March 1964

(Hansen, 1965; Whittow, 1980; Hays, 1981; Bolt, 1993)

The Alaskan earthquake struck on Good Friday when schools and businesses were closed. It occurred in a seismically active zone running westward along the coast of Alaska (Figure 10.4), parallel to the Aleutian Trench, and south of the Aleutian Islands arc. The earthquake was only one in a series of earthquakes in the area beginning 10 million years ago in the late *Pliocene*. In 1912 and 1934, earthquakes of magnitude 7.2 on the M_s scale occurred nearby. Other large earthquakes had occurred in neighboring areas in 1937, 1943, 1947 and 1958. These earthquakes were generated by the southward movement of Alaska over the Pacific Plate at a shallow angle of 20 degrees.

The 1964 earthquake was one of the strongest to be recorded since seismograph records were taken. Long wave vibrations lasted 4–7 minutes. The shock waves were so high in amplitude that seismographs ran off-scale all over the world, thus preventing accurate measurement of the earthquake intensity (reaching 8.6 and 9.2 on the M_s and M_W scales, respectively). The earthquake rang the earth like a bell, and set up seiching in the Great Lakes of America, 5000 km away. Water levels fluctuated in wells in South Africa on the other side of the globe. The ground motion was so severe that the tops of trees were snapped off. Earth displacements covered a distance of 800 km along the Daniel fault system parallel to the Alaskan coastline. Over 1200 aftershocks were recorded along this fault. Tsunami exceeded 10 m in height along the Alaskan coast and swept across the Pacific Ocean at intervals of one hour. The west coast of North America as far south as San Diego was damaged. In Crescent City, California, over 30 city blocks were flooded, resulting in $US7 million damage.

Maximum uplift and downthrow amounted to 14.4 and 3 m, respectively, the greatest deformation from an earthquake yet measured. In some locations, individual fault scarps measured 6 m in relief. Coastal and inland cliffs in south Alaska were badly affected by landslides. Land subsidence occurred in many places underlain by clays, and in some areas liquefaction was noted (Figure 10.5). Liquefaction is a process whereby firm but water-saturated material can be shaken and turned into a liquid that then flows downslope under the effect of gravity. About 60 per cent of the $US500 million damage was caused by ground failures. Five landslides caused $US50 million damage in Anchorage alone. Most of the ground failure in this city occurred in the Bootlegger Cove Clay, a glacial estuarine–marine deposit with low shear strength, high water content and high sensitivity to vibration. In the ports of Seward, Whittier, and Valdez, the docks and warehouses sank into the sea because of flow failures in marine sediments. Over the next nine hours, each of these towns was overwhelmed by tsunami 7–10 m in height.

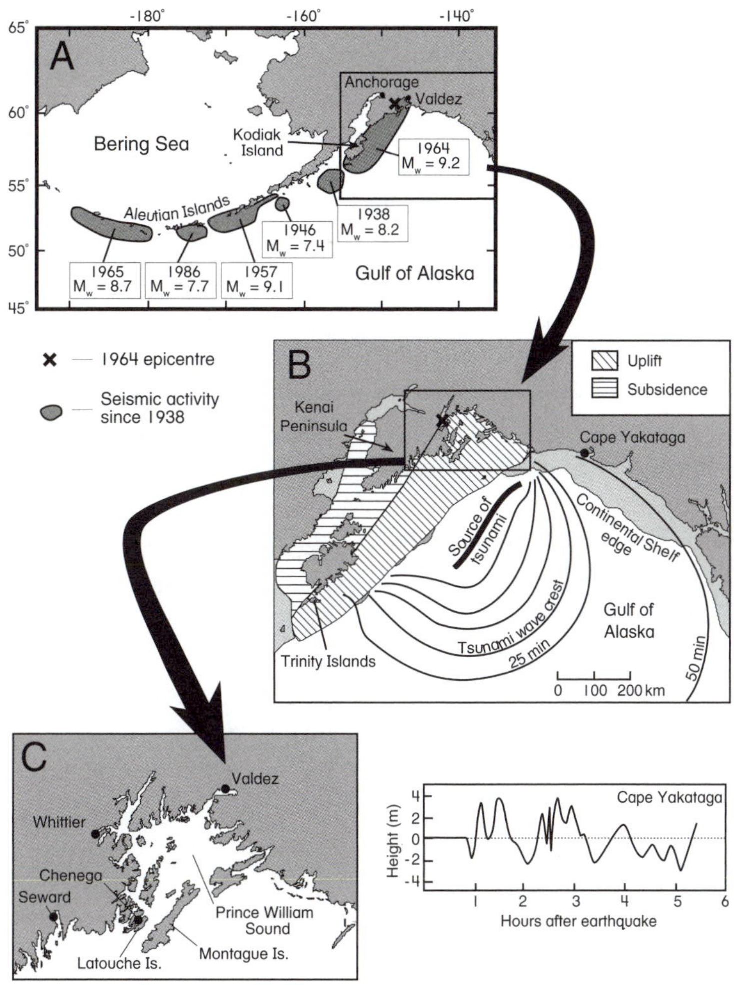

Fig. 10.4 Earthquake and tsunami characteristics of the Alaskan earthquake of 27 March 1964: A) Location of seismic activity since 1938; B) Gulf of Alaska land deformation caused by the earthquake and theorized open-Pacific tsunami wave front; C) Detail of Prince William Sound (based on Van Dorn, 1964; Pararas-Carayannis, 1998; and Johnson, 1999).

Fig. 10.5 Head scarp of one of numerous landslides that developed in Anchorage, Alaska, during the 27 March 1964 earthquake. Sand and silt lenses in the Bootlegger Cove Formation liquefied causing loss of soil strength (photograph courtesy of the United States Geological Survey).

The Californian earthquake hazard

(Keller, 1982; Oakeshott, 1983; Wood, 1986; Palm & Hodgson, 1992)

Of all seismic areas in the world, the San Andreas Fault running through southern California is one of the most potentially hazardous zones for earthquakes. Its many subsidiary branches pass by, or through, a region currently inhabited by an estimated 23 000 000 people (Figure 10.6). The most devastating earthquake in this region, in terms of property loss, was the San Francisco earthquake of 18 April 1906 that killed 498 – and maybe as many as 3000 – people. This quake was estimated at 8.25 on the M_s scale, and had a maximum vertical and lateral displacement of 2.5 m and 6 m, respectively. Fires, which broke out following the earthquake, were responsible for most of the damage. This earthquake was not unique. There have been earthquakes of similar size in recent geological times, with the earthquake at Fort Tejon on 9 January 1857 being the largest yet recorded in southern California. The Owen Valley tremor on 26 March 1872 is considered the third greatest earthquake to strike California in recent times. It produced surface faulting extending up to 150 km, with a maximum vertical displacement of 4 m. Other noteworthy earthquakes occurred in 1769 south-east of Los Angeles; on 10 June 1836 in the East Bay area of San Francisco; in June 1838 at Santa Clara; on 8 October 1865 in the Santa Cruz Mountains; on 21 October 1868 south of San Francisco; and on 24 April 1890 at Chittenden Pass.

The San Andreas Fault has been active since the *Jurassic*, 135 million years ago, and has accumulated 550 km of displacement. During the *Pleistocene* (last 2 million years) horizontal displacement has amounted to 16 km. Dating of charcoal deposited over the past few thousand years over various fault lines indicates that major earthquakes occur at intervals of 50–300 years. The Pacific Plate is moving slowly north-west relative to the North American Plate at the rate of 5 cm yr^{-1}. Most of the movement is sideways; but in a few places, there is a vertical component on the North American Plate side. In most places, this movement is continual; however, in some places, the plates become locked together and the movement is released in one sudden jerk. The San Francisco earthquake gave rise to the greatest lateral displacement – 6 m – yet measured along the San Andreas Fault. Even where slow creep takes place, earthquakes can occur.

In southern California around Los Angeles, there has been no major movement since the 1857 earthquake. The most recent earthquake, at Northbridge on 17 January 1994, measured only 6.7 on the M_s scale. Based upon geological evidence, this area together with the San Francisco area is overdue for a major earthquake greater than 8 on the M_s scale. Both cities exist in seismic gap areas, surrounded by zones that have experienced recent tremors in the past 30 years. At present, the probability of an earthquake occurring in these gaps is 50 per cent. One of the more overwhelming responses to the earthquake hazard in California is neglect. In the Parkfield area

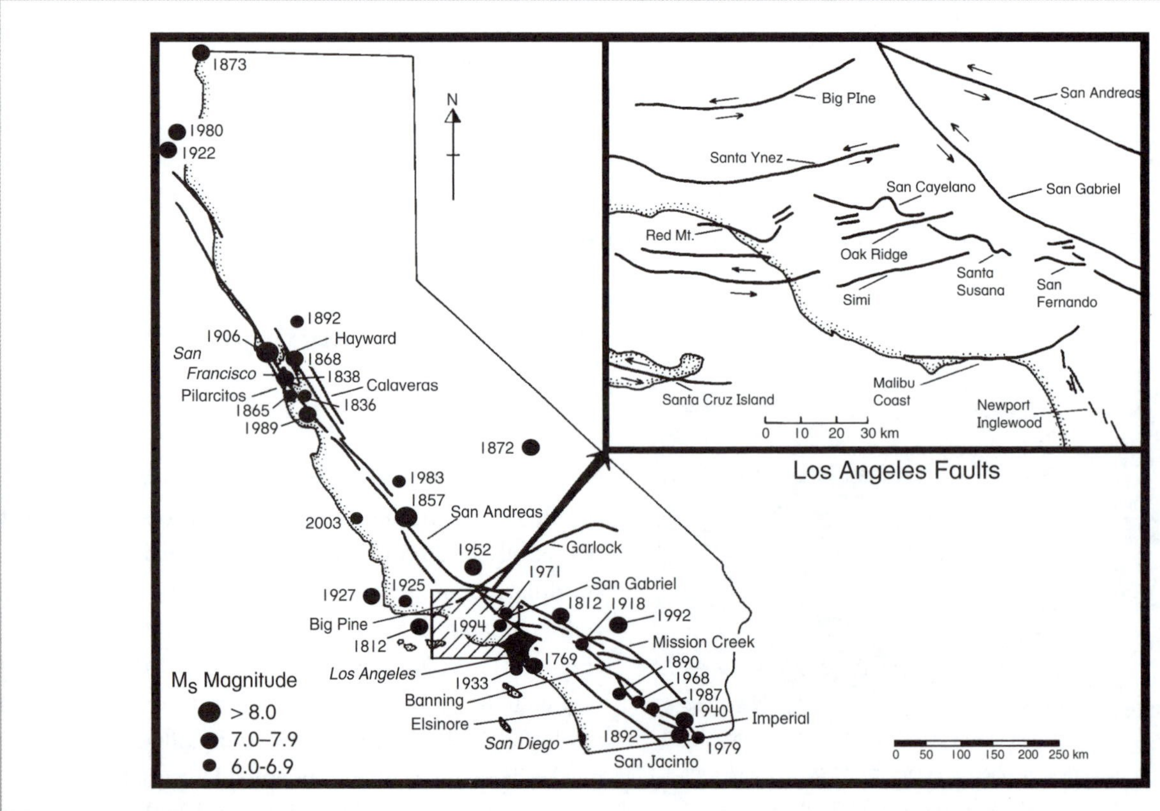

Fig. 10.6 Historically active fault systems of southern California and noteworthy earthquake epicenters for the last two centuries (adapted from Oakeshott, 1983 and Robinson, 2002).

of Los Angeles, where an earthquake with a 90 per cent probability of occurrence was forecast between 1985 and 1993, less than 20 per cent of the population purchased earthquake insurance and fewer than 17 per cent made their homes more earthquake resistant, despite widespread publicity about the likelihood of the event. Following the Loma Prieta earthquake of 1989, the increase in insurance in the most affected area was only 11 per cent, still leaving over 50 per cent of the population uninsured. The major populated segments of California lie in one of the world's most unstable zones for earthquakes and are poorly prepared for the consequences of these events.

A major earthquake in southern California would be catastrophic. Many past earthquakes in southern California produced little damage because they occurred in unpopulated areas. Many of these areas are now densely settled. The San Fernando earthquake of 1971, with an M_s magnitude of only 6.6, caused the collapse of public buildings, including two hospitals built to earthquake building code specifications. Sections of freeway overpasses built to the same standards also failed to withstand the shock waves. Subsequent to that earthquake, many public buildings and roadways were reinforced or rebuilt. Building codes have been tightened to account for the type of damage produced by that relatively small earthquake. Two dams in the area almost failed. One, the Lower Van Norman earth-filled dam, came within inches of collapsing and flooding 80 000 people in the lower San Fernando Valley. There are 226 dams in the San Francisco region alone with over 500 000 people living downstream. Many of these dams, together with others in southern California, have subsequently been strengthened. Some of these remedial measures have not worked. The 17 October 1989 Loma Prieta earthquake in San Francisco, measuring 7.1 on the M_s scale, collapsed a large section of the upper deck of the Nimitz Freeway in Oakland, even though the supports holding up this expressway were reinforced in the 1970s (Figure 10.7). Much of the housing construction in the San Francisco area since 1906 has occurred on unconsolidated landfill that could liquefy or amplify shock waves up to ten times, causing total destruction of buildings. This additional hazard was also brought home in the October 1989 earthquake, which caused the complete collapse of many houses in the Marina district for this reason.

Fig. 10.7 Aerial view of collapsed sections of the Cypress viaduct of Interstate Highway 880 (H.G. Wilshire, U.S. Geological Survey, <http://wrgis.wr.usgs.gov/dds/dds-29/>).

Not all of the branches of the San Andreas Fault have been mapped. For instance, the Diablo Canyon nuclear plant north of Los Angeles was built to withstand an earthquake of 8.5 on the M_s scale originating from a distant fault. However, during construction a major and active fault was discovered offshore within 5 km of the plant. While the plant's design was altered, there is still doubt whether or not it could withstand a major earthquake along the nearby fault. It is inevitable that the next major earthquake in southern California above 7 on the M_s scale will bring untold destruction. It has the potential to destroy large sections of major cities and to start numerous fires, which could turn into urban conflagrations similar to the 1906 earthquake. Numerous dams could fail causing major flooding and loss of life in downstream valleys, and the process of ground liquefaction, especially in the San Francisco area, could flatten suburbs in most of that city.

The Japanese earthquake hazard

(Hodgson, 1964; Holmes, 1965; Hadfield, 1992; EQE, 1995; Bardet et al., 1997)

The most notable earthquake in Japan in the last century occurred on 1 September 1923 in Sagami Bay, 50 km south-west of Tokyo (Figure 10.8). Known as the Great Kanto Earthquake, the crustal movements twisted the mainland in a clockwise direction, with a maximum horizontal displacement of 4.5 m and a vertical drop of 2 m. Until the Alaskan earthquake, these were some of the greatest crustal shifts recorded. Within the bay itself, changes in elevation were much larger owing to faulting, compaction, and submarine landslides. Measurements after the earthquake indicated that parts of the bay had deepened by 100–200 m with a maximum displacement of 400 m. The Tokyo earthquake immediately collapsed over a half million buildings and threw up a tsunami 11 m in height around the sides of Sagami Bay. However, these events were not responsible for the ultimately large death toll, because many of the collapsed buildings consisted of lightweight materials. The deaths resulted from the fires that immediately broke out in the cities of Tokyo and Yokohama, and raged for three days, destroying over 50 per cent of both cities. The fires rate as one of the major urban conflagrations in history – on the same scale as fires that destroyed London in 1666, Chicago in 1871, Moscow during the Napoleonic Wars in 1812, San Francisco during the 1906 earthquake, and Dresden and Hamburg during the Second World War. The outbreaks of fire were minor to begin with, but they occurred throughout both cities in large numbers, mainly because the earthquake occurred at lunchtime when many open fires were being used for cooking. Within half-an-hour, over 200 small fires were burning in Yokohama and 136 were raging in Tokyo. The rubble of collapsed buildings in the streets and cracked water mains compounded the difficulty in dealing with the fires. Firefighters could neither get to the fires nor obtain a reliable water source to extinguish them. Once the fires raged, the cluttered streets hampered evacuation. In Yokohama, 40 000 people fled to an open area around the Military Clothing Depot, which was subsequently engulfed in flames. All these people died from the searing heat or suffocation caused by the withdrawal of oxygen.

The spread of fire was aided initially by strong tropical cyclone winds. In the chapter on tropical storms, it was pointed out that the earthquake was probably triggered by the arrival of a tropical cyclone the previous day. The calculated change in load on the Earth's crust because of the pressure drop, and the increased weight of water from the storm surge, was 10 million tonnes km^{-2}. The fires were exacerbated by the development of strong winds generated by the cyclone in the lee of surrounding mountains. These

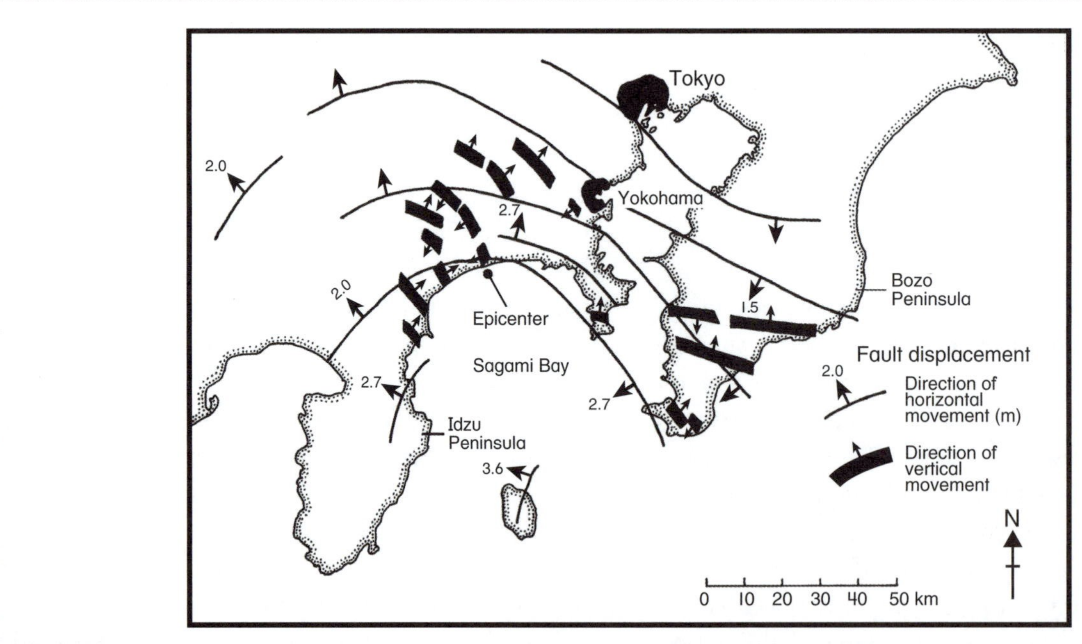

Fig. 10.8 Distribution of faulting for the Tokyo earthquake, 1 September 1923 (from Holmes, 1965).

föhn winds were desiccating. All descriptions of the fires allude to the strong winds, which swept flames out of control.

Since 1923, the Japanese economy has grown to be a crucial component of the global economy. The world has been waiting for the next large Tokyo earthquake. The death toll will be as high as 200 000 with up to 3.5 million people left homeless. The 1923 earthquake cost Japan 37.5 per cent of its gross national product (GNP). In 1987, a similar sized event would have cost 23 per cent of GNP or $US500–850 billion. This is more than Japan's entire overseas capital investment. The global insurance industry could cover only a small percentage of this loss and Japan would begin retrieving international investments to pay for reconstruction. Such a move would trigger a global recession that could last over five years. On 17 January 1995, a strong earthquake, measuring 7.2 on the M_s scale, struck the Kobe–Osaka region, the second most populated area in Japan. The earthquake – now known as the Great Hanshin Earthquake – killed 5500 people, injured 35 000 others and destroyed or badly damaged 180 000 buildings. Liquefaction caused the dockyards to sink below sea level and displaced supports for bridges. Shearing of inadequately designed or poorly constructed columns collapsed expressways, railway lines, and commercial premises. Building codes were designed to prevent structures crumbling during such an earthquake; yet, 4 per cent of buildings constructed to those codes suffered extensive damage and 2.5 per cent collapsed. Over 25 per cent of buildings higher than eight stories suffered severe damage. Fires consumed the crowded, older residential section of the city. Fewer than 5 per cent of homes were insured. Damage totalled $US100–150 billion or 13 per cent of the national budget, a staggering amount given the limited size of the earthquake. More significantly, the Nikkei index dropped 5.6 per cent, wiping the same amount off the share market. The fall of the share market spread globally. The Japanese economy, already in recession, stagnated further. Bureaucratic rigidity and archaic chains of command paralyzed relief efforts for several days. The response by the Self-Defense Forces was tardy, hampered by politics, and ineffectual. Everything about the Kobe earthquake indicates that a bigger event hitting Tokyo, as it inevitably must in the near future, will cause an even worse disaster scenario.

LIQUEFACTION OR THIXOTROPY

(Hansen, 1965; Scheidegger, 1975; Hays, 1981; Lomnitz, 1988; Bolt, 1993; Pinter et al., 2003)

Liquefaction is a process whereby relatively firm clay-free sands and silts can become liquefied and flow as a fluid. This can lead to the sinking of objects that they have been supporting. The mechanics of liquefaction will be discussed in more detail in Chapter 13. For now, liquefaction will be discussed qualitatively as an earthquake-induced hazard. The weight of a soil is supported by the grains, which touch each other. Below the watertable, spaces or *voids* in the soil fill with water. As long as the soil weight is borne by grain-to-grain contacts, the soil behaves as a rigid solid. However, if the pressure on water in the voids increases to the point that it equals or exceeds the weight of the soil, then individual grains may no longer be in contact with each other. At this point, the soil particles are suspended in a dense slurry that behaves as a fluid. It is this process, whereby a rigid soil becomes a liquid, that is termed liquefaction or thixotropy.

Liquefaction can be induced by earthquake-generated shear or compressional waves. Shear waves distort the granular structure of soil causing some void spaces to collapse. These collapses suddenly transfer the ground-bearing load of the sediment from grain-to-grain contacts to pore water. Compressional waves increase the *pore water pressure* with each passage of a shock wave. If pore water pressure does not have time to return to normal before the arrival of the next shock wave, then the pore water pressure increases incrementally with the passage of each wave (Figure 10.9). The process can be illustrated quite easily by going to the wettest part of the beach foreshore at low tide. On flat slopes, one can usually run easily along the sand. It is even possible to drive vehicles along this zone on some fine sand beaches. However, if you tap the sand in this zone slowly with your foot, it will suddenly become soft and mushy, and spread outwards. If the tapping is light and continual, the bearing strength of the sand will decrease and your foot will sink into the sand. All of the damaging earthquakes towards the end of the twentieth century – Loma Prieta, Northbridge, Kobe, and Izmit – were characterized by liquefaction.

Pore water pressure can return to normal only if there is free movement of water through the soil. Smaller particles inhibit water movement because

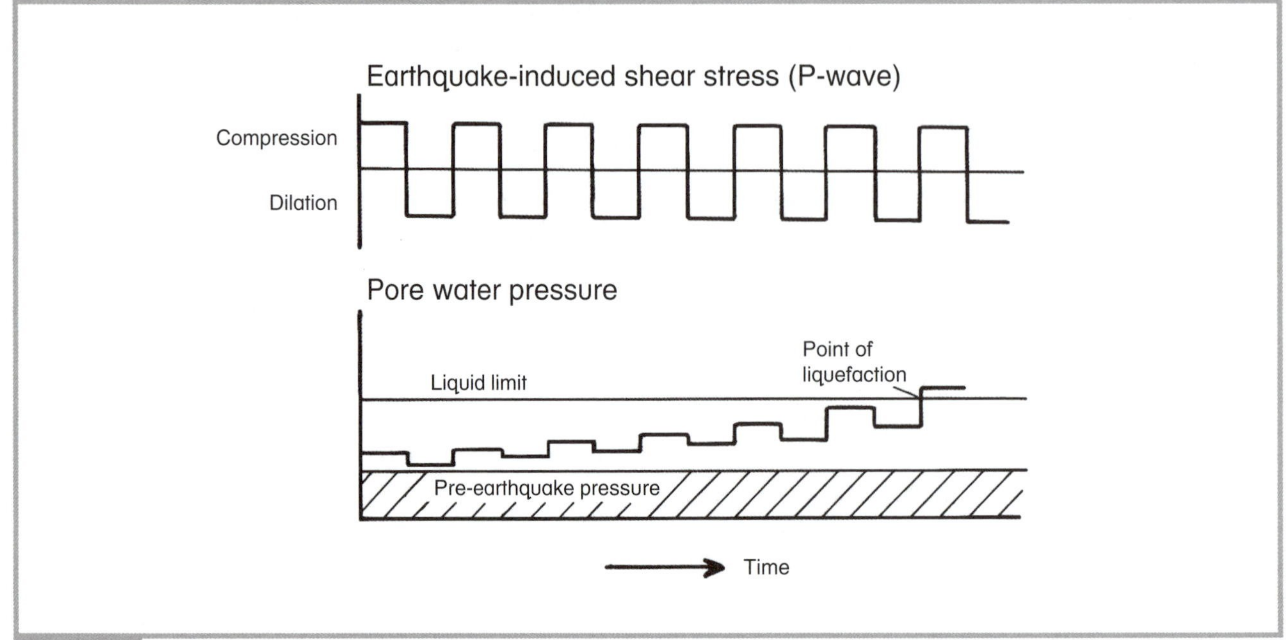

Fig. 10.9 Cumulative effect of compressional P-waves upon pore water pressure leading to liquefaction (after Bolt et al., 1975).

capillary water tension between particles is greater. This *cohesion* is so large with very fine silt and clays that the pore water pressure cannot reach the point of liquefaction. Medium- to fine-grained sands satisfy both criteria for liquefaction to occur. They are fine enough to inhibit rapid internal water movement, yet coarse enough that *capillary cohesion* is no longer relevant. Since these sand sizes are common in recently deposited river or marine sediments (in the last 10 000 years), liquefaction is an almost universal feature of earthquakes. The younger and looser the sediment deposit and the higher the watertable, the more susceptible the deposit is to liquefaction.

The build-up of pore water pressure can cause water to vent from fissures in the ground resulting in surface sand boils or mud fountains. Following the 7.6 M_s earthquake on 26 January 2001 in the Rann of Kachchh, India, the magnitude of groundwater evulsion was so great that it reactivated ancient river channels and formed shallow lakes over a 60 000 km^2 area. If pore water pressure approaches the weight of the overlying soil, then the soil at depth starts to behave like a fluid. Objects of any density will easily sink into this quicksand because the subsurface has virtually no *bearing strength*. Thus, liquefaction causes three types of failure: lateral spreads, flow failures, and loss of bearing strength. Lateral spreads involve the lateral movement of large blocks of soil as the result of liquefaction in subsurface layers, due to ground shaking during earthquakes. It occurs on slope angles ranging from 0.3–3.0 degrees, with horizontal movements of 3–5 m. The longer the duration of ground shaking, the greater the lateral movement and the more the surface layers will tend to fracture, forming fissures and scarps. During the 1964 Alaskan earthquake, floodplain sediments underwent this process, and lateral spreading destroyed virtually every bridge in the affected zone. Flow failures involve fluidized soils moving downslope, either as a slurry or as surface blocks overtop slurry at depth. These flows take place on slopes greater than 3 degrees. Some of these flows can move tens of kilometres at speeds in excess of 15 km hr^{-1}. More commonly, these flows can occur in marine sediment under water. Loss of bearing strength simply represents the transference of the ground load from grain-to-grain contacts to the pore water. Any object that has been sitting on this material and using it for support is then liable to collapse or sink. At present, the exact conditions responsible for earthquake-induced liquefaction have not been determined. Evidence indicates that the acceleration and number of shock waves control the process; however, prior history of the sediment cannot be ignored. Only mapping of subsurface material in earthquake-prone areas will delineate the area of risk from this process during earthquakes.

The effects of earthquake-induced liquefaction can be spectacular. For example, the Japanese earthquake at Niigata of 16 June 1964 registered 7.5 in magnitude on the M_S scale. The earthquake was located 60 km north of the city and, at Niigata, produced maximum ground accelerations of 0.16 per cent g – values that cannot be considered excessive. The city itself had expanded since the Second World War onto reclaimed land on the Shinano River floodplain. The earthquake shock waves did not destroy buildings. Rather, they liquefied unconsolidated, floodplain sediments, suddenly reducing their bearing strength. Large apartment blocks toppled over or sank undamaged into the ground at all angles (Figure 10.10). Afterwards, many buildings were jacked back upright, underpinned with supports, and reused.

The San Francisco earthquake of 1906 produced some evidence of liquefaction around the city. The breaking of water mains, which hampered fire suppression, was generated by liquefaction-induced lateral spreads, mainly near the harbor. Much of the development of the city since has occurred on top of estuarine muds with high water content. In addition, much of the harborside construction in southern California occurs on landfill, which can easily undergo liquefaction if water-saturated. The San Fernando earthquake in 1971 produced liquefaction in many soils and caused landslides on unstable slopes. Similar liquefaction problems were experienced during the October 1989 San Francisco earthquake. When a large earthquake strikes this city, as it inevitably will, liquefaction will be the major cause of property damage. The intensity of seismic waves does not have to be great to induce liquefaction – as illustrated in Figure 10.10, where the bearing capacity of soils underlying the buildings failed but the buildings remained virtually undamaged. However, the intensity of compressional shock waves triggering liquefaction may be considerably lower than the values experienced at Niigata. For instance, the Meckering, Western Australia, earthquake of 1968 induced liquefaction of sands underlying many parts of the expressway system in Perth, 100 km away from the epicenter. At Perth, the earthquake registered only 3 on the Mercalli scale. Not only did the edges of roadways sink 4 cm within hours of this seismic event, but the problem also persisted for weeks afterwards, eventually resulting in widespread damage to road verges.

Fig. 10.10 Liquefaction following the 16 June 1964, Niigata, Japan, earthquake. These workers' apartments suffered little damage from the earthquake, but toppled over because of liquefaction of unevenly distributed sand and silt lenses on the Shinano floodplain. Horizontal accelerations were as low as 0.16 per cent g. The buildings subsequently were jacked up and re-occupied (photograph courtesy of the United States Geological Survey, Catalogue of Disasters #B64F16-003).

The liquefaction process usually involves water movement in clay-free sand and silt deposits. In the loess deposits of China, it can involve air as the suspending medium. The Kansu earthquake of 16 December 1920 broke down the *shearing* resistance between soil particles in loess hills. The low *permeability* of the material prevented air from escaping from the loess, which subsequently liquefied; this resulted in large landslides that swept as flow failures over numerous towns and villages, killing over 200 000 people. The 1556 earthquake in Shensi Province may have generated similar loess failure, giving rise to a death toll in excess of 800 000.

Waterlogged muds can also amplify the long period components of shock waves. Earthquakes at Newcastle, Australia, in 1989 and in Mexico in 1985 provided clear evidence of this process at work. At Newcastle, the seismic wave from a weak tremor registering only 5.5 on the M_S scale amplified in river clays beneath the city. As a result, what was really only a small tremor still damaged over 10 000 buildings, creating a damage bill of $A1000 million. With the Mexican earthquake, damage was even more extensive. The Aztecs founded Mexico City in 1325 AD upon the deep, weathered ash deposits of a former lake bed, Lake Texcoco, later drained by the Spaniards. These deposits have weathered to form montmorillonitic clays with the ability to absorb water into their internal structure, thus increasing their water content by 350–400 per cent. From a geophysical point of view, these clays can be effectively modelled as water, even though they have a solid appearance. Today, Mexico City has grown to a population in excess of 12 million. Two earthquakes struck Mexico 400 km south-west of the city, offshore from Rio Balsas, on 19–20 September 1985. The first earthquake measured 8.1 on the M_s scale, while the second aftershock, 36 hours later, measured 7.5. Jalisco, Michoacan, and Guerrero, the three coastal states nearest the epicenter, suffered extensive damage, with the loss of about 10 000 people. Damage decreased sharply away from the epicenter. In Mexico City, peak accelerations corresponded to a calculated return period of 1 in 50 years. However, when the shock waves traversed the lake beds underlying Mexico City, they were amplified sixfold by resonance, such that the underlying sediments shook like a bowl of jelly under complex seiching. Surface gravity waves, akin to waves on the surface of an ocean, crisscrossed the basin for up to three minutes. These waves had wavelengths of about 50 m, the same as the base dimensions of many buildings, and caused the destruction of 10 per cent of the buildings in the city center, covering an area of 30 km^2. Of the 400 multistorey buildings wrecked, all but 2 per cent were between 6 and 18 stories high. Many had become pliable with age and had developed a fundamental resonance period of two seconds, the same as that obtained for peak accelerations of surface seismic waves. The buildings folded like a stack of cards. Close to 20 000 people were killed in Mexico City and the damage bill amounted to $US4.1 billion. The Mexican earthquake illustrates a fundamental lack in our knowledge regarding the nature of ground motion in sedimentary basins containing high water content, or alternating dry and wet sediment layers.

TSUNAMI

Description

(Wiegel, 1964; Shepard, 1977; Myles, 1985; Bryant, 2001)

Tsunami (both the singular and plural forms of the word are the same) are water wave phenomena generated by the shock waves associated with seismic activity, explosive volcanism, or submarine landslides. These shock waves can be transmitted through oceans, lakes, or reservoirs. The term tsunami is Japanese and means harbor (*tsu*) wave (*nami*), because such waves often develop as resonant phenomena in harbors after offshore earthquakes. Because some tsunami are preceded by a down-drawing of water followed by a rapid surge over the space of 30–120 minutes, the name *tidal waves* has been used to describe such phenomena. However tsunami have nothing to do with tides and this usage is now discouraged. Tsunami have a wavelength, a period, and a deep-water height; and can undergo shoaling, refraction and diffraction. Most tsunami originate from submarine seismic disturbances. The displacement of the Earth's crust by several metres during underwater earthquakes may cover tens of thousands of square kilometres, and impart tremendous potential energy to the overlying water. Tsunami are rare events, in that not all submarine earthquakes generate them. Between 1861 and 1948, only 124 tsunami were recorded from 15 000 earthquakes. Along the west coast of South America, 1098 offshore earthquakes have generated only 19 tsunami. This low frequency of occurrence may simply reflect the fact that most tsunami are small in amplitude and go unnoticed, or the fact that most earthquake-induced tsunami require a shallow focus seismic event greater than 6.5 on the M_s scale.

Submarine earthquakes have the potential to generate landslides along the steep continental slope that flanks most coastlines. In addition, steep slopes exist on the sides of ocean trenches and around the thousands of ocean volcanoes, seamounts, atolls, and guyots on the seabed. Because such events are difficult to detect, submarine landslides are considered a minor cause of tsunami. However, small earthquakes have generated tsunami and one mechanism includes subsequent tsunami landslides. A large submarine landslide or even the coalescence of many smaller slides has the potential to displace a large volume of water. Geologically, submarine slides involving up to twenty thousand cubic kilometres of material have been mapped. Tsunami arising from these events would be much larger than earthquake-induced waves. Only in the last thirty years has coastal evidence for these mega-tsunami been uncovered.

Tsunami can also have a volcanic origin. Of 92 documented cases of tsunami generated by volcanoes, 16.5 per cent resulted from tectonic earthquakes associated with the eruption, 20 per cent from *pyroclastic* (ash) flows or surges hitting the ocean, and 14 per cent from submarine eruptions. Only 7 per cent resulted from the collapse of the volcano and subsequent *caldera* formation. Landslides or avalanches of cold rock accounted for 5 per cent; avalanches of hot material, 4.5 per cent; *lahars* (mud flows), 3 per cent; atmospheric shock waves, 3 per cent; and lava avalanching into the sea, 1 per cent. About 25 per cent had no discernible origin, but probably were produced by submerged volcanic eruptions. The eruptions of Krakatau in 1883 and Santorini in 1470 BC were responsible for the largest and most significant volcano-induced tsunami.

There has been no historical occurrence of tsunami produced by a meteorite impact with the ocean. However, this does not mean that they are an inconsequential threat. Stony meteorites as small as 300 m in diameter can generate tsunami over 2 m in height that can devastate coastlines within a 1000 km radius of the impact site. The probability of such an event occurring in the next fifty years is just under 1 per cent. One of the largest impact-induced tsunami occurred at Chicxulub, Mexico, sixty-five million years ago at the Cretaceous–*Tertiary* boundary. While the impact was responsible for the extinction of the dinosaurs, the resulting tsunami swept hundreds of kilometers inland around the shore of the early Gulf of Mexico. Impact events are ongoing. Astronomers have compiled evidence that a large comet encroached upon the inner solar system and broke up within the last fourteen thousand years. The Earth has repetitively intersected debris and fragments from this comet. However, these encounters have been clustered in time. Earlier civilizations in the Middle East were possibly destroyed by one such impact around 2350 BC. The last rendezvous occurred as recently as 1500 AD; however, it happened in the southern hemisphere where historical records did not exist at the time. Only in the last decade has evidence become available to show that the Australian coastline preserves the signature of mega-tsunami from this latest impact event.

A partial geographical distribution of tsunami is provided in Table 10.2, while the location of tsunami in the Pacific since 47 BC are plotted in Figure 10.11. The worst regions for tsunami occur along the stretch of islands from Indonesia through to Japan and the coast of the Russia. Most tsunami in this region originate locally. Of 104 damaging tsunami in the past century, only nine have caused damage beyond their source regions. In the Pacific, these originate between Japan and the Kamchatka Peninsula inclusive, along the Aleutian Islands and south Alaskan coast, and along the coast of South America. For example, the Chilean earthquake in 1960 affected most of the coastline of the Pacific Ocean. Here, earthquakes greater in magnitude than 8.2 affect the entire Pacific Ocean and have a probability of occurrence of once every

Table 10.2 Percentage of tsunami in the world's oceans and seas.

Location	Percent
Atlantic east coast	1.6
Mediterranean	10.1
Bay of Bengal	0.8
East Indies	20.3
Pacific Ocean	25.4
Japan-Russia	18.6
Pacific east coast	8.9
Caribbean	13.8
Atlantic west coast	0.4

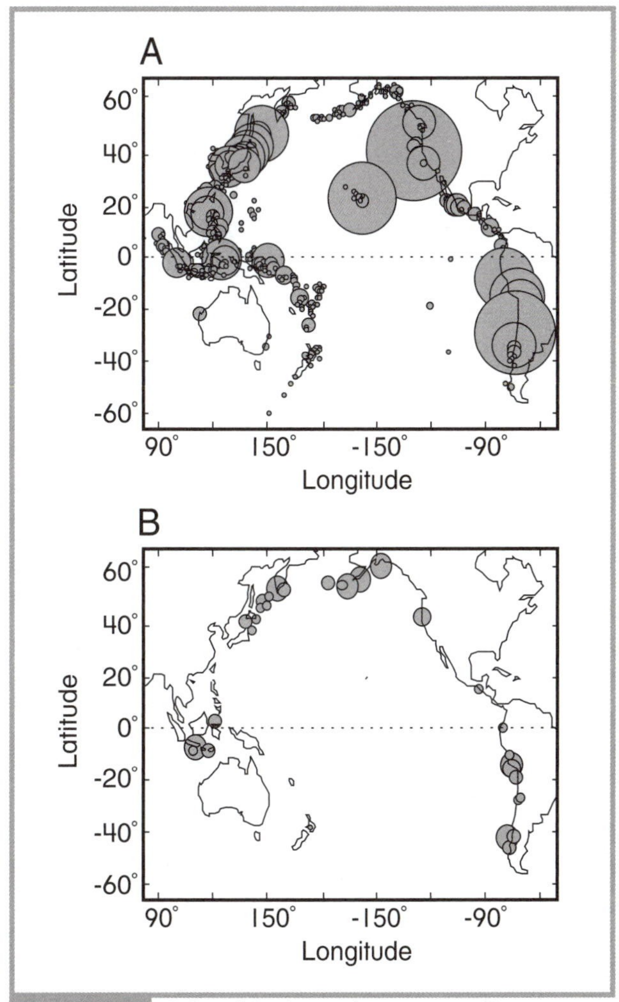

Fig. 10.11 A) Spatial distribution of tsunami since 47 BC; B) source areas for Pacific-wide events (adapted from National Geophysical Data Center and World Data Center A for Solid Earth Geophysics, 1984).

25 years. The high number of tsunami in Oceania includes those at Hawaii, which receives about 85 per cent of all tsunami affecting the Pacific. In fact, Japan and Hawaii are the two most tsunami-prone regions in the world, each accounting for 19 per cent of all tsunami measured. For this reason, Hawaii monitors closely all tsunami source regions in the Pacific.

Peru and Chile are most affected by local earthquakes. This coastline is also the only source area to affect the south-west Pacific region. Japan is mostly affected by locally generated tsunami, and directs its attention to predicting their occurrence locally. Since 684 AD, in the Japan region, 73 tsunami have resulted in over 200 000 deaths. Indonesia and the Philippines are also affected by locally originating tsunami. In the Philippines, recorded tsunami have killed 50 000 people, mainly in two single events in 1863 and 1976. Indonesia has experienced a similar death toll in recorded times; however, half this total has been caused by tsunami associated with local volcanism. The Atlantic coastline is virtually devoid of tsunami; however, the Lisbon earthquake of 1755 produced a 3–4 m wave that was felt on all sides of the Atlantic. The continental slope off Newfoundland, Canada, is seismically active and has produced tsunami that have swept onto that coastline. One recorded tsunami from this site reached Boston, with a height of 0.4 m, on 18 November 1929. One of the longest records of tsunami occurrence exists in the eastern Mediterranean. Between 479 BC and 1981 AD, 7 per cent of the 249 known earthquakes produced damaging or disastrous tsunami. Here, about 30 per cent of all earthquakes produce a measurable seismic wave.

Tsunami generated close to shore are not so large as those generated in deeper water. Tsunami have wave periods in the open ocean of minutes rather than seconds, and heights up to 0.5 m. Waves with this period will travel at speeds of 600–900 km hr^{-1} (166–250 m s^{-1}) in the deepest part of the ocean and 100–300 km hr^{-1} (28–83 m s^{-1}) across the continental shelf. In some cases in the Pacific Ocean, initial wave periods of 60 minutes have been recorded. Wave periods of this length in the open ocean will often degrade to less than 2.5 minutes in shallow water. Tsunami resonating in harbors typically have periods of 8–27 minutes. If the wave period is some harmonic of the natural frequency of the harbor or bay, then the amplitude of the tsunami wave can greatly increase over time. This resonance may cause the period of the wave to shift to higher wave periods (lower frequencies). Figure 10.12 plots typical tide gauge records or *marigrams* of tsunami at various locations in the Pacific Ocean. The concept that a tsunami consists of only one or two waves is not borne out by the records. Some wave periods range between 8 and 30 minutes and persist for over six hours. Wave characteristics are highly variable. In some cases, the waves consist of an initial peak that then tapers off in height exponentially over four to six hours. In other cases, the tsunami wave train consists of a maximum peak well back in the wave sequence.

The wavelength of an average tsunami of eight minutes is about 360 km (refer to Equation 8.3). As explained in Chapter 8, with this wavelength it is possible for the tsunami to feel the ocean bottom at any depth, and to undergo refraction effects. Refraction diagrams of tsunami give accurate prediction of arrival times across the Pacific, but not of run-up heights near shore. It has been shown that the spread of tsunami close to shore and around islands follows classic diffraction theory (Figure 8.3c). Figure 10.13 illustrates the movement of the Chilean tsunami wave of 23 May 1960, which wreaked havoc throughout the Pacific region. It clearly shows that the wave radiated symmetrically away from the epicenter, and was unaffected by bottom topography until islands or seamounts were approached. The island chains in the west Pacific clearly broke up the wavefront; however, much of Japan received the wave unaffected by refraction.

It is a very noticeable phenomenon that ships, anchored several kilometers out to sea, hardly notice the arrival of a tsunami wave. However, once the wave approaches shore, the wave very rapidly shoals and reaches its extreme height. This run-up height controls the extent of damage. Because most shorelines are relatively steep compared to the wavelength of the tsunami, the wave will not break or form a bore but will surge up over the foreshore (Figure 10.14). For example, the earthquake offshore from Gisborne, New Zealand, on 26 March 1947, generated a tsunami run-up 10 m high along a 13 km stretch of coast, while the eruption of Krakatau in 1883 generated a wave that reached elevations up to 40 m high along the surrounding coastline. The largest recorded earthquake-generated tsunami wave occurred in 1737 on the Kamchatka Peninsula when a 64 m high wave washed across the southern tip of the peninsula. By far

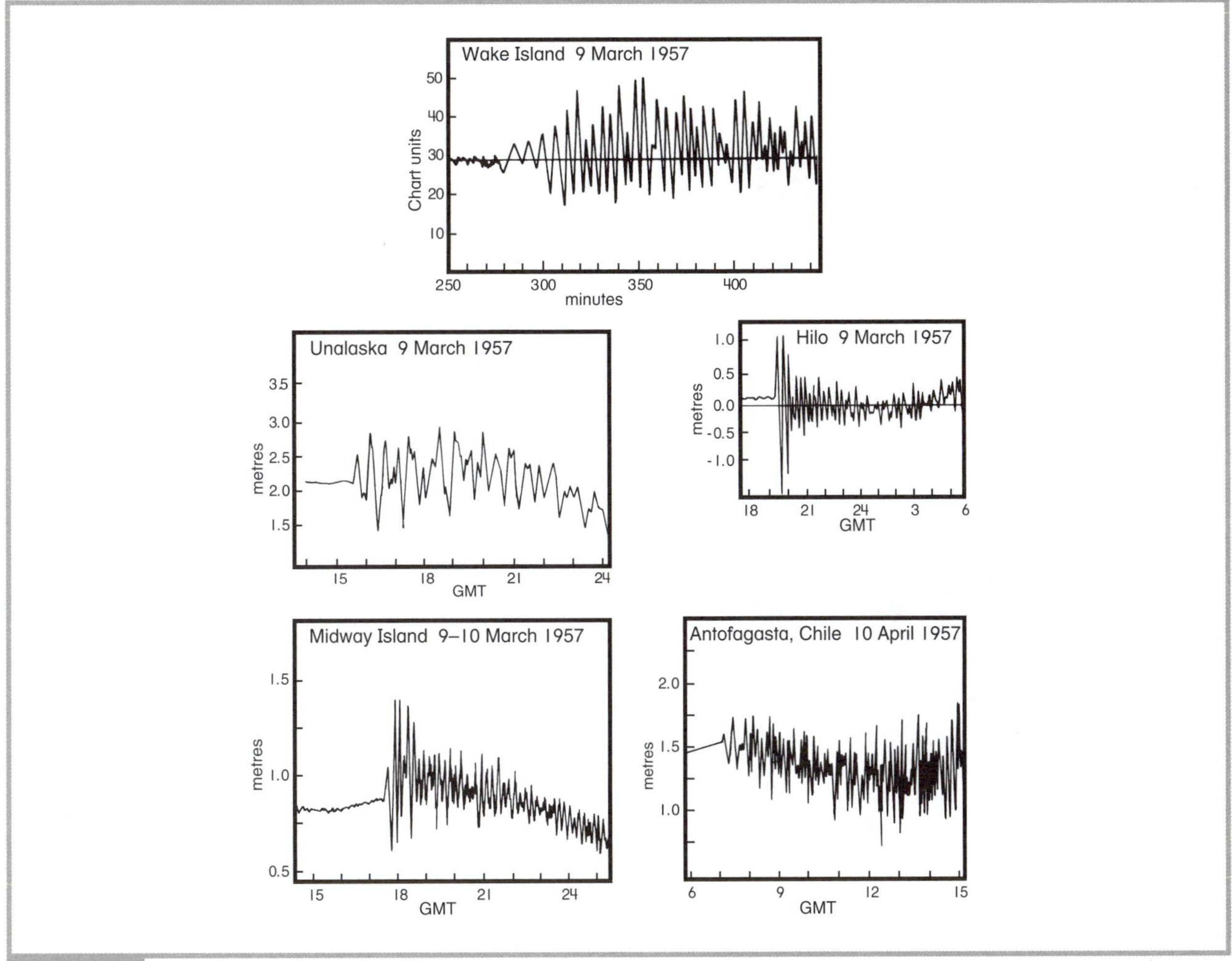

Fig. 10.12 Plots or marigrams of tsunami wave trains at various tidal gauges in the Pacific region (after Salsman, 1959; Van Dorn, 1959; Wiegel, 1964).

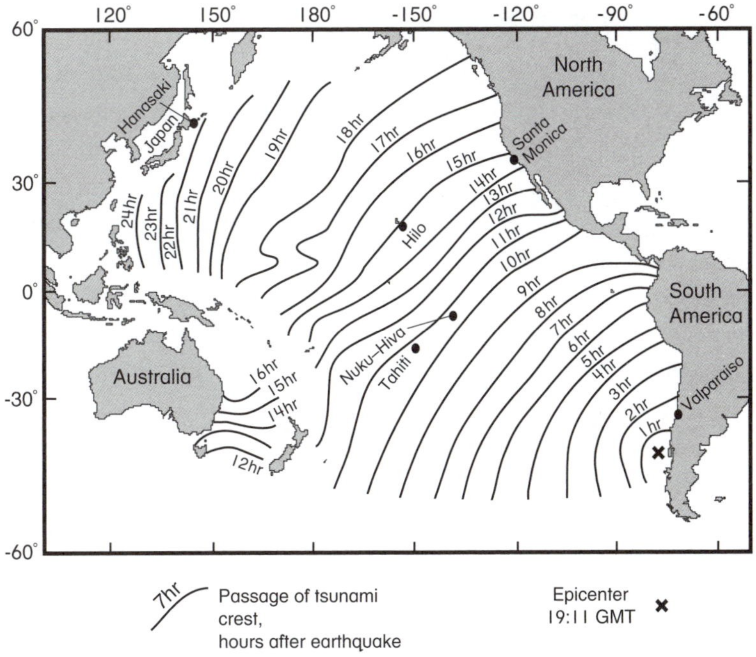

Fig. 10.13 Tsunami wave refraction pattern for 22 May 1960 Chilean earthquake in the Pacific Ocean (based on Wiegel, 1964 and Pickering et al., 1991).

Fig. 10.14 Sequential photographs of the 9 March 1957 tsunami overriding the backshore at Laie Point, Oahu, Hawaii. The tsunami was generated by an earthquake with a surface magnitude of 8.3 in the Aleutian Islands, 3600 km away. Fifty-four people died in this tsunami (photograph credit: Henry Helbush. Source: United States Geological Survey, Catalogue of Disasters #B57C09-002).

Fig. 10.15 The American gunship *Wateree* in the foreground and the Peruvian warship *America* in the background. Both ships were carried inland 3 km by a 21 m high tsunami during the Arica, Peru (now Chile) event of 13 August 1868. Retreat of the sea from the coast preceded the wave, bottoming both boats. The *Wateree*, being flat-hulled, bottomed upright and then surfed the crest of the tsunami wave. The *America*, being keel-shaped, was rolled repeatedly by the tsunami (photograph courtesy of the United States Geological Survey Catalogue of Disasters #A68H08-002).

the largest run-up height recorded is that produced by an earthquake-triggered rockfall into Lituya Bay, Alaska, on 9 July 1958. The resulting wall of water surged 524 m up the shoreline on the opposite side of the bay. Figure 10.15 dramatically illustrates the potential extent of flooding inland that such tsunami are able to generate. The American gunship *Wateree* and the Peruvian ship *America* were standing offshore during the Arica, Chile, tsunami of 13 August 1868. They were both swept 5 km up the coast, and 3 km inland over the top of sand dunes by two successive tsunami waves. The ships came to rest 60 m from a vertical cliff at the foot of the coastal range, against which the tsunami had surged to a height 14 m above sea level. Unfortunately, 25 000 deaths also resulted from this event along the coasts of Chile and Peru.

Run-up heights also depend upon the configuration of the shore, diffraction, characteristics of the wave, and resonance. Within some embayments, it takes several waves to build up peak tsunami wave heights. If the first wave in the tsunami wave train is the highest, then it can be dampened out. Figure 10.16 maps the heights of tsunami run-up around Hawaii for the 1946 Aleutian earthquake. The northern coastline facing the tsunami advance received the highest run-up. However, there was also a tendency for waves to wrap around the islands and strike hardest at supposedly protected sides, especially on the islands of Kauai and Hawaii. Because of refraction effects, almost every promontory also experienced high run-ups, often

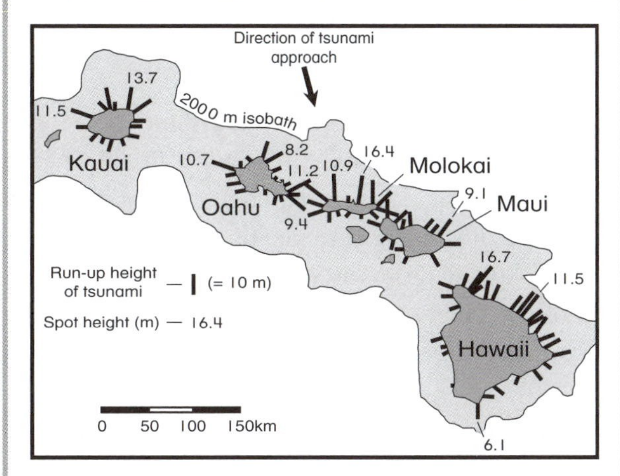

Fig. 10.16 Run-up heights of the 1 April 1946 tsunami around the Hawaiian Islands following an Alaskan earthquake (based on Shepard, 1977).

exceeding 5 m. Coastlines where offshore depths dropped off steeply were hardest hit, because the tsunami waves could approach shore with minimal energy dissipation. For these reasons, run-up heights were spatially very variable. In some places, for example on the north shore of Molokai, heights exceeded 10 m, while several kilometres away heights did not reach 2.5 m.

Tsunami magnitude scales

(Abe, 1979, 1983; Horikawa & Shuto, 1983; Hatori, 1986; Shuto, 1993; Geist, 1997)

Earthquake-generated tsunami are associated with events registering more than 6.5 on the M_s scale. Two-thirds of damaging tsunami in the Pacific region have been associated with earthquakes of magnitude 7.5 or greater. Tsunami can be scaled according to the size of their run-up. The following scale, known as the Imamura–Iida scale, was defined using approximately one hundred Japanese tsunami between 1700 and 1960:

$$m_{II} = \log_2 H_{rmax} \qquad (10.1)$$

where m_{II} = Imamura–Iida's tsunami magnitude scale (dimensionless)

H_{rmax} = maximum tsunami run-up height

On the Imamura–Iida scale, the biggest tsunami in Japan – the Great Meiji Sanriku tsunami of 1896, which had a run-up height of 38.2 m – had a magnitude of 4.0. The other great Sanriku tsunami of 1933, which ranks as the second largest recorded tsunami in Japan, had a magnitude of 3.0. Japanese tsunami have between 1 and 10 per cent of the total energy of the source earthquake. This can be related to a tsunami's magnitude using the Richter scale (Table 10.3). Only earthquakes of magnitude 7.0 or greater are responsible for significant tsunami waves in Japan, with run-up heights in excess of 1 m. However, as an earthquake's magnitude rises above 8.0, the run-up height and destructive energy of the wave dramatically increase. A magnitude 8.0 earthquake can produce tsunami run-up of between 4 and 6 m in height. Intensities of 8.75 generated the 38.2 m wave run-up of the Great Meiji Sanriku tsunami.

The Imamura–Iida magnitude scale has acquired worldwide usage. However, because the maximum run-up height of a tsunami can be so variable along a coast, Soloviev proposed a more general scale as follows:

$$i_s = \log_2 (1.4\, \overline{H}_r) \qquad (10.2)$$

where i_s = Soloviev's tsunami magnitude (dimensionless)

$\overline{H}_r$ = Mean tsunami run-up height along a stretch of coast (m)

This scale and its relationship to both mean and maximum tsunami wave run-up heights are summarized in Table 10.4. Neither the Imamura–Iida nor the Soloviev scales relate transparently to earthquake

Table 10.3 Earthquake magnitude, tsunami magnitude and tsunami run-up heights in Japan (based on Iida, 1963).

Earthquake Magnitude Richter Scale	Tsunami Magnitude	Maximum run-up (m)
6	–2	<0.3
6.5	–1	0.5–0.75
7	0	1.0–1.5
7.5	1	2.0–3.0
8	2	4.0–6.0
8.25	3	8.0–12.0
8.5	4	16.0–24.0
8.75	5	>32.0

Source: Based on Iida, 1963.

magnitude. For example, both tsunami scales contain negative numbers and peak around a value of 4.0. Tsunami waves clearly carry quantitative information about the details of earthquake-induced deformation of the seabed in the source region. Knowing the tsunami magnitude, M_t, it is possible to calculate the amount of seabed involved in its generation using the following formula:

$$M_t = \log_{10} S_t + 3.9 \tag{10.3}$$

where S_t = area of seabed generating a tsunami (m^2)

Table 10.4 Soloviev's scale of tsunami magnitude (source: from Horikawa and Shuto, 1983).

Tsunami magnitude	Mean run-up height (m)	Maximum run-up height (m)
-3.0	0.1	0.1
-2.0	0.2	0.2
-1.0	0.4	0.4
0.0	0.7	0.9
1.0	1.5	2.1
2.0	2.8	4.8
2.5	4.0	7.9
3.0	5.7	13.4
3.5	8.0	22.9
4.0	11.3	40.3
4.5	16.0	73.9

The above scales imply that the size of a tsunami should increase as the magnitude of the earthquake increases. This is true for most *teleseismic tsunami* in the Pacific Ocean; however, it is now known that many smaller earthquakes can produce large, devastating tsunami. The Great Meiji Sanriku earthquake of 1896 and the Alaskan earthquake of 1 April 1946 were of this type. The Sanriku earthquake was not felt widely along the adjacent coastline. Yet, the tsunami that arrived thirty minutes afterwards produced run-ups that exceeded 30 m in places and killed 22 000 people. These types of events are known as *tsunami earthquakes*. Submarine landslides are thought to be one of the reasons why some small earthquakes can generate large tsunami; but this explanation has not been proven conclusively. Presently, it is believed that slow rupturing along fault lines causes tsunami earthquakes.

Disaster descriptions

(Cornell, 1976; Myles, 1985; Bryant, 2001)

Over the past two thousand years there have been 462 597 deaths attributed to tsunami in the Pacific region. Of these deaths, 95.4 per cent occurred in events that killed more than one thousand people each. The number of deaths is recorded in Table 10.5 for each of the main causes of tsunami, while the events with the largest death tolls are presented in Table 10.6.

Table 10.5 Causes of tsunami in the Pacific Ocean region over the last 2000 years.

Cause	Number of events	% of events	Number of Deaths	% of deaths
Landslides	65	4.6	14,661	3.2
Earthquakes	1,171	82.3	390,929	84.5
Volcanic	65	4.6	51,643	11.2
Unknown	121	8.5	5,364	1.2
Total	1,422	100.0	462,597	100.0

Source: National Geophysical Data Center and World Data Center A for Solid Earth Geophysics (1998) and Intergovernmental Oceanographic Commission (1999).

Table 10.6 Large death tolls from tsunami in the Pacific Ocean region over the last 2000 years.

Date	Fatalities	Location
22 May 1782	50,000	Taiwan
27 August 1883	36,417	Krakatau, Indonesia
28 October 1707	30,000	Nankaido, Japan
15 June 1896	27,122	Sanriku, Japan
20 September 1498	26,000	Nankaido, Japan
12 August 1868	25,674	Arica, Chile
27 May 1293	23,024	Sagami Bay, Japan
4 February 1976	22,778	Guatemala
29 October 1746	18,000	Lima, Peru
21 January 1917	15,000	Bali, Indonesia
21 May 1792	14,524	Unzen, Ariake Sea, Japan
24 April 1771	13,486	Ryukyu Archipelago
22 November 1815	10,253	Bali, Indonesia
May 1765	10,000	Guanzhou, South China Sea
16 August 1976	8,000	Moro Gulf, Philippines

Source: National Geophysical Data Center and World Data Center A for Solid Earth Geophysics (1998) and Intergovernmental Oceanographic Commission, (1999).

Tectonically generated tsunami account for the greatest death toll, 84.5 per cent, with volcanic eruptions accounting for 11.2 per cent – mainly during two events, the Krakatau eruption of 26–27 August 1883 (36 417 deaths) and the Unzen, Japan, eruption of 21 May 1792 (14 524 deaths). The number of fatalities has decreased over time and is slightly concentrated in South-East Asia, including Japan. The biggest tsunami of the twentieth century occurred in Moro Gulf, Philippines, on 16 August 1976 where 8000 people died. The largest total death toll is concentrated in the Japanese Islands where 211 300 fatalities have occurred. Two events affected the Nankaido region of Japan on 28 October 1707 and 20 September 1498, killing 30 000 and 26 000 people, respectively. The Sanriku Coast of Japan has the misfortune of being the heaviest populated, tsunami-prone, coast in the world. About once per century, killer tsunami have swept this coastline, with two events striking within a forty-year time span. On 15 June 1896, a small earthquake on the ocean floor 120 km south east of the city of Kamaishi sent a 30 m wall of water crashing into the coastline and killing 27 122 people. The same tsunami event was measured 10.5 hours later in San Francisco on the other side of the Pacific Ocean. In 1933, disaster struck again when a similarly positioned earthquake sent ashore a wave that killed three thousand inhabitants. Deadly tsunami have also affected Indonesia and the South China Sea. In the South China Sea, recorded tsunami have killed 77 105 people, mainly in two events in 1762 and 1782. Indonesia has experienced a comparable death toll (69 420 deaths) over this period with the largest following the eruption of Krakatau in 1883.

While earthquakes are responsible for most destructive tsunami, the Santorini and Krakatau volcanic eruptions were probably the most devastating volcanic events. The Krakatau eruption of 27 August 1883, which will be discussed in more detail in the next chapter, sent out a tsunami wave measured around the world. The wave rounded the Cape of Good Hope in South Africa 6000 km away, and was recorded along the English Channel 37 hours later. On the other side of the Pacific Ocean, water levels were affected along the west coast of Panama and in San Francisco Bay. The waves that rounded the world were probably associated with a substantial atmospheric shock wave, as some tsunami were detected in bodies of water that were not connected to each other. For instance, it is difficult to see how the tsunami wave could have got through the island archipelagos of the west Pacific to register at San Francisco. In addition, a 0.5 m oscillation was measured in Lake Taupo, in the center of New Zealand, coincidentally with the passage of the atmospheric shock wave. The Santorini eruption of 1470 BC generated a tsunami that must have destroyed all coastal towns in the eastern Mediterranean. The Santorini crater is five times larger in volume than that of Krakatau, and twice as deep. On adjacent islands, there is evidence of *pumice* stranded at elevations up to 250 m above sea level. The initial tsunami waves may have been 60 m in height as they spread out from Santorini.

The Lisbon earthquake of 1 November 1755, which is possibly the largest earthquake known (9.0 on the M_s scale), resulted in the lower town being submerged under a tsunami 15 m in height. The wave raised tide levels 3–4 m above normal in Barbados and Antigua, West Indies, on the other side of the Atlantic. Tsunami also affected the west coast of Europe and the Atlantic coast of Morocco. The Spanish port of Cadiz, as well as Madeira in the Azores, was also hit by waves 15 m high, while a 3–4 m high wave sank ships along the English Channel. Water level oscillations were also noticed in the North Sea. The Caribbean has also witnessed its share of devastating tsunami. Noted events occurred in 1842, 1907, 1918, and 1946. However, the worst tsunami recorded destroyed Port Royal, Jamaica, in June 1692. The earthquake that triggered the tsunami collapsed the city, sending much of it sliding into the sea because of liquefaction. The resulting tsunami flung ships standing in the harbor inland over two-storey buildings. Because of the small size of the town, the death toll was fortunately only 2000.

The most active area seismically producing tsunami is situated along the eastern edge of the Nazca Plate, along the coastlines of Chile and Peru. This region has been inundated by destructive tsunami at roughly 30-year intervals in recorded history – in 1562, 1570, 1575, 1604, 1657, 1730, 1751, 1819, 1835, 1868, 1877, 1906, 1922 and, most recently, 1960. The tsunami events of 21–22 May 1960 were produced in Chile by over four dozen earthquakes with magnitudes up to 8.5 and 9.6 on the M_s and M_W scales, respectively. A series of tsunami waves, spread over a period of 18 hours, took over 2500 lives across the Pacific, and produced property damage in such diverse places as Hawaii, Pitcairn Island, New Guinea, New Zealand, Japan, Okinawa, and the Philippines (see Figure 10.13).

The tsunami resulted from sudden land displacement of 2–4 m along the fault during the earthquakes. Within 15 minutes of the faulting, three large waves devastated coastal towns in Chile, where over 1700 people lost their lives. The arrival of the first wave was predicted at Hilo, Hawaii, five hours in advance to an accuracy of one minute. Unfortunately, 61 people, mainly sightseers, were killed as the wave hammered the Hawaiian Islands. When the wave reached Japan 24 hours later, it still had enough energy to reach a height of 3.5–6 m along the eastern coastline. Five thousand homes were washed away, hundreds of ships sank and 190 people lost their lives. In Hawaii, this event illustrated the folly of some humans in the face of disaster. Despite plenty of warning, only 33 per cent of the residents evacuated the affected area in Hilo. Over 50 per cent evacuated only after the first wave arrived, and 15 per cent stayed behind even after large waves had beached. Many of those killed were spectators who went back to see the action of a tsunami hitting the coast.

Prediction in the Pacific region

(Sokolowski, 1999; Bryant, 2001; International Tsunami Information Center, 2004)

The most devastating ocean-wide tsunami of the past two centuries have occurred in the Pacific Ocean. Following the Alaskan tsunami of 1946, the United States government established a tsunami warning system in the Pacific Ocean under the auspices of the Seismic Sea Wave Warning System. In 1948, this system evolved into the Pacific Tsunami Warning Center (PTWC). Warnings were initially issued for the United States and Hawaiian areas; but, following the 1960 Chilean earthquake, the scheme was extended to all countries bordering the Pacific Ocean. Until 1960, Japan had its own warning network, believing at the time that all tsunami affecting Japan originated locally. The 1960 Chilean tsunami proved that large submarine earthquakes in the Pacific Ocean region could spread tsunami ocean-wide. The Pacific Tsunami Warning System was significantly tested following the Alaskan earthquake of 1964. Within forty-six minutes of that earthquake, a Pacific-wide tsunami warning was issued. This earthquake also precipitated the need for further international cooperation on tsunami warning in the Pacific. The United States National Weather Service currently maintains the Center. Currently, twenty-five countries cooperate in the Pacific Tsunami Warning System, in one of the most successful disaster mitigation programs in existence. The countries involved include Canada, the United States and its dependencies, Mexico, Guatemala, Nicaragua, Colombia, Ecuador, Peru, Chile, Tahiti, Cook Islands, Western Samoa, Fiji, New Caledonia, New Zealand, Australia, Indonesia, Philippines, Hong Kong, People's Republic of China, Taiwan, Democratic People's Republic of Korea, Republic of Korea, Japan and the Russian Federation. An additional ten countries or dependencies receive PTWC warnings. Many of the primary countries also operate national tsunami warning centers, providing warning services for their local area.

The Pacific Tsunami Warning System relies on any earthquake 6.5 or greater on the Richter scale registering on one of thirty-one seismographs outside the shadow zones of any P- or S-waves originating in the Pacific region (Figure 10.17). These stations are operated by the Center itself, the West Coast and Alaskan Tsunami Warning Center, the United States Geological Survey's National Earthquake Information Center, and various international agencies. Once a suspect earthquake has been detected, information is relayed to Honolulu where requests for fluctuations in sea level on tide gauges are issued to member countries operating sixty tide gauges scattered throughout the Pacific. These gauges can be polled in real time so that warnings can be distributed to one hundred dissemination points with three hours' advanced notice of the arrival of a tsunami. For any earthquake with a surface magnitude of 7 or larger, the warnings are distributed to local, state, national, and international centers. Administrators, in turn, disseminate this information to the public, generally over commercial radio and television channels. The National Oceanic and Atmospheric Administration (NOAA) Weather Radio system provides direct broadcast of tsunami information to the public via very high frequency (VHF) transmission. The US Coast Guard also broadcasts urgent marine warnings on medium frequency (MF) and VHF marine radios. Once a significant tsunami has been detected, its path is then monitored to obtain information on wave periods and heights. These data are then used to define travel paths using refraction–diffraction diagrams calculated beforehand for any possible tsunami originating in any part of the Pacific region.

The United States supplements these warnings using six seabed transducers that have been installed

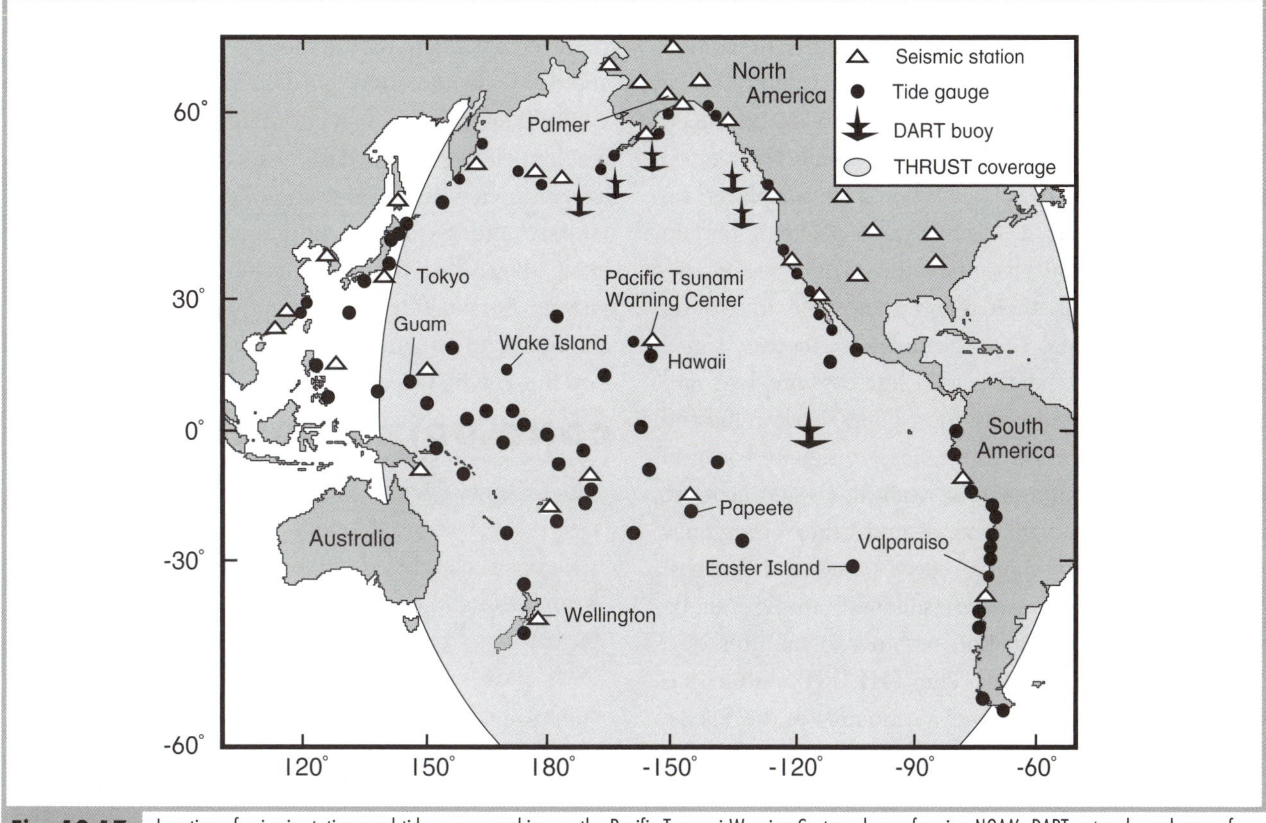

Fig. 10.17 Location of seismic stations and tide gauges making up the Pacific Tsunami Warning System, buoys forming NOAA's DART network, and area of potential coverage of the THRUST satellite warning system (sources: Bernard, 1991; González, 1999; <http://vishnu.glg.nau.edu/wsspc/tsunami/HI/Waves/waves00.html> [cited 2000, now defunct]

and linked to satellites for rapid communication (Figure 10.17). The project is known as the Deep-Ocean Assessment and Reporting of Tsunami (DART). The bottom transducers can detect a tsunami with heights of only 1 cm in water depths of 6000 m. The United States also warns its west coast separately. The West Coast/Alaskan Tsunami Warning Center (WC/ATWC) in Palmer, Alaska, was established after the 1964 Alaskan earthquake to provide tsunami warnings to coastal areas of Alaska following local earthquakes. In 1982, the Center's mandate was extended to include the coasts of California, Oregon, Washington, and British Columbia. Alarms are triggering by any sustained, large earthquake monitored at eight seismometers positioned along the west coast of North America. Regional warnings can be issued within fifteen minutes of the occurrence of any earthquake having a surface magnitude greater than 7.

Localized earthquakes pose dilemmas in many countries. Separate warning systems exist for Hawaii, Russia, French Polynesia, Japan and Chile. The Russian warning system was developed for the Kuril–Kamchatka region following the devastating Kamchatka tsunami of 1952. This system is geared towards the rapid detection of the epicenter of coastal tsunamigenic earthquakes, because some tsunami here take only 20–30 minutes to reach shore. In French Polynesia in 1987, a system was developed by the Polynesian Tsunami Warning Center at Papeete, Tahiti, to detect both near- and far-field tsunami. The system uses the automated algorithm TREMORS and calculates the expected tsunami height from any epicenter in the Pacific. Near-field tsunami are a threat in Japan, especially along the Sanriku coastline of north-eastern Honshu where only 25–30 minutes lead time exists between the beginning of an earthquake and the arrival at shore of the resulting tsunami. The P-wave for any local earthquake can be detected within seconds using an extensive network of high-frequency and low-magnification seismometers. Within the next two minutes, the height of the tsunami along the adjacent coastline can be predicted using graphic forecasting models based upon the size of prior earthquakes and the distance to their epicenters. Finally, Chile, being the source of many teleseismic tsunami, does not rely upon the Pacific Tsunami Warning System, because this does not give a long enough lead time in which to evacuate the coast. Project THRUST (Tsunami Hazards Reduction Utilizing Systems Technology) was established offshore from Valparaiso, Chile, in 1986 to provide advanced warning (within two minutes) of locally generated tsunami along this coastline. When a sensor placed on the seabed detects a seismic wave above a certain threshold, it transmits a signal to the GEOS geostationary satellite, which then relays a message to ground stations. The signal is processed and another signal is transmitted via satellite to a low-cost receiver and antenna operating twenty-four hours a day, located along a threatened coastline. This designated station can be pre-programmed to activate lights and acoustic alarms, and to dial telephones and other emergency response apparatus when it receives a signal. For a cost of $US15 000, a life-saving tsunami warning can be issued to a remote location within two minutes of a tsunamigenic earthquake. The THRUST system has the potential to be used over a wide area of the Pacific (Figure 10.16).

The Pacific Tsunami Warning System is not flawless. The risk still exists in many island archipelagos along the western rim of the Pacific, that local earthquakes could generate tsunami too close to shore to permit advanced warning. The 7.8 magnitude (M_s scale) Philippines earthquake of 16 August 1976, in the Moro Gulf on the south-west part of the island of Mindanao, generated a local tsunami 3.0–4.5 m high that reportedly killed 8000 people. The event was virtually unpredictable because the earthquake occurred within 20 km of a populated coastline. Similarly, the PNG tsunami of 17 July 1978, which killed 2200 people, also escaped the notice of the Pacific Tsunami Warning System. Finally, our knowledge of tsunami is rudimentary for many countries and regions in the Pacific Ocean. While travel-time maps have been drawn up for source regions historically generating tsunami – for example, the coasts of South America, Alaska and Kamchatka Peninsula – the entire Pacific Rim coastline has not been studied. This was made apparent on 25 March 1998 when an earthquake with a surface magnitude, M_s, of 8.8 occurred in the Balleny Islands region of the Antarctic directly south of Tasmania, Australia. Because of the size of the earthquake, a tsunami warning was issued; but no one knew what the consequences would be. The closest tide gauges were located on the south coasts of New Zealand and Australia. The Pacific Tsunami Warning Center in Hawaii had to wait to see if any of these gauges reported a tsunami before they issued warnings further afield. While that may have helped residents in the United States or Japan, it certainly was little comfort to residents living along coastlines facing the Antarctic in the antipodes. In cities such as Adelaide, Melbourne, Hobart and Sydney, emergency hazard personnel knew they were the 'mine canaries' in the warning system. Fortunately, the Antarctic earthquake was not conducive to tsunami and no major wave propagated into the Pacific Ocean.

CONCLUDING COMMENTS

The seismic risk for many locations in the world can now be assessed to the point that many buildings and structures can be designed to withstand damage from tremors. However, the seismic risk in many locations, even in developed countries or ones where earthquakes have been recorded for centuries, cannot be assessed completely. For example, the seismic risk for Australia is very much a product of that country's distribution of cities (Figure 10.2). In North America, the assessment of seismic risk in the eastern part of the continent is still deficient because major earthquakes occur so infrequently that, even after centuries of settlement, the full extent of the hazard is not known (Figure 10.2). This chapter has not highlighted recent earthquakes but concentrated upon two large events, the Alaskan earthquake of 1964 and the Tokyo earthquake of 1923, to illustrate the force that can be released in a tremor and the destruction that can result. The Alaskan earthquake was one of the largest events in the twentieth century. Even though it occurred in a relatively isolated part of the world, its magnitude and subsequent tsunami attracted world attention. The Tokyo earthquake is noteworthy because of the appalling death toll and the manner in which those deaths occurred. The descriptions of these events could easily have been set within southern California, a region where earthquakes with the same magnitude as the Alaskan one can easily occur, and a region where the effects upon dense human settlement might be as dramatic. In discussing the Californian earthquake hazard, it was hoped to highlight the extent of the hazard and some of the inadequacies in people's attempts to minimize future damage and loss of life. Despite decades of retro-fitting of structures to withstand earthquakes, engineering expertise is only recently coming to terms with the fact that the effects of seismic waves upon structures has been underestimated by a factor of 2–4. The San Francisco and Kobe earthquakes of 1989 and 1995, respectively, triggered this reassessment. There are extensive regions prone to large earthquakes where people must live and adjust to this threat.

Finally, very little discussion was given to the threat of tsunami along coastlines where historical records are minimal. The coastline of Australia illustrates this point. This country has been scrutinized by written observations for only the past 230 years. Historically, the largest tsunami wave height measured on a tide gauge – 1.07 m – occurred at Sydney in May 1877. Observations exist on the north-west coast of Australia for run-up heights of 4 m following the Krakatau eruption in 1883 and of 6 m in August 1977 following an Indonesian earthquake. Yet, Aboriginal legends originating from coastal regions frequently note the occurrence of large tsunami. Detailed fieldwork examining a range of signatures left by tsunami in coastal deposits and the erosion of bedrock across Australia indicate that tsunami have run up to 35 km inland on the north-west coast and topped headlands 130 m high south of Sydney. The extent and magnitude of this evidence could be caused only by a comet or meteorite impact with the ocean. These events are not ancient. Radiocarbon dating indicates that these mega-tsunami occurred within the last 300–500 years. Both Aboriginal and Maori legends also substantiate a cosmogenic event within this time frame. Our knowledge of tsunami as a hazard is deficient in terms of both cause and extent.

REFERENCES AND FURTHER READING

Abe, K. 1979. Size of great earthquakes of 1837–1974 inferred from tsunami data. *Journal Geophysical Research* 84: 1561–1568.

Abe, K. 1983. A new scale of tsunami magnitude, M_t. In Iida, K. and Iwasaki, T. (eds) *Tsunamis: Their Science and Engineering*. Reidel, Dordrecht, pp. 91–101.

Bardet, J.P., Idriss, I.M., O'Rourke, T.D., Adachi, N., Hamada, M., Ishihara, K. 1997. *North America–Japan Workshop on the Geotechnical Aspects of the Kobe, Loma Prieta, and Northridge Earthquakes Quakes*. Report to National Science Foundation, Air Force Office of Scientific Research and Japanese Geotechnical Society.

Bernard, E.N. 1991. Assessment of Project THRUST: past, present, future. *Proceedings of the Second UJNR Tsunami Workshop*, Honolulu, Hawaii 5–6 November 1990, National Geophysical Data Center, Boulder, pp. 247–255.

Bolt, B.A. 1993. *Earthquakes*. W.H. Freeman and Co., New York.

Bolt, B.A., Horn, W.L., MacDonald, G.A., and Scott, R.F. 1975. *Geological Hazards*. Springer-Verlag, Berlin.

Bryant, E. 2001. *Tsunami: The Underrated Hazard*. Cambridge University Press, Cambridge.

Cornell, J. 1976. *The Great International Disaster Book*. Scribner's, New York.

EQE 1995. *The January 17, 1995 Kobe earthquake*. <http://www.eqe.com/publications/kobe/kobe.htm> (URL defunct as of 2004)

Geist, E.L. 1997. Local tsunamis and earthquake source parameters. *Advances in Geophysics* 39: 117–209.

Giardini, D., Grünthal, G., Shedlock, K. and Zhang, P. 1999. *Global Hazard Seismic Map*. Global Seismic Hazard Assessment Program, UN/International Decade of Natural Hazard Reduction.

González, F.I. 1999. Tsunami! *Scientific American*, May: 44–55.

Hadfield, P. 1992. *Sixty Seconds That Will Change the World: The Coming Tokyo Earthquake*. Tuttle, Boston.

Hansen, W.R. 1965. Effects of the Earthquake of March 27, 1964 at Anchorage, Alaska. *United States Geological Survey Professional Paper* No. 542-A.

Hatori, T. 1986. Classification of tsunami magnitude scale. *Bulletin Earthquake Institute* 61: 503–515 (in Japanese).

Hays, W.W. 1981. Facing geologic and hydrologic hazards: Earth-science considerations. *United States Geological Survey Professional Paper* No. 1 240-B: 54–85.

Hodgson, J.H. 1964. *Earthquakes and Earth Structure*. Prentice-Hall, Englewood Cliffs, New Jersey.

Holmes, A. 1965. *Principles of Physical Geology*. Nelson, London.

Horikawa, K. and Shuto, N. 1983. Tsunami disasters and protection measures in Japan. In Iida, K. and Iwasaki, T. (eds) *Tsunamis: Their Science and Engineering*. Reidel, Dordrecht, pp. 9–22.

Iida, K. 1963. Magnitude, energy and generation of tsunamis, and catalogue of earthquakes associated with tsunamis. Proceedings of the Tsunami meetings associated with the Tenth Pacific Science Congress. *International Union of Geodesy and Geophysics Monograph* 24: 7–18.

Intergovernmental Oceanographic Commission 1999. *Historical tsunami database for the Pacific, 47 BC–1998 AD*. Tsunami Laboratory, Institute of Computational Mathematics and Mathematical Geophysics, Siberian Division, Russian Academy of Sciences, Novosibirsk, Russia, <http://tsun.sscc.ru/HTDBPac1/> (URL defunct as of 2004)

International Tsunami Information Center 2004. <http://www.prh.noaa.gov/itic/more_about/itic/itic.html>

Johnson, J.M. 1999. Heterogeneous coupling along Alaska-Aleutians as inferred from tsunami, seismic, and geodetic inversions. *Advances in Geophysics* 39: 1–106.

Keller, E.A. 1982. *Environmental Geology* (3rd edn). Merrill, Columbus, Ohio, pp. 133–167.

Lomnitz, C. 1988. The 1985 Mexico earthquake. In El-Sabh, M.I. and Murty, T.S. (eds) *Natural and Man-Made Hazards*, Reidel, Dordrecht, pp. 63–79.

Myles, D. 1985. *The Great Waves*. McGraw-Hill, New York.

National Geophysical Data Center and World Data Center A for Solid Earth Geophysics 1984. *Tsunamis in the Pacific Basin 1900–1983*. United States National Oceanic and Atmospheric Administration, Boulder, Colorado, map 1:17 000 000.

Oakeshott, G.G. 1983. San Andreas Fault: Geologic and Earthquake History. In Tank, R.W. (ed.) *Environmental Geology*. Oxford University Press, Oxford, pp. 100–107.

Palm, R. and Hodgson, M.E. 1992. *After a California Earthquake: Attitude and Behavior Change*. Geography Research Paper No. 233, University of Chicago.

Pararas-Carayannis, G. 1998. *The March 27, 1964 Great Alaska Tsunami*. <http://www.geocities.com/CapeCanaveral/Lab/1029/Tsunami1964GreatGulf.html>

Pickering, K.T., Soh, W., and Taira, A. 1991. Scale of tsunami-generated sedimentary structures in deep water. *Journal of the Geological Society London* 148: 211–214.

Pinter, B., Gobron, N., Verstraete, M.M., Mélin, F., Widlowski, J-L., Govaerts, Y., Diner, D.J., Fielding, E., Nelson, D.L., Madariaga, R., Tuttle, M.P. 2003. Observing earthquake-related dewatering using MISR/Terra satellite data. *EOS, Transactions, American Geophysical Union* 84: 37, 43.

Ritchie, D. and Gates, A.E. 2001. *Encyclopedia of Earthquakes and Volcanoes*. Facts on File, New York.

Robinson, A. 2002. *Earthshock*. Thames & Hudson, London.

Salsman, G.G. 1959. The tsunami of March 9, 1957, as recorded at tide stations. *U.S. Coast and Geodetic Survey Technical Bulletin* No. 6.

Scheidegger, A.E. 1975. *Physical Aspects of Natural Catastrophes*. Elsevier, Amsterdam.

Shepard, F.P. 1977. *Geological Oceanography*. University of Queensland Press, St Lucia.

Shuto, N. 1993. Tsunami intensity and disasters. In Tinti, S. (ed.) *Tsunamis in the World*. Kluwer, Dordrecht, pp. 197–216.

Sokolowski, T.J. 1999. The U.S. West Coast and Alaska Tsunami Warning Center. *Science of Tsunami Hazards* 17: 49–56.

United States Geological Survey 2002. *Most destructive known earthquakes on record in the world*. <http://neic.usgs.gov/neis/eqlists/eqsmosde.html>

Van Dorn, W.G. 1964. Tsunamis. *Advances in Hydrosciences* 2:51–46.

Whittow, J. 1980. *Disasters: The Anatomy of Environmental Hazards*. Pelican, Harmondsworth.

Wiegel, R.L. 1964. *Oceanographical Engineering*. Prentice Hall, Englewood Cliffs, New Jersey.

Wood, R.M. 1986. *Earthquakes and Volcanoes*. Mitchell Beazley, London.

Volcanoes as a Hazard

CHAPTER 11

INTRODUCTION

(Bolt et al., 1975; Tazieff & Sabroux, 1983)

Of all natural hazards, volcanoes are the most complex. Whereas a tropical cyclone has a predictable structure, or a drought can generate a predictable sequence of events in rural communities, such predictions cannot be made in respect of volcanoes. There is a multitude of volcanic forms, and each event appears unique in the way that it behaves, and the physical and human consequences it produces. This chapter will examine the different types of volcanoes and the secondary phenomena associated with their occurrence. It will conclude with a detailed description of some of the more spectacular volcanic disasters that have occurred in recorded history.

Volcanoes are conduits in the Earth's crust through which gas-enriched, molten silicate rock magma reaches the surface from beneath the crust. The origin of magma is still debated, but it is generally believed from seismic evidence that the mantle is partially liquefied 75–300 km below the Earth's surface. There are two types of magma. The first type consists of silica-poor material from the mantle, and forms basaltic volcanoes. The second type consists of silica-rich material originating from either the melting of the crust in subduction zones or the partial differentiation of liquefied mantle material. This second type forms 'andesitic' volcanoes, the largest group of which rings the western Pacific Ocean, where the Pacific Plate is subducting beneath the Eurasian Plate (Figure 9.1). Basaltic lavas have temperatures of between 1050 and 1200°C, while andesitic ones are 100–150°C cooler. For each 500°C decrease in temperature, fluidity decreases tenfold. For this reason, and because of their higher silica content, andesitic lavas are 200–2000 times more viscous than basaltic lavas. Andesitic lavas are also gas-rich, with 50–90 per cent of the gas consisting of water assimilated from dissolved crustal material. This high water content and high *viscosity* makes andesitic volcanoes, located over supposed subduction zones, much more explosive than basaltic volcanoes sitting over hot spots in the center of crustal plates.

Magma is not necessarily extruded at the Earth's surface by pressure originating in the mantle below. Instead, lava penetrates through tens of kilometres of weakened, fractured crust because of gas pressure. All magmas, be they andesitic or basaltic, contain various amounts of dissolved gas. These magmas can become oversaturated in gas due to crystallization of the magma, its cooling down, decreases in hydrostatic pressure, or enrichment of volatile molecules from outside the magma. The gas in oversaturated lavas begins to vesiculate or form bubbles, a process that immediately makes the magma lighter or more buoyant. If the magma is forced quickly upward by pressure from below, then the *hydrostatic pressure* acting on the magma to dissolve these bubbles can decrease rapidly, causing a chain reaction of new bubble formation. If the magma is gas-rich, as andesitic lavas are, and the chain reaction is rapid enough, an explosive eruption can occur. Once explosive magma reaches the atmosphere, it is fragmented into small particles by the gas and by the force of the eruption. These pieces are called *pyroclastic ejecta* or tephra and are classified according to size as follows: bombs (liquid, boulder-sized blobs), *blocks* (angular, solid boulders), *scoriae* (cinder material of any size), *lapilli* (fragments 2–60 mm in diameter), and sands and ash (< 0.1 mm). Depending upon the size and distance hurled, much of this material falls back to the ground around the vent to build up a cone.

The study of the gas phase of magma is crucial in determining whether the volcano will become explosive. If the magma is rising slowly, a small quantity of gas will continuously escape through vents to the surface through *fumaroles*. This movement tends to be slow because the viscosity of the magma inhibits bubble release. Bubbles, once formed, tend to stick to liquid–solid interfaces no matter how buoyant they become. These gases also have sufficient time to dissolve in groundwater. The monitoring of groundwater or escaping gas can give an accurate indication of the maturity of the magma, and its closeness to the surface. Some of these *geochemical* techniques for prediction were discussed in Chapter 10.

In some volcanic eruptions, only this *degassing* process initially affects the upper magma. As this magma is forced out, the magma immediately beneath degasses explosively. In this way, there may not be one big explosion, but a series of continual blasts, each lasting a few seconds to minutes, over several days. Degassed magma may also pour out of the volcano as lava, which (depending upon viscosity, output and the slope of the ground) can flow from a few meters to several hundreds of kilometers from the vent. All lava soon solidifies to form hard rock, although extremely thick lava beds may take years to cool down completely to surrounding environmental temperatures. There are three types of basaltic lava that can be ejected from a volcano: pahoehoe, aa, and block lava. Pahoehoe is basaltic lava that solidifies at the top, but which is still fed from below by pipe vesicles. Such lava has a smooth, wrinkled surface, is less than 15 m thick and flows at rates of a few metres per minute. Aa forms as a thin river of lava, less than 10 m thick, with a spiky, cinder-like topping. Block lava consists solely of andesite or rhyolite up to 300 m thick, and has a surface coating consisting of smooth-sided fragments. It has the slowest traveling speed of all lava flows, moving only a few metres per day. Both aa and block lava behave in all respects like a river and can be bordered by levee banks.

The total energy of a volcanic event can be broken down into four processes as follows:

- energy released by volcanic earthquakes,
- energy necessary to fracture the overburden,
- energy expended in ejecting material, and
- energy expended in producing atmospheric shock waves and, occasionally, tsunami.

Volcanic tremor and earthquake energy can be assessed in similar ways to any other earthquake event. These methods of measuring energy have been discussed under earthquake magnitude in Chapter 10. Earthquakes of magnitude 7 or larger on the Richter scale have been produced by major explosive eruptions, and contain more than 4×10^{15} joules of energy. The energy required to break up overburden is typically greater than 1×10^{11} joules. Energy spent in ejecting material consists of kinetic energy, potential energy, and thermal energy. These terms can be calculated for most volcanic eruptions, and indicate that volcanic explosions release as much energy as large nuclear devices – typically 1×10^{15} joules. The energy used in generating the atmospheric shock wave can be expressed as follows:

$$E_s = 1.25 \times 10^{13} \sin\phi \; 0.5 At^2 \qquad (11.1)$$

where E_s = energy to generate atmospheric shock wave
$\sin \phi$ = distance from source of the explosion in degrees on the Earth
A = amplitude of shock wave in millibars
t = wave period in seconds

A similar type of equation as 11.1 can be used to measure the energy contained in any tsunami wave generated by an eruption. These shock waves require energy expenditure greater than 1×10^{15} joules. In total, volcanic eruptions contain energy in excess of 1×10^{16} joules. Some of the larger eruptions, such as Krakatau in 1883, or Tambora in 1815, contained energy in excess of 1×10^{18} joules (see Figure 11.1 for

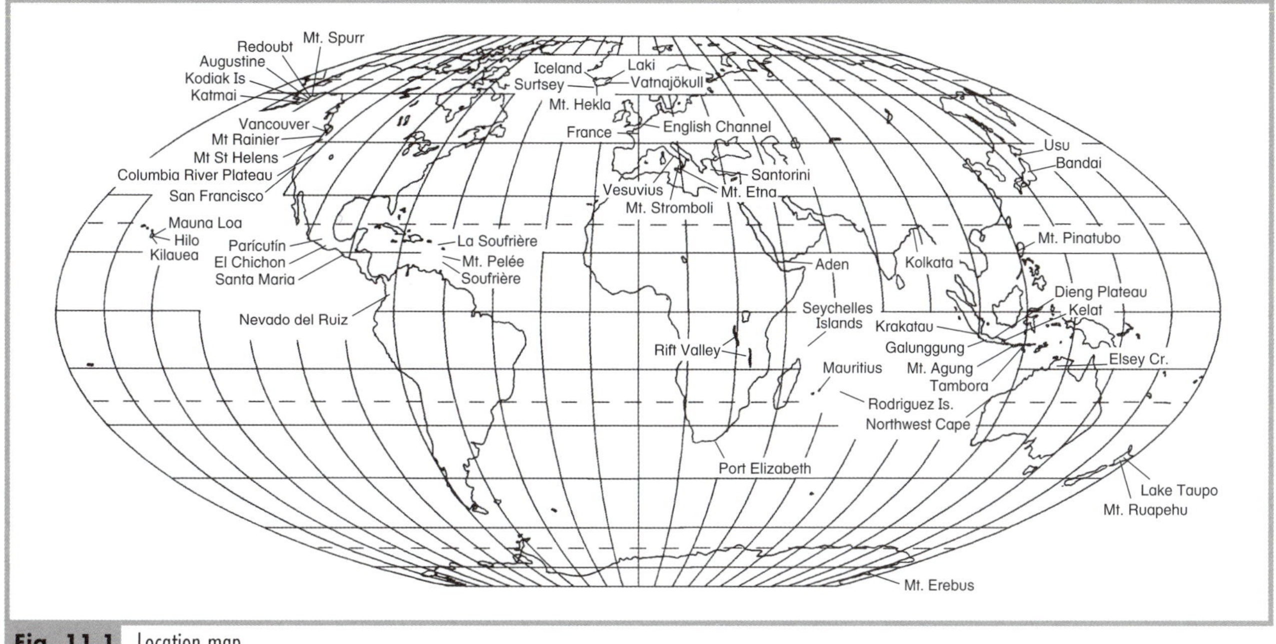

Fig. 11.1 Location map.

all major placenames mentioned in this chapter and not located on individual site maps). By calculating the amount of energy partitioned among these various processes, it is possible to classify the type of volcano and its magnitude. This is of little help in predicting volcanic events, but it can give information on the type of eruption that has occurred. For instance, eruptive volcanoes expend considerable energy in easily measured atmospheric shock waves and tsunami. Knowledge of these values can be used to evaluate the overall energy of the volcano. In addition, the continual monitoring of energy expenditure, and the way it is partitioned, can be used to predict whether a volcano is dying down or increasing in activity.

TYPES OF VOLCANIC ERUPTIONS

(Fielder & Wilson, 1975; Blong, 1984; Wood, 1986)

While most classification schemes refer to a specific historical event characterizing an eruption sequence, it should be realized that each volcanic eruption is unique, and may over time take on the characteristics of more than one type. There are two types of non-explosive volcanoes. Volcanoes producing *flood lavas* are the least hazardous of all eruptions, because they occur very slowly, permitting plenty of time for evacuation. Flood lavas are basaltic in composition. At present, such eruptions are rare, occurring only in Iceland, where they happen with sufficient warning to permit safe evacuation. In the geological past, some volcanoes have had enormous magma chambers that emptied and flooded landscapes over vast distances in a short period. Some flood lavas have laid down deposits 500 m thick over distances of 300 km. The largest such deposits on Earth (flood lavas also occur on other planets and their moons) make up the Deccan Plateau in India, the island of Iceland, the Great Rift Valley of Africa, and the Columbia River Plateau in the United States. 'Hawaiian'-type eruptions are very similar to flood lavas. However, at times, they can produce tephra and faster moving, thin lava flows. Hawaiian eruptions tend to be aperiodic, building up successive deposits over time. These can form large cones that, in the case of the Hawaiian Islands, can be tens of kilometres high and hundreds of kilometres wide at their base.

Explosive eruptions can be more complex. The simplest form begins with the tossing out of moderate amounts of molten debris. 'Strombolian' volcanoes, named after Mt Stromboli in Italy, throw up fluid lava material of all sizes. Most eruptions of this type toss out bombs a few hundreds of metres into the air, rhythmically, every few seconds in events that can go on for years. They can also produce moderate lava flows. Strombolian volcanoes are characterized by a very symmetrical cinder cone that grows in elevation around the vent. Strombolian volcanoes also represent an intermediate stage between basaltic and acidic magma extrusions. 'Vulcanian'-type eruptions are more acidic, and give off large blocks of very viscous, hot magma with few if any flows. This very viscous material may be tossed up to heights of 10 km (40 km in extreme cases). Activity is explosive for periods ranging up to several months. These volcanoes build up tephra and block cones. Usually the lava freezes in vents between eruptions, so that the strength of the plug ultimately determines the amount of pressure that can build up before the next eruption. 'Surtseyan' eruptions are violently explosive because hot, fluid magma encounters seawater. These types of eruptions are termed 'phreatomagmatic' and make up 8.8 per cent of all volcanic events. Because of the interaction of magma and water, large clouds of very fine dust are produced that can drift several kilometres upward into the atmosphere. Coarser particles tend to form rings around the vent, rather than cones or large ash sheets. The rubbing together of fine dust particles builds up static electricity, leading to numerous and spectacular lightning displays around the rising column of ejecta.

'Plinian' events are more explosive, with wide dispersal of tephra. Over 1 km^3 of magma may be driven straight up, at velocities of 600–700 m s^{-1}, to heights of 25 km by a continuous gas jet and thermal expansion over the course of several hours. As the magma chamber slowly empties, the character of the magma changes and the eruption becomes starved for material. Usually, the volcano then collapses internally under its own weight to form a caldera. The Mt Vesuvius eruption of 79 AD was of this type, and will be described later in more detail. While such events have been dramatic on Earth, on other planets with lower gravity and thinner atmospheres, Plinian eruptions have been truly cataclysmic.

If the eruptive ash column from an explosive volcano collapses under the effect of gravity, it can produce a *debris avalanche* of ash and hot gas that can race downslope at speeds reaching 60 km hr^{-1}. This type of eruption is termed a 'Peléean' eruption after the explosion of Mt Pelée in 1902. The debris clouds are called *pyroclastic flows* or *nuées ardentes*, and the resulting debris deposits are termed *ignimbrites*. The large death toll of Vesuvius in 79 AD can be attributed to an associated pyroclastic flow event. Pyroclastic flows will be examined later in this chapter when secondary volcanic hazards are examined in detail. The most destructive and largest Peléean eruptions are termed the 'Katmaian' type, after the Katmai eruption at the eastern end of the Alaska Peninsula on 6 June 1912. This eruption ejected ten times the material of the 1902 Mt Pelée event. As the magma chamber emptied, a 2500 m high mountain collapsed into a caldera 5 km across and 1 km deep. The eruption was heard 1200 km away, with dust being deposited over a 1500 km distance. The pyroclastic flow from this eruption flooded into a valley 20 km long, leaving an estimated 11 km^3 of hot, fused ash spread to a depth of 250 m.

Peléean eruptions should not be used to refer to all volcanic events that generate hot ash flows. Many destructive pyroclastic flows have not originated as ash columns collapsing under the effect of gravity, but have been preceded by a ground or basal surge caused by the lateral eruption of the volcano. This type of explosion is termed a 'Bandaian' eruption. The 1980 eruption of Mt St Helens in the United States was of this type, which is more properly known as a *base surge*. Because the force of the eruption shoots out particles at velocities in excess of 150 km hr^{-1}, the blasts have been likened to cyclones of ash. Objects in the paths of such hot and violent blasts are destroyed by sandblasting. Base surges often precede a large and slower moving pyroclastic flow, and represent the most destructive of all types of volcanic eruptions.

VOLCANIC HAZARDS

(Bolt et al., 1975; Whittow, 1980; Hays, 1981; Tazieff & Sabroux, 1983; Blong, 1984; Wood, 1986; Ritchie & Gates, 2001)

There are many hazardous phenomena produced directly, or as secondary effects, by volcanic eruptions. The association of earthquakes and tsunami with volcanoes has already been covered in Chapter 10, and will be discussed in detail for particular events at the end of this chapter. Excluding these two phenomena, it is possible to group volcanic hazards into six categories as follows:

- lava flows,
- ballistics and tephra clouds,
- pyroclastic flows and base surges,
- gases and acid rains,
- lahars (mudflows), and
- glacier bursts (*Jökulhlaups*).

Lava flows

Since the sixteenth century, 20 km^2 of land per year has been buried by lava flows. There are about 60 lava flow events per century that result in severe social and economic disruption. The main reason for this disruption is the fact that lava flows in volcanic areas are very fertile, and have attracted intense agricultural usage and dense settlement. Re-eruption of volcanoes in these areas has, thus, led to the destruction of this agricultural land and to loss of life. Much of this destruction can be prevented simply by mapping and avoiding those areas most frequently inundated by lava flows. Techniques also exist to modify flow behavior, especially for pahoehoe and aa flows.

Based upon the more than 1000 flows that have occurred historically, low-viscosity flows have a median length of 4.1 km and an average thickness of 10 m, while high-viscosity flows have a median length of 1.3 km and an average thickness of 100 m. There is a 1 per cent probability that high- and low-viscosity lava flows will exceed, respectively, 11 km and 45 km in length. The fastest moving, hottest and most mobile flows are the low-viscosity Hawaiian- and Icelandic-type eruptions. These flows would appear to be dangerous; however, they are thin and often consist of pahoehoe-type lava, which cools very rapidly. Pahoehoe lavas may develop large diameter lava tubes extending several kilometres under the flow. These can provide conduits for subsequent lava venting once the original flow has cooled at the surface. Such flows can travel great distances, but are the easiest to divert. Aa flows are thicker and more viscous. They tend to channelize to depths up to 30 m. Both aa and block lava flows are more difficult to divert.

The characteristics of flows can vary depending upon temperature, viscosity, *yield strength*, and expulsion rate. Because of the high temperatures of flows (880–1130°C), all carbon materials such as cloth, paper and wood are easily ignited upon contact with lava. Once temperatures drop below 800°C, a skin forms on the magma that effectively inhibits further heat loss. Lavas have been known to have high internal temperatures 5–6 years after they have stopped flowing. As temperature decreases, viscosity increases, and it is this factor that stops most flows. While the velocity of a flow depends upon gravity or the slope of the terrain, the rate at which lava is expelled also controls the speed of movement. Most expulsion rates range between 15 and 45 m s^{-1}, but values above 1000 m^3 s^{-1} have been measured. The distance reached by lava is a direct function of this expulsion rate, such that the distance of travel in kilometers is about three times the expulsion rate measured in m^3 s^{-1}. The actual velocity of flows is in a range of 9–100 km hr^{-1}, with average values around 30 km hr^{-1}.

While most documented flows rarely exceed 4 km^3 in volume, some have caused impressive damage. The Laki fissure eruption of 1783–1784 in Iceland covered 550 km^2, had a length of 88 km and a total volume of 12.3 km^3. This eruption caused melting of glaciers and flooding of valuable farmland. The sulfurous vapors from the lava hung over the countryside for weeks. The 'haze famine' that resulted killed 20 per cent of Iceland's population, and wiped out 50 per cent of its cattle and 75 per cent of its sheep and horses. Extensive flows sweeping down the slopes of Mt Vesuvius engulfed the town of Bosco Trecase at its base in 1906 and the town of San Sebastiano in 1944. Lava from Mt Etna in 1669 reached the wall of the feudal city of Catania 25 km away, eventually overtopping the walls and moving slowly through the city. In 1928, the town of Mascali was also destroyed by lava from Mt Etna. Parícutín in Mexico, in 1946, buried 2400 hectares of forest and agricultural land before completely covering the town of San Juan Parangaricutiro, 5 km away from the main vent. Eruptions from Mauna Loa and Kilauea in Hawaii have consistently flooded agricultural land, destroyed small villages surrounding the cones, and threatened the major city of Hilo on several occasions. In January 2002, lava from Mt Nyiragongo in the Democratic Republic of the Congo buried the town of Goma under 1.5 m of lava and caused over 500 000 people to flee for their lives.

The easiest way to stop lava movement is to enhance the processes that slow it down. Flow depends upon yield strength, which can be increased by lowering the temperature of the melt, increasing the rate of gas escape from the flow, stirring, and seeding the lava with foreign nuclei. In some cases, increasing the yield strength of the flow may only turn a fluid flow into a more viscous one, making it thicker and more difficult to stop. Because of the insulating properties of solid lava, cooling by water is not always effective. It has been observed that lava flows can continue to move even when submerged under the sea. One notable exception appears to be the attempts in 1973 to stop a block lava flow, which threatened the town of Vestmannaeyjar in Iceland. Water was intensively pumped for months onto, and just behind, the moving lava front, thus increasing the viscosity of the magma and slowing its rate of progress. Agitation of the flow can be difficult to achieve except by bombing. This was tried on one tube feeding a pahoehoe flow that threatened Hilo, Hawaii, in 1935. The flow was disrupted enough that it turned into a more viscous aa flow, and the tube became blocked. The bombing might not have been the reason for the flow termination, because it was already in the final stages of expulsion. However, there are still contingency plans for bombing lava flows if they threaten Hilo in the future. The sides of a flow can also be breached, thus diverting lava to a safer path. However, what one person considers safe, another may consider a threat. This problem beset the unfortunate inhabitants of Catania in the 1669 eruption of Mt Etna. The citizens of the walled city made the first recorded attempt to alter a lava flow. They clad themselves in wetted animal skins, and dug into one of the lava levees controlling the route of the flow towards their city. Unfortunately, the diversion sent the flow towards the town of Paterno. Five hundred enraged residents of this latter city raced up the slopes to attack the citizens of Catania, and terminated their efforts after winning a pitched battle. The lava flow clogged the breach, proceeded to Catania, overflowed its 20 m high walls, and inundated large sections of the town. A similar attempt in Hawaii in 1942, rather than diverting the flow, only sent it on a parallel path.

Barrier construction has also been used to divert aa flows, or give more time for effective evacuation. Most lava flows will pond behind the flimsiest obstacle, and then simply overtop it, as eventually occurred in Catania. Barriers should not be used to stop a flow, but simply to divert it. Because lava flows can become very viscous as they cool down (106 times more viscous than water), they may only be flowing over very low grades and exerting very little force. They can thus be diverted successfully by putting guiding barriers diagonally in their paths, as long as the diversion route is steep. However, the rapid movement of some flows rules out time-consuming barrier construction. Even construction of barriers long distances in front of the advancing lava may be impractical, because the flow could easily stop naturally before reaching the barrier. Barriers must also consist of denser material than the lava, be attached to the ground, or be three times wider at their base than the flow thickness itself. Less dense material will simply be buoyed up by the lava and incorporated into the flow. If barriers only divert a flow, there is always the problem of finding a safe diversion path that does not affect someone else's property or life. Only additional study in the mechanics and behavior of lava flows will permit barrier construction to be used effectively.

Ballistics and tephra clouds

(Prabaharan, 2002)

The eruption of Vesuvius in 1944 illustrates another major hazard of volcanic eruptions. At that time the Allied war effort in the area was severely hampered by the bombing of airfields, not by the Germans, but by liquid lava blobs tossed out by the volcano. Strombolian-, Vulcanian-, Surtseyan-, and Plinian-type eruptions all shoot out ash to heights greater than 30 km. The larger particles consist of boulder-sized blobs of fluid magma and remnant blocks of the volcano walls. Measured velocities are in the range of 75–200 $m\ s^{-1}$, and maximum distances are obtained with trajectory angles of 45°. Measured distances for boulder material rarely exceed 5 km, although material weighing 8–30 tonnes has been projected over distances of 1 km or more. This type of debris can be voluminous and hot, and can fall over a small area. It can also be extremely destructive. The density of projectiles varies considerably so that the kinetic energy of impact is wide-ranging. Larger bombs can behave exactly as human-made projectiles in their explosive effect when hitting the ground. Panoramic photographs of vegetated landscapes subject to sustained projectile bombardment look exactly like war bombing scenes.

The production of tephra or ash can also be destructive (Figure 11.2). The eruption of Mt Hekla, Iceland, in 1947 ejected 100 000 $m^3\ s^{-1}$ of material, dropping ash as far away as Finland two days later. Tephra can rise at velocities of 8–30 $m\ s^{-1}$, with drift rates downwind of 20–100 $km\ hr^{-1}$. While much of this dust falls out locally, fine material < 0.01 mm in size can be thrown up to heights in excess of 27 000 m into the stratosphere. Here, residence times may be two or more years. The eruption of Tambora in Indonesia on 5–10 April 1815 is the largest recorded tephra eruption. It blasted 151 km^3 of material into the atmosphere as fine ash, which was responsible for a cooling of the Earth's surface temperature by 0.5–1.0°C. As already mentioned in Chapter 9, volcanic dust exerts a dramatic control on the Earth's temperature, with many of the changes in temperature over the last three centuries correlating well with fluctuations in volcanic dust production.

Fig. 11.2 Tephra ejecta from Mt St Helens, 22 July 1980. Pyroclastic flows generated the lower ash clouds to the right. Note the spreading of the ash cloud at the tropopause and the fallout of coarse particles downwind (photograph courtesy of the United States Geological Survey).

The area covered by tephra deposits can be considerable because most tephra falls from atmospheric suspension within a short distance of the volcano. Fine ash settling out from Krakatau in 1883 covered an area of 800 000 km^2. This volcanic material can be extremely fertile once incorporated into the soil as a fine dust; however, it can have severe environmental consequences. Only a small proportion of volcano-related deaths can be attributed directly to ash fallout, nevertheless, ash fallout 70–80 km from Krakatau was still hot enough to burn holes in clothing and vegetation. Tephra can also contain toxic fluoride compounds, which are poisonous to animals attempting to eat contaminated fodder. The 'haze famine' of 1784 in Iceland, and the death by famine of 80 000 people following the eruption of Tambora in 1815, can be attributed to the destruction of vegetation by tephra. The glassy tephra from Mt St Helens in 1980 had a severe impact on road conditions and on motor vehicles. Tephra from the 1982 eruption of Galunggung volcano in western Java almost brought down, in separate incidents, two 747 passenger jets bound for Australia. The initial volcanic eruption had not been noticed nor reported to aviation authorities. The first plane to succumb was a British Airways 747 flying between Singapore and Australia, when it entered the ash cloud at 11 300 m. All four engines cut out, one after the other, sending the plane into a 13-minute silent descent, through 7300 m, before the pilot could restart three of the engines. Impact with the ash pitted the cockpit windscreen so badly that the pilot had to land the plane by peering out a side window. Two weeks later, a Singapore Airlines 747 flying the same route strayed into another ash cloud and dropped 2500 m before the pilot regained control. Only after this second incident was a warning issued for planes to avoid the area. Since then more than 80 commercial airplanes have flown into clouds of volcanic ash. Mt Pinatubo in 1991 damaged 20 commercial planes in transit at a cost of $US100 million. Explosions on Redoubt, Galunggung, Mt Spurr, and Mt Pinatubo sent clouds of ash into the atmosphere that damaged planes 150, 200, 1200, and 1740 km, respectively, from the site of the eruptions.

Pyroclastic flows and base surges

(Branney & Zalasiewicz, 1999)

Mention has already been made of the collapsing ash columns – termed 'nuées ardentes' or 'pyroclastic flows' (Figure 11.3) – and the basal surges produced by the lateral explosion of volcanoes. Base surges are akin to the lateral blast waves accompanying a nuclear explosion, and can pick up considerable dust in addition to carrying the ash originating from the volcano itself. The Bandai eruption of 1888 saw a 300 m high volcano pulverized and converted to a debris avalanche in this manner. The blast wave can reach speeds in excess of 150 km hr^{-1}, and the dust can sandblast objects. The Taal eruption in the Philippines in 1965 ablated 15 cm of wood from the sides of trees within 1 km of the eruption. The basal surge may be followed by a pyroclastic flow. The Mt St Helens eruption followed this sequence. The classic description of a pyroclastic flow is 'an avalanche of an exceedingly dense mass of hot, highly charged gas and constantly gas-emitting fragmental lava, much of it finely divided, extraordinarily mobile, and practically frictionless, because each particle is separated from its neighbors by a cushion of compressed gas' (Perrett, 1935). In pyroclastic flow, fine particles are suspended and transported by turbulent whirlwinds of gas. While appearing as choking clouds of ash, the particles are so widely spaced that they rarely collide. As these turbulent eddies pass over a spot, they can deposit alternating layers of coarse and fine sediment as the velocity of the current waxes and wanes in a similar fashion to gusts of wind. The destructive force of the flow can

Fig. 11.3 Beginning of a pyroclastic flow or nuée ardente on Mt Ngauruhoe, New Zealand, in January 1974. Tephra has been blasted vertically only to collapse under its own weight. The pyroclastic flow consists of ash suspended in hot carbon dioxide, and can be seen as the rapidly moving cloud closely hugging the ground to the left in the photograph (photograph courtesy of the United States Geological Survey, Catalogue of Disasters #0401-09j).

thus vary over short distances. The Mt Vesuvius pyroclastic flow of 79 AD knocked over walls and statues in some places, while short distances away furniture inside buildings was left undisturbed. Where the cloud meets the ground, conditions are different. Sediment concentrations increase and sediment particles ranging in size from silts to boulders undergo billions of collisions. The momentum of the flow is equalized between the particles and the flowing current of air. In some cases, the grains may flow independent of any fluid under a phenomenon known as granular flow. Granular flow tends to expel coarse particles to the surface; however, if fluid moves upwards through the flow then a process called fluidization may allow sediment particles in dense slurries to move as a fluid and remain unsorted. Different processes probably operate at different levels in pyroclastic flows. Turbulence lifts finer particles into the current higher up, while at ground level it enhances fluidization. Similarly, the falling-out of particles from slurries near the ground entraps smaller particles into a deposit while expelling water. The latter process also enhances fluidization. From time to time, turbulent vortices penetrate to the bed allowing the deposition of alternating layers of fine and coarse material. The resulting product is a disorganized deposit containing a wide range of particle sizes with evidence of layering.

Not only can the blast from the flow be destructive, but the flow can also be extremely hot. Pyroclastic flows are also called 'glowing avalanches' because of the presence of heated debris within the flow. The Mt St Helens flow produced temperatures around 3500°C near the source and 50–200°C near the margins of the flow. These temperatures were sustained for less than two minutes. Temperatures in the 1902 Mt Pelée flow were as high as 1075°C, as shown by the partial melting of gold coins and bottles. Temperatures can vary as much as velocities do. For example, the Mt Vesuvius flow of 79 AD was hot enough to carbonize wood in places, while food in other locations was left uncooked. The deposits from pyroclastic flows can remain hot for considerable periods of time. Temperatures in excess of 400°C were found within one deposit from the Augustine, Alaska, eruption of 1975, several days after the event.

Pyroclastic flows can also travel at high speeds because they behave like a density flow under the effects of gravity. Velocity measurements of small ash flows range between 10 and 30 m s^{-1}, while larger flows can obtain speeds of 200 m s^{-1}. At these velocities, the flows can override high topographic barriers. The Mt Pelée flow of 1902 refracted around topographic obstacles, and appeared to increase in velocity seaward. Both basal surges and pyroclastic flows are exceedingly deadly because of their speed, and the fact that they are mixed with large amounts of hot, toxic gases.

Historically, pyroclastic flows have covered substantial areas. The Katmai eruption in Alaska in 1912 – one of the biggest flows measured – covered an area of 126 km^2, while the Mt St Helens pyroclastic flows extended over 600 km^2. The Rabaul eruption, 1400 years ago, produced a flow 2 m deep over

1200 km^2. Eruptions in the Taupo volcanic zone of New Zealand are the largest to have occurred on Earth. There is geological evidence that Taupo has discharged pyroclastic material at the rate of 1 km^3 min^{-1}. The 186 AD eruption of Taupo produced 60 100 km^3 of tephra, an amount five times greater than that produced by Krakatau in 1883. The tephra covered all of the North Island of New Zealand with at least 10 cm of ash, while intense pyroclastic flows overran 1000 m high ridges around the crater and flooded an area of 15 000 km^2.

From ignimbrites – deposits resulting from pyroclastic flows – there is some evidence that the flows have climbed ridges 500–1000 m high up to 50 km from the source of eruptions. At these high velocities, the flows can also be erosive, gouging out channels in less resistant sediments. Ignimbrites produced by Peléean volcanoes are distinct from ash fallout deposits resulting from Plinian eruptions. Because particles settle from suspension in air, tephra deposits from the latter tend to be well-sorted and spread evenly over the landscape. Ignimbrites on the other hand consist of particles of all sizes that have been transported with air as the interstitial medium. They are made up of volatile and gas-rich magmas of a rhyolitic or andesitic nature, are chaotically sorted, and tend to pool in depressions because their flow is controlled by gravity. The heat in a pyroclastic flow can weld ignimbrite material together, whereas Plinian deposits remain unconsolidated. Some ignimbrites tend to have an inverted size grading in which finer material is overlain by coarser blocks. For this to happen, the cloud must be very dense, travel at high velocities and be thin. Such a process is called *shear sorting*, and is analogous to large lumps in a sugar bag coming to the surface as the bag is shaken. Alternatively, these well-sorted deposits may represent evidence of base surges, which usually produce thin, well-sorted beds less than 10 m thick. Ground surges, thus, appear to precede a large and slower moving pyroclastic flow, resulting in chaotically sorted ignimbrites being deposited on top of well-sorted ground-surge deposits. Because well-sorted ignimbrites arc plentiful, basal surge events may be more common than previously thought.

Gases and acid rains

(Symons et al., 1988)

Many pyroclastic flows were believed at first to consist solely of hot gases. While this has been found to be untrue, gases cannot be ruled out as a hazard. On average, explosive volcanoes eject 4 × 10^6 tonnes of sulfur dioxide into the atmosphere annually. This is highly variable from year to year. For example, the Mt Pinatubo eruption of 15 June 1991 released 20 × 10^6 tonnes of sulfur dioxide in the space of a few weeks. The more explosive the eruption, the greater the transport of gas on adsorbed particles. Passively, degassing volcanoes annually release another 9 × 10^6 tonnes of sulfur dioxide. While appearing voluminous, these amounts represent only 5–10 per cent of the annual anthropogenic flux to the atmosphere. However, emissions of sulfur dioxide by volcanoes can have significant local and regional effects, mainly on crops. Carbon monoxide emissions are more dangerous, being toxic to mammals in low quantities, and inhibiting leaf respiration in plants. Even CO_2 can prove harmful. Volcanoes can produce large quantities of this gas (20–95 per cent of the gas discharge) which, being denser than air, can pool in depressions and suffocate livestock. On 21 August 1986, these gases, plus sulfur dioxide and cyanide, were released by a landslide from the bottom sediments of Lake Nios, a dormant volcanic crater in Cameroon, Africa. Clouds of deadly gases, reaching concentrations of 20–30 per cent CO_2, hugged the ground and flowed under gravity into topographic lows killing all animal life in their path. Over 1700 people died within a matter of minutes, while 10 000 survivors suffered skin burns. A gas discharge in neighboring Lake Monoun, Cameroon, killed 37 people in 1984, while a similar disaster on the Dieng Plateau in Java in 1978 killed 180 people.

Many gases react with water vapor under the high temperatures of volcanic eruptions to form hydrochloric, sulfuric, carbonic, and hydrofluoric acids. Passively degassing volcanoes release annually into the atmosphere 0.05–4.7 million tonnes of fluorine and 0.3–10 million tonnes of chlorine as hydrofluoric and hydrochloric acid, respectively. Icelandic eruptions are notoriously high in fluorides. The hydrofluoric acid content in the Hekla, Iceland, eruption of 1970 reached 1700 ppm. Notable eruptions containing high amounts of HCl have been Agung, Bali (1.5 million tonnes), in 1963; Augustine, Alaska (525 000 tonnes) in 1976; Soufrière, Guadeloupe (1 million tonnes) in 1979; and Mt Erebus in the Antarctic (370 000 tonnes) in 1983. In the troposphere, acid can be precipitated out of the atmosphere through condensation, scavenged by ash, or dissolved in water-rich eruption plumes. Above 6000°C, hydrofluoric acid forms calcium fluorosilicate that adheres to the glassy surfaces of tephra particles. Despite this culling, acids can still pose a hazard over a large area. The Katmai eruption of 1912 dispersed a sulfuric acid rain that damaged clothes hung out to dry in Vancouver, 2000 km to the south. The Laki, Iceland, eruption in 1783–1784 filled the skies over most of the north Atlantic, western Eurasia, and the Arctic with a dry sulfuric fog that led to a cold winter. Explosive eruptions can also inject chlorine in the form of HCl directly into the stratosphere where it can react with ozone. Tambora in 1815 and Krakatau in 1883 were explosive eruptions, ejecting a minimum of 2.1 and 3.6 million tonnes of chlorine as HCl, respectively. The eruption of El Chichon, Mexico, in 1982 released 40 000 tonnes of HCl into the stratosphere between 20 and 40°N latitudes.

Lahars

(Newhall & Punongbayan, 1996)

One of the more unusual, though still hazardous, phenomena produced by volcanoes consists of lahars, or mudflows, that can occur at the time of the eruption (primary lahars) or several years afterwards (secondary lahars). About 5.6 per cent of volcanic eruptions in the last 10 000 years have produced mudflows at some time. Primary lahars can be generated by pyroclastic flows or by eruptions of crater lakes. In the former case, a pyroclastic flow can easily entrain water from streams and rivers as it moves down topographic lows. In the process, the gas-rich flow is slowly converted to a fast moving, heated mudflow as more water is entrained in the mix. The Toutle River lahars (see Figure 11.4) from Mt St Helens in 1980 had this origin. Volcanoes with crater lakes can produce mudflows at the time of any eruption, if the crater lake is ruptured. The size of the mudflow is then related directly to the volume of water in the lake. The 1919 eruption of Mt Kelat on Java expelled water from a crater lake, covering 200 km^2 of farmland and killing over 5000 people. Attempts have been made to control the Kelat situation by drilling tunnels through the walls of the crater to lower the level of the lake. The first attempt was in 1929, when Dutch engineers reduced the lake volume from 21 to 1 million m^3. A subsequent eruption in 1951 blocked the tunnels and, even after repairs, an eruption in 1966 generated lahars that took hundreds of lives. Indonesian engineers have since replaced the tunnels and drained the lake completely to negate the threat of future lahars.

Secondary lahars are caused by rain falling on freshly deposited, uncompacted tephra. Such water-soaked material is very unstable, and can move downslope as a mudflow that entrains all loose debris in its path. If acids have not been leached out of the tephra, then lahars can become acidic enough to cause serious burns. Such flows have covered enormous areas. One

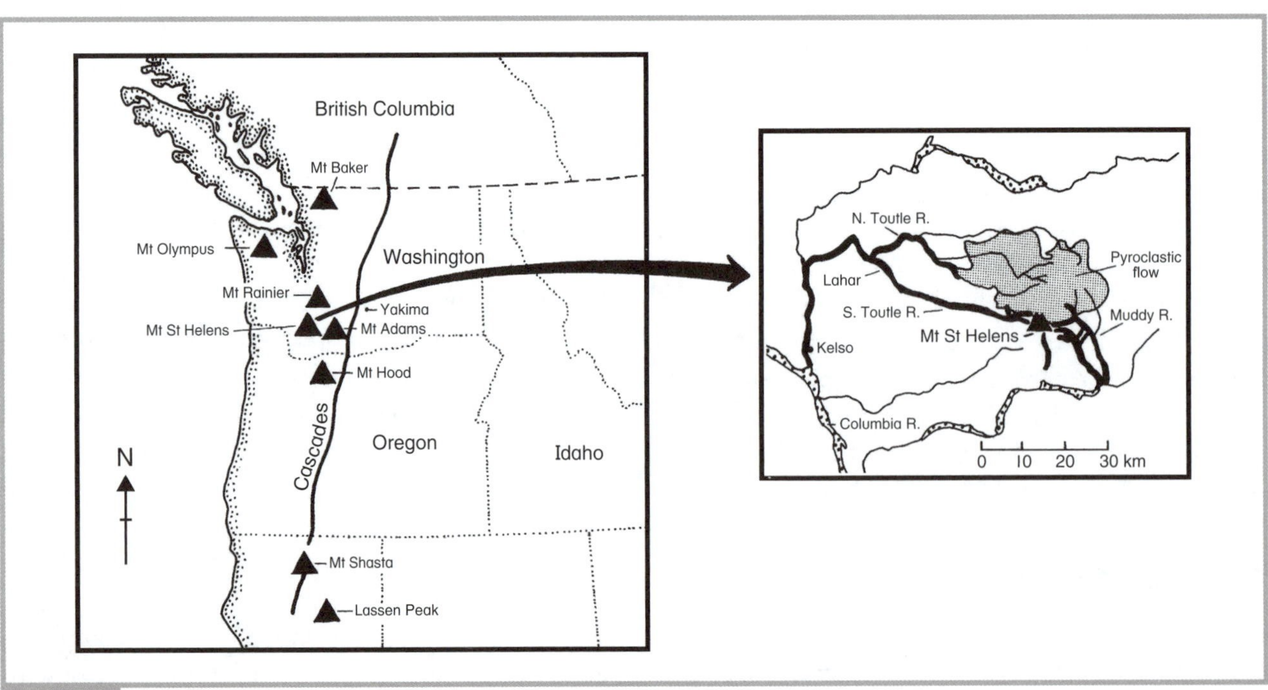

Fig. 11.4 Location map of Mt St Helens (adapted from Blong, 1984).

on Mt Rainier in Washington State 5000 years ago had a volume of 1.9 × 10^9 m^3 and flowed 60 km from the summit. Other lahars preserved in the geological record in the United States have extended over 5000–31 000 km^2. Secondary lahars following the Plinian eruption of Mt Pinatubo in the Philippines in June 1991 left more than 50 000 persons homeless, and disturbed the livelihoods of 1 400 000 people in 39 towns and four large cities over an area of 1000 km^2 (Figure 11.5). The eruption was one of the largest of the twentieth century and culminated on 15 June in an explosive eruption that caused the summit to collapse into a 2.5 km diameter caldera. Pyroclastic flows exceeded 5–6 km^3 and spread up to 16 km from the volcano over an area of 400 km^2. The final eruption discharged 3.4–4.4 km^3 of ash. On the day of the climactic eruption, Typhoon Yunya passed close to the volcano. First its winds scattered tephra to a thickness of 10–33 cm over an area of 2000 km^2, second its rain soaked into the ash and caused many buildings to collapse and, third, runoff turned the pyroclastic flows into enormous lahars. The following monsoon rains continued the threat, repeatedly generating hot lahars that buried towns and agricultural land under 5–30 m of ash, destroyed bridges, and cut roads. A typical lahar was 2–3 m deep and 20–50 m wide. It consisted of 50 per cent ash moving as slurry at velocities of 4–8 m s^{-1} with peak discharges of 200–1200 m^3 s^{-1}. A few lahars reached speeds of 11 m s^{-1}, with peak discharges of 5000 m^3 s^{-1}. Lahars were triggered by as little as 6 mm of rainfall in half-an-hour. By the end of the monsoon season, these lahars had moved about 0.9 km^3 of sediment and buried 300 km^2 of lowland. The 1992 monsoon season yielded over 2000 mm of rain that generated lahars that moved 0.5–0.6 km^3 of sediment. Typically, 1000 m^3 of ash was removed from each square kilometer of upland watershed by each millimeter of rainfall. Each subsequent monsoon season reiterated the process but with diminishing volumes of sediment being transported. Because of the installation of warning systems, the lahars after the main eruption killed few people.

Fig. 11.5 Lahar at Bamban following the eruption of Mt Pinatubo in June 1991. The lahar has cut the main road linking major cities in the north. As well, it has caused vertical accretion of the river channel. To prevent sediment-laden waters spilling across the floodplain and into the town, levees have been built to contain the lahar (photo by Cees J. van Western, International Institute for Aerospace Survey and Earth Sciences, The Netherlands. Taken from NOAA's National Geophysical Data Center hazard photos <http://www.ngdc.noaa.gov/seg/hazard/slideset/36/36_728_slide.shtml>).

Because lahars are restricted to topographic lows, they can be predicted in advance. The build-up of tephra on a volcano's slopes can be easily measured, so that steps can be taken to evacuate people in threatened areas. After the Usu eruption in 1977 in Japan, it was calculated that tephra deposits thicker than 0.5 m, on slopes longer than 300 m and steeper than 17–18°, would require only 20 mm of rain at a rate of 10 mm hr^{-1} to initiate lahars. The 86 lahars that occurred afterwards closely followed these predictions. In New Zealand, lahars threaten the Whakapapa skifield situated on the north-west slopes of Mt Ruapehu. Here, seismometers have been installed to detect small earth movements indicating the start of a lahar. These trigger alarms on the ski-fields giving skiers 5–10 minutes to ski out of valleys. While lahars are very common and can reach velocities in excess of 20 km hr^{-1}, they can be prevented in some cases. Unfortunately, lahars can also be initiated by erosion of old tephra deposits years, or decades, after the original eruption has taken place. For instance, Mt Rainier was affected by a lahar in 1947 even though no volcanic eruption has been recorded historically.

Glacier bursts or Jökulhlaups

(Halldórsson & Brandsdóttir, 1998)

Of particular danger during a volcanic eruption is the melting of glaciers or snowfields by hot lava or gases. The water can be subsequently heated and mixed with mud debris to form a steaming hot lahar. These events are termed jökulhlaups, an Icelandic word meaning 'glacier bursts'. Jökulhlaups are restricted mainly to Iceland and the Andes; however, they can occur elsewhere. For example, the Mt St Helens lahars were probably exacerbated by melting of snow and ice by hot gases. In November 1985, over 20 000 people were killed by a glacier burst mudflow that swept off the Nevado del Ruiz volcano in Colombia. This was the worst such event in the twentieth century. Jökulhlaups in Iceland are caused by continuous melting of the Myrdalsjökull or Vatnajökull icecaps, which lie overtop active volcanoes. The Vatnajökull glacier caps the 40 km^2 Grimsvötn caldera, which continually emits hot gases. The thickness of these glaciers prevents heat escape, while their weight prevents water from flowing out of the caldera. Meltwater builds up slowly beneath the ice until water depths are sufficiently deep to float the glaciers. Jökulhlaups have dominated Iceland since first reported in the twelfth century. Until the twentieth century, they occurred every 5–15 years. During the twentieth century, large jökulhlaups took place in 1903, 1913, 1922, 1934, 1938 and 1996. Smaller events now occur 2–3 times per decade. Discharges during a jökulhlaup create some of the largest floods on Earth. In 1918, the discharge of a jökulhlaup from the Myrdalsjökull icecap exceeded three times the discharge of the Amazon, the largest flowing river in the world. The Katla volcano in Iceland has had glacier bursts from the Vatnajökull glacier exceeding 92 000 $m^3 s^{-1}$, with a total volume in excess of 6 km^3. The latter floods over time have formed an outwash plain 1000 km^2 in area. The last major jökulhlaup occurred at Grimsvötn on 5 November 1996. Flow peaked at 45 000 $m^3 s^{-1}$ as 3 km^3 of heated water discharged over three days across the southern outwash plain of Iceland. Blocks of ice 15 m in diameter and weighing up to 1000 tonnes knocked out bridges and destroyed the main highway.

Water discharge can also accompany volcanic eruptions in areas without snow or ice. The 1902 eruptions of Mt Pelée and Soufrière in the West Indies were both accompanied by massive flooding of dry rivers. The 1947 eruption of Mt Hekla, Iceland, led to a discharge of 3 000 000 m^3 of water, which could not be attributed solely to melting of snow or ice. These latter events suggest that water may be expelled from the watertable by escaping gases during some volcanic eruptions. Nor do these discharges have to accompany the eruption. In 1945, the Ruapehu volcano in New Zealand erupted, emptying its crater lake. This lake slowly refilled inside a newly formed crater that became dammed by glacial ice. In 1953, an ice cave formed that began lowering the crater. On Christmas Eve of that year the crater wall suddenly collapsed near the cave causing the lake to drop 6 m in two hours at the rate of 900 $m^3 s^{-1}$. The floodwater swept down the Whangaehu River picking up mud and boulders and forming a lahar. Two hours later, at 10:00 pm, the debris-laden water swept away a section of the Tangiwai Rail Bridge, three minutes before the Wellington–Auckland night train arrived. The train plunged through the gap, killing 151 people.

VOLCANIC DISASTERS

Volcanoes are visually one of the most spectacular natural hazards to occur, and probably one of the most devastating in terms of loss of human life. Even the Sahelian droughts, or earthquakes that completely flatten cities, leave more survivors than dead. Volcanoes have wiped out entire populations. There were only two survivors out of a population of 30 000 in the town of St Pierre, Martinique, following the Mt Pelée eruption of 1902. Four volcanic events stand out in the historic record. Three of these, Mt Vesuvius, Krakatau, and Mt Pelée, are significant because of either the enormity of the eruption, or the resulting death toll. The other, Mt St Helens, stands out because it represents a major eruption beginning a period of volcanic activity at the end of the twentieth century.

Santorini, around 1470 BC

(Yokoyama, 1978; Pichler & Friedrich, 1980; Kastens & Cita, 1981; Cita et al., 1996; Pararas-Carayannis, 1998)

The prehistoric eruption of Santorini, also known as Stronghyli – the round island – occurred about 1470 BC off the island of Thera, north of Crete in the southern Aegean Sea (Figure 11.6). Probably the biggest volcanic explosion witnessed by humans, it is also one of the most controversial because legend, myth, and archaeological fact are frequently intertwined and distorted in the interpretation of events. The eruption has been linked to the lost city of Atlantis (described by Plato in his *Critias*), to destruction of Minoan civilization on the island of Crete 120 km to the south (Figure 11.6), and to the exodus described in the Bible of the Israelites from Egypt. Certainly, Greek flood myths refer to this or similar events that generated tsunami in the Aegean. Plato's story of Atlantis is based on an Egyptian story that has similarities with Carthaginian and Phoenician legends. There is no

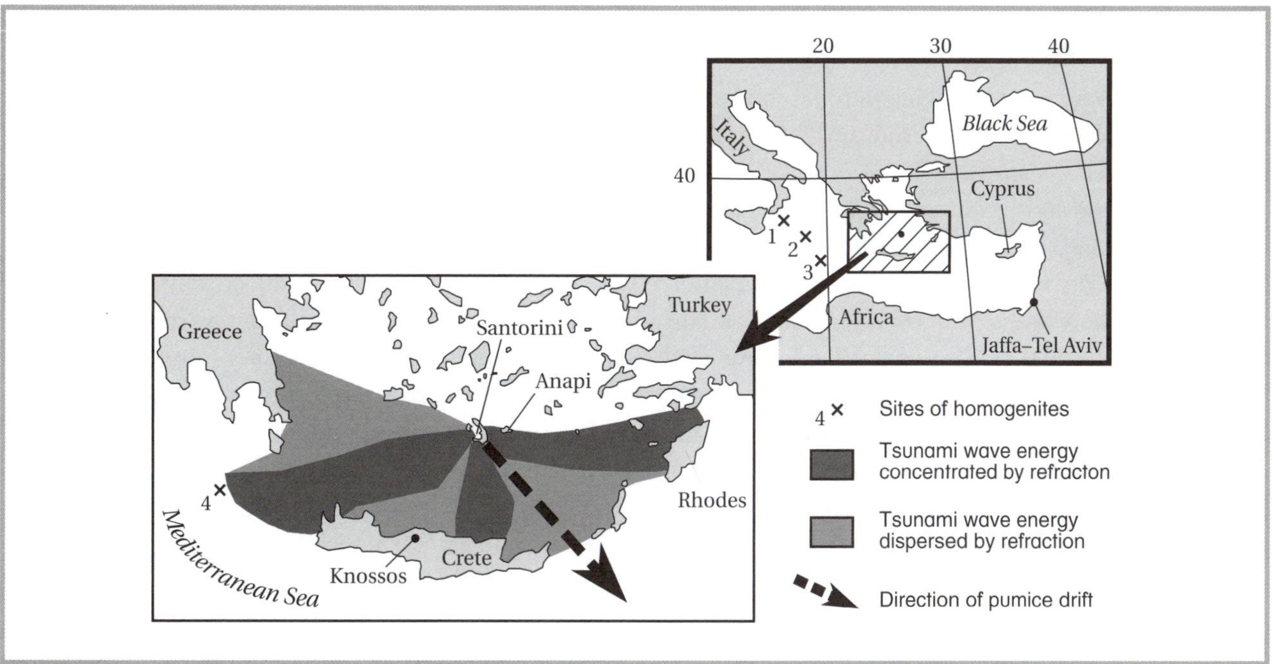

Fig. 11.6 Eastern Mediterranean region affected by the Santorini eruption around 1470 BC. The numbers refer to sites where homogenites exist: 1) Calabrian Ridge, 2) Ionian Abyssal Plain, 3) Bannock Basin, and 4) Mediterranean Ridge. Refraction patterns based on Kastens and Cita (1981).

doubt that the great Bronze Age Minoan Empire declined around the time of the eruption.

Santorini is part of a volcanic island chain extending parallel to the coast of Asia Minor. The volcano has erupted explosively at least twelve times during the last two hundred thousand years. Its height has been reduced over this time from a single mountain 1500 m high to a submerged caldera ringed by three islands less than 500 m in height. The timing of the last major eruption around 1470 BC is disputed. It has been linked to the plagues of Egypt (Exodus 6:28–14:31) and the exodus of the Israelites from that country. In 1 Kings 6:1, the exodus is dated as occurring 476 years before the rule of King Solomon, which began in 960 BC, putting the Exodus around 1436 BC. Other evidence indicates that the exodus occurred in 1477 BC. Both dates encompass the reign in Egypt of either Hatshepsut or her son Tuthmosis III of the eighteenth dynasty. The transition between the two rulers is known as a time of catastrophes. In the biblical account, the river of blood may refer to pink ash from the Santorini eruption preceding the explosion. This ash, after it was deposited, would easily have mixed with rainwater and flowed into any stream or river, coloring it red. The three days of darkness possibly refer to tephra clouds blowing south across Egypt at the beginning of the eruption. The darkness was described as a 'darkness, which could be felt'. Egyptian documents around 1470 BC refer to a time of prolonged darkness and noise; to a period of nine days that 'were in violence and tempest: none ... could see the face of his fellow', and to the destruction of towns and wasting of Upper Egypt. Acidity in Greenland ice cores indicates that the major eruption occurred in 1390 ± 50 BC, although radiocarbon dating on land suggests an age around 1450 BC or 1470 BC. Dendrochronology based on Irish Bog Oak and Californian Bristlecone Pine puts the age of the event as early as 1628 BC. At this latter time, Chinese records report a dim sun and failure of cereal crops because of frost. Large volcanic events cool temperatures globally by as much as 1°C over the space of several years. The range of dates may not be contradictory because there is evidence that Santorini may have erupted several times over a time span of two hundred years.

The eruption around 1470 BC had four distinct phases. The first was a Plinian phase with massive pumice falls. This was followed by a series of basal surges producing profuse quantities of pumice up to 30 m thick on Santorini. The third phase was associated with the collapse of the caldera and production of pyroclastic flows. About 4.5 km^3 of dense magma was ejected from the volcano producing 10 km^3 of ash. The volume of ejecta is similar in magnitude to that produced by the Krakatau eruption in 1883. The ash drifted to the east-south-east, but did not exceed 5 mm thickness in deposits on any of the adjacent islands, including Crete. The largest thickness of ash measured in marine cores appears to originate from pumice that floated into the eastern Mediterranean. It is possible at this stage that ocean water made contact with the magma chamber and produced large explosions that generated tsunami. The final phase of the eruption was associated with the collapse of the caldera in its south-west corner. The volcano sank over an area of 83 km^2 to a depth of 600–800 m. This final collapse produced the largest tsunami, directed westward (Figure 11.6). It is estimated that the original height of the tsunami was 46–68 m, and maybe as high as 90 m. The average period between the dozen or more peaks in the wave train was fifteen minutes.

Evidence of the tsunami is found in deposits close to Santorini. On the island of Anapi to the east, sea-borne pumice was deposited to an altitude of 40–50 m above present sea level. Considering that sea levels at the time of the eruption may have been 10 m lower, this represents run-up heights greater than those produced by Krakatau in the Sunda Strait. On the island of Crete, the wave arrived within thirty minutes with a height of approximately 11 m. Refraction focused wave energy on the north-east corner of Crete, where run-up heights reached 40 m above sea level. In the region of Knossos, the tsunami swept across a 3 km wide coastal plain reaching the mountains behind. The backwash concentrated in valleys and watercourses, and was highly erosive. Evidence for the tsunami is also found in the eastern Mediterranean on the western side of Cyprus, and further away at Jaffa–Tel Aviv in Israel. At the latter location, pumice has been found on a terrace lying 7 m above sea level at the time of the eruption. However, the tsunami wave here had already undergone substantial defocusing because of wave refraction as it passed between the islands of Crete and Rhodes. The greatest tsunami wave heights occurred west of Santorini. The wave in the central Mediterranean Sea was 17 m high; while closer to Italy over the submarine Calabrian Ridge it was 7 m high. Bottom current velocities under the wave crest in these regions ranged from 20 to 50 cm s^{-1} – great enough to entrain clay- to gravel-sized particles. The maximum pressure pulse produced on the seabed by the passage of the wave ranged from 350 to 850 kdynes cm^{-2}. Spontaneous liquefaction and flow of water-saturated muds is known to occur under pressure pulses of 280 kdynes cm^{-2} and greater.

Vesuvius (25 August 79 AD)

(Bolt et al., 1975; Whittow, 1980; Blong, 1984; Sigurdsson et al., 1985)

Mt Vesuvius lies on the south-east corner of the bay of Naples on the west coast of Italy (Figure 11.7). The Roman towns of Pompeii and Herculaneum were situated at the base of the south-western slopes, 5 km from the mountaintop. Both towns were prosperous regional centers serving as summer retreats for wealthy Romans. Mt Vesuvius, prior to its eruption, consisted of a single, flat-topped cone with a small crater. The Romans recognized it as a volcano, but there is no Roman or Etruscan record of its ever being active. In 63 AD, a major earthquake struck the region with its epicenter at Pompeii. Both the towns of Pompeii and Herculaneum were severely damaged, and by the time of the eruption in 79 AD, only minor rebuilding had been completed in Pompeii.

The description of the eruption comes from the eye-witness account of Pliny the Younger, who recounts the death of his father during the eruption. The writings of Cassius provide descriptions of the destruction of the two cities. Pliny the Elder was the commander of the Roman fleet at Misenum, 30 km to the west at the entrance to the Bay of Naples. On 24 August, Vesuvius began to erupt, accompanied by violent earthquakes. Pliny sailed to the scene to rescue inhabitants at the base of the mountain. He could not get near the shore because of a sudden retreat of the shoreline, and his boats were covered in a continuous ash fall. He sailed that night to Stabiae to stay with a friend. Unfortunately, Stabiae was downwind from Pompeii and, during the night, the ash fall became so deep that people realized they would have to flee for their own safety. Pliny the Elder reached the coast during the daylight hours of 25 August, but the ash fall was so heavy that it still appeared to be night. As the magma chamber emptied, the top of the volcano collapsed inward, consuming 50 per cent of its volume and producing a caldera 3 km in diameter. In the early hours of 25 August, a series of six surges and pyroclastic flows swept the area. The first surge overwhelmed Herculaneum. All remaining inhabitants met a grisly death from asphyxiation as they sheltered on the beach. The fourth surge reached Pompeii at about 7:00 am. The last two surges were the largest, and reached the towns of Stabiae and Misenum. At Stabiae, Pliny the Elder died on the beach, apparently from a heart attack. It

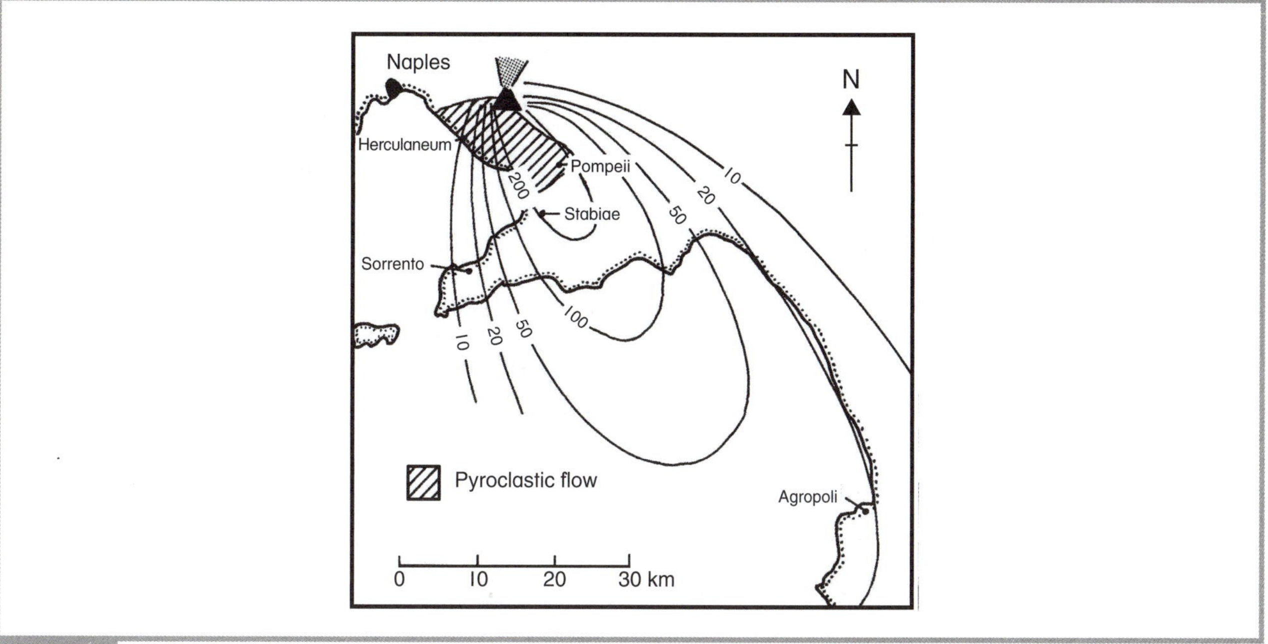

Fig. 11.7 Location map of Mt Vesuvius (Blong, ©1984, with permission Harcourt Brace Jovanovich Group, Australia).

took three days before it was light enough to recover his body. The pyroclastic ash falls buried Stabiae and Pompeii to depths of 2 and 3 m, respectively. Most of the residents of Pompeii appear to have fled, but 10 per cent were literally caught dead in their tracks by the pyroclastic flows that climaxed the eruption. Herculaneum was also affected by the tephra fallout but, in the latter stages of the eruption, was engulfed by a mudflow 20 m thick. This lahar appears to have been caused by groundwater expulsion as the caldera collapsed.

The eruption marks one of the first volcanic disasters ever documented. The sequence of events at Vesuvius has been labelled a Plinian-type eruption after Pliny's descriptions. Many of the buildings in Pompeii and Herculaneum protruded above the ash. The buildings were ransacked of valuables, dismantled for building materials, and subsequently buried by continued eruptions. The ash reverted to rich agricultural land and Pompeii was forgotten. It was not until 1699 that Pompeii was rediscovered and excavated. Vesuvius has continued to erupt aperiodically until the present day. Subsequent major eruptions occurred in 203, 472, 512, 685, 787 and five times between 968 and 1037 AD. It then remained dormant for 600 years until 1630, when it produced extensive lava flows that killed 700 people. The last major eruption in 1906 produced tephra up to 7 m thick on the north-east side of the mountain. About 300 people were killed, mainly under collapsing roofs. If the 79 AD eruption recurred today, over 1 500 000 people would be affected.

Krakatau (26–27 August 1883)

(Verbeek, 1884; Latter, 1981; Self & Rampino, 1981; Nomanbhoy & Satake, 1995; Winchester, 2003)

The eruptions of Tambora and Krakatau in the nineteenth century make Indonesia, with its dense population, one of the most hazardous zones of volcanic activity in the world. Krakatau lies in the Sunda Strait between Sumatra and Java, Indonesia (Figure 11.8). The Javanese *Book of Kings* describes an earlier eruption that generated a sea wave that inundated the land and killed many people in the northern part of Sunda Strait. Krakatau had last been active in 1681 and, during the 1870s, the volcano underwent increased earthquake activity. In May of 1883, one vent became active, throwing ash 10 km into the air. By the beginning of August, a dozen Vesuvian-type eruptions had occurred across the island. On 26 August, loud explosions recurred at intervals of ten minutes, and a dense tephra cloud rose 25 km above the island. The dust particles formed nuclei for water condensation and muddy rain fell on adjacent islands. The explosions could be heard throughout the islands of Java and Sumatra. In the morning and later that evening, small tsunami waves 1–2 m in height swept the strait striking the towns of Telok Betong on Sumatra's Lampong Bay, Tjaringin on the Java coast

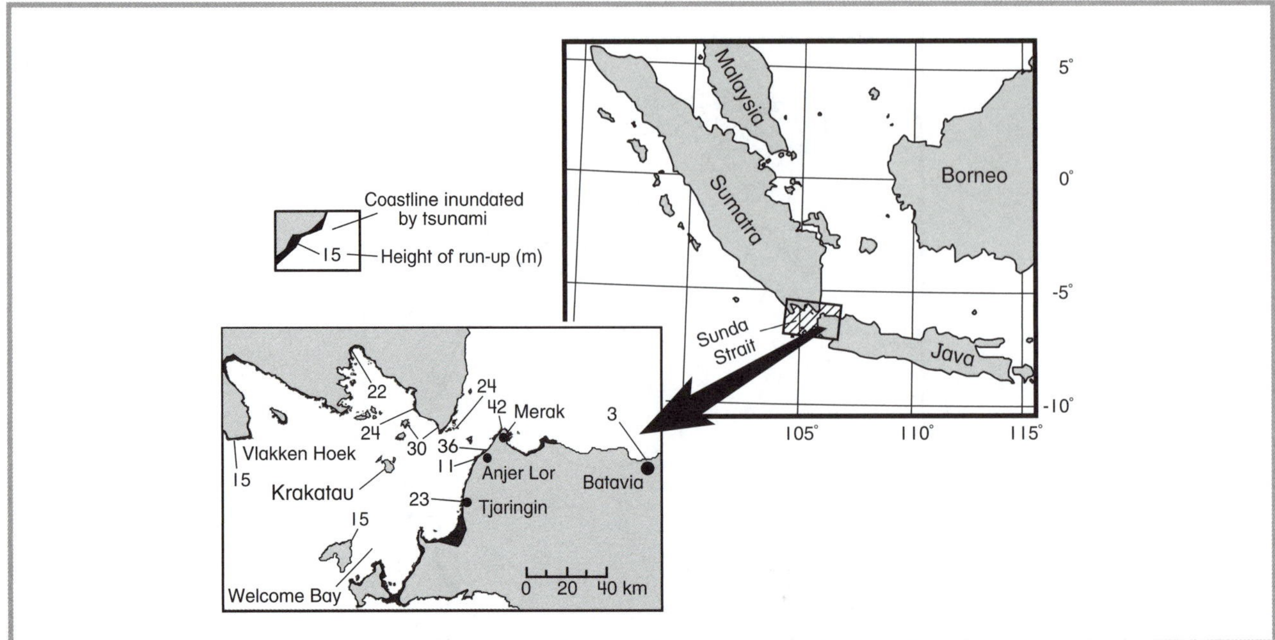

Fig. 11.8 Coastline in the Sunda Strait affected by tsunami following the eruption of Krakatau on 26–27 August 1883 (based on Verbeek, 1884; Blong, 1984 and Myles, 1985).

north of Pepper Bay and Merak. On the morning of 27 August, four horrific explosions occurred. The first explosion, at 5:28 am, destroyed the 130 m peak of Perboewatan, forming a caldera that immediately infilled with seawater and generated a tsunami. At 6:44 am, the 500 m high peak of Danan exploded and collapsed, sending more seawater into the molten magma chamber of the eruption and producing another tsunami. A third explosion occurred at 8:20 am. The fourth blast, at 9:58 am, tore apart the remaining island of Rakata. Including ejecta, 9–10 km^3 of solid rock was blown out of the volcano. About 18–21 km^3 of pyroclastic deposits spread out over 300 km^2 to an average depth of 40 m. Fine ash spread over an area of 2.8×10^6 km^2 and thick pumice rafts impeded navigation in the region up to five months afterwards. A caldera formed, 6 km in diameter and 270 m deep, where the central island had once stood. This final blast was the largest sound ever heard by modern humans, and was recorded 4800 km away on the island of Rodriguez in the Indian Ocean, and 3200 km away at Elsey Creek in the Northern Territory of Australia. Windows were shattered 150 km away. The atmospheric shock wave traveled around the world seven times. Barometers in Europe and the United States measured significant oscillations in pressure over nine days following the blast. The total energy released by the third eruption was equivalent to 840×10^{15} joules.

The blasts were accompanied by a cloud of ash that rose to a height of 30 km. The debris from this cloud formed deposits with a volume of 13 km^3. On the island remnants, 60 m of ignimbrite were deposited on top of 15 m of tephra. Large rafts of pumice ejected by the volcano blocked the Sunda Straits. Tephra fell over an area of 300 000 km^2 and turned day into night 200 km downwind for forty-eight hours. Dust encircled the globe within two weeks of the eruption, mainly within the zone of tropical easterlies. Within three months, dust had spread into the northern hemisphere, covering most of the United States and Europe. In France, solar radiation dropped 20 per cent below normal and, three years afterwards, was still 10 per cent below normal at the ground surface. Spectacular, prolonged sunsets were observed everywhere, caused by high altitude dust scattering incoming solar radiation as it entered the Earth's atmosphere at low angles. The first occurrence of such sunsets sparked false alarms to fire departments in the eastern United States as residents reported what appeared to be distant, but horrific, fires.

The two pre-dawn blasts each generated tsunami that drowned thousands in the Sunda Strait. The fourth blast-induced wave was cataclysmic and devastated the adjacent coastlines of Java and Sumatra within 30 to 60 minutes. The coastline north of the eruption was struck by waves with a maximum run-up height of 42 m (Figure 11.8). The tsunami penetrated 5 km inland over low-lying areas. The largest wave struck the town of Merak. Here, the 15 m tsunami rose to 40 m because of the funnel-shaped nature of the bay. The town of Anjer Lor was swamped by an 11 m high wave, the town of Tjaringin by one 23 m in height, and the towns of Kelimbang and Telok Betong were each struck by a wave 22–24 m high. In the latter town, the Dutch warship *Berouw* was carried 2 km inland and left stranded 10 m above sea level. The highest run-up reached 42 m above sea level at Merak in the north-east corner of the Strait. Coral blocks weighing up to 600 tonnes were moved onshore. Within the Strait, eleven waves rolled in over the next fifteen hours, while at Batavia (now Jakarta) fourteen consistently spaced waves arrived over a period of thirty-six hours. Between 5000 and 6000 boats were sunk in the Strait. In total, 36 417 people died in major towns and three hundred villages were destroyed because of the tsunami.

Within four hours of the final eruption, a 4 m high tsunami arrived at Northwest Cape, Western Australia, 2100 km away. The wave swept through gaps in the Ningaloo Reef and penetrated 1 km inland over sand dunes. Nine hours after the blast, three hundred riverboats were swamped and sunk at Kolkata – formerly Calcutta – on the Ganges River, 3800 km away. The wave was measured around the Indian Ocean at Aden on the tip of the Arabian Peninsula, Sri Lanka, Mahe in the Seychelles Islands, and on the island of Mauritius. The furthest this tsunami wave was observed was 8300 km away at Port Elizabeth, South Africa. Tsunami waves were measured over the next thirty-seven hours on tide gauges in the English Channel, the Pacific Ocean, and in Lake Taupo in the center of the North Island of New Zealand – where a 0.5 m oscillation in lake level was observed. Around the Pacific Ocean, tide gauges in Australia, Japan, San Francisco, and Kodiak Island measured changes of 0.1 m up to twenty hours after the eruption. Honolulu recorded higher oscillations of 0.24 m with a periodicity of thirty minutes.

The tsunami in the Pacific have been attributed to the atmospheric pressure wave, because many islands effectively obscure the passage of any tsunami from the Sunda Strait eastward. The atmospheric pressure wave also accounts for seiching that occurred in Lake Taupo, which is not connected to the ocean. Finally, it explains the long waves observed along the coasts of France and England when the main tsunami had effectively dissipated its energy in the Indian Ocean. The generation of tsunami in Sunda Strait and the Indian Ocean has been attributed to four causes: lateral blast, collapse of the caldera that formed on the north side of Krakatau Island, pyroclastic flows and a submarine explosion. Lateral blasting may have occurred to a small degree on Krakatau during the fourth explosion; however, its effect on tsunami generation is not known. During the final explosion, Krakatau collapsed in on itself forming a caldera about 270 m deep and with a volume of 11.5 km^3. However, modelling indicates that this mechanism underestimates tsunami wave heights by a factor of three within Sunda Strait. Krakatau generated massive pyroclastic flows. These flows probably generated the tsunami that preceded the final explosion. At the time of the final eruption, ash was ejected into the atmosphere towards the north-east. Theoretically, a pyroclastic flow in this direction could have generated tsunami up to 10 m in height throughout the strait; however, the mechanism does not account for tsunami run-ups of more than 15 m in height in the northern part of Sunda Strait. The pyroclastic flow now appears to have sunk to the bottom of the ocean and traveled 10–15 km along the seabed before depositing two large islands of ash. The 40 m high run-up measured near Merak to the north-east supports this hypothesis. The tsunami's wave height corresponds with the depth of water around Krakatau in this direction. Water was simply expelled from the seabed by the pyroclastic flows. As well, the fourth explosion of Krakatau, at 9:58 am, more than likely produced a submarine explosion as ocean water came in contact with the magma chamber. A submarine explosion could have generated tsunami 15 m high throughout the Strait. If the explosion had a lateral component northward, as indicated by the final configuration of Krakatau Island, then this blast, in conjunction with the pyroclastic flow, would account for the increase in tsunami wave heights towards the northern entrance of Sunda Strait.

Mt Pelée (8 May 1902)

(Bolt et al., 1975; Whittow, 1980; Blong, 1984; Scarth, 2002)

The year 1902 was not a good year for the residents of the West Indies or the surrounding Caribbean (Figure 11.9). The Pacific coast of Guatemala was struck by a strong earthquake on 18 January and again

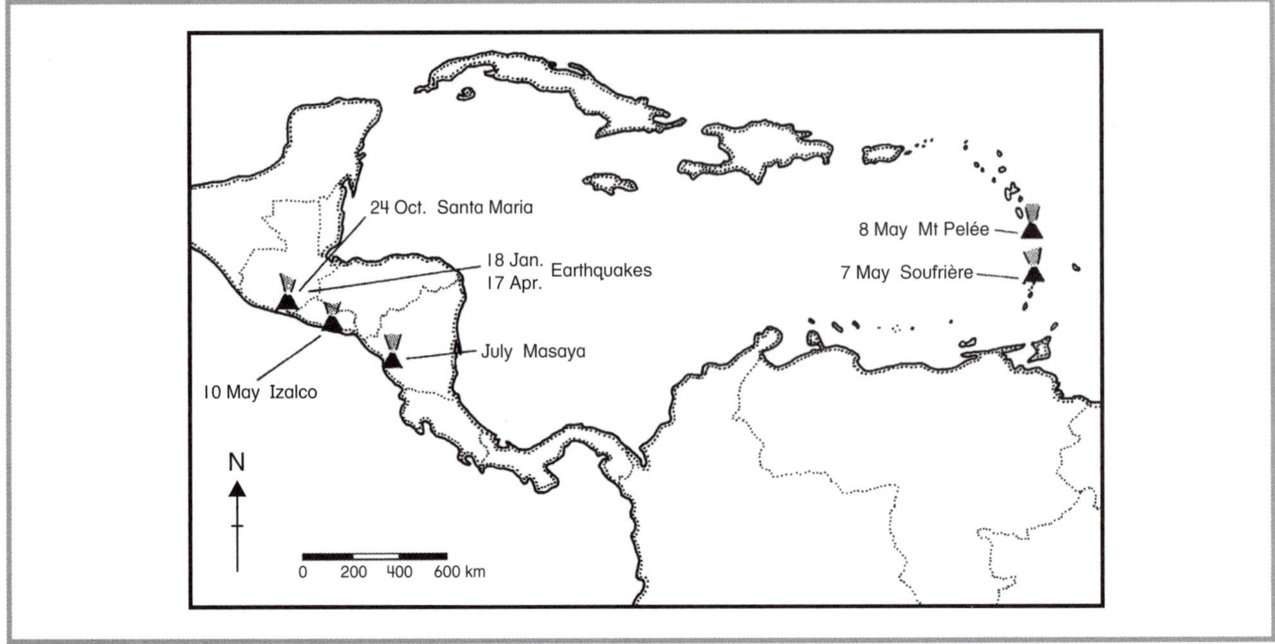

Fig. 11.9 Location map of volcanoes and earthquakes in the Caribbean region in 1902.

on 17 April. On 20 April, Mt Pelée began to erupt on the island of Martinique and on 7 May Soufrière, on the nearby island of St Vincent, erupted killing 2000 people. Izalco in El Salvador erupted on 10 May, Masaya in Nicaragua erupted in July, and Santa Maria in Guatemala exploded on 24 October. The latter eruptions mainly caused property damage. The worst eruption in terms of loss of life was the explosion of Mt Pelée on 8 May. It ranks as one of the most catastrophic natural eruptions witnessed.

The eruption of Soufrière on St Vincent, 160 km south of Martinique, was a prelude to the Mt Pelée disaster. Soufrière became active in 1901 and, in April of 1902, earthquake activity was severe enough that most residents were evacuated to the southern part of the island. On 6 May, the volcano started emitting steam, and the Rabaka Dry and Wallibu Rivers became torrents of muddy water. At 2:00 pm on 7 May, Soufrière generated a pyroclastic flow that destroyed most of the northern part of the island and killed 1565 people. Fortunately, the evacuations avoided a higher death toll.

Mt Pelée began to erupt on 20 April 1902. The mountain was conical in shape, with a notch on the south-west side that led into a topographic low called the Rivière Blanche (Figure 11.10). This valley descends steeply for about 1.5 km and then bends sharply westward. The town of St Pierre lies on the coast 3 km south of this bend. St Pierre was one of the most prosperous cities in the Caribbean. Its business centered on the export of rum made from local sugar cane. The eruption was preceded by a long series of natural warnings foreboding imminent disaster. Lava was found in the crater shortly after the initial eruption and, in the last week of April, the volcano belched ash and sulfurous fumes so intense that birds dropped dead from the sky. Reports were issued that the volcano was still safe, and that St Pierre was protected from any lava flows by the ridge between it and the Rivière Blanche. On the night of 3 May, Mt Pelée entered a new stage of eruption. The Roxelane River, running through St Pierre, turned into a torrent of mud, knocking out the power supply in the city. Fissures opened up on the slopes, and sent out steam and boiling mud that killed 160 people in the town of Ajoupa-Bouillon. About 10 000 people then fled the slopes and crowded into St Pierre. Biblical-type plagues then afflicted the city and countryside. Snakes and insects driven from the mountain by the heat and gas emissions invaded the city's suburbs. More than 100 deadly pit vipers entered the north of the city. The army was called out to shoot them, but not before over 50 people and 200 animals had been killed. At the sugar plantation at the mouth of Rivière Blanche, ants and deadly centipedes (the latter 0.3 m long) invaded the mill buildings, swarmed up the legs of horses and the workmen, and bit them.

By 5 May, the crater at the top of the mountain had filled with muddy water. On 5 May, this was blasted

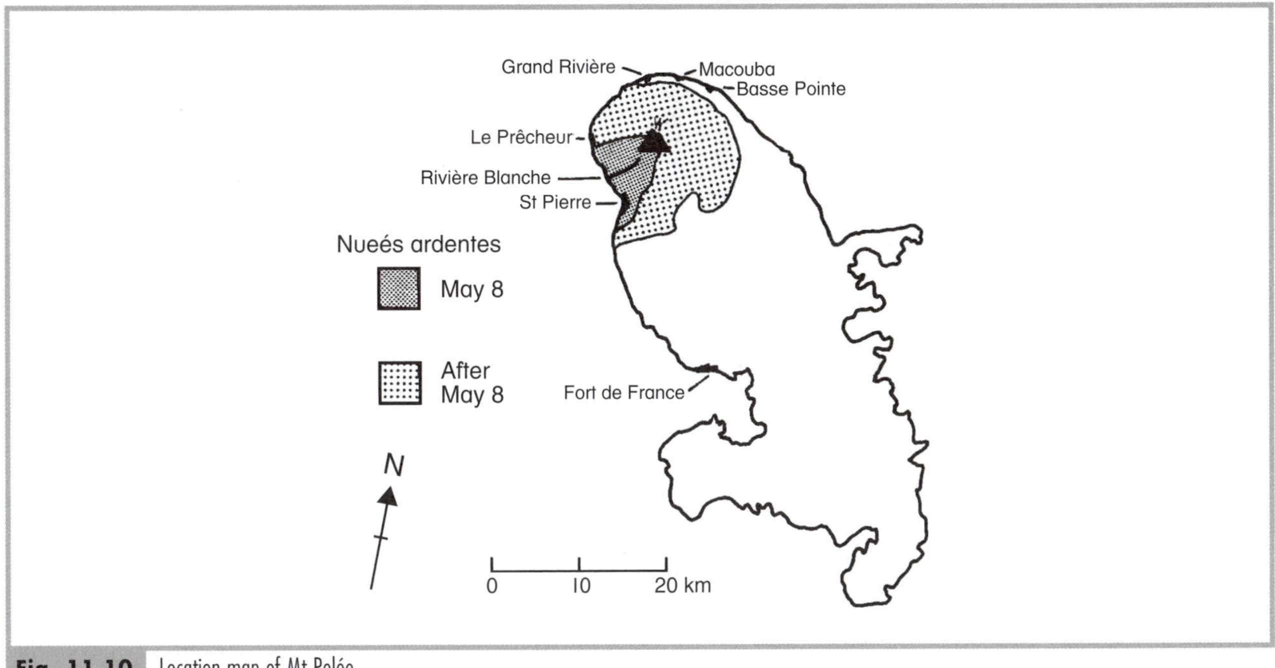

Fig. 11.10 Location map of Mt Pelée.

out, sending a large mudflow down the Rivière Blanche to the coast. The lahar killed 25 people. When the 35 m high wall of mud hit the ocean, it generated a tsunami that crashed into the lower part of St Pierre, killing 100 people there. On 6 May, Mt Pelée issued tephra that was so voluminous that it crushed roofs and clogged streets. All the rivers around the mountain then flooded for the next two days as the mountain began roaring and blowing out pumice. Hot gases rose to the surface, expelling groundwater from the watertable. On 7 May, Mt Soufrière on the adjacent island of St Vincent erupted and reports soon circulated that this eruption had released the pressures building up under Mt Pelée. Then Mt Pelée quietened, supporting this rumor.

During these events, the government, headed by Governor Mouttet, struggled to make sense of the eruptions. Scientific advice was sought, but there was no expert living on the island. Rumors were dispelled by press releases stating that there was no threat. Comparisons were made with Krakatau, which had erupted nineteen years earlier. Because Mt Pelée was not near the ocean, the risk of tsunami was dismissed. Earthquakes were viewed as the main hazard. No one realized that the nuées ardentes that had begun on 7 May heralded an unfamiliar phenomenon that posed far more of a threat. Without any adequate knowledge of what to expect next, Mouttet – in what must be rated as one of the most foolhardy acts in history – took his wife and most of the senior members of his administration to St Pierre on the evening of 7 May to allay fears of impending doom amongst the citizens. Troops were ordered to begin patrolling the streets the following morning to prevent panic. Those citizens who wanted to leave St Pierre had already done so.

At 8:00 am on 8 May, a series of four violent explosions shot dust to heights of 15 km. A second cloud was blasted as a basal surge, horizontally to the south-west. This cloud immediately flowed down the Rivière Blanche. When it came to the bend in the river valley, it overrode the bank, and part of the flow, traveling at velocities in excess of 160 km hr^{-1}, swept through the city of St Pierre at 8:02 am. Eyewitness accounts, from the few people who had fled the city early that morning, describe the pyroclastic flow as a hurricane of flame. Temperatures within the gas cloud were as high as 1075°C. Walls of buildings were blown down, a three-ton statue of the Virgin Mary was tossed 12 m, 18 of the 20 ships in the harbor were sunk, and most of the standing city set ablaze. The hot blast exploded a rum distillery and ignited rum, which then flowed through the streets completing the incineration of the city (Figure 11.11). The dust cover averaged only 30 cm in thickness throughout the city, but it engulfed the area in complete darkness. Towards the mouth of the Rivière Blanche, pyroclastic deposits thickened to 4 m. Up to 30 000 people, including Governor Mouttet, died within two minutes. Many had clothing

Fig. 11.11 The city of St Pierre, Martinique, after the 8 May 1902 eruption of Mt Pelée. The pyroclastic flow, which destroyed the city, swept over the ridge at the top of the photograph. The waterfront had been wrecked by a tsunami caused by a lahar several days previously (photograph from a book by A. Lacroix, courtesy of the Geological Museum, London).

stripped from them by the force of the blast. Others were grotesquely disfigured, either as their body fluids boiled and burst through their skin, or as their muscles contracted in spasms as they fought for air. There were only two survivors, one of whom, Auguste Ciparis, was a condemned prisoner in the local jail. He suffered severe burns and later had his sentence commuted. Subsequent blasts on 19 May and 20 August swept across most of the northern and western slopes, causing further destruction. The last explosion killed 2000 people in five mountain villages. Of all historical volcanic disasters, the residents of the city of St Pierre and the island of Martinique appear to have suffered one of the most devastating eruptions known.

Mt St Helens (18 May 1980)

(Hays, 1981; Lipman & Mullineaux, 1981; Keller, 1982; Blong, 1984; Coates, 1985)

The Mt St Helens eruption is notable for several reasons. Firstly, it was the first large explosive eruption to occur in the world for several decades. Secondly, it heralded a period of substantial volcanic activity at the beginning of the 1980s. Thirdly, it had associated with it several hazard phenomena. Finally, it was the largest eruption to occur in the conterminous United States in recorded history. Mt St Helens is situated in the Cascade Mountains of Washington State (Figure 11.4). It had not had a major eruption for 123 years and, compared to Mt Rainier, was considered a relatively minor volcano. However, research in the mid-1970s began to indicate that ash found in the region originated mostly from Mt St Helens. On 20 March 1980, a swarm of micro-earthquakes around Mt St Helens signaled renewed activity, which culminated in a small eruption on 27 March sending clouds of ash 6 km skyward. Harmonic tremors began on 3 April, signaling the movement of magma into the magma chamber below the volcano. By this time, over 70 million m^3 of tephra had been ejected. Prior to the eruption, the north side of the volcano bulged more than 150 m at a rate of 1.5 m per day.

At 8:30 am on 18 May, an earthquake of magnitude 5.1 on the Richter scale started landslides in the bulge region. Immediately, the volcano erupted and sent out a lateral explosion, which was felt 425 km away. A dark cloud of ash, containing 2.7 km^3 of material weighing 520 million tonnes, rose to a height of 23 km over the next nine hours. The height of the mountain was reduced by 500 m (Figure 11.12). Over 500 km^2 of forest were devastated by the resulting base surge that reached temperatures of 2600°C. The blast overrode ridges over 10 km away. In some places, debris deposits accumulated to depths of 150 m. Snow and ice on the mountain were melted by the blast and formed lahars that rushed into Spirit Lake, filling the lake to depths of 60 m. The lahars then moved over 50 km down the north and south forks of the Toutle River, reaching Kelso on the Columbia River. Lahars also flowed down into the Muddy River to the east. In all, over 300 km of roads and 48 road bridges were extensively damaged. The lahars and fine ash that settled in the surrounding area were soon carried into the Columbia River, where the 180 m wide shipping canal was reduced from a depth of 12 m to one of 4.3 m. Only 60 people lost their lives, mainly because the government had ordered evacuations as the intensity of the eruption increased. However, some of this death toll included people who were permitted back into the area only days before the eruption. Tephra fallout became the worst nuisance after the initial eruption. While most of the dust was blasted out the side (Figure 11.12), ash fell to the ground in eastern North America and over the Atlantic. Within 17 days, ash had encircled the globe at a height of 9–12 km, at the top of the troposphere. Some ash moved into the stratosphere, but this amount was minor. Temperatures were estimated to have decreased by 0.5°C for a few weeks directly downwind because of the reduced

Fig. 11.12 The north side of Mt St Helens after the 18 May 1980 eruption. The crater has not formed by collapse, but by a lateral blast. The material in front represents the pyroclastic debris from that blast (photograph courtesy of James Ruhle and Associates, Fullerton, California).

incoming solar radiation, but the worldwide effect on climate was virtually irrelevant. The greatest effect occurred within 700 km of the eruption. Ash accumulated to depths of 10–15 mm, 150 km downwind of the eruption. It also made driving conditions in and around the area treacherous for five days because of low visibility and slippery roads. Speed limits had to be reduced to 8–10 km hr^{-1} to prevent accidents. The deposition of ash proved a major clean-up problem in urban centers. The small town of Yakima, Washington, with a population of 50 000, took ten weeks to remove half-a-million tonnes of ash, a task that cost $US2 million. The fine dust clogged air filters on cars, got into brakes and wrecked motors. Dust fallout on electrical equipment caused equipment failure, while crop yields noticeably decreased because the ash coated leaves and lowered the efficiency of photosynthesis.

Given the magnitude of the event, the media hype, and the reaction by the public and government, the true economic effects of the disaster tended to be overestimated. Washington State estimated the damage at $US2700 million, and Federal Congress appropriated just under $US1000 million for disaster relief. In effect, the eruption caused $US844 million damage. Clean-up cost $US270 million; agricultural losses amounted to $US39 million; property damage, mainly to roads and bridges, cost $US85 million; and commercial timber losses amounted to $US450 million. The eruption sparked renewed research interest into other volcanoes along the Cascade and Sierra Nevada ranges. Mt St Helens also highlighted the fact that the previous 60 years worldwide were generally quiescent volcanically.

CONCLUDING COMMENTS

There is no doubt that the Earth is experiencing one of the most intense periods of volcanism in the last 10 000 years. This period began at the beginning of the seventeenth century, concomitant with global cooling that peaked in the Little Ice Age. The volcanic events of the latter half of the twentieth century must be viewed in this context, rather than as freak eruptions of supposedly dormant volcanoes. The eruptions of El Chichon, Mt St. Helens, and Mt Pinatubo are the most dramatic examples of this reawakening. While Mt Pinatubo stands out as a dominant event, there is little realization that it was but one of three major volcanic eruptions in 1991–1992. The other two were Mt Hudson in Chile and Mt Spurr in Alaska. Since then, volcanic activity has become quiescent. Whether or not this represents a resumption of inactivity similar to that between the eruption of Caribbean volcanoes in 1902 and Mt Agung in 1961 is academic. However, since the eruption of Mt St Helens in 1980, renewed volcanic activity and associated phenomena have taken more than 24 000 lives. In our present era, volcanic eruptions are pervasive, unpredictable, and deadly.

REFERENCES AND FURTHER READING

Blong, R.J. 1984. *Volcanic Hazards: A Sourcebook on the Effects of Eruptions*. Academic Press, Sydney.

Bolt, B.A., Horn, W.L., MacDonald, G.A., and Scott, R.F. 1975. *Geological Hazards*. Springer-Verlag, Berlin.

Branney, M and Zalasiewicz, J. 1999. Burning clouds. *New Scientist* 17 July: 36–41.

Cita, M.B., Camerlenghi, A., and Rimoldi, B. 1996. Deep-sea tsunami deposits in the eastern Mediterranean: new evidence and depositional models. *Sedimentary Geology* 104: 155–173.

Coates, D.R. 1985. *Geology and Society*. Chapman and Hall, New York.

Fielder, G. and Wilson, I. 1975. *Volcanoes of the Earth, Moon and Mars*. Elek, London.

Halldórsson, M.M. and Brandsdóttir, B. 1998. *Subglacial volcanic eruption in Gjálp, Vatnajökull, 1996: The jökulhlaup*. <http://www.hi.is/~mmh/gos/vat-update.html>

Hays, W.W. 1981. Facing geologic and hydrologic hazards: Earth-science considerations. *United States Geological Survey Professional Paper* 1240-B: 86–109.

Kastens, K.A. and Cita, M.B. 1981. Tsunami-induced sediment transport in the abyssal Mediterranean Sea. *Geological Society of America Bulletin* 92: 845–857.

Keller, E.A. 1982. *Environmental Geology* (3rd edn.) Merrill, Columbus, Ohio.

Latter, J.H. 1981. Tsunamis of volcanic origin: summary of causes, with particular reference to Krakatau, 1883. *Bulletin Volcanologique* 44: 467–490.

Lipman, P.W. and Mullineaux, D.R. (eds). 1981. The 1980 Eruptions of Mount St Helens, Washington. *United States Geological Survey Professional Paper* No. 1250.

Myles, D. 1985. *The Great Waves*. McGraw-Hill, New York.

Newhall, C.G. and Punongbayan, R.S. (eds) 1996. *Fire and Mud: Eruptions and Lahars of Mount Pinatubo, Philippines*. Philippine Institute of Volcanology and Seismology, Quezon City and University of Washington Press, Seattle.

Nomanbhoy, N. and Satake, K. 1995. Numerical computations of tsunamis from the 1883 Krakatau eruption. *Geophysical Research Letters* 22: 509–512.

Pararas-Carayannis, G. 1998. *The waves that destroyed the Minoan Empire (Atlantis)*. <http://www.geocities.com/CapeCanaveral/Lab/1029/.html> (URL defunct as of 2004)

Perrett, F.A. 1935. Eruption of Mt Pelée 1929–32. *Carnegie Institute of Washington Publication* No. 458.

Pichler, H. and Friedrich, W.L. 1980. Mechanism of the Minoan eruption of Santorini. In *Thera and the Aegean World: 2*. Proceedings of the Second International Scientific Congress, Santorini, Greece, August 1978, pp. 15–30.

Prabaharan, D.J. 2002. Fear of flying: Assessing the risks of volcanic ash clouds for aviation. *GIS User* 50: 22–23.

Ritchie, D. and Gates, A.E. 2001. *Encyclopedia of Earthquakes and Volcanoes*. Facts on File, New York.

Scarth, A. 2002. *La Catastrophe: Mount Pelée and the Destruction of Saint-Pierre, Martinique*. Terra Publishing, Harpenden.

Self, S. and Rampino, M.R. 1981. The 1883 eruption of Krakatau. *Nature* 294: 699–704.

Sigurdsson, H., Carey, S., Cornell, W., and Pescatore, T. 1985. The eruption of Vesuvius in 79 AD. *National Geographic Research* 1: 332–387.

Symons, R.B., Rose, W.I., and Reed, M.H. 1988. Contribution of Cl- and Fl-bearing gases to the atmosphere by volcanoes. *Nature* 334: 415–418.

Tazieff, H. and Sabroux, J.C. (eds). 1983. *Forecasting Volcanic Events*. Lange and Springer, Berlin.

Verbeek, R. D. M. 1884. The Krakatoa eruption. *Nature* 30: 10–15.

Whittow, J. 1980. *Disasters: The Anatomy of Environmental Hazards*. Pelican, Harmondsworth.

Winchester, S. 2003. *Krakatoa: The Day the World Exploded: 27 August 1883*. Viking, New York.

Wood, R.M. 1986. *Earthquakes and Volcanoes*. Mitchell Beazley, London.

Yokoyama, I. 1978. The tsunami caused by the prehistoric eruption of Thera. In *Thera and the Aegean World: 1*. Proceedings of the Second International Scientific Congress, Santorini, Greece, August 1978, pp. 277–283.

Land Instability as a Hazard

CHAPTER 12

INTRODUCTION

One of the most widespread natural hazards is the unexpected and sometimes unpredictable movement of unconsolidated weathered material (*regolith*) or weathered rock layers near the Earth's surface. Landslides and avalanches, while historically not renowned for causing as large a death toll as other natural disasters such as tropical cyclones or earthquakes, have had just as dramatic an impact on property and lives. The sudden movement of slope material is as instantaneous as any earthquake event but it is a more widespread problem. In any moderate- to high-relief region subject to periods of high rainfall, slippage of part or all of the regolith downslope is probably the most common hazard. Nowhere is this problem more prevalent than in cold regions underlain by *permafrost* or ground ice. Of a slower nature, and just as widespread a hazard, is land *subsidence*. While much of a land surface may be stable or even flat, there is a wide range of natural processes that can generate ground collapse. Another important aspect of land instability is the multitude of factors that can trigger ground movement. Almost all of the hazards presented in this book can generate secondary land instability problems. In many cases, associated landslides have contributed significantly to the large death toll from earthquakes and cyclones. Even droughts can exacerbate ground instability through the process of repeated drying and wetting of expansive clays. This can lead to surface deformation and eventual destruction of structures with insufficient foundations.

In this chapter, a broad overview of the wide range of land instability hazards will be presented. Firstly, the basic principles of soil mechanics will be described to show how surface material becomes unstable, and to point out what factors exacerbate ground failure. Secondly, each of the main types of land instability will be examined in turn, together with a description of some of the major disasters. To limit the coverage of such a wide topic as land instability, direct, human-induced and ice-related (*cryogenic*) factors will be discussed only at a cursory level.

SOIL MECHANICS

(Young, 1972; Chowdhury, 1978; Finlayson & Statham, 1980; Goudie, 1981; Aune, 1983; Bowles, 1984)

Stress and strain

Consider a body of soil with mass ω sitting on a slope of angle α. This mass is affected by gravity and tends to move downslope. The effect of gravity is directly related to the angle of the slope as follows:

$$\text{effect of gravity on a slope} = \omega \sin \alpha \quad (12.1)$$

where ω = mass

α = the slope angle

Any force (gravity is by far the most important) that tends to move material downslope is termed a stress (Figure 12.1). If a building is built on the slope, the weight of the building adds more stress to the soil. Additional weights on a slope are termed 'stress increments'. Stress increments can consist of rain, soil moved from upslope, buildings, or any other mass. While gravitational stress is most obvious, there are other stresses that can exist in the regolith. Molecular stress arises with the movement of soil particles or even individual molecules. It is associated with such phenomena as swelling and shrinking of *colloids* (clay soil particles) on wetting and drying, thermal expansion and contraction, and the growth of ice crystals. Biological stress refers to the stress exerted on a soil body by the growth of plant roots or the activities of animals.

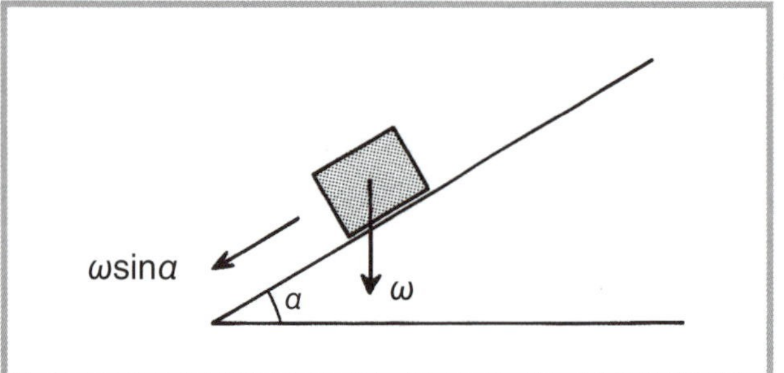

Fig. 12.1 Effect of gravity on an object with mass ω on a slope with angle α.

The effect of stress upon a soil or a regolith is called *strain*. Strain is not directly equated with stress but incorporates the additional factor of slope angle. Strain may not uniformly occur in the soil body, but may be restricted to joints where fracturing will eventuate. Strain can affect inter-particle movements, or act on the overall soil column. In combination, these strains result from what is labelled 'net shear stresses'.

Friction, cohesion and coherence

One of the forces of resistance against the movement of material downslope is friction. Friction is the force that tends to resist the sliding of one object over, or against, another. For regolith material sitting on a firm base, friction is due to irregularities present at the contact between these two materials no matter how smooth this contact may appear to be. The irregularities tend to interlock with each other and prevent movement. Generally, the greater the weight of an object pressing down upon the contact, the greater the amount of friction. While friction is dependent on the degree of roughness of a surface, it is independent of the area of contact between a body of regolith and the underlying substrata. Small areas of soil material tend to fail at the same angles as larger areas. Figure 12.1 illustrates the application of stress to an object on a slope. There is a resisting force against movement due to friction. Sliding will commence when the applied stress exceeds maximum frictional resistance.

Friction is expressed as a coefficient μ and this coefficient is equal to the tangent of the slope angle at which sliding just begins:

$$\tan \alpha_\mu = (\omega \sin\alpha_\mu)(\omega \cos\alpha_\mu)^{-1} \quad (12.2)$$

where α = angle of plane sliding friction
μ = coefficient of friction

The equations of movement of objects can be expressed in terms of critical frictional resistance ($R_{f\text{crit}}$) and critical applied force (F_{crit}). Failure commences when the critical applied force exceeds the critical frictional resistance as follows:

$$F_{\text{crit}} > R_{f\text{crit}}$$

where $R_{f\text{crit}} \sim \omega \tan\alpha_\mu$ (12.3)

This equation implies that as the weight of an object approaches zero the force required to move it downslope approaches zero. For non-rigid objects (unconsolidated material) this is not so, because as the weight of the regolith approaches zero, there is still an additional force resisting downslope movement. This force is termed cohesion, defined as the bonding that exists between particles making up the soil body. The above relationship, now including cohesion, can be expressed as follows:

$$F_{crit} > R_{f\,crit} \quad (12.4)$$

where $R_{f\,crit} \sim \omega \tan\alpha_\mu + c$
c = cohesion

Strictly, the term cohesion refers to chemical or physical forces between clay particles. The term *coherence* is used to describe the binding of soil particles of all sizes, as a group or mass, due to capillary cohesion by water at the interface between individual grains, by chemical bonds or by *cementation*. In cementation, the chemical bonds are considered primary and, thus, are very strong. Common cementing agents include carbonates, silica, alumina, iron oxide, and organic compounds. Cementation can also occur because of compaction, especially if material comes under pressure from above. The compaction process gives stability to materials on slopes. With certain grain shapes, especially with clays, it is possible to realign grains so that they interlock effectively, increasing coherence and the resisting force to movement. Chemical bonds consist mainly of oppositely charged electrical fields that develop on the surfaces of large molecules, especially clay minerals. These attracting charges are called *Van der Waals' bonds*, which for clay minerals remain active even when the clay particles or colloids are moved relative to each other. For clays, this gives rise to plasticity, which will be described later.

Figure 12.2 presents the type of bonds that can exist depending upon particle size and the relative strength of bonding. Van der Waals' bonds are restricted to material less than 0.03 mm or 30 microns in diameter. As the material gets smaller, the bonding strength increases considerably until it reaches values of 1 kg cm^{-2} for sizes less than 0.001 microns. As grain size increases, capillary cohesion (mainly due to water) becomes dominant. Capillary cohesion is also very important in bonding clay minerals, creating forces three orders of magnitude stronger than Van der Waals' bonds. The only way that coherence of material coarser than 1 cm can be achieved is by compaction and cementation.

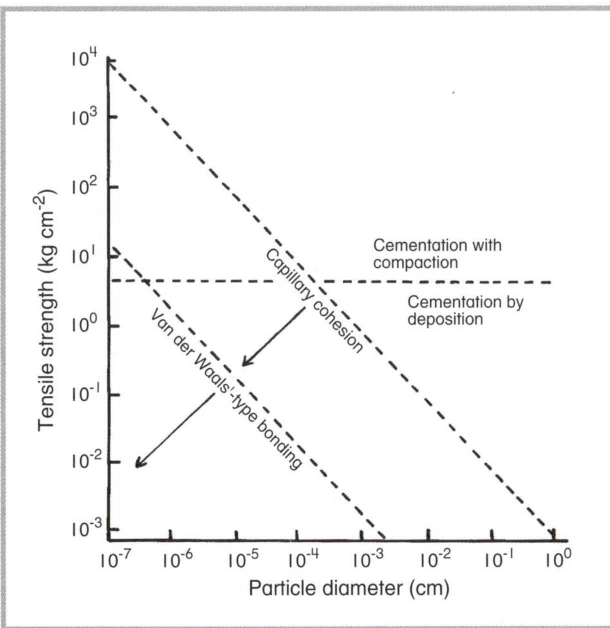

Fig. 12.2 The strength of bonding on particles of different sizes (Finlayson & Statham, ©1980 with permission Butterworths, Sydney).

The effect of capillary cohesion can be demonstrated easily. Damp sand can be shaped into almost vertical walls without any sign of failure. As the sand dries out and the cohesive tension of water at sand grain interfaces is removed, the sand pile begins to crumple, eventually reaching the angle of repose for loose sand, which is only 33°. Cohesion in damp sand is produced totally by the effect of water tension between sand grains, and gives wet sand a remarkable degree of stability. Note, however, that the sand cannot be too wet, otherwise liquefaction occurs. In other words, if more water (or any other material for that matter) is added to the sand, such that the pore spaces in the material are filled and the capillary thickness of water at the grain interfaces is exceeded, then the cohesion of the material approaches zero and the material begins to behave as a fluid. This process will be discussed in more detail later.

SHEAR STRENGTH OF SOILS: MOHR–COULOMB EQUATION

The way that soil particles behave as a group or mass (coherence) depends not only upon the inner cohesion of the soil particles but also upon the friction generated between individual soil grains. The latter characteristic is internal friction or shearing resistance. How much shear stress a soil or regolith can withstand is given by the following equation, termed the *Mohr–Coulomb equation*:

$$\tau_s = c + \sigma \tan\phi \qquad (12.5)$$

where τ_s = the shear strength of the soil
c = soil cohesion
σ = the normal stress (at right angles to the slope)
ϕ = the angle of internal friction or shearing resistance

The Mohr–Coulomb equation is represented diagrammatically in Figure 12.3. Note that this equation is in a form similar to Equation 12.4 except for two differences. Firstly, in the Mohr–Coulomb equation the critical force for movement is determined by the stress normal to the ground surface, rather than by the weight piled on a slope.

Secondly, the angle of the slope has been replaced by the angle of shearing resistance, which represents the angle of contact between particles making up

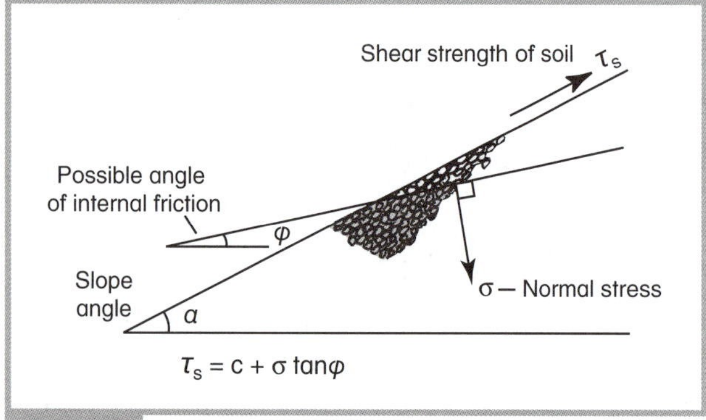

Fig. 12.3 Schematic representation of the Mohr–Coulomb equation. Note that the angle ϕ is not necessarily the slope angle, but the angle of internal friction within the slope mass.

unconsolidated material. Loosely compacted material tends to have a lower angle of shearing resistance than very compressed or compacted material. All unconsolidated material tends to fail at internal angles less than the slope angle upon which it is resting. The Mohr–Coulomb equation can also be used to define the *shear strength* of a unit of rock resting on a failure plane and the susceptibility of that material to landslides. In this case, the formula must be modified to include the characteristics of bedding planes. If the stress applied to the soil exceeds the shear strength, then the material will fail and begin to move downslope.

Pore-water pressure

Almost all material, no matter what its state of consolidation, contains pores or voids, which may be filled with air or, more often, water. If the water does not adhere solely to individual particles as capillary water, but completely fills voids forming a watertable, then it will flow freely under a pressure head. If the soil is incompressible, a rise in the watertable will cause the pressure of water in voids at depth to increase. Furthermore, some of the added pressure may be taken up by the grain-to-grain contacts in the soil. If an external load is applied to a soil mass on a slope in the form of additional water, or overburden, then the pore-water pressure will build up in that mass and water will be expelled at weak points. These weak points occur along fracture lines, bedding planes, or at the base of the regolith, where pore-water pressure is highest. As water drains from a soil body, the pore-water pressure will decrease.

The increase in pore-water pressure in the Mohr–Coulomb equation reduces the effective resistance of the soil body or regolith as shown by the following equation:

$$\tau_s = c + (\sigma - \xi) \tan \phi \qquad (12.6)$$

where ξ = pore-water pressure

In this case, *normal stress* σ is reduced by the pore-water pressure Í. If overburden is added to a soil, it will immediately increase pore-water pressure. Unless the excess pore-water pressure is reduced through drainage of water, the critical stress on that soil will exceed the critical resistance of the soil, causing slope failure. If, on the other hand, wet material is added upslope, there is no change in the stress being applied to the soil. However, the water in this material can drain into the soil downslope, increasing progressively over time its pore-water pressure. Failure of the slope may happen sometime after the overburden was emplaced upslope. A slope that has been stable under existing load and watertable conditions can also become unstable if drainage patterns are changed in the surrounding area. Dams raise watertables locally – a fact that may cause subsequent slope failure in adjacent areas some time after the dam is filled. Construction of buildings on slopes with septic systems can also increase the pore-water pressure, leading to subsequent failure. The source of water need not come solely from septic tanks. It could also come from lawn watering and other domestic discharges.

Earthquakes can increase the pore-water pressure of a soil with each passage of a compression shock wave. As the pore-water pressure increases, it may not be reduced fast enough by discharge of water from the ground before the next compression wave sweeps past. The effective stress in the material thus decreases with each shock wave until the pore-water pressure is equal to the normal stress in the soil (Figure 10.9). At this point liquefaction will result. In unconsolidated sediment, the smaller the particle size, the more water movement is inhibited because of capillary cohesion. Very fine silt and clays thus should be susceptible to liquefaction during earthquakes. However, depending upon the degree of consolidation, weathering, or other factors, cohesion in these materials may exceed the increases in pore-water pressure during the passage of earthquake shock waves. Thus, liquefaction tends to occur best in medium- to fine-grained sands that have not completely compacted. Because these types of sediments are widespread, especially in marine environments or on river floodplains, liquefaction is an almost universal feature of moderate-to-large earthquakes.

There is an exception to the above occurrence of liquefaction. Some sodium *cation*-rich clays – termed *quick clays* – will liquefy if salt is leached by fresh water. This scenario often occurs with glacially deposited marine clays that are subsequently elevated above sea level and flushed with fresh water. The salt or sodium chloride deposited with the clays acts as electrolytic glue, adhering to the clay particles and providing cohesiveness and structure to the clay matrix. If the sodium is leached out or replaced by calcium – a procedure that can be performed even by the mobilization of free calcium in cement foundations into the surrounding clay subsoil – the clay dramatically loses its cohesiveness. These clays then become very responsive to shock waves and liquefy during moderately intensive earthquakes. Such a process occurred in the Bootlegger Cove clay underlying Anchorage during the Alaskan earthquake of 1964 (see Figure 12.4 for the location of major placenames mentioned in this chapter). The process can also be instigated by the flushing of evaporite lake clays or highly weathered clays such as montmorillonite, and is certainly exacerbated by irrigation, pipe leakage, or simple lawn watering. The latter circumstances exist today in many parts of southern California including the San Joaquin Valley, southern coast ranges, Mojave Desert, and Los Angeles Basin – all regions that are seismically active.

Rigid and elastic solids

Depending upon how stress and strain are related, soil bodies may behave in four ways: as rigid solids, as elastic solids, as plastics, or as fluids. The process of liquefaction determines the point at which a soil body behaves as a fluid. The concept of a rigid solid is also easily explained. No matter how much stress is applied to a solid, that solid is considered rigid until the stress exceeds the strength of the material. At this point, the solid either deforms or fractures. The rate at which the stress is applied determines how the solid will behave. For instance, toffee candy, when given a sharp jolt, will fracture. In this case, the toffee can be considered a rigid solid. If, however, gentle pressure is applied to the toffee, it may simply deform. In this case, the toffee is not rigid under this type of stress. A solid may be considered *elastic* if a stress is applied that results in slow, continuous deformation proportional to the applied stress before fracturing. The deformation at this stage is reversible, hence the description of the body as an elastic. Earth materials behave elastically as long as the stress changes applied are small enough.

Plastic solids

The relationship between the moisture content of a soil and its volume permits three factors to be determined: shrinkage, plasticity, and liquid limits. These three terms are labelled *Atterberg limits*. They refer mainly to properties of clays and form a continuum, illustrated schematically in Figure 12.5. The shrinkage limit is defined as the moisture content of a soil at which point the soil stays at a constant volume upon drying. Above the shrinkage limit, soil material will behave as a rigid or elastic solid because it consists of

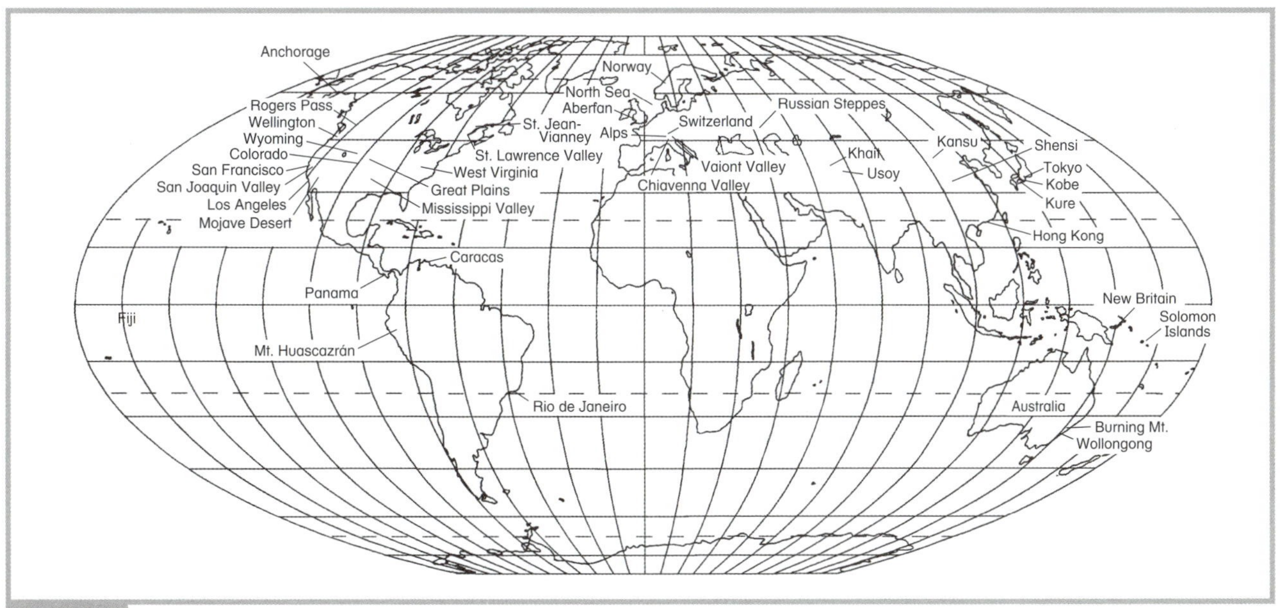

Fig. 12.4 Location map.

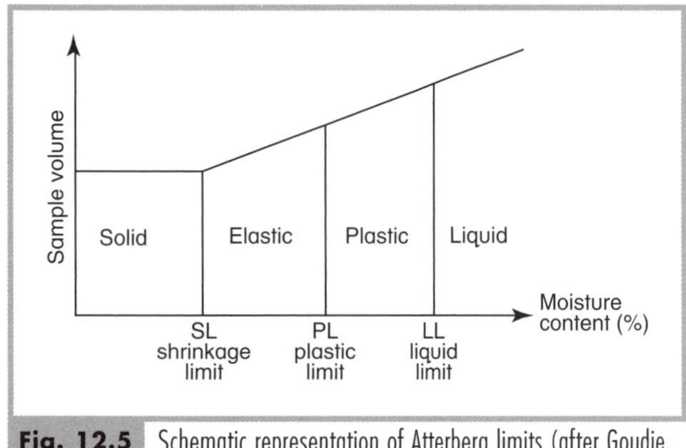

Fig. 12.5 Schematic representation of Atterberg limits (after Goudie, 1981).

interlocked assemblages of grains and particles that possess a finite strength, termed the *yield limit*. However, if a force is applied to that material such that the rate of deformation is proportional to the amount of applied stress above this yield limit, then the material is termed a 'plastic solid'. The difference between an elastic solid and a plastic solid depends upon how the material reacts after the stress is released. In an elastic solid, yield strength has not been exceeded, and the solid will tend to return to its original configuration after deformation. In a plastic solid, the deformation is irreversible because the yield limit has been exceeded. The yield or plastic limit is defined by the minimum moisture content at which point that material, usually clay, can be molded. It is also the point where the angle of shearing resistance in clays approaches zero as moisture content increases. This limit defines plasticity and, for clays, this state can be maintained over a wide range of moisture contents. As long as clay has a moisture content below the plastic limit, it will support objects. However, when the plastic limit is reached, then the bearing strength of the clay is greatly reduced and it will begin to deform. The liquid limit defines the moisture content at which the clay flows under its own weight. When the liquid limit is reached, clay behaves like a fluid and easily flows downslope. This point also defines when liquefaction occurs, because the material has no shear strength.

Plastic and shrinkage limits are very much a function of the type of clay material. There are three main types of clays, dependent upon the degree of weathering of feldspars or other easily weatherable minerals. Feldspars progressively weather to *montmorillonite*, *illite*, and finally *kaolinite*. Montmorillonite [$(Mg,Ca)O.Al_2O_35SiO_2.nH_2O$] is composed of one alumina layer sandwiched between two silica layers. The layers are bound together by Van der Waals' bonds. The clay minerals have a large residual negative charge because of a charge imbalance in their structure. This leads to the extensive addition of water molecules to the structure of the clay. As a result, the cation exchange, plasticity, and swelling capacity of such clays are high, making them especially susceptible to deformation and swelling. As the clays hydrate through progressive chemical weathering, they begin to lose these characteristics. Illite [$KAl_2(OH)_2(AlSi_3(O,OH)_{10})$], a daughter product of montmorillonite, contains a higher proportion of silicon atoms, inducing a high, net negative charge between layers. Potassium ions are attracted to these sites in the crystal lattice and bind the silica layers together, thus reducing their swelling capacity and plasticity. Kaolinite [$Al_2O_3.2SiO_2.2H_2O$] is a highly weathered clay mineral in which all potassium ions have been stripped from the lattice. These clay minerals consist of a layer of silica and alumina bonded together by Van der Waals' bonds or hydrogen ions. Both bonds are strong. There is little substitution of other atoms or molecules possible in the mineral; hence, it has low plasticity and low swelling capacity.

CLASSIFICATION OF LAND INSTABILITY

Introduction

(Sharpe, 1968; Varnes, 1978; Finlayson & Statham, 1980; Crozier, 1986)

Land instability can best be described by setting the various types of failure into a general classification. At present, land subsidence can be considered as a special case of land instability because it is more related to the behavior of the substratum than it is to the stress applied to material on a slope. Most classifications of land instability are based upon the type of material and the type of movement. There are five types of movement: falls, topples, slides, lateral spreads and flows. Material can consist of bedrock, consolidated soil and regolith, loose debris, various mixtures of sediment and water, and pure water in the form of snow or ice. Unfortunately, many of the classifications vary in their emphases. Some are based on geotechnical aspects, reflecting an engineering orientation; while others center on processes and morphology, reflecting a geomorphological perspective. Each classification has tended to bring in different terminology that at times is confusing and contradictory.

A useful classification in terms of material and movement is shown in Table 12.1. This classification has since been modified to include topples and lateral spreads. The scheme emphasizes the composition of the material being moved, but it and its modification clearly do not include time. From the hazard point of view – in terms of warning, human response, and prevention – these classifications can be limited. A temporal classification that gives some concept of the speed of movement of instability can be more attractive. In Table 1.4, which ranks hazard characteristics and impacts mentioned in this text, some emphasis is placed on the suddenness of the event. One such temporal classification is shown in Figure 12.6. It is immediately obvious that this classification includes expansive soils, which are ignored in most engineering and morphometric classifications. Figure 12.6 is used in this chapter, not because it is better than

Table 12.1 United States Highway Research Board Landslide Committee classification of mass movements (from Leopold et al., 1964).

Type of movement \ Type of material		Bedrock →			Soil
Falls		Rockfall			Soil fall
Slides	No. of Units: Few	Rotational ↔	Planar ↔		Rotational
		Slump	Block glide		Block slump
	↓ Many		Rockslide	Debris slide	Lateral spread
Flows		All unconsolidated			
		Rock fragments →	Sand or silt →	Mixed →	Plastic
	Moisture content: Dry	Rock fragment flow	Sand run; Loess flow		
			Rapid earth flow	Debris avalanche	Slow earth flow; Creep
					Solifluction
	↓ Wet		Debris flow: sand or silt	Rubble flow: dirty snow avalanche	Mud flow

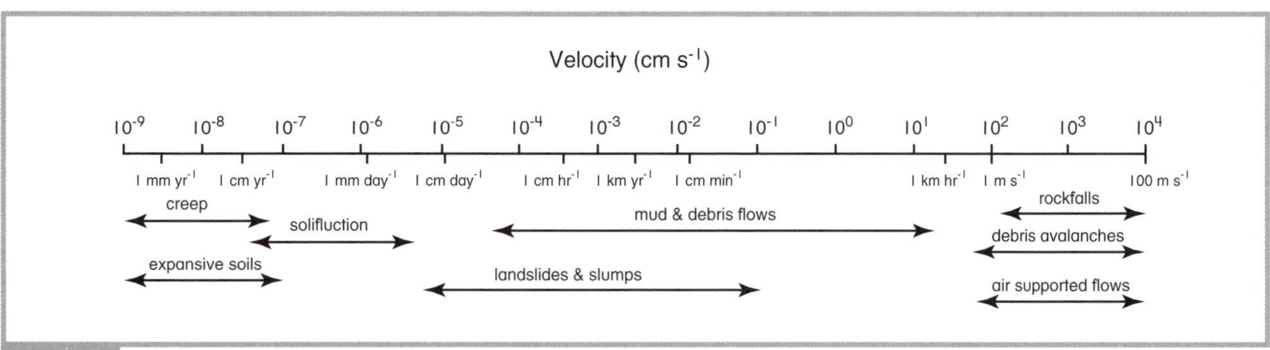

Fig. 12.6 Classification of land instability based upon rate of movement (derived from Finlayson & Statham, 1980).

the other classifications, but because it emphasizes time and includes expansive soils – the latter in the long term causing the most expensive type of land instability.

Expansive soils

(Hays, 1981)

Expansive soils annually cause more than $US3000 million damage (1989 dollar value) to roads and buildings in the United States. This cost exceeds the combined, annual damage bill from all climatic hazards in this country. Because the process works so slowly, damage is not always obvious. The actual cost to society may in fact exceed $US10 000 million annually. Fifty per cent of the damage occurs to highways and streets, while 14 per cent occurs to family dwellings or commercial buildings. Of the 250 000 homes built on expansive clays in the United States each year, about 10 per cent will undergo significant damage and 60 per cent will undergo minor damage in their life span. Figure 12.7 maps the extent of expansive soils in the conterminous United States. Over a third of the country is affected. The extent of the problem is not unique to the United States: it occurs on a similar scale in Australia and on the Russian steppes, where soil and geological conditions are similar. In Australia, 38 per cent of homes in the southern half of the continent show evidence of cracking, with again 10 per cent developing serious damage over their life span. During the drought of 2002–2003, the damage bill reached $US600 million.

Expansive soils are produced mainly by clays derived from two major groups of rocks. The first group consists of aluminum silicate minerals in volcanic material that decomposes to form montmorillonite. The second group consists of shales containing this clay mineral. In Australia, both these rock types are very prevalent because large areas of the eastern half of the continent were subject geologically to volcanism. Additionally, slow evolution of the Australian landscape has permitted the widespread accumulation of large quantities of montmorillonite clay as *cracking clays* on inland river systems.

Expansion usually takes place when water penetrates the lattice structure of clay minerals. Two conditions must be met before swelling can take place. First, it must be possible for a volume change to occur and, second, water must be present. An increase in soil moisture content of only 1–2 per cent is sufficient to cause expansion. For the volume change to be made, sufficient clay must be present. A thick layer of expansive clay has greater potential volume change than a thin layer. If the load on the clay is high, then compressive forces may exceed the expansive force exerted by clay upon wetting. Because this compressive load decreases towards the ground surface, the presence of

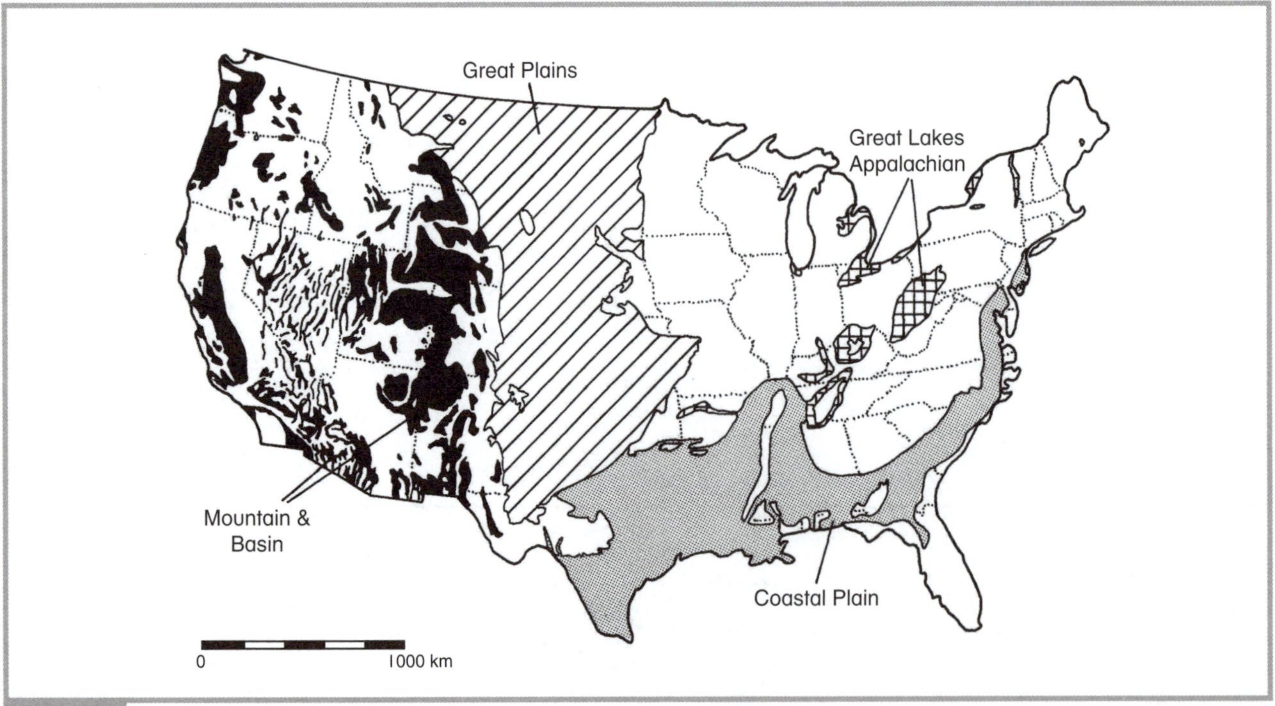

Fig. 12.7 The extent of expansive soils in the United States (adapted from Hays, 1981).

expansive clays in the upper soil layers increases the effect of swelling and shrinking. Large buildings positioned on expansive clays will not usually be affected because the weight of the structure can prevent expansion. Extensive surficial cracking due to desiccation is also indicative of expanding clays. In Australia, the presence of *gilgai* (undulations in the soil surface with wavelengths of 1–2 m) is indicative of the problem.

People can enhance the effect of expanding clay by disrupting ground drainage and modifying vegetation. Drainage from a house gutter or septic tank into an area of expansive clays can cause that material to expand, while the drier material under the house does not. This differential expansion can affect foundations closest to the wetted soil. The removal of vegetation can also result in increased soil moisture because evapotranspiration from trees or shrubs is negated. In built-up areas where rainfall variation is high, expanding clays can undergo repetitive cycles of swelling and shrinking, hence putting unequal stress on foundations over time. The process is quite common in Australia where drought and flooding rains alternate between El Niño and La Niña events. This is illustrated schematically for my house in Figure 12.8. During droughts, the soil underneath my house will dry out more slowly than that in the open. The exposed soil shrinks and the outer foundations of the house tend to settle before the inner ones. After rainfall, the exposed soil is wetted and expands. A wetting front then proceeds under the house causing expansion over time towards the centre of the house. Trees around a house can accelerate the drying effect during drought because of their higher evapotranspiration, while lawn watering can accelerate the wetting effect. Cracking of interior walls, especially during a change from drought to wet or vice versa, is very common. Floors can become creaky as pilings sink differentially due to variations in clay or moisture content.

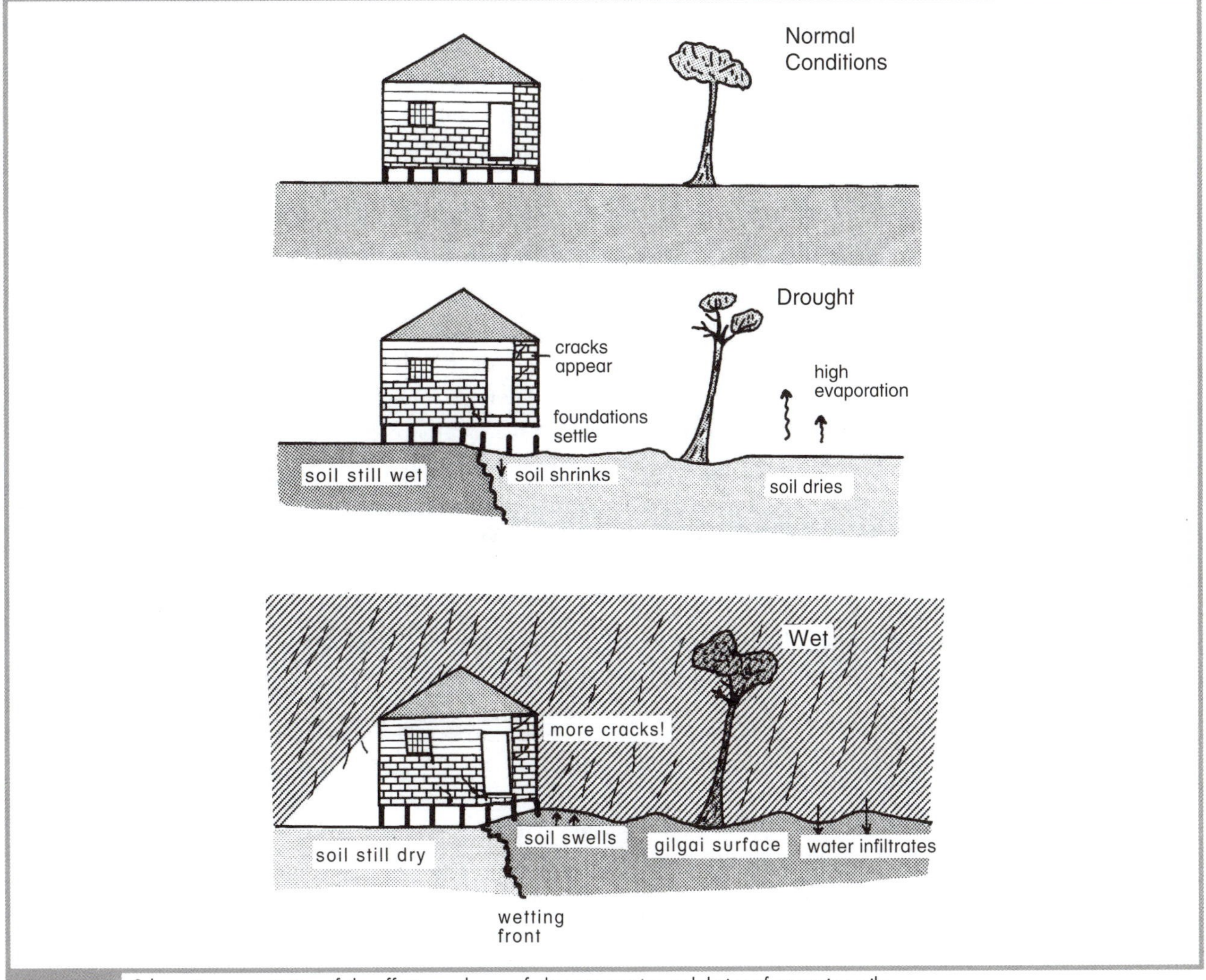

Fig. 12.8 Schematic representation of the effect on a house of alternate wetting and drying of expansive soils.

The most efficient way to reduce damage caused by expansive soils is to avoid them. In Australia, however, this is almost impossible. Other methods to minimize damage include removing the expanding soil (especially at the surface); applying heavy loads to the surface; preventing water access to the site; pre-wetting the soil and preventing its drying out forever afterwards (although damp soil attracts termites); or ensuring that foundations are sunk deep enough to minimize the effects of near-surface swelling and shrinking. It is also possible to change the ionic character of the soil by adding hydrated lime, $Ca(OH)_2$, to the surface of an expanding clay. The lime lowers the exchange capacity of the clay. Water, thus, cannot be substituted into the internal lattice structure of this clay mineral as efficiently, and expansion is reduced. Note that the addition of lime to a soil subject to cracking is not recommended, if that soil is also quick clay. Liming dramatically decreases this latter soil's cohesiveness.

Creep and solifluction

(Leopold et al., 1964; Sharpe, 1968; Young, 1972; Finlayson & Statham, 1980)

Under sustained shearing stresses, all soil and rock materials on slopes exhibit viscous behavior, which is termed creep. Where the melting of ground ice exacerbates the movement, the process is termed solifluction. Rates of creep are not substantial and rarely exceed 1–2 mm yr^{-1}, although velocities over 100 mm yr^{-1} have been measured on slopes as steep as 40°. Because creep is more active towards the surface of a soil profile, the rate of movement and the resulting displacement decrease with depth. This is partly due to compaction and increased loading at depth. If material has some long-term instability, shearing planes or surfaces will develop in the region of maximum shear stress. This point usually occurs at the base of the soil profile between the B and C horizons, or along fractures (bedding planes) in the underlying weathered bedrock. A stone line, above which banding in the soil occurs, and below which imperceptible movement takes place, often shows this. Material in this zone will shear until its strength is reached. If further shearing eventuates upward in the profile, the displacement becomes cumulative and, hence, greatest towards the surface.

Creeping of a regolith downslope can also be accomplished by alternating expansion and shrinking of a soil. The most obvious mechanism for this process is the presence of expanding clays under the influence of seasonal inequalities in rainfall. The process can also be accomplished by freezing and thawing of water in the soil, and by salt crystallization. Each time expansion happens, the soil tends to be pushed upward at right angles to the slope (Figure 12.9). This process weakens the coherence of the soil mass. When shrinkage or thawing eventuates, the soil settles back to its original position; however, gravity will tend to move the material slightly downslope. Creep rates should be proportional to the tangent of the angle of the slope, and all material should move downslope. In practice, it has been found that the rates of movement on a constant slope can vary greatly over small distances, to the point of being random. There are also measurements indicating that creep movement can occur upslope. This result implies that expansion and contraction do not operate normal to the slope. In this regard, a slope must be considered three-dimensional, with movement possible in any direction including upslope and laterally. This aspect has already been recognized in periglacial environments, where lateral sorting of sediment dominates the micro-morphological evolution of the landscape. In environments where a distinct thawing season exists after the soil has been frozen, or where rainfall is seasonal, creep rates will vary seasonally. The process is fastest in spring, when temperatures are still cool enough for the ground to re-freeze at night, and during the wet season, when distinct seasonal inequalities in rainfall exist.

Creep in clay does not always depend upon expansion and contraction. If clay is wet enough to become plastic, it will deform under load in the downslope direction. As the rate of creep depends upon the weight of overburden, deformation may occur only at depth in the profile and not at the surface. Creep, in clays at depth

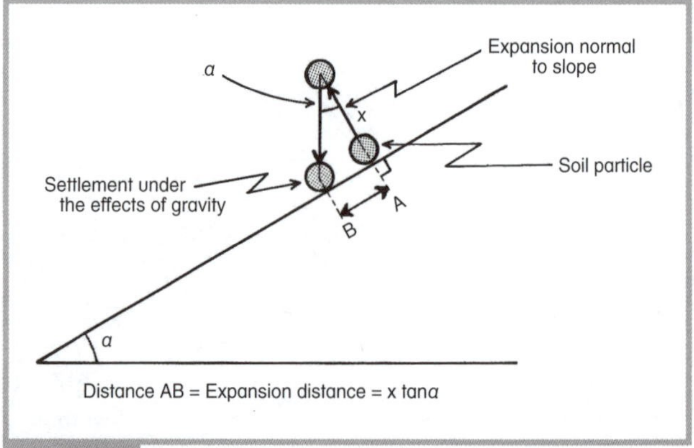

Fig. 12.9 Mechanism for the downhill movement caused by soil creep of material on a slope.

in a particular zone, can cause the clay minerals to realign themselves, reducing shear strength in this layer, which subsequently becomes the location for development of a shear plane. The cohesiveness of clay decreases with increasing moisture content; so, continual creep in clays has its highest rate where the clay is moist for the longest duration. Creep processes are not restricted solely to clays or to locations with expansive soils. Because creep operates under the effect of gravity, any disturbance to a soil on a slope will result in downhill movement of material. Animal burrowing, penetration by plant roots, tree collapse, and simple trampling by animals or people can be the cause of this disturbance. While the short-term disturbance seems minor, aggregated over time these effects can have significant influence on near-surface creep rates.

The process of solifluction involves additional factors. Solifluction is restricted to environments where ground freezing takes place. The expansive force initiating creep is due to moisture freezing in the soil, because frozen water occupies a greater volume than the equivalent weight in liquid form. Solifluction can produce accelerated rates of downslope movement of material on slopes as gentle as 1°. Freezing in soil can also concentrate water at a freezing front. Large volumes of water can be drawn to this front by capillary action, forming ice lenses that can constitute up to 80 per cent by volume of the soil mass. The melting of all or some of this ice can increase pore-water pressures and form shear planes for slope failure. In areas underlain by permafrost, where seasonal melting prevails in an active surface zone, solifluction can totally dominate a landscape such that material on all slopes appears to be in a continual state of rapid movement downslope. The discussion of solifluction and its geomorphic ramifications are beyond the scope of this text; but it should be realized that ground ice and solifluction processes represent a serious hazard to construction and transport in Arctic regions of Canada, Alaska, and Russia. Most mountain regions are also cool enough, because of their high elevation, to be affected by solifluction. About 20 per cent of the world's landmass is susceptible to these deleterious agents.

Mud and debris flows

(Sharpe, 1968; Bolt et al., 1975; Cornell, 1976; Whittow, 1980; Innes, 1983; Wieczorek et al., 2001)

A flow is any slope failure where water constitutes a major component of the material and a major controlling factor in its behavior. Flows can be subdivided into two categories, based upon the type of material. If clays dominate, the flow is termed a 'mudflow' or 'earth flow' in the United States and a 'mudslide' in the United Kingdom. Note that this term incorporates lahars mentioned in the previous chapter. The term 'lahar' is more restrictive in that it is a mudflow originating in volcanic ash. If the range in particle size is highly variable, then the flow is termed a *debris flow*. Generally, the material in a flow is water-saturated and unconsolidated. Not only can water permeate through the material, but it can also be absorbed by the sediment. For this reason, most flows represent the further downslope movement of material that has already undergone some sort of failure. For instance, a rockfall that disintegrates into smaller pieces while falling will produce an unconsolidated debris mantle when it comes to rest. Over time, this debris can chemically weather, breaking down into smaller particles. When wetted, this material may absorb water, increase in weight, and begin to fail again at lower slope angles.

Flows occur where high water content increases pore-water pressure in the deposit, a process that decreases shearing strength. Flows usually begin moving along a basal shear plane. Rates of movement can be as low as 1–2 m yr^{-1}, or as high as 600 m or more per year. Fastest movement occurs during the wettest months. Mudflows do not necessarily move at consistent rates; they commonly surge. In southern California, surge rates of 3–4 m s^{-1} have been observed in large mudflows. Typically, even with lahars, velocities average less than 20 km hr^{-1}. Slow-moving earth flows have caused property damage, but minimal loss of life, in Czechoslovakia, England, Switzerland, and Colorado. In periglacial environments dominated by solifluction, mudflows are a common occurrence.

Coarse debris flows are much more difficult to move. Fine material must be present to aid water retention. The debris flow may consist of very low volumes of sediment moving downslope at rates of several kilometers per hour. Coarser debris tends to get extruded to the surface and to the sides during movement, such that coarse-sized levee banks are built to the sides and front of the moving deposit. Typically, debris flows require slopes of 30–40° for movement to start, but flow continues over slopes of 12–20° (Figure 12.10). Note that this type of flow does not include large, catastrophic events moving coarse debris

Fig. 12.10 Widespread debris flows initiated on slopes of 12–30° in the Kiwi Valley area, Wairoa, New Zealand, by intense rainfall in February 1977. Rainfall intensities of 127 mm in 2 hours, and 252 mm in 5.5 hours, were recorded (photograph courtesy of the Hawke's Bay Catchment Board, Napier, New Zealand).

at higher velocities. Such an event is termed a debris avalanche and will be discussed in detail below. When mudflows and debris flows stop, excess water tends to drain from the deposit, forming either lobate or planar fans. In some cases, where subsequent filtering out of sediment has happened, flow deposits can be mistaken for alluvial fan deposits. The thickness of the deposit is inversely proportional to the fluidity of the deposit, and to the slope angle at rest.

Under extreme wetness, both mudflows and debris flows can liquefy, whereupon the weight of material is borne by pore-water pressure and not grain-to-grain contacts. This problem is particularly severe in marine clays laid down following the retreat of the last major glaciation. Subsequent desalinization forms quick clays, which are very susceptible to liquefaction. The St Lawrence Valley in eastern Canada was effectively flooded by the ocean during the waning phases of continental glaciation, with the subsequent deposition of extensive, thick clay beds susceptible to earth flow and liquefaction. Mudflows with volumes in excess of 20 million m^3 have been quite common in this region. The St Jean-Vianney, Quebec, failure in 1971 took 35 lives. Similar types of deposits have been recorded in Scandinavia and Russia. In 1893, at Vaerdael, Norway, 112 people were killed by one such failure.

Unstable mine tailings can also generate debris flows. The Aberfan, Wales, disaster of 21 October 1966 was one of the more tragic disasters of this nature. A 250 m high, fine-grained, rain-soaked coal tailings pile failed and turned into a debris flow, which swept through a school killing 116 children, the entire juvenile population of the town. A similar flow in a coal mining waste dump in West Virginia killed 118 people in 1972. The Aberfan disaster illustrated the unsafe nature of coal tailings in the United Kingdom – over 25 per cent were subsequently found to be in a similar condition of instability.

Nothing in recent history matches the Caracas, Venezuela, floods of 15–17 December 1999. Centuries ago, Caracas was deliberately built inland, behind the 2700 m high El Ávila coastal range, to hide the city from invaders. It has grown to a city of 5 million people plagued by the worst characteristics of Third World urban centers: slums with high densities of 25 000 km^{-2} built in hazard-prone areas, haphazard construction, lack of planning, and unenforceable building codes. Hillsides with 80 per cent slopes have been built upon. The disaster began during a month of rain, culminating in over 900 mm falling in a 72-hour period between 15 and 17 December. The rain triggered landslides, mudflows and, worst of all, debris flows, killing an estimated 30 000 people. While the rains were excessive, probability analysis has shown that 300 mm or more of rain in 24 hours could be experienced once in 25 years. Similar floods had occurred in 1798, 1912, 1914, 1938, 1944, 1948, and 1951. Geomorphic evidence indicated debris flows were common. The rains of 1999 reactivated nine large alluvial fans on the coastal side of the El Ávila range and created 28 slides containing as much as 58 500 m^3 of material. Unfortunately, the fans had been urbanized (Figure 12.11). The rains also stripped steep slopes of their regolith and moved boulders with dimensions as large as $11.3 \times 5.0 \times 3.5$ m, and volumes of 1310 m^3, downstream. Debris flow velocities reached 14.5 m s^{-1}. Under these conditions, buildings up to eight storeys high were destroyed. The extent of the disaster was awesome. The volume of sediment moved in individual debris flows reached 1.9 million m^3, amongst the largest ever recorded. However, these volumes are at least an order of magnitude smaller than the largest triggered by volcanic explosions, eruptions, or earthquakes. Over 8000 homes and 700 apartment buildings were destroyed or damaged. The damage bill totalled $US1.8 billion.

Fig. 12.11 Aerial view of erosion and debris flow deposition during the Caracas, Venezuela, urban flood of December 1999. The deposit is up to 6 m in thickness and contains 1.9 million m^3 of sediment. Gaps denote areas where buildings were removed by the floods (photo by Lawson Smith, United States Geological Survey, <http://pubs.usgs.gov/of/2001/ofr-01-0144/Venezuela/tnhires/pages/image030.htm>).

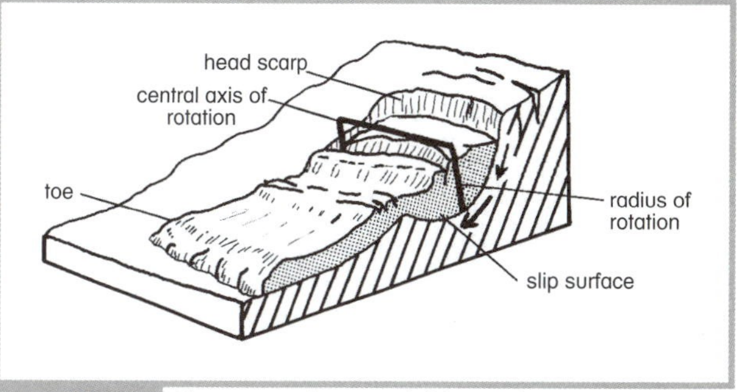

Fig. 12.12 The process of rotational sliding (Whittow, 1980, © and reproduced with the permission of John Whittow, Department of Geography, University of Reading).

Landslides and slumps

Mechanics

(Leopold et al., 1964; Chowdhury, 1978; Whittow, 1980)

Slope failure causing landslides can be modelled using the Mohr–Coulomb equation. Because many slides occur along a distinct shear plane, this equation can be modified to include the effects of shear planes within an unconsolidated deposit, bedding planes and joints in weathered or unweathered bedrock, and the presence of a watertable. Landslides, therefore, tend to happen in two different types of material, consisting of either bedrock or unconsolidated sediment, usually clay. In the simplest case, where the Mohr–Coulomb equation is easily applied, landslides move downslope parallel to the line of failure. However, landslides can undergo rotation as well. In this case, the failure develops slowly as a *slump*. Figure 12.12 illustrates this process of rotational sliding or slumping. While the mechanics of rotation are the same as a simple planar slide, the movement of the material is determined by the arc of rotation centered at a point outside the slip material. Rotational slides can be discussed in terms of a head scarp, where the material has separated from the main slope; a slip surface, upon which the material rotates; and a toe, which consists of the debris from the failure. Rotational slides may also generate transverse cracks across the body of the slipped mass and tensional cracks above the head scarp. These tensional cracks can be lines of failure for future slides. The toe of the slide controls the terminal point of the failure. Interference with the toe of the slide can also initiate further failure.

Causes

(Sharpe, 1968; Finlayson & Statham, 1980; Crozier, 1986)

With landslides and slumps, not only is the stress applied to the area of potential failure important but the behavior of the toe also becomes crucial to the timing of the slide. One of the major causes of landslides is undercutting of the basal toe in some manner. The causes of landslides in any form are thus multitudinous. Geological or topographical factors condition the location of a potential slide. If weathered or unweathered bedrock contains inherent lithological or structural weakness, then these areas will favor slide failure. Lithologically favorable material includes leached, hydrated, decomposed, chloritic, or micaceous rocks, shales, poorly cemented sediments, or unconsolidated material. Additionally, if this material is well-rounded or contains clay, then it will flow more freely under pressure. Stratigraphically favorable conditions for landslides include massive beds overlying weaker ones, alternating permeable and impermeable beds, or clay layers. Structurally favorable conditions include steeply or moderately dipping foliations, cleavage, joint, fault, or bedding planes. Rock that is strongly fractured, jointed, or contains parallel alignment of grains because of crushing, folding, faulting, earthquake shock, columnar cooling or desiccation is also likely to fail. Internal deformation structures resulting from tectonic movement, solution or underground excavation will aid slope failure, especially if such features do not reach the surface. Any water-saturated, porous lens within a regolith will also tend to form a zone of preferred shearing. Topographically, any cliffed or steep slope is also susceptible to land sliding, as are areas of block faulting, previous landslides or subsidence, and artificial excavation. Finally, any area that has been denuded of its soil-retaining vegetation because of deforestation, overgrazing, cultivation, fires, or climatic change has the potential for sliding.

Initiating causes of landslides are just as plentiful. The removal of basal support provides the easiest way to start a landslide. Natural agents such as undercutting by running water, waves, wind, and glacial ice are prime candidates. Extrusion of material from the base of a slope is also effective. This can take the form of outflow of plastic material within the deposit, washout of fines, and melting of ice. The material at the base could also change its characteristics through water absorption due to flooding, solution of soluble material such as limestone or salt, or chemical alteration. In developed areas, people have now become the major cause of toe instability on slopes through mining, excavation, quarrying, basal construction, and road, rail and canal works. Overloading of the slope material is also a common means of initiating failure. Overloading can occur by saturating the slope with rainwater or water from upslope streams or springs, or by loading the slope with snow or debris from upslope instability. Humans can also initiate overloading by dumping spoil or by building foundations and structures on the slope.

Slope failure can also be induced by reducing internal coherence in the slope material. Increased lubrication due to heavy rainfall or runoff filtering through a regolith is a common mechanism for triggering failure, either immediately or some months afterwards. Cracking of the slope material through partial failure, desiccation, earthquakes, or internal movement can also allow water to penetrate more easily into the slope material. Blockage of drainage, usually by raising the base of the watertable, can reduce shear resistance internally. As already mentioned, the activities of humans through drainage modification, deforestation, overgrazing, or water discharge from septic systems and domestic usage also reduce internal coherence through lubrication. Earthquakes and volcanoes can either crack slope material, or reduce coherence through the process of liquefaction. Even thunderstorms can trigger rockfalls and landslides because of the vibrations that are transmitted through slopes. Similarly, humans now perform activities that vibrate the ground and could trigger landslides. These activities include vehicle movements, blasting, pile-driving, drilling, and seismic work.

A number of processes can cause slope failure due to prying or wedging. Water freezing in cracks, increased pore-water pressure after heavy rain, expansion of soil material because of hydration, oxidation, carbonation or the presence of swelling clays, tree root growth, and temperature changes are the main mechanisms. Finally, strains in the earth due to sudden changes in temperature, atmospheric pressure, or the passage of earth tides are also factors that could trigger landslides. The above conditions favoring landslides may act in concert to trigger instability. As a result, slope failure can happen instantaneously, irregularly but cumulatively over time, or with considerable

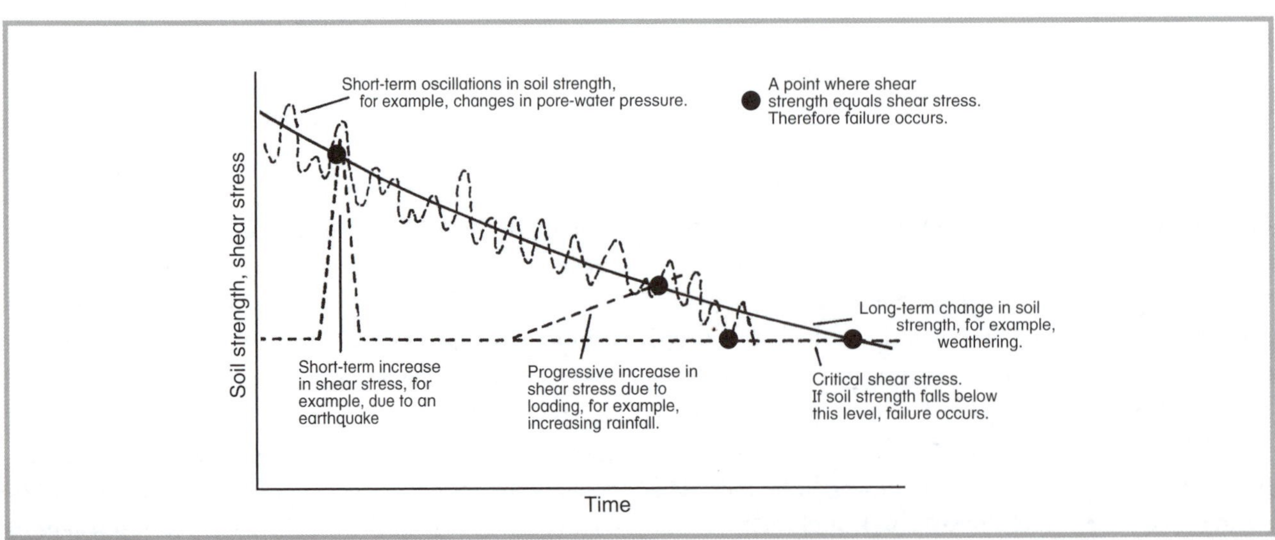

Fig. 12.13 The complexity of landslide initiation over time (Finlayson & Statham, 1980 © with permission Butterworths, Sydney).

temporal lag after the occurrence of a triggering event. Figure 12.13 illustrates this complexity in landslide initiation. Ultimately, failure will eventuate when the shear strength of the soil is exceeded by a critical shear stress. The critical shear stress can fluctuate slowly or dramatically over time, while the shear strength of the material can oscillate because of seasonal factors, or decrease slowly because of any of the above conditions. Slope failure may happen at a number of points in time. If stress dramatically increases, for instance because of heavy rainfall, and exceeds soil strength, then the cause of the landslide is obvious. However, soil strength could be decreasing, or the stress on a slope increasing slowly and imperceptibly over time, to the point that the cause of the landslide is not definable. Thus, with any landslide, there is randomness as to the timing and cause of the failure.

Landslide disasters

(Bolt et al., 1975; Cornell, 1976; Whittow, 1980; Coates, 1985; Blong & Johnson, 1986)

Landslide disasters are difficult to separate from other types of land instability, and from some of the larger disasters that trigger the event. For instance, tropical cyclones, apart from drowning people, usually bring very heavy rainfalls; this not only increases pore-water pressure in potentially unstable material but, through added weight, also increases stress in slope deposits. Some of the worst natural disasters have occurred because of landslides triggered by earthquakes. The Chinese earthquakes in Shensi province in 1556, and at Kansu in 1920, caused failure in loess deposits, collapsing homes dug into the cliffs. Urban areas seem to be most affected by landslides. In Rio de Janeiro in 1966, record-breaking rainfall in January and March caused catastrophic landslides that struck shantytowns around the mountains in the city. Over 500 people were killed in the slides and another 4 000 000 were affected by disrupted transportation and communications. In the following two-year period, over 2700 people were killed in the Rio de Janeiro area by landslides and other slope instability events that afflicted over 170 km^2. In Hong Kong, a tropical cyclone in 1976 dropped 500 mm of rain in two days, triggering landslides on steep slopes and killing 22 people. Similar slides there in 1966 killed 64 people. Many of the slides occurred where dense urban sprawl encroached upon steep slopes and undermined the toe of unstable sediments. Japan is also particularly vulnerable to typhoon-generated slides near urban areas. Kobe was hit in 1939 by rain-induced landslides that killed 461 people and damaged 100 000 homes. In 1945, comparable landslides killed 1154 people in Kure. In 1958, Tokyo was struck by a typhoon that generated over 1000 landslides and killed 61 people.

In underdeveloped, heavily vegetated regions, especially in the tropics, mega-landslides play a major role in the downslope movement of material. In 1935, New Guinean landslides cleared 130 km^2 of vegetated slopes. The Bialla earthquake of 10 May 1985, on the island of New Britain in Papua New Guinea, triggered a mega-landslide in the Nakanai Range that dramatically infilled the Bairaman River (Figure 12.14). About 12 per cent of slopes in New Guinea are subject to landslide denudation every century. Elsewhere in the tropics, an area of 54 km^2 was cleared on slopes in Panama in 1976. Tropical Cyclone Wally, which struck Fiji in April 1980, and Cyclone Namu, which passed over the Solomon Islands in 1986, both generated hundreds of landslides with an average volume of material moved in excess of 500 m^3 per hectare. Such landslides had not been witnessed in these areas in the previous 50 years. The cyclones effectively destabilized the landscape enough to ensure continued mass movements triggered by less severe rainfalls for years to come.

Fig. 12.14 Mega-landslide on the northern slope of the Nakanai Range, New Britain, Papua New Guinea. This slide was triggered by the magnitude 7.0 Bialla earthquake on 10 May 1985. The slide occurred in Miocene limestone and extensively infilled the Bairaman River Valley (photograph by Dr Peter Lowenstein, courtesy of C.O. McKee, Principal Government Volcanologist, Rabaul Volcanology Observatory, Department of Minerals and Energy, Papua New Guinea Geological Survey).

The ubiquitous nature of landslides is illustrated with reference to the conterminous United States in Figure 12.15. By far the most prevalent zone for landslides occurs in the Appalachian Mountains in the eastern part of the continent. Second in importance are the Rocky Mountains, particularly in the states of Colorado and Wyoming. The coastal ranges along the Pacific Ocean form the third most hazardous zone. In January 1982, heavy rains triggered more than 18 000 landslides in the San Francisco Bay region, causing 25 deaths and more than $US100 million in damage. It was the third biggest disaster in this region after the 1906 San Francisco and 1989 Loma Prieta earthquakes. The threat is even greater in the Los Angeles region. Landslides in the Los Angeles area killed 200 people on 2 March 1938. In all, landslides in some form affect about one-seventh of the United States. While most of the failures happen in mountainous areas, they are not restricted to these zones. The Mississippi Valley has a low, but significant, risk, as do plateaus on the southern Great Plains where weak shales are found. The United States is not unique in this respect. Landslides are a common geomorphic process in most countries with steep or raised terrain.

Rockfalls

(Scheidegger, 1975; Voight, 1978)

Rockfalls are one of most rapid land-instability events. They are generally restricted to unvegetated, vertical, rock cliff-faces but can occur in certain unconsolidated sediments that permit vertical slopes to develop. Generally, the vertical faces must be continually maintained by erosion of talus and screes that form at the base of cliffs. If this debris is permitted to accumulate, then the vertical face becomes buried. The debris is removed through chemical breakdown, washing out of fines and slippage. There is, thus, a balance between the rate of cliff retreat controlled by the occurrence of rockfalls, and the rate of removal of debris at the base. The production of talus does not have to occur catastrophically, but instead debris can simply accumulate through the continual addition of material as it is loosened from the cliff-face by weathering, rainfall, small shock waves, or frost action.

The preservation of a vertical face is dependent upon the characteristics of the bedrock. Generally, less resistant rock such as shales will not be able to develop cliff-faces, because the rock is susceptible to weathering and slippage. Situations where there are massive beds of sandstone or limestone, or where these resistant beds overlie weaker beds, are ideal for the formation of cliffs. However, massive beds without much jointing do not erode easily, and hence do not produce rockfalls. For example, the escarpments along the east Australian coastline have faces up to 100 m or more high in places, and appear to be retreating at rates as low as 1 mm yr^{-1} without significant rockfalls. Significant joints or vertical failure planes should also

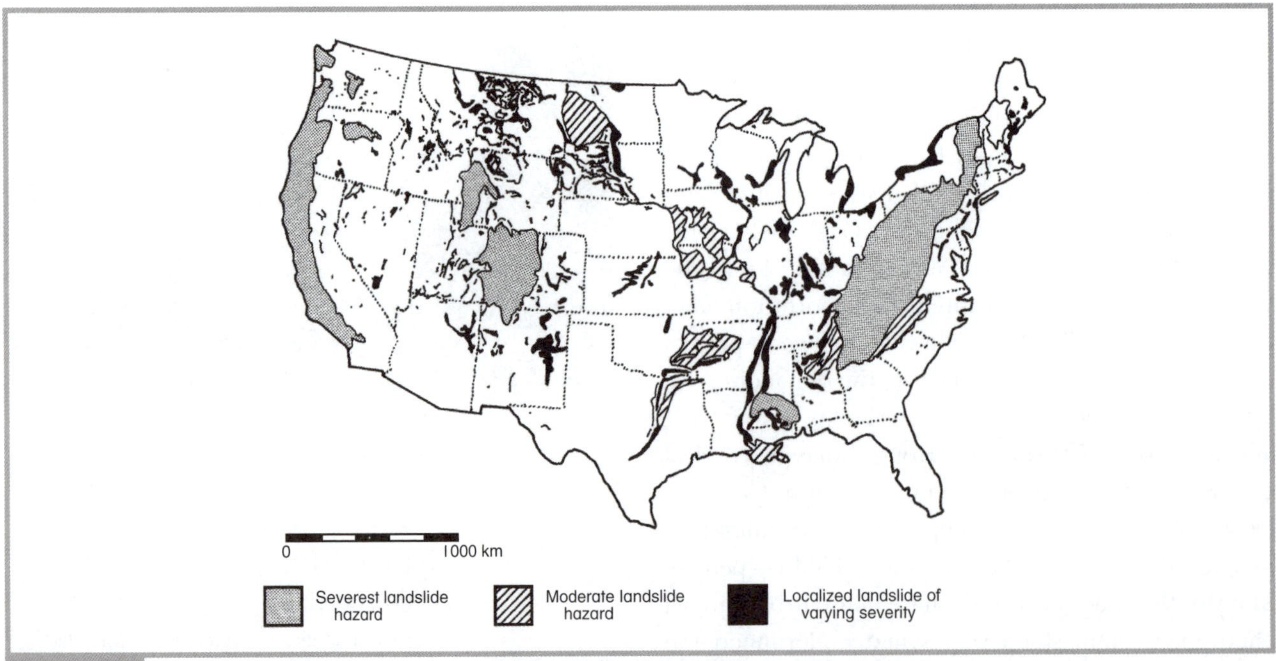

Fig. 12.15 Extent of landslides in the coterminous United States (after Hayes, 1981).

be present for the occurrence of large rockfalls. Even jointing is not a guarantee that rockfalls will eventuate. Blocks separating from cliffs along vertical lines of weakness must also undergo disintegration for rockfalls to take place. Otherwise, they simply glide downslope intact.

Globally, rockfall is the major process for the removal of debris from cliffs in scarped or steep mountainous regions affected by periglacial activity. While many of the factors that trigger landslides initiate rockfalls, the majority occur because of earthquakes. Many of the types of landslides discussed above, especially those occurring in the European Alps, began as rockfalls. In deeply snow-covered mountains, rockfalls are a major triggering mechanism for avalanches.

Debris avalanches

(Varnes, 1978; Voight, 1978)

Debris flows and mudflows can take on catastrophic proportions several orders of magnitude greater in size and speed than described above. The term 'debris avalanche' is more appropriate to these types of flows. Many of these events begin as landslides or rockfalls. All appear eventually to entrain a variety of particle sizes and behave as a fluid flow. The high velocities appear to be achieved by lubrication provided by a cushion of air entrapped beneath the debris. The flow literally travels like a hovercraft. In some instances, there is little evidence of water involved in the avalanche and, in other cases, masses of debris appear to have traveled with minimal deformation or re-alteration. By far the most destructive debris avalanches in recent times occurred around Mt Huascarán, Peru, in 1962, and again in 1970. The 1962 event started as an estimated 2 million m^3 of ice avalanched from mountain slopes. This ice mixed with mud and water, and turned into a much larger mudflow with a volume of 10 million m^3. It swept down the Rio Shacsha Valley killing 4000 people, mainly in the town of Ranrahirca. Material, some of which consisted of boulders over 2000 tonnes in weight, traveled down the valley at 100 km hr^{-1} and 100 m up valley slopes. Having survived that event – and believing that another, similar, catastrophe was unlikely – residents of the area then experienced an earthquake on 31 May 1970, which caused the icecap of Mt Huascarán, together with thousands of tonnes of rock, to cascade down the valley. This debris flow entrained boulders the size of houses. As it traveled down the valley at speeds of 320 km hr^{-1}, it picked up more debris each time it crossed a glacial moraine. At the bottom of the valley, the flow had incorporated enough fine sediment and water to become a mudflow with a 1 km wide front. Eyewitness accounts describe the flow as an enormous, 80 m high wave with a curl like a huge breaker as it approached the town of Yungay. Ridges over 140 m in height were overridden, and boulders weighing several tonnes were tossed up to 1 km beyond the rims of the flow. Within four minutes, 50–100 million m^3 of debris completely obliterated Ranrahirca and the much larger town of Yungay. The flow then continued down the Rio Santa reaching the Pacific Ocean 160 km downstream. Over 25 000 lives were lost. Only those who had raced to the tops of ridges survived. The disaster ranks with the Mt Pelée eruption of 1902 for the completeness of its destruction.

Many large landslides described as such in the literature should in fact be classified as debris avalanches. Almost all have originated in steep mountain areas. For instance, the Bialla slide referred to above as a megaslide, and shown in Figure 12.14, is technically a debris avalanche. Historically, many large landslides fit into this category. At Elm, Switzerland, on 11 September 1881, about 10 million m^3 of mountain slipped onto the town, burying it to a depth of 7 m and killing 115 people within 55 seconds. A debris avalanche of similar size at Goldau in Switzerland, in 1806, buried four villages and killed 457 people. On 4 September 1618, debris avalanches in the Chiavenna Valley in Italy killed over 2400 people. Larger disasters have also taken place. Reportedly, the total population of 12 000 in the town of Khait, Tajikistan, was killed in 1949 by converging avalanches triggered by earthquakes in the Pamir mountains. The largest amount of debris ever involved in a recorded failure occurred in the same mountains at Usoy in 1911. Approximately 2.5×10^9 m^3 of rock failed, damming the Murgab River and forming a lake 284 m deep and 53 km long. The debris avalanche was triggered by an earthquake registering 7 on the M_s scale that destroyed the town of Usoy, killing 54 inhabitants. A failure one-tenth this size, containing 0.25×10^9 m^3 of soil and rock, fell into a reservoir across the Vaiont Valley of Italy on 9 October 1963. The impact sent a wave of water 100 m high over the top of the dam and down the valley, killing 2000 people.

Air-supported flows (avalanches)

(Bolt et al., 1975; Scheidegger, 1975; Whittow, 1980; Wolman, 2003–2004)

Avalanches are commonly thought of in terms of ice or snow cascading down slopes; however, they are not restricted to frozen water. Snow avalanches are part of a phenomenon of debris movement involving the interstitial entrapment of air. Any material may be involved, including hot gases entrained within volcanic tephra (as described for pyroclastic flows in Chapter 11), mixtures of sediment and water in debris avalanches (as described in the previous section), and collapsing loess (as occurred during the Shensi, China, earthquake of 1556). The material in an avalanche flow is supported by air rather than by grain-to-grain contacts. Thus, avalanches are analogous to the liquefaction process whereby air, rather than water, fills voids and sustains the weight of the flow. Even if the avalanche makes contact with the ground, or contains a high proportion of liquid water, the internal air content ensures that coherence is minimal, and friction so low that the avalanche can obtain high ground velocities. Without turbulence, it is possible for velocities to reach in excess of 1200 m s^{-1}. However, avalanches rarely exceed 300 m s^{-1} because of the disruption to movement by turbulence. Unfortunately, turbulence permits most avalanches to entrain loose debris up to the size of boulders, a process that can cause substantial erosion along part of its path. The swift movement ensures that such flows are very powerful, and that they are preceded by a shock wave, which for snow avalanches may exert forces greater than 0.5 tonnes m^{-2}, as shown in Figure 12.16. The internal turbulence within a snow avalanche tends to generate swirls of debris with velocities in excess of 300 km hr^{-1}, greater than the movement of the avalanche itself. Forces of 5–50 tonnes m^{-2} can be generated internally, and are of sufficient magnitude to uproot trees and move large structures.

If the debris in the flow has liquefied, and if velocities are high and the flow thin, it is possible for shear sorting to occur internally in the flow. Shear sorting tends to produce internal sorting of debris such that larger objects are expelled to the top and sides of the flow. Basal surges in volcanoes behave in this fashion, and produce ash deposits that are well-sorted with the coarsest material towards the top of the deposit. The debris avalanche that wiped out Yungay in Peru in 1970 preferentially moved boulders towards the top and sides of the flow as it underwent shear sorting; however, it is not certain whether water or air was involved in the sorting process. The results of shear sorting enhance the prospect of survival for people caught in snow avalanches. Because human bodies are relatively large compared to the rest of the debris in the flow, they tend to move to the top of the avalanche under shear sorting. Rescuers simply probing the

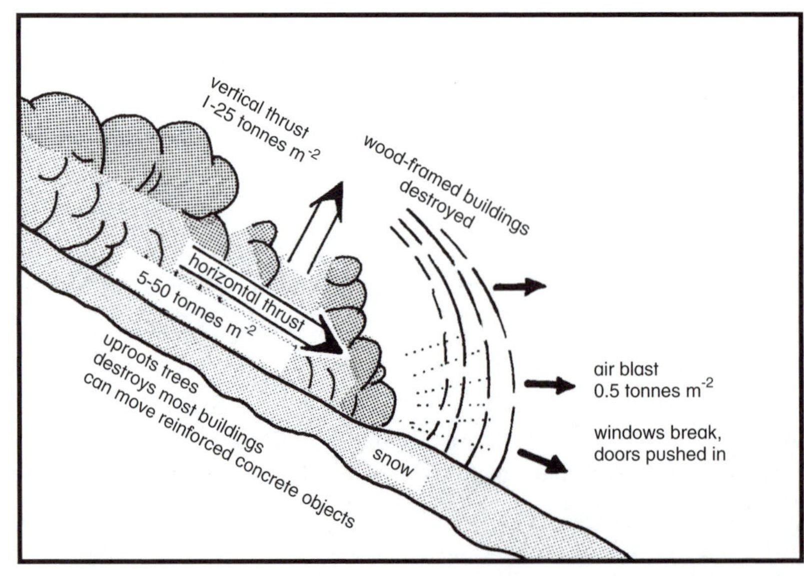

Fig. 12.16 Process and impact forces generated by a snow avalanche (Whittow, 1980, © and reproduced with the permission of John Whittow, Department of Geography, University of Reading).

upper layers of the avalanche deposit find and rescue many people buried by snow avalanches.

The word 'avalanche' has always conjured up images of snow cascading down mountains at high speed and wiping out unsuspecting alpine villages. This image is based upon accumulative disasters in the European Alps. Snow avalanches are also a common hazard in the Himalayas, Andes, and Rocky Mountains, especially in spring. One of the more recent disasters happened in 1978 at Col des Mosses, Switzerland, when an avalanche overwhelmed a ski-lift and killed 60 people. Even the Prince of Wales was almost killed by an avalanche that overtook his skiing party in 1987. Two of the largest disasters have occurred during wars. Hannibal's 218 BC invasion of Rome across the Italian Alps in winter was thwarted by snow avalanches that killed 15 000–18 000 of his men. During the First World War, snow avalanches – many of which were deliberately triggered by shelling – killed around 40 000 Italian and Austrian troops fighting in the Italian Alps. Other disasters have been less costly, but just as spectacular. Between 1718 and 1720 in Switzerland, four separate snow avalanches took over 250 lives. The largest Swiss avalanche, in 1584, took 300 lives. In 1910, an avalanche at Wellington, Washington, in the Rocky Mountains, swept through two trains killing 96 people. In the same year at Rogers Pass, British Columbia, avalanches killed 62 people. As recently as 2000–2001, avalanches killed 176 people worldwide.

In recent years, snow avalanche disasters have decreased with the advent of monitoring techniques and remedial measures designed to negate the hazard. For instance, most zones have been mapped in the populated mountainous areas where snow avalanches are likely to occur. The fact that avalanches tend to recur in the same places has aided this mapping. Sharp increases in temperature, after a winter of heavy snowfall or during a spring accompanied by rainfall, produce ideal conditions for avalanches. Studies of the dynamics of snow and ice on slopes have allowed these conditions to be identified to a high degree of accuracy. Sensitive micropenetrometers driven into snow at 2 cm s^{-1} can produce detailed profiles of snow hardness and identify layers of loose snow underlain by snow that has metamorphosed into ice called depth hoar that forms prime conditions for avalanching. In milder environments, refreezing of melted slush – called 'zarame-yuki' in Japan – can produce similar conditions. In the North American Rockies and the European Alps, army field cannon or explosives are used to trigger snow avalanches before they can build up to catastrophic dimensions. These techniques have proved effective in keeping passes free of large disruptive flows, and in protecting ski slopes in resort areas. Where road and train routes must be kept open, avalanche chutes are used to divert frequently recurring flows. These chutes simply consist of roofs that permit the avalanche to pass harmlessly over the road or railway. Near settlements, where these techniques are impractical, other techniques for braking, deflecting or disrupting the flow have been established. Upward-deflecting barriers can be constructed along preferred avalanche pathways to convert a ground-based avalanche into a slower, less forceful, airborne one. Concrete blocks can be arranged on slopes to divert avalanches away from settlements. The judicious alignment and reinforcement of buildings facing avalanche-prone slopes can also negate damage and prevent loss of life.

SUBSIDENCE

(Sharpe, 1968; Bolt et al., 1975; Coates, 1979, 1985; Whittow, 1980; Hays, 1981)

Land subsidence is a major type of land instability that does not depend upon many of the basic mechanisms outlined in this chapter so far. Rather, it is due to factors that include the effects of humans through water and oil extraction, and mining activities. The proper discussion of these latter factors is beyond the scope of this text. In this section, only the natural causes and mechanisms of land subsidence will be discussed.

Land subsidence is important at several scales of magnitude, spatially and temporally. For instance, a small sinkhole caused by limestone solution is not significant except to a local landowner; however, uniform subsidence of several square kilometres can become severely disruptive to a much wider community. In a few minutes, the Alaskan earthquake of 1964 produced regional, tectonically induced subsidence along the south coast of Alaska that had a profound effect upon the landscape. This sudden change in the relative location of land and sea resulted in severe economic problems in harbors. At the same time, there are locations – such as the south coast of England – where long-term subsidence of the crust has been occurring since the demise of the last continental icesheets, and as far back as the Pliocene. While coastal erosion and flooding can result from this regional subsidence, it is possible for humans to adjust to these changes.

Natural mechanisms causing land subsidence fall into three groupings: chemical, mechanical and tectonic. Chemical agencies of land subsidence mainly involve solution by groundwater of limestone, halite, gypsum, potash, or other soluble minerals. Because limestone is very extensive, the development of *karst* topography can lead to sudden ground subsidence over large areas. This is a particular problem in the south-eastern and mid-western parts of the United States, areas that are underlain by extensive limestone deposits. In Australia, subsidence features are widespread across the Nullarbor Plain; however, the region is neither densely settled nor intensely farmed, so that karst subsidence has little effect on people. A rarely recognized chemical cause of subsidence is the natural ignition (or, more recently, industrial vandalism) of coal seams such as has occurred at Burning Mountain, north of Scone, New South Wales, Australia.

Mechanical factors involve the removal, by extrusion, of material at depth in the soil. This can include the ejection of quick clays. However, by far the most effective process is the melting of ice lenses in permafrost areas. Because living and decayed vegetation provides an insulating blanket against the penetration of heat into permanently frozen ground, its removal will often result in catastrophic and irreversible melting of ground ice and subsequent surface collapse. The emplacement of heat conduits, such as foundations and telephone poles, will also permit the transfer of heat from the surface to depth where melting will occur. Even heat from a building can conduct into permafrost to cause melting and subsidence unless it is prevented from doing so by an intervening layer of insulation. The effect of subsidence in permafrost regions has major implications for landform development and the survival of fragile, cold-climate ecosystems.

Tectonic factors involve warping of the Earth's crust. The Earth is an elastic body that can be deformed by the weight of glaciation, and that will return to its original shape after the ice load is removed. Glaciation causes forebulging some distance in front of the ice front. In these regions following glacial retreat, the crust settles instead of rebounding. Large sections of the southern coastline of England, including London and northern Europe, are affected by this process, and are presently undergoing subsidence at rates in excess of 2–3 mm yr^{-1}. This is being exacerbated by the long-term downwarping of the southern North Sea Basin. If there is a threat of sea level rise in the next century, then it is possible for this rise to be exacerbated by tectonic subsidence in these regions. Subsidence can also be caused by ground loading. This is an obvious factor where major rivers are presently depositing large volumes of sediment on shallow continental shelves. Much of the coastline surrounding the Mississippi Delta is subsiding because of continual, long-term deformation of the crust due to the deposition of sediment debouched from the Mississippi River.

CONCLUDING COMMENTS

Land instability hazards encompass the most diverse range of processes and phenomena of any hazard discussed in this text. They also occur over the greatest time span, from soil creeping downslope at rates of millimetres per year, to the various types of avalanches reaching speeds in excess of 300 km hr^{-1}. In many respects, land instability is the most commonly occurring hazard. For instance, solifluction affects 20 per cent of the Earth's landmass, while expansive soils and landslides impact on 33 per cent and 14 per cent, respectively, of the conterminous United States. However, the topic of land instability hazards is not an easy one to cover. The fact that these hazards have attracted equal attention from engineering- and geology-related disciplines has led to a confusion of terms and a discipline-specific emphasis. This has meant that some important phenomena such as expansive soils are relegated in treatment to disciplines, in this case soil science, that do not always attract the attention of natural hazard researchers. It has also made exchange between disciplines difficult, not only because of the clash over terminology, but also because engineers tend to centre discussion around mathematics while geologists and geomorphologists are more interested in processes and morphology. Despite attracting the attention of a wide range of researchers, many aspects of land instability remain unclear. For instance, little is known about the mechanics of debris avalanche movement. This has meant that our ability to predict the occurrence of many land instability hazards is far from effective. For example, research in the Wollongong region of Australia would indicate that landslides should be prevalent whenever the monthly rainfall total exceeds 600 mm. This condition was met for the month of April in both 1988 and 1989. While numerous artificial embankments failed, the number of landslides to occur on natural slopes did not reach these predictions. Finally, land instability hazards are frequently associated with other natural hazards. Flows, landslides, and avalanches can all occur as secondary phenomena following tropical cyclones and other large-scale storms, earthquakes, and volcanic eruptions. Even drought affects soil stability because of its control on the frequency of soil contraction and expansion. For these reasons, land instability hazards are secondary phenomena and cannot be ranked as important as floods, storms, earthquakes, or volcanoes.

REFERENCES AND FURTHER READING

Aune, Q.A. 1983. Quick clays and California's clays: no quick solutions. In Tank, R.W. (ed.) *Environmental Geology*. Oxford University Press, Oxford, pp. 145–150.

Blong, R.J. and Johnson, R.W. 1986. Geological hazards in the southwest Pacific and southeast Asian region; identification, assessment, and impact. *BMR Journal of Australian Geology and Geophysics* 10: 1–15.

Bolt, B.A., Horn, W.L., MacDonald, G.A., and Scott, R.F. 1975. *Geological Hazards*. Springer-Verlag, Berlin.

Bowles, J.E. 1984. *Physical and Geotechnical Properties of Soils* (2nd edn). McGraw-Hill, Sydney.

Chowdhury, R.N. 1978. *Slope Analysis*. Elsevier, Amsterdam.

Coates, D.R. 1979. Subsurface influences. In Gregory, K.J. and Walling, D.E. (eds) *Man and Environmental Processes*. Dawson, Folkestone, pp. 163–188.

Coates, D.R. 1985. *Geology and Society*. Chapman and Hall, New York.

Cornell, J. 1976. *The Great International Disaster Book*. Scribner's, New York.

Crozier, M.J. 1986. *Landslides: Causes, Consequences and Environment*. Croom Helm, London.

Finlayson, B. and Statham, I. 1980. *Hillslope Analysis*. Butterworths, London.

Goudie, A. (ed.). 1981. *Geomorphological Techniques*. Allen and Unwin, London.

Hays, W.W. 1981. Facing geologic and hydrologic hazards: Earth-science considerations. *United States Geological Survey Professional Paper* 1240-B: 54–85.

Innes, J. L. 1983. Debris flows. *Progress in Physical Geography* 7: 469–501.

Leopold, L.B., Wolman, M.G., and Miller, J.P. 1964. *Fluvial Processes in Geomorphology*. Freeman, San Francisco.

Scheidegger, A.E. 1975. *Physical Aspects of Natural Catastrophes*. Elsevier, Amsterdam.

Sharpe, C.R.S. 1968. *Landslides and Related Phenomena: A Study of Mass-Movements of Soil and Rock*. Cooper Square, New York.

Whittow, J. 1980. *Disasters: The Anatomy of Environmental Hazards*. Pelican, Harmondsworth.

Wolman, D. 2003–2004. Charge of the ice brigade. *New Scientist* 20–27 December 2003–3 January 2004: 44–46.

Varnes, D.J. 1978. Slope movement types and processes. In Schuster, R.L. and Krizek, R.J. (eds) *Landslides: Analysis and Control*. National Research Council Transportation Research Board Special Report No. 176: 11–33.

Voight, B. (ed.). 1978. *Rock Slides and Avalanches I: Natural Phenomena*. Elsevier, Amsterdam.

Young, A. 1972. *Slopes*. Oliver and Boyd, Edinburgh.

Wieczorek, G.F., Larsen, M.C., Eaton, L.S., Morgan, B.A. and Blair, J.L. 2001. Debris-flow and flooding hazards associated with the December 1999 storm in coastal Venezuela and strategies for mitigation. *U.S. Geological Survey Open File Report 01-144*, <http://pubs.usgs.gov/of/2001/ofr-01-0144/>

PART 3

SOCIAL IMPACT

Personal and Group Response to Hazards

CHAPTER 13

INTRODUCTION

The mitigation and survival of any natural hazard ultimately depends upon the individual, family, or community. An individual has the choice to heed warnings, to prepare for an impending disaster, and to respond to that event. In the end, the individual, family unit or local community endures the most of a disaster in the form of property loss, injury, loss of friends or relatives, or personal death. If these small social units could recognize the potential for hazards in their environment, and respond to their occurrence before they happened, then there would be minimal loss of life and property. Unfortunately, not everyone heeds advice or recognizes warning signs, not necessarily because of stupidity, but for very important personal and socio-economic reasons. The reaction of individuals, families, and community groups after a disaster also determines their ability to survive physically and mentally. Some individuals and families can overcome all bureaucratic obstacles in rebuilding. Others end up with shattered lives despite all the support available to them. This chapter will describe and account for some of these personal and group reactions.

BEFORE THE EVENT

(Burton et al., 1978)

Warnings and evacuation

People react differently to their perception of a hazard. While some may not know that an earthquake is about to occur until they hear the rumbling and feel the shaking of the ground, most will know that they live in an earthquake-prone area. For example, there are 23 000 000 people living along the San Andreas Fault in southern California. The bulk of this population knows that a major earthquake is overdue, but continues to live there despite the risk. People are faced with the same decision in Wellington, New Zealand, where the occurrence of a major earthquake could destroy most of the city at any time. Wellington is expanding along its fault zone. In both cases, it is the 'sense of place' or 'home' that overrides all common sense about the threat of the hazard. People born in an area tend to want to stay there. Home is familiar, it has people one knows and associates with at a personal level. There is some sort of historic identity with the area, which is difficult to give up, and which is defended against change. There is a need to maintain links with the past or with ancestors, to sustain one's roots. People require an association with place and history. No threat of a hazard will make them leave, and they will rationalize the threat to minimize its occurrence. Residents of San Francisco refer to the 'Great Fire of 1906' rather than the 'Great Earthquake of 1906'. In these people's minds, great fires are rarer than great earthquakes and, hence, a similar disaster would be less likely to recur. If a place is attractive climatically, has an alluring lifestyle, or promises better economic opportunities, then it will continue to grow despite any calamity that can be associated with it. That is why people continue to move to California, that is why Mexico City has grown so rapidly since 1960, and that is why Wellington, New Zealand, continues to grow.

In the short term, even in the above locations, a warning of imminent disaster can also invoke a casual reaction bordering on disregard. For example, during the 1983 Ash Wednesday bushfires in Australia, some people nonchalantly sat around pubs escaping the heat of the day, while they watched one of the worst conflagrations witnessed by humanity bearing down on them (Figure 13.1). A non-evacuation reaction may be macho-based bravado, a public display of fearlessness, a normal psychological response of a low-anxiety individual, a religious taboo, or even the sensible thing to do. An example of a religious taboo occurred in Bangladesh (East Pakistan) during the 1970 cyclone surge. At the time the cyclone struck many women were prohibited, by established religious practice, from going outside for the period of a month. Even if residents had been given warning of the storm surge, few women would have evacuated and broken that taboo. Non-reaction to imminent disaster can occur for other reasons. Warnings of disaster imply evacuation, and that may appear too radical a departure from everyday life to be acceptable. Peer-group pressure also may be at work. If you are the only one in your neighborhood to evacuate, then you may be ridiculed or made to look silly. This has always been a human response to warnings of doom. For example, in 1549, a cadi (judge) in eastern Persia predicted an earthquake for his city. After trying to persuade his friends to leave their homes for the open, and after spending time alone in the cold outside, he eventually returned home only to be killed along with 3000 others in the ensuing earthquake (Michaelis, 1985).

Fig. 13.1 Drinkers at a pub in Lorne, Victoria, casually watching a bushfire bearing down on them on Ash Wednesday, 16 February 1983 (photograph © and reproduced courtesy of *The Age*, Melbourne). This fire and others turned into one of the greatest conflagrations ever witnessed.

Impending disasters are also times of worry, fear, and confusion. Most people at Hilo, Hawaii – reacting to sirens warning of the approach of a tsunami following the Chilean earthquake of 23 May 1960 – were confused (Lachman et al., 1961). Only 41 per cent of people evacuated to safer locations following the sirens. Twenty-seven per cent of the population, including some who evacuated, did not know what the sirens meant, while 46 per cent thought they were only a preliminary warning. More worrying was the fact that no mechanism existed to inform people it was safe to return. Of the sixty-one people killed by the tsunami, many believed it was safe to return after the first few waves had arrived at shore. The overt decision to 'stay put' may represent a subliminal mechanism for coping with the unknown and the unexpected. Generally, those of low socio-economic status, ethnic minorities, and women respond least to warnings. Older people find such change difficult to handle physically and mentally. For example, one old man died in the Mt St Helens eruption because he refused to move despite all the warnings. He had lived there all his life, and nothing was going to make him move. During tropical cyclones in the United States, the militia has evacuated people, at gunpoint, from coastal areas. Old couples died in the 1983 Australian bushfires along the coast south of Melbourne, because they had retired to those communities and there was nowhere else for them to go. If their retirement home were to be destroyed by fire, they would stay behind to save it, or 'go down with the ship'.

This reaction to imminent disaster can be explained in psychological terms. High-anxiety people perceive events occurring around them as confirming their anxiety. They tend to take action and in fact be psychologically conditioned for survival. Similarly, people who believe that they are in control of their life will also take action to survive. In contrast, people who are low-anxiety or fatalistic believe their lives are controlled externally. This difference also relates to variation in socio-cultural groups. For example, as mentioned in Chapter 3, Sims and Baumann (1972) found that there was a higher incidence of death due to tornadoes in the south of the United States than in the north. People in Alabama distrust the Weather Bureau and its warnings. They await their fate and, as a result, there is a greater incidence of death there because of tornadoes. On the other hand, people in Illinois believe they are in control of their own lives, are technologically orientated, and take precautions against the threat of tornadoes. They better survive the risk as a result.

For some, the decision to evacuate is an economic one. Any evacuation costs money, no matter who organizes it or carries it out. An evacuation involves loss of income for most of the community. For example, fishermen in the Caribbean do not evacuate during cyclones. Staying behind with their boat is more acceptable than evacuating, knowing that it is going to cost them money to move, and loss of income for the time they have moved. Loss of income may be unacceptable if your income is paying for an expensive mortgage, or keeping you from the moneylender. In that case, evacuation will be contemplated only if the hazard is a certainty.

The 'cry wolf' syndrome lulls others. For example, even though the residents of Darwin in 1974 had up to three days' warning of the arrival of Cyclone Tracy, only 30 per cent took precautions. This was because a cyclone had approached Darwin in the previous week, and had veered away. False cyclone threats are always happening in Darwin. Tracy also had the misfortune of occurring just before Christmas. Preparing for Tracy meant interrupting Christmas celebrations for a cyclone that might never come. Evacuation meant going inland, where there were no facilities, and the possibility of being trapped by heavy rains. Similarly, in the United States, tornado warnings are ignored because they are issued for very large areas, when their probability of occurrence in part of that area is small. The Weather Bureau assumes that the total population will respond to a warning. However, so few people have actually experienced a tornado that the majority of the population fails to take any precautionary measures. Civil defense groups in the United States now make more dramatic pleas for evacuation. For instance, after a newscaster has broadcast a tropical cyclone warning in the weather report, the head of the civil defense organization may interrupt further broadcasting, identify who they are and, in an authoritative voice, state that the cyclone threat is now real and people should take immediate action to evacuate for their own personal safety. Of course, this did not occur before Hurricane Andrew made landfall in 1992.

There are also those who are quite willing to evacuate, even though the signs of impending disaster are not obvious to the public. Many people in touch with nature, such as North American Indians, have left areas of disaster weeks in advance, responding to subtle signs shown by insects and animals. For example, some Indians in the Mt St Helens area packed up and left before the eruption, because they noticed animals doing the same thing. Indian tribes in southern California have left flood-prone areas in the midst of drought, because they noticed insects and ground burrowing animals seeking higher ground. Several weeks later flash flooding occurred when the drought broke.

Evacuation can also be dangerous. It may seem advantageous to flee the landfall of an approaching cyclone and, certainly – if warning is given of an earthquake – it is wise to leave buildings and head for open spaces. The same type of evacuation might also seem appropriate if a wildfire is swooping towards your home at 20 km hr^{-1}, as happened in Australia's 1983 Ash Wednesday bushfires. However, a study by Wilson and Ferguson (1985) of houses occupied at Mt Macedon showed that it was better not to flee this bushfire. Only 44 per cent of unattended houses survived compared to 82 per cent of occupied houses. This evidence was ignored during the devastating bushfire that swept through Canberra 20 years later. Six people lost their lives at Mt Macedon; however, this could not be attributed to the fact that people remained at home to fight the fires. Evacuation during the fire was downright deadly. Other evacuees clogged roads, smoke made driving impossible, and some cars ended up trapped in exposed areas, where death from heat radiation was a risk. Similarly, it is unwise to try to flee approaching tornadoes in cars. Studies in the United States have shown that 50–60 per cent of deaths and injuries from tornadoes are caused by flying glass from broken windscreens.

In some cases, where the disaster occurs suddenly, evacuation may not be a viable option. During most earthquakes or landslides, individuals have only a few seconds to react. Personal decisions must be deliberately made against a background of seemingly organized flight by tens or hundreds of people. The decision to 'buck the trend' becomes mentally more difficult and, in certain cases, physically impossible. As an expert, I know that standing in the open during a lightning storm is dangerous. Yet I have stood immobile watching a junior soccer match under such circumstances, too shy to march onto the field and order everyone off because of an imminent threat. Most individuals, be they sane or not, demonstrate remarkable composure at the threat of a nature hazard.

Preparedness, if warned

In many of the above cases, people who stay put and do not prepare for disaster are the exception. For example, in the United States, millions of people have evacuated from the path of a hurricane in a timely and orderly fashion. The threat of personal loss of life, or loss of close family, outweighs the risk of staying. For most inhabitants of cyclone-threatened areas in Australia, the approach of the cyclone means tying down loose objects in the home and yard, stocking three days' supply of food and water, and taping windows. Such precautions are taken seriously and initiated individually.

Preparation for imminent disaster by individuals centers around four areas: self, family, property, and community – in that order. Organizing can range from packing one's clothes and toothbrush to mental preparation. The latter usually includes some sort of catharsis such as conveying apprehension or fear to oneself, family or friends, or taking a sedative or alcoholic drink. In some cases, it may take the form of prayer for divine aid or intervention. Mental preparation appears to be a first priority in meeting any hazard. For those with immediate family, the second priority is ensuring their safety, preparedness, and mental stability. Attempts may be made to check the safety of close family living in the same threatened area, or communicating assurances of safety to distant relatives.

Next, there is a direct concern for property. Usually, this involves the immediate home either well before, or immediately preceding, any disaster. For example, McKay (1983) found following the 1983 Ash Wednesday fires that the degree of preparation for bushfires in the Adelaide Hills varied throughout the area. Fewer than 20 per cent of people did nothing. Fifty-four per cent were involved in either temporary or permanent measures. Temporary measures involved such preventative tasks as clearing gutters of leaves, and cutting away undergrowth around buildings. Permanent measures involved structural changes to the house, landscape modifications, or the construction of a sprinkler system. These remedies required little maintenance, but were more expensive. There was no socio-economic relationship between the number and types of measures used. However, people who rented their accommodation tended to perform fewer measures. About 60 per cent of people carrying out preparations did so after seeking advice from neighbors, friends, or relatives. Only 9 per cent obtained information from public bodies and 11 per cent from the Country Fire Brigade. Drabek (1986) also found a similar trend for many other hazards worldwide. Obviously, people trust neighbors or relatives more for advice than they trust government bodies with expertise. Significantly, people who had neighbors who had experienced bushfires, or people who themselves were affected in the previous few years by fires, took greater precautions. In Australia, people's enthusiasm for implementing preventative measures against bushfires may wane with the memory of the last fire. As the threat increases, large segments of the community may be unprepared for bushfires. Overcoming this may require an annual advertising program, making people aware of the continuing threat. A similar pattern has been found in the United States with people's degree of preparedness to cyclones.

During any tragedy, victims usually attempt to save their home. If there is little hope that the house can be saved, then personal belongings or mementos such as photographs, trophies, or presents are rescued first. If there is time, then more expensive, transportable property is saved. Reaction to pets can also be included in this behavior. Sometimes considerable risk can be expended with people dying in attempts to save pets or livestock. For example, floods in Australia have seen farmers sheltering dairy herds in their house. Evacuation of 40 000 families in Mississauga, Ontario, Canada, following a gas leak in 1979, also involved 30 000 pets. If preparation for an imminent hazard can be completely carried out, and the victim perceives that such preparation has enhanced their chance of survival, or preserved personal property, the shock of the event is lessened and recovery after the event is more successful.

DEALING WITH THE EVENT AND ITS AFTERMATH

Response during the event

Response during the event centers on the same four areas focused on by people preparing for a disaster. It is doubtful if anyone going through a tropical cyclone will be worrying about the state of the American or Australian economy. Almost all descriptions of human behavior during a natural disaster refer to a lack of panic. People may be scared but, in the face of adversity, most remain outwardly calm. Hysteria is rarely displayed. In an extensive search of the literature, Drabek (1986) could find no evidence that disasters generated mass fear or panic. Generally, the media overtly reinforced a panic myth by exaggerating incidents of non-rational behavior. If panic exists, it usually represents a rapidly reinforced, collective response to flee some very real threat; or panic occurs simply because people in a rapidly developing disaster situation, such as a fire, spend critical time confirming the threat, thus leaving little time to flee.

It is a human instinct to try to protect one's family during disaster. This instinct tends to override all fear. One mechanism ensuring maximum protection is for family members to remain close or huddled together during life-threatening situations. Tropical cyclone victims in Australia usually find themselves huddling together for refuge in the toilet, because in new houses this room is best designed to withstand high winds. During Cyclone Tracy in 1974, huddling environments were not always conducive to survival. One family who had huddled together in bed in the main bedroom found that they were exposed to the full force of the storm as the house disintegrated around them. The bed began to drift downwind across the floor towards the edge of the house and came to rest only when one of the legs fell through the floor. Earthquake victims are often found huddled together in the crush of fallen buildings. In miraculous situations, such closeness often permits an adult to save a child. For instance, a father, fleeing the debris flow of the Mt Huascarán earthquake in Yungay, Peru, in 1970, managed to throw his two children up a slope to safety just as the flow overtook and killed him. Following the Nevado del Ruiz lahar of 1985 in Columbia, a child was found trapped in a building in mud up to her neck, propped up by her father who had suffocated to death.

The above examples illustrate another basic response during disaster, and that is the innate instinct for survival. The 'miracle babies' of the 1985 Mexican earthquake survived in the maternity ward of a collapsed hospital for up to two weeks after the earthquake because their bodies went into a natural state of hibernation that conserved fluids and energy. Victims – some severely wounded – have been dug out of collapsed mines, earthquake rubble, bombed buildings, and avalanches after similar periods of incarceration. Almost all survived because of a determination to live, a feeling that it was too early to die, a feeling that there was still more to be done in life. The Tarawera eruption of 10 June 1886, in New Zealand, buried Maori villages under more than 2 m of ash and mud, killing 156 people. A 100-year-old survivor, Tuhoto of the Arawa tribe, was found alive and in good health after being buried in his house for four days. He, in fact, told his rescuers to go away because the gods were taking care of him. He died ten days later under hospital care. The girl trapped in the lahar mud in Columbia in November 1985 met a similar fate. She remained cheerful and survived for several days stuck in the mud, only to die just before being pulled free. The physical devastation greeting the first outsiders to arrive in Darwin after Cyclone Tracy led many to expect more than the 64 deaths that actually occurred. Over 90 per cent of homes were totally destroyed and many families endured Tracy completely exposed to the elements. The small number of casualties underscores the ability of people to endure more hardship, deprivation, undernourishment, and shock than is normally believed possible. It is only after the event that survivors must cope with the psychological trauma that inevitably sets in.

Humans may simply be curious to see what is going on during a disaster. This behavior is not to be confused with sightseeing, which involves people who have not experienced the disaster wanting to see what has happened. For example, during the passage of the eye of a hurricane, almost everyone wants to go outside and look around. You and your family appear to be uninjured, but you wonder, 'What does the house look like, where's the dog, how are the neighbors doing?' When a flood crest affects the Mississippi River system in America, people who could be flooded are always wandering down to the river to look at the river's height. During eruptions, Mt Vesuvius has always attracted local sightseers curious to find out what is going on. Even residents of St Pierre journeyed up to Mt Pelée in 1902 to see what was happening a few days prior to the eruption. Ten thousand people flocked to San Francisco beaches after the 1964 Alaskan earthquake to see the tsunami coming in. The death toll in America during that tsunami event, and the one that hit Hawaii in 1946, resulted from mainly curious (and probably uninformed) residents returning home to evaluate the damage too early.

Maybe there is some euphoria to living through natural disasters or record-breaking events. The news is reporting the heaviest one-day rainfall on record and you wander outside to stand in it. I can still remember as a child awaiting for days the approach of Hurricane Hazel in southern Ontario, because no one in the area had ever experienced a tropical cyclone. When Hazel finally struck, there was a sense of relief that it had arrived at last. The same thrill of expectation goes through anyone who is witnessing for the first time the initial stages of a ground tremor. Then reality sinks in. The tremor becomes a major earthquake; the nonchalant attitude to a cyclone becomes a nightmare as the house disintegrates around you; the flood hits home as rumors of friends being wiped out become facts. The disaster has struck, and new emotions and needs arise. Mechanisms for cleaning up after the disaster must be carried out, and reconstruction and recovery instigated. At this stage, the emphasis shifts slowly away from the individual towards social groups.

Death and grief

One of the immediate aspects of a disaster is coping with the injured and the dead. Often the search for survivors and rescuing the injured become major components of large-scale natural disasters, requiring international cooperation. While rescue operations tend to take precedence over coping with the dead, dead bodies pose special problems, both hygienically and legally. The health aspect of dead, but unburied, bodies is obvious. Decaying corpses, human or animal, breed disease and must be cleared away within hours or days if epidemics are to be avoided. The more challenging problem occurs in trying to define people as legally dead. The problem is minor in the case of unidentifiable bodies. Forensic science can be used to match bodies against reports of missing people. After time, it will simply be classified as a 'John Doe' and buried as such. The more difficult legal problem occurs when a person is reported as missing, and a body cannot be found. Legally, western society presumes that such a person is still alive. They could be wandering around with amnesia; they may have taken the opportunity to flee an unbearable family or work situation in the confusion of the disaster; or they might be responsible for the disaster and are fleeing conviction. Unless there is a body, the person is considered – for up to seven years – to be alive. A will cannot be read, nor an estate settled, within this statutory period. In extreme cases, where a will has not been written, it is possible for a person's estate to be sealed off. A dead person's car cannot be moved from a driveway; a house is locked up; and business assets frozen. Not only can such action be financially trying for the surviving relatives, but the presence of the dead person's possessions in the immediate vicinity is also a constant reminder that they are missing.

In all societies, the trauma of death is resolved by a burial or memorial service. It allows family members and close friends to publicly show feelings of grief. Mentally, it permits an individual to accept the death and get on with life. Psychologically and sociologically, funeral services and death rites are essential to overcome issues associated with disaster fatalities. If the disaster is large-scale and the death toll significant, then society organizes mass burials and memorial services. An official period of national mourning may also be instigated following large disasters. Total strangers who have never known any of the dead or had any association with the affected communities may find that they want to attend some sort of funeral. Even in Australia, where less than 10 per cent of the population regularly attends church services, special services had to be arranged to fulfill this need following both Cyclone Tracy and the Ash Wednesday bushfires.

Possessions and homes

While the death of close family may seem to be the worst outcome of a disaster, the destruction of property may leave longer-lasting scars. No funeral service is held for a wrecked home, or a burnt photograph. Following the Ash Wednesday bushfires, people stated that the worst aspect was losing, firstly, personal photographs; secondly, prized trophies or awards; thirdly, property that had some sort of link to dead parents, children, or friends. Some even vocalized that the loss of possessions was worse than losing close family. Following a bushfire, nothing can replace old family photographs or a melted-down trophy. Almost everyone has seen pictures of people returning home after a natural disaster to pick through the ruins to find some memento. The pictures often show people sitting afterwards in complete despair or shock when nothing can be found. The personal belongings give some sense of continuity to life. They also provide a sense of history – a connection with past events or ancestors. As a survivor of Hurricane Andrew in 1992 said, 'You work all your life, and for what? Insurance took good care of me, but I lost my home. Not just my house and everything in it, but my home' (Santana, 2002). Survivors have gone to extreme lengths to try to recover some of these possessions. Relatives are contacted to provide copies of significant photographs, and antique shops may be searched to obtain replicas of prized objects.

Many people immediately commit themselves to rebuilding in exactly the same location. Often a duplicate of the original house is built. This was the case for many people rebuilding after the Ash Wednesday bushfires. Homes were re-established in their entirety to designs that offered little protection against repeat bushfires. The need to rebuild exact replicas can be taken to extremes. For example, following the San Francisco earthquake of 1906, the city rejected a modern design in favor of a street layout and a building program that would, as quickly as possible, recreate the original city. Sometimes investment in assets conditions reconstruction. For instance, following the Hawke's Bay, New Zealand, earthquake of 3 February 1931, the town of Hastings was rebuilt to its original, narrow street plan, even though total destruction permitted an opportunity to rebuild wider streets. The decision to rebuild in this manner was made at the adamant insistence of one councilor, who owned the only shop with any part of its structure still standing. The councilor wanted to save money by not having to rebuild a new shop, which he would have to do if the streets were widened.

Feelings about property can be quite intense. For example, following the autumn coastal storms of 1974 around Sydney, Australia, I was walking with colleagues along remnants of the beach in front of a series of houses propped up by makeshift supports. As we dodged waves surging up amongst the rocks, one resident furiously yelled that we were on private property and that he would 'get the police after us' for trespassing. This incident highlights a second stage of response following a natural disaster. After the initial impact, and after all one's resources have been utilized to ensure safety of the family, the anger sets in. For some people, the response manifests itself in long-term psychological shock that alters life forever.

Anti-social behavior

No group of people is so disdained as the sightseer or the looter. Looting, while rare, does occur. However, it also tends to be exaggerated by the media. In the United States, the prospect of looting traditionally has been so great following disasters, that the National Guard is often mobilized to block off affected areas and ordered to shoot looters on sight. Residents have to show evidence that they live in an area, and must restrict their activities to the street in which they live. In Australia, looting is a reality after most disasters. People were arrested following both Cyclone Tracy and the Ash Wednesday bushfires, but no one was ever jailed. The Lismore flood of 1954 saw looters taking whatever they could carry from the business district, including a rescue boat that was used openly to transport stolen goods.

A real problem for rescue and relief organizers after large disasters is coping with the sightseers. While people may just want to see what a natural disaster such as a cyclone or bushfire can do, the victims often perceive sightseeing as an intrusion on their privacy by the nosy, or as an opportunity by those who survived unscathed to gloat over their luck and the victims' misfortune. In Australia, sightseeing can take on mammoth proportions, not only by the public but also by the politicians and bureaucrats. Almost every cabinet minister in the Whitlam government made a fact-finding journey to Darwin after Cyclone Tracy. The same planes that evacuated survivors returned nearly as full with government sightseers. Following the Ash Wednesday bushfires, which occurred in the middle of the 1983 election campaign, almost all party leaders made the journey to affected areas – not, as they took pains to state in front of the cameras, for political reasons, but to see if they could offer genuine help.

People's misfortune can also breed extraordinarily ghoulish behavior. For example, an old lady was robbed in busy downtown Toronto, Canada, in broad daylight, after being blown to the ground in January 1978 by wind gusts in excess of 160 km hr^{-1}. As she lay pinned to the sidewalk, a more agile motorist, in what at first appeared to be an act of chivalry, stopped his car and ran to assist her. Before stunned onlookers could react, he snatched her purse, hopped back into his car and sped away. Michaelis (1985) reports that after the Naples earthquake of November 1980, smartly dressed people in flashy cars were seen leaving the scene with orphaned babies and young children, abducted for themselves or to be sold for adoption. Relief goods in Naples were also pilfered and resold on the black market. Holthouse (1986) also reports that during the 1918 Mackay, Queensland, cyclone and subsequent massive flood, boat-owners were seen offering to rescue people stranded on rooftops and in trees on payment of a substantial fee: no fee, no rescue. Following Cyclone Zoe, which struck the impoverished Solomon Islands of Tikopia, Anuta and Fatutaka on 28 December 2002 with winds of 300 km hr^{-1}, the policemen crewing a rescue boat demanded an allowance of $A1250 before they would leave the capital Honiara to travel 1000 km with urgently needed relief supplies. In the longer term, disasters attract strangers and hangers-on, the type of people you were warned about as a child. As Hauptman (1984) so aptly puts it, 'the sort of trashy people who follow disasters . . . people who could not hold a job any place else'. Just like vultures flocking to encircle a dying body, these people flock to victimize survivors, who often let their guard down following a disaster. They have to because they are now reliant upon total strangers for assistance. The worse their financial plight, the more reliant they become. Moreover, for every honest stranger dispensing cash and materials for relief, there is a dishonest one disguised as a contractor or household merchant attempting to steal it back.

Perhaps the most gruesome aspects of human behavior are those that arise because of severe famine following drought. Societies have ritualized human sacrifice on a grand scale in attempts to appease gods believed responsible for a drought. Couper-Johnston (2000) describes some of the more elaborate rituals established by the Aztecs in Mexico in response to ENSO and La Niña events. In the drought El Niño years around 1450, the Aztecs were sacrificing 250 000 victims annually to curry favor with the rain god Tlaloc. Men were beheaded at the tops of pyramids built especially for such sacrifices, woman danced in rituals before being slaughtered, and children were drowned after having their nails extracted so that priests could interpret the signs of forthcoming weather from their cries. Another response to impeding death from famine is to sell your children and yourself into slavery. Captains of ships recruiting labor for cotton and sugar cane plantations in Australia, Hawaii, and Fiji in the nineteenth century would scurry to South Pacific islands on the news of drought, because they could either steal a weakened population or coerce them into slavery with offers of food and water. Indenture is similar to slavery. In Mexico, successive droughts found 60 per cent of the rural population locked permanently into forced labor by the beginning of the twentieth century.

Finally, when all else fails, people resort to cannibalism to survive. The behavior is ubiquitous. Following crop failure in 1315–1317 in western Europe, people in Ireland waited at the fringes of funerals, then dug up the bodies from churchyards and ate them. In Germany, criminals were snatched from the gallows to be eaten, often before they had died. Couper-Johnston (2000) describes other examples of cannibalism. In the drought of 1200–1201 in Egypt, cannibalism became so rampant that people walking in the streets were at risk of being snatched by hooks lowered from the windows above. Doctors ate their patients and guests were invited to dinner by friends – never to be seen again. When two parents were caught with a small roasted child and burnt publicly for their crime, the populace ate one of the bodies the next day. Cannibalism has often become entrenched in societies experiencing repeated drought. For example, during the drought of 1640–1641 in China, the Ming dynasty tolerated markets in human flesh.

Resettlement

In almost all cases of a major natural disaster, people want to go back and resettle in the same place they were living beforehand. This may seem odd if the place happens to lie in a major earthquake zone such as the San Andreas Fault, or a coastal area subject to recurring tsunami. There are a number of reasons why people return home, even when that environment has been shown to be hazardous.

There may simply be no alternative place to live or occupation to undertake. People cannot go elsewhere if they are employed in agriculture or fishing and they do not know any other career. Additionally, they are experts in farming or fishing in that particular area, and cannot afford to lose that expertise. For instance, people farming the rich soils surrounding volcanoes may want to return after an eruption because they know that the tephra will increase crop yields. The residents of the town of Sanriku on the east coast of Japan, which has been struck repeatedly by large tsunami, have always returned because they know the fishing conditions along that section of coast. The latter example also illustrates that, in some cases, people return to the same place because there may be nowhere less hazardous for them to go. The citizens of Managua, Nicaragua, also found themselves in that situation after the 1972 earthquake. The whole country is seismic and no other location is necessarily safer. In Nicaragua or Japan, movement from one area of the country to another does not necessarily increase a person's sense of security.

Evacuation can also mean loss of assets. By returning, people can at least salvage something. Residents at Terrigal, north of Sydney, were caught in that predicament following the Australian coastal storms of 1974. The value of houses, perched precipitously at the edge of dunes eroded by storm waves, collapsed. Residents returned to these homes to try to protect them from further erosion. Fortunately, for some, housing prices climbed back to normal and they were able to sell out with minimal loss. Unfortunately, for others, storms in 1978 encroached further upon properties, and land had to be resumed for token sums by the local council. Change after a disaster also raises legal complications, especially if governments attempt to restructure cities and cut across existing private or commercial property boundaries. For traditional peoples, a high value is placed upon traditional land rights. All of these examples show that humans are territorial, and may not want to surrender property no matter what has happened to it.

There are also psychological benefits in reconstructing at the site of a disaster. By rebuilding as fast as possible, people can forget the actual event. They also keep busy at something they are familiar with, and can master. This keeps their minds off the disaster, builds up self-esteem, and overcomes the feelings of hopelessness and despair. It has already been pointed out that the city of San Francisco was rebuilt almost exactly as it had been before the 1906 earthquake and, after the 1983 bushfires in Victoria, some home-owners rebuilt exact replicas of their original home. In the case of San Francisco, the decision to rebuild in a hazardous area leads to outright denial of the hazard. Residents of San Francisco will point out that San Francisco is as safe as any other area in southern California.

A loyalty to local history, and a commitment to a locality, also draws people back to a disaster scene. Rebuilding signals a loyalty to friends or relatives who might have been killed. Their lives were not lost in vain, and they will be remembered. For example, following the Colombian lahar tragedy, the whole affected area was declared a national shrine by the government in memory of the 20 000 people buried beneath the mud. Within months, survivors were resettling the area. This sense of loyalty also includes the concept of ancestor attachments. Many people want to return to their homeland. In some cases, this relates to ethnicity, in other cases, to respect for ancestor links. Indian tribes in the Americas and the Aboriginal groups in Australia have equally strong attachments to the land. To be removed and then not be allowed to return to an ancestral homeland makes life meaningless.

Religious beliefs or morals are sometimes used to dismiss the hazardousness of an area. Many people believe that disasters are due to the wrath of a vengeful or angry god. Disasters do not occur because areas are hazardous, but because the people living there did something wrong. For instance, the Lisbon earthquake of 1755 was perceived at the time to be punishment for the wickedness of its citizens, to be God's answer to the cruelty of the Portuguese Inquisition. Accounts of the destruction of the town of St Pierre, Martinique, after Mt Pelée erupted, mention the fact that the town had a reputation for being permissive. Prophets of doom have predicted the imminent destruction of San Francisco by an earthquake because of its large homosexual population. In Papua New Guinea, some members of the Orokaivan group rationalized the 1951 Mt Lamington eruption as payback for the killing of missionaries. Hymns, gifts, offerings, and prayers can placate an angry god who can cause disaster. Hazardous environments in Indonesia and the Philippines are often viewed as being imbued with a divine spirit that can be quietened in this fashion.

Similarly, such beliefs can also dictate against resettlement. For example, Maoris in New Zealand attributed the 1886 eruption of Mt Tarawera to an act of revenge by an alleged sorcerer. This supposed sorcerer was, in fact, Tuhoto, the same man referred to earlier who was found alive after being buried for four days. Tuhoto had visited some friends close to the site of the eruption earlier in the year and, following his visit, the child of a chief had died. The child's grandmother accused Tuhoto of bewitching the child. When Tuhoto heard about the accusation, he was so angered that he called upon the god of volcanoes to wipe out the whole village. Following the eruption, no Maori would have anything to do with Tuhoto. The site of the eruption and the surrounding devastated area were considered taboo, and to this day are avoided by Maoris.

Resettlement may also be rationalized by a belief that modern-day technology can prevent destruction in future disasters. In southern California, great faith is put into reinforcing older buildings and expressway overpasses against collapse. Minor earthquakes in 1971 and 1989 have shown that not all this 'technological strengthening' works. New Zealand is pursuing the idea of building large buildings on rubber bungs that can absorb earthquake shock waves and minimize destruction. Certainly, in areas where this has been done, earthquakes are not so destructive as they are in regions where buildings are not built to earthquake standards. The large death tolls of the Agadir, Morocco, earthquake of 1960, the Armenian earthquake of 1988, and the Bam, Iran earthquake of December 2003, attest to this latter fact – even though these events were of lesser magnitude than recent Californian ones with low death tolls.

Moving elsewhere after a disaster may also be too disruptive to a wider segment of society. Evacuation of large numbers of people represents a major task that can put pressure on housing, government, and consumer supplies in other parts of a country. By containing the survivors, efforts can be efficiently directed to a single area. For example, the evacuation of Darwin following Cyclone Tracy (described in Chapter 2) was not effective. It caused disruption to the people who were evacuated. Had the numbers been larger, it would have severely disrupted the economy in the south. There is an inertia effect resisting the relocation of so many people. Large communities attract a communications and business infrastructure that represents a significant capital investment that cannot be easily abandoned or relocated. For instance, Messina, Italy, after the earthquake of 1908, rather than being abandoned, was rebuilt in the same area because it was a significant port. Tangshan, China, provides an important link between Beijing and Tienshen on the coast. For this reason, it had to be quickly rebuilt following the 1976 earthquake. The only concession to the disaster was the rebuilding of the city as a series of satellite cities spread out to minimize damage in any future earthquake. Some countries such as Japan and China have more centralized economies, where the government decides the movement and location of people. There is no personal choice to leaving or resettlement. The government makes the decision, and either enforces or encourages it.

A final reason why people may choose not to resettle elsewhere is that the disaster may appear to be a one-off or exceedingly rare event. In this case, alternatives to resettlement in the same area do not arise. For example, tropical cyclones affecting southern Ontario, Canada, are relatively rare events. Hurricane Hazel in 1954 represented a 1:200-year event with major flooding in Toronto. While areas prone to such infrequent flooding have been set aside in Toronto as open space, and hazard planning takes account of this type of flooding, no plans for massive evacuation have been established in case another cyclone strikes. The event is just too rare to be of significant concern. The same is true of some earthquakes and most volcanoes. Historically, some of the largest earthquakes in the United States have occurred at locations such as New Madrid, Missouri, and Charleston, South Carolina, which are not considered seismic. An earthquake disaster movie set in South Carolina would not be believable. Nor does the disaster have to be infrequent to be dismissed. It just has to be perceived as infrequent. The slopes of Mt Vesuvius in Italy, and the town of St Pierre, Martinique, have been resettled despite the fact that the local volcanoes are still considered dangerously active. This perception of infrequency can form very quickly, because people have short memories of disaster events. For example, within six months of the 1974 Brisbane River floods in Australia, property values in flooded areas had returned to their pre-flood levels.

Myths and heroes

The myth that people panic during disasters has already been dispelled in this chapter. There are a number of other myths about human response during disasters that should also be dismissed. The first myth is that people experiencing the calamity appear confused, are stunned, and helpless. If anything, disasters bring out the strength in people. Survivors, far from being helpless, are almost immediately back into the rubble trying to save people and property. At this stage in disaster relief, the victims are more than capable of organizing themselves and performing the rescue work. The only things they may lack are the tools to carry out the tasks efficiently. Secondly, it is generally believed that disasters bring out anti-social behavior, stories of mothers dropping their babies to go after a few seeds of grain spilled from a passing truck, of men pushing children from lifeboats on sinking ships. Such behavior is rare, or else it is often the product of media distortions and bears little resemblance to reality. While there may be anti-social behavior – for example the ghoulish conduct of baby-snatchers in Naples mentioned above – survivors are generally very sociable. Disasters kindle community spirit and altruistic acts in people and businesses. In Australia, there are examples of total strangers in new cars driving up to flood and bushfire victims, tossing them the keys and walking away. In the 1994 Sydney fires, a major supermarket chain opened up its warehouses and trucked away its soft drinks to be off-loaded free at the fire-fronts.

Another aspect about disasters, which may or may not be mythical, is the role of heroes. People know that if they ever have the misfortune of being caught in a disaster and require help, there will be someone trying to save them. In the United States, forest fires make heroes. For example, the actions of the crew on the St Paul and Duluth train to Hinckley, Minnesota, during the fire of 1894 were certainly gallant. Despite decreasing visibility due to smoke, the train was casually driven into the town – only to be met by 150 fleeing townspeople, closely pursued by a wall of flame. The engineer decided to evacuate people to a marsh 7 km back up the line, and throttled in reverse at full speed. However, the fire overtook the train, consuming one passenger car after another. As flames licked through cracks and windows, the conductor calmly used a fire extinguisher to put out fires on women's dresses. In the cab, the situation was worse. Both the engineer's and fireman's clothes caught fire, and the cab began to burn. The glass in the cab window burst from the heat, cutting the engineer's jugular vein; the controls caught fire; the seats began to smolder; and the cabin lamp melted. Several times both men collapsed in the heat, but managed to keep the train moving. As the train reached the marsh and erupted into flame, both passengers and crew flung themselves into the mud where they sheltered for four hours. The fireman, however, after carrying the bleeding engineer to safety, returned to the train, uncoupled the engine and moved it to safety, while the conductor struggled back to the nearest station to warn the next train coming through. The crew's actions saved over 300 people.

In Australia, no one would doubt that Major-General Stretton's work in the first weeks after Cyclone Tracy was heroic. His organizational skills stand out as superhuman. He was able to mobilize evacuations and ensure that the city was cleaned up, laying the foundations for the reconstruction that was to follow. He made decisions that appeared innovative and effective. He was given an impossible task to make order out of chaos; and he succeeded without any additional casualties. Stretton rose to the occasion, but that is what was expected of him. He had a military background, and was in charge of the National Disasters Organisation in Canberra. His training, experience, and position aided him in performing his assignment.

Yet, there have been situations where common people with few of these attributes have also spontaneously risen to the occasion. When rescuers arrived at the scene two days after the 4 February 1976 earthquake in Guatemala, they found an old man in charge. He had organized rescue, reassured survivors, and when the aid came in, he took charge of the distribution of food and supplies. The other survivors deferred to his wisdom whenever a decision or action was required. In a way, hero stories reinforce society's sense of security. Nowhere was this altruism more evident than in the terrorist attacks on the World Trade Center on 11 September 2001. One gentleman pushed a wheelchair-bound employee down 65 flights of stairs to safety. Another chose to stay behind with a handicapped person who could not flee. When the transcripts of telephone calls made from the doomed building were released two years later, they showed employees carrying out their duties professionally. As a New York Port Authority spokesperson said, 'these people were heroes until the end'.

ADDITIONAL IMPACTS

(Dalitz, 1979; Reser, 1980)

Emotional problems

'Why did it have to happen to me? I've worked hard all my life and the house is gone. There stands George's house next door. He built it after winning the lotto. The shrubs are stolen, and he was at the pub all through the fire. Look, his fence isn't even scorched.'

This scenario describes well the reaction that begins to set in after a disaster. While humans are quite resilient during disaster, a state of shock or numbness that borders on self-pity sets in afterwards. These feelings inevitably turn to anger within 24–48 hours: maybe anger at oneself for not preparing better for the event, maybe anger at a neighbor who better survived the disaster, maybe anger at some larger organization such as the Weather Bureau for not giving enough warning about the approach of a cyclone, maybe anger at some unidentified group such as the conservationists who prevented prescribed burning. The object may be real or imaginary. It makes no difference. One just becomes angry. The outlet may be physical or verbal abuse. A colleague of mine walked into a landslide disaster area, within the anger time frame, to take pictures, and left with a bleeding nose after being punched in the face. Commonly, people must talk about the event to anyone who will listen. The press preys upon this characteristic to get pictures and material on the disaster, and often records this anger stage. In some cases, the press becomes the object of the resentment. Relief and social workers are aware of the fact that if they remain around the scene of a disaster long enough, they must expect to be on the receiving end of some of the verbal attacks.

For some, a disaster elicits a victim–helper syndrome. Large natural disasters such as earthquakes can completely destroy the physical fabric of a community. Survivors become very dependent upon outside aid for their daily requirements such as shelter, clothes, food, and services. Often the victims have no resources left for reconstruction. They may be suffering from injury or shock after losing family members. These people become victims of the disaster. Without anything that is familiar, without any ability to cope or power to direct their own lives, they succumb to lethargy and establish a complete dependence upon the rescuer. This disaster syndrome is not that common, being most pronounced following traumatic catastrophes for only a minority of people. In some cases, unless aid is directed to re-establishing a normal existence, the victims can become permanently locked into a dependency syndrome. They become disaster refugees. For example, the Ethiopian government at the end of the Sahel drought of 1984 forcibly repatriated refugees to relief camps in their northern homelands, because they had become so dependent upon emergency aid that they were hesitant to return to what they still perceived as drought-ravaged farms. A significant number of evacuees of Cyclone Tracy, after a year of living in the southern states of Australia where the standard of living was better, also found it difficult deciding to return home to Darwin.

The evacuation of survivors causes other psychological problems. For instance, after Cyclone Tracy, most residents were initially evacuated. However, husbands were permitted back to rebuild. Families were often separated for up to a year. When finally reunited, parents had drifted apart socially, and a gulf had arisen between fathers and children who had not seen each other for months. In this case, the resulting tragedy of Darwin was not the initial death toll from the cyclone; it was the fact that over 50 per cent of families who had experienced the disaster separated within a decade. Nor was this phenomenon restricted to evacuees. Families in the south, who had volunteered their homes as temporary billets, were faced with increased stress and – for some individuals – tempted into an extra-marital affair with younger guests. Evacuees, especially those people who were self-employed, also experienced loss of income. This was not necessarily a problem for government workers or employees of benevolent national companies, who were relocated to southern offices. Those who were not so lucky faced unemployment or found temporary work. For some, this alternative employment became so attractive that they were reluctant to return to Darwin and face an uncertain future. Single-parent families left in the south during the reconstruction also faced difficulties with children. Because of the trauma and changed surroundings, younger children underwent regressive behavior. Without a father-figure, rejected by their peers as odd, and considered transient by the community, school-aged children found it difficult adjusting to a strange environment and fitting into the community's social activities. Many developed antisocial behavioral traits, became delinquent, or took up drugs and alcohol to cope. The problems were especially acute for younger children (see Table 13.1). Behavior of these non-returnee children significantly differed from similarly aged children who either had returned to Darwin or, for some reason, had not left.

Within Australia, this situation is not unique to Cyclone Tracy. The drought of 1982–1983 drove many families off the land into rural towns and the capital cities, where society was more permissive than generally experienced on isolated farms. Many children

Table 13.1 Frequency of psychological problems amongst young survivors of Cyclone Tracy, Darwin, 1974.

Trait	Stayers %	Returnees %	Non-Returnees %
fear of rain and wind	19.8	24.2	29.3
fear of dark	10.8	13.2	11.5
fear of jet noise	4.5	7.9	15.5
clinging to mother	6.3	5.3	12.6
bed-wetting	2.7	6.8	7.6
thumb-sucking	0.0	1.6	2.0
temper tantrums	2.7	1.1	7.2
fighting	0.9	3.2	4.9
deliberate vandalism	2.7	1.1	3.4
injuries	0.0	2.6	5.5
diseases	9.9	5.8	10.9

Source: Western and Milne, 1979.

found the availability of drugs, which at the time were beginning their dramatic spread beyond the capital cities, irresistible. At the time of the Live Aid Concert for Ethiopia, Molly Meldrum (the Australian organizer of that relief campaign) observed that we were more concerned in Australia with people's plight in the Sahel than we were about the unrecognized or deliberately ignored drug problems of children driven into cities by the Australian drought. Significant adolescent problems still existed in country towns in 1989, six years after that drought, with many young people drifting to capital cities or resort towns and contributing to delinquency rates in those places.

As time passes, the shock of a natural hazard and the trauma of death also lead to psychiatric problems. The Aberfan, Wales, disaster, while not strictly a natural disaster, illustrates this phenomenon well. Failure of a 250-metre high, coal tailings pile killed 116 children, the entire juvenile population, in a manner of seconds. The whole population was personally affected and underwent psychological trauma. Over 158 residents subsequently sought psychiatric treatment. A similar, but less severe, situation arose in communities badly affected by Australia's 1983 Ash Wednesday bushfires. The number of people requiring psychiatric help to cope with the disaster and its after-effects increased significantly afterwards. The number of suicides rose in the year following a particularly bad tornado in Wichita Falls, Kansas, in 1979.

Just as work stress over time can lead to increased illness and absenteeism in workers, so the culmination of life changes, such as those resulting from a natural disaster, can lead to psychosomatic illness. Psychosomatic illness is any physical disease, such as ulcers, depression, or arthritis that owes its origin to a psychological problem. Scales have been constructed to foreshadow the likelihood of an individual undergoing a psychosomatic illness. Death in the immediate family, a household move, the birth of a child, a change in job, or a divorce all contribute equally to the chance of physical illness in the following year. Again, non-returnees from the evacuation of Darwin in 1974 suffered a significant increase in stress-related illnesses. Further study has shown that it was not necessarily non-returnees that suffered more; it was the people who wanted to return but could not (Kearney & Britton, 1985). In this latter group, there was a 50 per cent increase in the number of people who were nervous, depressed, or worried about the future, and a 100 per cent increase in the number of people complaining about bowel trouble or who were using sedatives.

The prolonged prominence of psychosomatic illness may not only destabilize a family unit in the short term but, if not resolved, have profound long-term ramifications for society. Where one or more family members suffer from a psychological disorder, the chance of that disorder appearing in offspring is significantly increased. Natural disasters can also breed a cycle of sociological and psychological disorder, affecting a future generation who has never actually experienced the event. Ten years after Cyclone Tracy, social workers in Brisbane were able to recognize cyclone-traumatized victims, who had been evacuated to this city, as a distinct subgroup.

The impression should not be given that all natural disasters produce copious psychological problems. A natural disaster may simply apply stress to an already-stressed person or family. The disaster may simply be 'the straw that broke the camel's back'. The evacuees from Darwin who did not return, and who developed psychological problems, may have been those people in Darwin already suffering problems, who would have left the city permanently some day even if Cyclone Tracy had not struck. Even if the families were stable units, the long-term isolation in unfamiliar surroundings and crowded accommodation was an abnormal social pressure. The study by Luketina (1986) of the 1984 Southland floods in New Zealand indicated that people evacuated for periods similar to the Darwin evacuees also underwent exceptional stress. However, after life returned to normal, very few admitted to, or sought help for, psychological problems attributed to the floods. The exceptions to this occurred with the elderly – some of whom became confused by the events – with those already going through the stages of marriage breakdown, and with children. Immediately after the floods, 7 per cent of children from flooded homes were exhibiting behavioral changes of concern to parents and teachers. However, the majority of these problems disappeared as life normalized after several months.

There is a sequence of psychological adjustment following a disaster. Foremost, the majority of disaster victims appear to be resilient and adaptive to the disaster. For example, in the Brisbane floods of 1974 the community shunned disaster relief. Fewer than 2 per cent of the victims subsequently sought help from a counseling service. Of course, in many of these events there was no large loss of life nor substantial injury. A significant criterion for disaster-induced psychosis appears to be the occurrence within a short period of major life changes. If the disaster generates these changes, or adds to ones that have already occurred, then an individual or family will be more likely to be affected psychologically by that disaster.

Critical incidence stress syndrome

A more worrying problem with disasters is the fact that anyone who encounters a tragedy spawning death or injury can be affected by that tragedy for the rest of their lives. The experience may be as innocuous as being a bystander to the tragedy. For instance, during the Waterloo, London, subway station fire of 1987, trains coming into the station stopped and the doors remained closed. Then, the trains pulled swiftly away, despite people on the platform desperately banging on windows and doors trying to get in to escape the inferno. For the stunned passengers on the trains, the stop lasted only 30 seconds, but the experience will be with them for the rest of their lives. The closer one is to the disaster, the more permanent and damaging the effect. Sadly, the closest people are the heroes. Whereas a hero may have been awarded society's everlasting gratitude and should be basking in glory, many in fact suffer from extreme depression and eke out shattered lives. The pattern is now known as *Critical Incidence Stress Syndrome*, and is recognized by rescue workers and heroes as a severe, long-term psychological response to disaster. The pattern is characterized by extreme depression and unfounded guilt feelings. For example, the policeman who repeatedly risked his life to save commuters in the King's Cross subway fire, in London in 1987, cannot ascertain why he has survived while others died; he now lives a timid life, afraid to go outdoors. The passenger who became a human bridge permitting fellow travelers to crawl to safety after the *Herald of Free Enterprise* ferry sank at Zeebrugge, Belgium, in 1987, lives a life without direction, even though he has been given every bravery award his country can offer. Another policeman, who dragged a burning man from the Bradford stadium fire in 1985, finds himself so emotionally weakened that he cannot help crying, years after the incident, over trivial matters. Even Major-General Stretton was affected by guilt and depression because he believed his evacuation orders after Cyclone Tracy had shattered survivors' lives. As with the victims of many disasters who receive psychiatric counseling within minutes of being rescued, professional counselors are now debriefing rescue workers following their rescue work. In New South Wales, Australia, these debriefing sessions take the form of team counseling and, depending upon the magnitude of the disaster and the response, can continue for months afterwards. The policy is not to let rescue workers discuss the disaster clannishly in an unstructured environment such as the local pub. Instead, counseling is directed towards relieving initial feelings of anger and guilt that may not only destroy that person's life but their abilities as a rescue worker. Where death is involved, there is no rosy conclusion to a disaster. Anyone who has witnessed the tragedy, be they victim, bystander, rescue worker or hero, can potentially have their lives irreparably harmed by the experience.

REFERENCES AND FURTHER READING

Burton, I., Kates R.W., and White, G.F. 1978. *The Environment as Hazard*. Oxford University Press, Oxford.

Couper-Johnston, R. 2000. *El Nino: The Weather Phenomenon That Changed the World*. Hodder and Stoughton, London.

Dalitz, E.R. 1979. Personal reactions to natural disasters. In Heathcote, R.L. and Thom, B.G. (eds) *Natural Hazards in Australia*. Australian Academy of Science, Canberra, pp. 340–351.

Drabek, T.E. 1986. *Human System Responses to Disaster: An Inventory of Sociological Findings*. Springer-Verlag, New York.

Hauptman, W. 1984. On the dryline. *Atlantic Monthly*, May: 76–87.

Holthouse, H. 1986. *Cyclone: A Century of Cyclonic Destruction*. Angus and Robertson, Sydney.

Kearney, G.E. and Britton, N.R. 1985. Insurance response in disaster. *Report of Proceedings of Research Workshop on Human Behaviour in Disaster in Australia*. Natural Disasters Organisation, Mt Macedon, pp. 300–324.

Lachman, R., Tatsouka, M., and Bonk, W.J. 1961. Human behaviour during the tsunami of May, 1960. *Science* 133: 1405–1409.

Luketina, F. 1986. The psychological effects of floods. *Soil and Water* 22(1): 20–24.

McKay, J.M. 1983. Community adoption of bushfire mitigation measures in the Adelaide Hills. In Healey, D.T., Jarrett, F.G. and McKay, J.M. (eds) *The Economics of Bushfires: The South Australian Experience*. Oxford University Press, Melbourne, pp. 116–131.

Michaelis, A.R. 1985. Interdisciplinary disaster research. *Report of Proceedings of Research Workshop on Human Behaviour in Disaster in Australia*. Natural Disasters Organisation, Mt Macedon, pp. 325–346.

Reser, J.P. 1980. The psychological reality of natural disasters. In Ollier, J. (ed.) *Response to Disaster*. Centre for Disaster Studies, James Cook University, Townsville, pp. 29–44.

Santana, S. 2002. Remembering Andrew. *Weatherwise* 55 July/August: 14–20.

Sims, J.H. and Baumann, D.D. 1972. The tornado threat: coping styles of the north and south. *Science* 176: 1386–1392.

Western, J.S. and Milne, G. 1979. Some social effects of a natural hazard: Darwin residents and cyclone 'Tracy'. In Heathcote, R.L. and Thom, B.G. (eds) *Natural Hazards in Australia*. Australian Academy of Science, Canberra, pp. 488–502.

Wilson, A.A.G. and Ferguson, I.S. 1985. Fight or flee? A case study of the Mt. Macedon bushfire Pt 2. *Bushfire Bulletin* 8(1 & 2): 7–10.

Epilogue

CHAPTER 14

CHANGING HAZARD REGIMES

(Houghton et al., 1995, 1996; Bryant, 2001; Nott, 2003)

Throughout this book the commonness of natural hazards has been continually emphasized. For example, the eruption of Krakatau in 1883 was not an unusual event. It was matched by earlier eruptions of similar magnitude and will be witnessed again. This argument smacks of uniformitarianism, but it is not meant to support this concept. Rather, it acknowledges that our existing realms of natural hazards fit a magnitude–frequency distribution that can be described, and from which the probability of occurrence of future extremes can be predicted. This is not necessarily how natural hazards behave over time. Existing hazard regimes are not immutable. They can change. My present research into mega-tsunami illustrates this point. In the historic record, tsunami in Australia have not exceeded 1.07 m on tide gauges, did not have run-ups of 4 m above sea level nor penetrate more than 1 km inland. Over the past fifteen years, evidence has been found along Australia's New South Wales coastline for large tsunami that have gone inland up to 10 km, transported boulders the size of boxcars up 30 m high cliffs (Figure 14.1) and swept over headlands 130 m high. Nor is the evidence restricted to the New South Wales coastline. In north-western Australia, one event penetrated 35 km inland in the Great Sandy Desert. These mega-tsunami have occurred at periodic intervals over the last 8000 years. They were generated by comet or meteorite impacts with the surrounding oceans – because no known earthquake or volcanic eruption has produced similar magnitude evidence. Astronomers believe that a large comet entered the inner solar system about 15 000 years ago. It broke up, and the Earth periodically sweeps through its debris trail in late June or mid-November each year. This is the Taurids meteorite

Fig. 14.1 Imbricated boulders stacked against a 30 m high cliff face on the south side of Gumgetters Inlet, New South Wales, Australia. Some of these boulders are the size of a boxcar. Note the person circled for scale.

stream. The Earth passes through the concentrated head of this stream every 2520 years with another minor concurrence every 1100 years. The Earth's last encounter occurred around 1500 AD. This is when the last mega-tsunami affected the New South Wales coastline. Australian Aboriginal and Maori legends describe a comet shower that can be dated to this time. We have also found a recent impact crater, generated by a comet about 1 km in diameter, off the south coast of Stewart Island, New Zealand. Currently, the Earth is not affected by large or plentiful objects in the Taurids. As a result, Australia – together with much of the globe – exists presently within a quiescent epoch where tsunami are generated mainly by earthquakes or volcanoes. In Australia, the dichotomy is that field evidence for mega-tsunami does not match a limited historical record devoid of such events. The dichotomy can be resolved knowing that tsunami are very much characteristic of a changing hazard regime.

Other climatic and geological hazard regimes have also changed. For example, since 1700 the Earth has experienced more and larger volcanic eruptions than at any time during the last 10 000 years. We are also witnessing the termination of the Little Ice Age, which appears to follow a 1500-year climate cycle. During cold phases, storms are more frequent and intense in northern Europe (Figure 3.12), and temperatures more extreme. For example, in the 1420s, the Burgundy region of France – during one of the periods of the Middle Ages when temperatures were cooling rapidly – witnessed six of the coldest years in 1000 years. However, embedded within this cooling was the warmest summer experienced for a millennium. This summer was not just slightly above normal, but exceptionally so. The grape harvest in that year occurred at the beginning of August instead of during the second or third week of September.

Finally, the extreme 1 in 100 year rainfalls that many regions of the world have been experiencing may simply represent the occurrence in time of part of a much more intense rainfall regime. Wollongong, Australia, mentioned in this text, is a region where extreme rainfalls of 200–400 mm in 24 hours are common. Embedded in rainfall records for this region are rainfalls of 840 mm in 9 hours, and 1000 mm within one month. These rainfalls do not plot on the straight line in probability of exceedence diagrams. They represent a different climate regime conducive to extreme floods. This aspect is not unique to Wollongong. Across Australia, there is geomorphic evidence for floods that are greater than any that could be produced by maximum probable rainfalls. Nor is Australia unique in the world. The increase in flooding globally (Figure 6.10) is evidence that rainfall (or drought) regimes are changeable.

The effects of global warming, whether caused by the termination of the Little Ice Age or by enhanced Greenhouse gas concentrations, can also be accommodated within the concept of a changing hazard regime. Climatologists acknowledge that the Earth's climate is not consistent over time. At the height of the Last Ice Age 22 000 years ago, temperatures fluctuated dramatically, rising and falling 4–6°C over the space of a few years. These 'flickerings' are probably subdued under our present climate regime, but they appear when climate gets colder. The fluctuations in the 1420s in France were probably evidence of this fundamental feature of Pleistocene climate.

The warming that is occurring presently and is hypothesized to increase represents a change in hazard regime. The potential effect of changes in carbon dioxide induced by human activity has been the subject of scientific enquiry for the past 155 years. In 1861, John Tyndall suggested that carbon dioxide changes in the atmosphere might be responsible for changes in climate. In 1896, Svante Arrhenius proposed that a doubling of carbon dioxide might warm the atmosphere by 5°C. In 1938, Callender showed how carbon dioxide and water vapor absorb long wave radiation at different wavelengths. He suggested that anthropogenic production of carbon dioxide would alter the natural balance between incoming solar radiation and outgoing long wave radiation, and cause global warming that would exceed natural variations in historical times. The seminal conference that catalyzed world attention on the significant ramifications of enhanced 'Greenhouse' gas concentrations in the atmosphere was the WMO-sponsored conference of 9–15 October 1985 in Villach, Austria. A policy statement released by this conference pointed out that most current planning and policy decisions assumed a constant climate – an immutable hazard regime – when, in reality, increases in 'Greenhouse' gases by the 2030s would warm the globe between 1.5 and 4.5°C, leading to a sea level rise of 20–140 cm. The only thing that has changed since then is the realization that numerous gases – CO_2, CH_4, N_2O, CO, O_3 and CFCs – are implicated in these predictions. While present temperatures have been matched during the Mediaeval Warm Period (centered on the thirteenth century), there is no match in the geological record to the current rapid increase in Greenhouse gases. Thus, the climatic consequences of increased Greenhouse gases are speculative and uncertain.

MODERN CONSEQUENCES OF NATURAL HAZARDS

This book has been concerned with natural hazards that are within the present realm of possibility: that we have lived with, will continue to experience, and hopefully can survive. The basic purpose of the text is to present the reader with a clear description of the mechanisms producing climatic and geological hazards. An attempt has been made to enliven that presentation with historical descriptions and supportive diagrams. There are four themes that are indirectly addressed throughout the text. It is worthwhile here briefly summarizing these.

Firstly, natural hazards are predictable. This is especially applicable to climatic hazards. With satellite monitoring, short-range computer forecasting models and the wealth of experience of trained personnel and scientists, most climatically hazardous events can be forecast early enough to give the general population time to respond to any threat. Whether or not these warnings are believed is most relevant in today's world. The fact that the British Meteorological Office ignored an extra-tropical storm warning issued by their counterparts in France; that residents in Darwin were more interested in partying for Christmas when Cyclone Tracy was barreling down on them; or that the duty officers in Sydney ignored their own amateur storm-watch observers as the worst hailstorm on record overtook the city, does not negate the fact that most climatically hazardous events are being predicted. It only illustrates that a society, its organized bureaucracy, or its individuals have a major influence in determining how effectively warnings of, and preparations for, a natural disaster are accepted.

Also pointed out in this text is the fact that the degree of prediction is strongly influenced by regular astronomical cycles, such as the 11- and 22-year sunspot and the 18.6-year lunar cycles. Sunspot numbers have tended to obscure the more significant prognostic variable, namely solar geomagnetic activity. These avenues of forecasting have been strongly criticized. Yet, frequencies equating with the return period of sunspots appear in enough climatic and geophysical data that solar geomagnetic cycles cannot be ignored as a precursor for natural hazard events. Similarly, the correlation of rainfall and drought cycles with the 18.6-year lunar cycle should not be ignored. The coincidence of drought, across most of the grain producing regions of the world, with the M_N lunar tidal component now demands that this cycle be taken seriously.

Secondly, the occurrence of natural hazards is increasing in frequency: if not in terms of the number of events, then in terms of economic cost. The Munich Reinsurance Company, especially since 1960, can substantiate this trend through payout claims. This point is well-illustrated for volcanic activity this century. After the Caribbean eruptions of 1902, volcanism leading to large death tolls virtually ceased for half-a-century. A similar quiescence is evident in climatic records. The 1930s and 1940s were exceedingly passive as far as storm events were concerned. Air temperatures globally tended to be warm, and weather patterns did not fluctuate strongly. This does not imply that there were no climatic disasters. After all, this period witnessed the dust bowl years on the Great Plains of the United States, and Hitler's armies experienced a Russian winter that was only matched in severity by that which afflicted Napoleon in 1812. Beginning about 1948, climate globally began to become more extreme. This variability has increased substantially since 1970. For example, since 1980 Australia has withstood two of the largest regional flood events on record – sandwiched between three of the largest 1 in 200 year droughts recorded. In the United Kingdom, in the late 1980s, temperatures seasonally tended to oscillate from the coldest to the warmest on record. This enhanced variability is prominent in both climatic and geological phenomena. In terms of natural hazard events, we presently live in a period that may be experiencing a shift in hazard regime.

Thirdly, natural hazards are pervasive in time and space. A simple plot of the location of the hazards mentioned in this book will illustrate this fact. Normally, we do not expect droughts to strike England, tornadoes to sweep through Edmonton, Canada, or snow to fall in Los Angeles. However, such events have happened in the latter half of the twentieth century. Part of the reason so-called unusual events are occurring in places where they are not expected may lie in the fact that we have lived through a passive period for hazards, as mentioned above. Another explanation is the fact that, as we develop longer geophysical records, extreme events are more likely to appear in those records. Additionally, technological advances in communications have dramatically shrunk our spatial awareness to the point that today's unusual weather or geological event is everyone's experience. Such developments have heightened people's awareness of natural hazards. I believe that the long-term prognosis, at any location, is for the occurrence of a wide range of hazards that have not necessarily been detected in the historical record. We are experiencing a shift in regime for a number of hazards. It is not unusual for an earthquake of 7.2 on the Richter scale to occur in the supposedly aseismic area of Tennant Creek, Australia (September 1988), nor for tornadoes to sweep through New York City (July 1989). These extreme events, while not common in these areas, are certainly within the realm of possibility. There is a pressing need for the serious assessment of extreme hazards in many parts of the world. For example, volcanic activity is possible in Australia even though there are no active volcanoes, and tropical cyclones could affect an area like southern California. In the former case, volcanoes have erupted in Australia in the past 10 000 years. In the latter case, southern California lies close to an area generating tropical cyclones. Natural hazards, and in some cases very unexpected hazards, are simply more common than perceived.

Finally, the pervasiveness of so many hazards, plus ones that we have not discussed, should not be viewed pessimistically. When a natural disaster occurs, and is splattered across our television screens, we tend to believe that few survive. In fact, it is surprising that so few people die or are injured. There have been only a few disasters in the past 150 years that have wiped out most of a population. Mt Pelée in 1902 left two survivors out of 30 000; the Mt Huascarán debris flow of May 1970 killed most of the population of Yungay, Peru, while the November 1985 eruption of the Nevado del Ruiz volcano in Colombia killed most of the people in the path of its subsequent lahar. Consider the odds of survival for other large natural disasters mentioned in this text: Cyclone Tracy killed 64 people out of a population of 25 000 in Darwin on 25 December 1974; only one out of every four people died in the horrific Tangshan earthquake in China in 1976; the death toll in the Tokyo earthquake of 1923 was 140 000 people out of a population well in excess of 1 million. Human beings are very resilient and apt at surviving and, in particular, are gifted at coping with disasters. These aspects were emphasized in Chapter 13. Any text on natural hazards tends to emphasize, as does this one, the sensational, negative, or gruesome aspects of disasters. However, it is only appropriate to end this book by noting people's remarkable ability during calamities to rescue, survive, and recover from such events.

REFERENCES AND FURTHER READING

Bryant, E. 2001. *Tsunami: The Underrated Hazard.* Cambridge University Press, Cambridge.

Houghton, J.T., Meira Filho, L.G., Bruce, J., Hoesung Lee, Callander, B.A., Haites, E., Harris, N. and Maskell, K. (eds) 1995. *Climate Change 1994: Radiative Forcing of Climate Change and an Evaluation of the IPCC IS92 Emission Scenarios.* Cambridge University Press, Cambridge.

Houghton, J.T., Meira Filho, L.G., Callander, B.A., Harris, N., Kattenberg, A. and Maskell, K. (eds) 1996. *Climate Change 1995: The Science of Climate Change*. Cambridge University Press, New York.

Nott, J. 2003. The importance of prehistoric data and variability of hazard regimes in natural hazard risk assessment: example from Australia. *Natural Hazards* 30: 43–58.

Select Glossary of Terms

A

Adiabatic Lapse Rate. The rate at which air temperature decreases with elevation because of the decreasing effect of gravity. The result is an increase in air volume without any change in the total amount of heat. The normal adiabatic lapse rate is –1 °C for each 100 metres of elevation: but the saturated adiabatic lapse rate, when air cools more slowly, is 0.4–0.9°C per 100 metres. If a parcel of air is moved towards the ground the reverse process occurs and the air begins to warm.

Agglutinate. The tendency for high-pressure cells originating from the poles to stagnate and stack up at a particular location over mid-latitude oceans. The Bermuda or Azores High represents one of these locations at which stagnation tends to occur. *See* Mobile Polar High.

Albedo. The degree to which short wave solar radiation is reflected from any surface or object.

Andesite. Volcanic rock containing the minerals andesine, pyroxene, hornblende, or biotite. In a melted state, such material contains a high amount of dissolved gas.

Aperiodically. Recurring but not with a fixed regularity. For instance, the first day of each month occurs aperiodically because there are not the same number of days in every month.

Aseismic. Characterizing a region that historically has not experienced earthquakes.

Atterberg Limits. These limits define the moisture content of a soil at the points where it becomes elastic, plastic, and liquid.

B

Barrier Island. A sandy island formed along a coastline where relative sea level is rising or has risen substantially in the past. Shoreward movement of sediment cannot shift the island landward fast enough, so the island becomes separated from the mainland by a lagoon or marsh.

Basaltic. Referring to the most common liquid rock produced by a volcano, containing mainly magnesium, iron, calcium, aluminum, and some silicon. It forms a hard, but easily weathered, fine-grained, dark-grey rock. Volcanoes producing magma with these constituents are classified as basaltic.

Base Surge. The sudden eruption of a volcano laterally rather than vertically.

Bathymetry. Depths below sea level.

Bearing Strength. The amount of weight that a soil can support, including its own weight, before the soil deforms irrevocably.

Benioff Zone. The zone at a depth of 700 km associated with earthquakes along a subduction zone.

Biomass. The total weight of all living organisms.

Bistable Phasing. Over two to three centuries in many locations, droughts and wet periods tend to alternate every 9.3 years. In some cases, a drought is suddenly followed, nine years, later by another drought before the alternating cycle is re-established. This is bistable phasing.

Blizzard. Any wind event with velocities exceeding 60 km hr^{-1}, with temperatures below –6 °C, and often with snow being blown within tens of metres of ground level.

Blocking High. A high-pressure system whose normal eastward movement stalls, blocking the passage of subsequent low- or high-pressure systems and deflecting them towards the pole or equator. Mobile polar highs are high-pressure cells that have either run out of momentum in their movement away from the poles or have not been displaced by another high.

Blocks. A type of lava that consists of angular, boulder-sized pieces of solid rock blown out of an erupting volcano.

Bombs. (1) *See* East-Coast Lows. (2) Boulder-sized pieces of liquid lava blown out of an erupting volcano.

Boreal. (1) Pertaining to the northern hemisphere. (2) Northern hemisphere forest, consisting mainly of conifers and birch.

Bruun Rule. A relationship stating that a sandy beach retreats under rising sea level by moving sand from the shoreline and spreading it offshore to the same thickness that sea level has risen.

C

Caldera. The round depression formed at the summit of a volcano caused by the collapse of the underlying magma chamber that has been emptied by an eruption.

Capillary Cohesion. The binding together of soil particles by the tension that exists between water molecules in a thin film around particles.

Catastrophists. A group of people, little challenged before 1760, who believed that the Earth's surface was shaped by cataclysmic events that were acts of God.

Cation. An ion or group of ions having a positive charge.

Cavitation. A process, highly corrosive to solids, whereby water velocities over a rigid surface are so high that the vapor pressure of water is exceeded and bubbles begin to form.

Cementation. The process whereby the spaces or voids between particles of sediment become filled with chemical precipitates such as iron oxides and calcium carbonate, which effectively weld the particles together into rock.

Chaparral. The name given to dense, often thorny, thickets of low brushy vegetation, frequently including small evergreen oaks This vegetation is found in the Mediterranean climate of the United States south-west.

Chernozem. A very black soil rich in humus and carbonates forming under cool, temperate, semi-arid climates.

Clustering. The tendency for events of a similar magnitude to be located together in space or time.

Coherence. (1) Of soils: the binding together of soil particles due to chemical bonding, cementation or tension in thin films of water adhering to the surface of particles. (2) Of time series: the tendency for a phenomenon to be mirrored at another location over time. For instance, wet periods in eastern Australia often occur simultaneously with a heavy monsoon in India.

Cohesion. The capacity of clay soil particles to stick or adhere together because of physical or chemical forces independent of inter-particle friction.

Colloids. Fine clay-sized particles 1–10 microns in size, usually formed in fluid suspensions.

Conflagration. A particularly intense fire with a heat output of 60 000–250 000 kW m^{-1} along the fire front.

Continental Shelf. That part of a continental plate submerged under the ocean, tapering off seaward at an average slope of less than 0.1 degrees, and terminating at a depth of 100–150 m in a steep drop to the ocean bottom.

Convection. The transfer of heat upward in a fluid or air mass by the movement of heated particles of air or water.

Convective Instability. A parcel of air continuing to rise when lifted if its temperature is always greater than the surrounding air – a process that is guaranteed if condensation occurs releasing heat into the parcel.

Coriolis Force. The apparent force caused by the Earth's rotation, through which a moving body on the Earth's surface deflects to the right in the northern hemisphere and to the left in the southern hemisphere.

Correlation. The statistical measure of the strength of a relationship between two variables.

Cracking Clays. Soils consisting of clays that have a high ability to expand when wetted, and to shrink and, hence, crack when dried.

Creep. The imperceptibly slow movement of surface soil material downslope under the effect of gravity. *See* also Tectonic Creep.

Cretaceous. The geological time period 135–65 million years BP, which was characterized by sudden, large extinctions of many plant and animal species.

Critical Incidence Stress Syndrome. The condition affecting rescuers who perform superhuman, selfless feats of heroism or who are involved with death during a disaster. Such people often suffer severe guilt, become reclusive, and generally broken in spirit for months or years after the event.

Crust. The outermost shell of the Earth, 20–70 km thick under continents, but only 5 km thick under the oceans. It consists of lighter, silica-rich rocks, which flow on denser liquid rock making up the mantle.

Cryogenic. Any process involving the freezing of water within soil or rock.

Cumulonimbus. A dense cloud consisting of a lower dark layer that is potentially rain-bearing, and a multiple white fluffy crown reaching the top of the troposphere. Such cloud forms by thermal convection and is characteristic of thunderstorms.

Cycles, cyclic or cyclicity. Recurring at regular intervals over time: for instance, sunspot numbers, which recur at 10–11 year intervals.

Cyclogenesis. The process forming an extra-tropical, mid-latitude cell or depression of low air pressure.

D

Debris Avalanche. A flow of debris of different particle sizes that takes on catastrophic proportions. Volumes of 50–100 million cubic metres traveling at velocities of 300–400 km hr^{-1} have been measured.

Debris Flow. Any flow of sediment moving at any velocity downslope in which a range of particle sizes is involved.

Deflation. The removal by wind, over a long time period, of fine soil particles from the ground surface.

Degassing. The sudden or slow release into the atmosphere of gas dissolved in melted rock.

Degradation. The slow loss from a soil of humus, minerals, and fine particles essential to plant growth; or the contamination of a soil by salts or heavy minerals that are toxic to plant growth.

Desertification. The process whereby semi-arid land capable of supporting enough plant life for grazing or extensive agriculture becomes infertile because of increased aridity, degradation, or overuse.

Dilation. Rock in the Earth's crust will tend to crack and expand in volume when subjected to forces capable of fracturing it.

Dust Bowl. Part of the American southern Great Plains, centered on Oklahoma, which experienced extensive dust storms during the drought years of 1935–1936 at the height of the Great Depression.

Dust Veil Index. A subjectively defined index that attempts to measure the relative amounts of dust injected into the stratosphere by volcanic eruptions. The index is reference to a value of 1000 for the eruption of Krakatau in 1883.

E

East-Coast Lows. Very intense storms that develop in mid-latitudes on the seaward, eastern sides of continents. Such lows develop over warm ocean water in the lee of mountains and are dominated by intense convection, storm waves, and heavy rainfall. Events where wind velocities reach in excess of 100 km hr^{-1} within a few hours are called 'bombs'.

Easterly Wave. In the tropics, pressure tends to increase parallel to the equator with easterly winds blowing in straight lines over large distances. However, such airflow tends to wobble, naturally or because of some outside forcing. As a result, isobars become wavy and the easterly winds develop a slight degree of rotation that can eventually intensify over warm bodies of water into tropical cyclones.

El Niño. The name given to the summer warming of waters off the coastline of Peru at Christmas time (from Spanish, meaning 'The Child').

Elastic. A property of any solid that deforms under a load and then tends to return to its original shape after the load is released.

Electric Resistivity. The degree to which a solid body can conduct an electric current.

Electrosphere. The zone of positively charged air at 50 km elevation, the strength of which determines the frequency of thunderstorms.

Electrostatic. The generation of electrical charges on stationary particles.

Endemic. A term used to describe a disease that exists naturally within a specific region because of its adaptation to climate and the presence of vectors and reservoirs. If the vectors and reservoir species return or controls are relaxed, then the disease can easily become re-established. *See also* Vectors, Reservoirs.

Energy. The capacity of any system to do work. Energy can be described as 'kinetic', which is the energy any system possesses by virtue of its motion; as 'potential', which is the energy any system possesses because of its position; and 'thermal', which is the energy any system possesses because of its heat.

ENSO. Abbreviation for the words 'El Niño–Southern Oscillation', referring to times when waters off the coast of Peru become abnormally warm and the easterly trade winds blowing in the tropics across the Pacific Ocean weaken or reverse.

Epicenter. The point at the surface immediately above the location of a sudden movement in the Earth's crust generating an earthquake.

Episodic. The tendency for natural events to group into time periods characterized by common or loosely connected features. Weak cyclicity or regularity may be implied in this term.

Equinox. The time occurring twice yearly, around 21 March and 21 September, when the apparent motion of the sun crosses the Earth's equator. This results in all parts of the Earth's surface receiving equal amounts of daylight.

Eucalyptus. A genus of tree, which is mainly evergreen, is rich in leaf oil, flammable, drought-resistant, and capable of growing in nutrient-deficient soil.

Eustatic. Worldwide sea level change.

Evaporite. A sediment deposit that has formed from the residue left after the evaporation of salt water.

Evapotranspiration. The amount of moisture lost by plants by evaporation to the atmosphere, mainly through leaf openings or stomata. This amount is always greater than that lost from equivalent, unvegetated surfaces.

Expansive Soils. Soils consisting of clay particles – mainly montmorillonite – that absorb water when wetted and thus expand in volume.

Extra-Tropical Low or Depression. A mid-latitude, low-pressure cell with inwardly and upwardly spiraling winds. Unlike tropical cyclones, the cell can develop over land as well as water, usually in relation to the polar front and with a core of cold air.

Eye. The center of a tropical cyclone (or tornado), where air is descending, surrounded by a wall structure created by a sharp decline in pressure and characterized by intensely inwardly spiraling winds.

F

Fault. A planar zone where one part of the Earth's crust moves relative to another. A fault is categorized according to the direction of movement in the horizontal or vertical plane (see Figure 9.8). Transcurrent faults are fault lines approximately at right angles to major lines of spreading crust in the oceans.

Feedback. The situation where one process reinforces (positive) or cancels (negative) the effect of another. For instance, a microphone placed close to an amplifier catches its own sound waves and rebroadcasts them louder. This is positive feedback.

Fetch. The length of water over which wind blows to generate waves. The longer the fetch, the bigger the wave.

Firestorm. Under intense burning, air can undergo rapid uplift and draw in surrounding air along the ground to replace it. This movement can generate destructive winds at ground level in excess of 100 km hr^{-1}.

Flood. A large discharge of water flowing down – if not outside – any watercourse in a relatively short period compared to normal flows.

Flood Lava. Lava flows occurring over a very large area.

Floodplain. Flattish land adjacent to any watercourse built up by sediment deposition from water flowing outside the channel during a flood.

Föhn Wind. Air rising over a mountain and undergoing condensation releases latent heat of evaporation and, thus, cools adiabatically at a slow rate of about 0.5°C per 100 m. When this air then descends downslope on the leeward side, it tends to warm at the faster, dry adiabatic lapse rate of 1°C per 100 m. The resulting wind is much warmer and drier than on the windward side.

Foreshocks. The small seismic shocks preceding a major earthquake or volcanic event.

Freezing Rain. Light rain falling through a layer of cold air or upon objects with a temperature below freezing will tend to freeze or adhere to those objects. Over time the accumulated weight breaks power lines and tree branches.

Friction. The force due to the resistance of one particle sliding over another because of a weight pressing irregularities on the surface of both particles together.

Fumarole. A hole or vent from which volcanic fumes or vapors issue.

G

General Air Circulation. Equatorial regions have an excess of heat energy relative to the poles because incoming solar radiation exceeds outgoing long wave radiation. To maintain a heat balance over the Earth's surface, this excess heat flows mainly in air currents from the equator to the pole.

Geochemical. Involving chemical reactions on, or inside, the Earth.

Geo-Electric Activity. Processes involving the transmission or induction of electrical currents along the surface of, or through, the Earth.

Geomagnetic Activity. Fluctuations in the Earth's magnetic field over space and time. The strength of this field and its variation can be changed by solar activity.

Geophysical. Dealing with the structure, composition, and development of the Earth, including the atmosphere and oceans.

Geosyncline. An area of sediment accumulation so great that the Earth's crust is deformed downward.

Gilgai. The tendency for soil, affected by alternating wetting and drying, to overturn near the surface. The process can result in local relief up to 2 m.

Glacier Burst. The sudden release of huge volumes of water melted by volcanism under a glacier and held in place by the weight of ice until the glacier eventually floats.

Graupel. Aggregated ice or hail coated in supercooled water.

Greenhouse Effect. The process whereby certain gases in the atmosphere, such as water vapor, carbon dioxide and methane, transmit incoming short wave solar radiation, but absorb outgoing long wave radiation, thus raising surface air temperature.

Gumbel Distribution. The plot of the magnitude of an event on a logarithmic scale against the average length of time it takes for that event to occur again.

Gyres. Large rotating cells of ocean water that often completely occupy an ocean basin. The Gulf Stream forms the western arm of one such gyre in the north Atlantic.

H

Haboob. The sand or dust storm resulting from a Harmattan wind. The term is used in Arabia, North Africa and India. *See* Harmattan.

Hadley Cell. Heated air at the equator rises and moves poleward in the upper atmosphere. It cools through the loss of long wave radiation and begins to sink at about 20–30° north and south of the equator, whereupon part of the air returns to the equator. This forms a large, semi-permanent vertical cell on each side of the equator. Postulated by George Hadley in 1735.

Hale Sunspot Cycle. Sunspot numbers peak every 10–11 years. Every alternate peak is larger, forming a 22-year cycle, a phenomenon first noted by the American astronomer George Hale. The Hale cycle is, in fact, the full magnetic sunspot cycle.

Harmattan. A dry, dusty trade wind caused by mobile polar highs moving from the east over North Africa. These are the source of dust storms.

Hectare. Standard international unit measuring area. 1 hectare = 1000 m^2 = 2.471 acres.

Hectopascal. Standard international unit, abbreviated as hPa, measuring air pressure: 1 hPa = 1 millibar.

Hindcasting. The procedure whereby wave characteristics can be calculated from the wind direction, strength and duration derived from weather charts.

Holocene. The most recent part of the Quaternary, beginning 15 000 years BP, when the last major glaciation terminated.

Howling Terrors. An Australian term for small, destructive, tropical cyclones with an eye diameter of less than 20 km – also called 'kooinar' by Aborigines.

Hydrostatic. Describing pressure in groundwater exerted by the weight of water at levels higher than a particular location on a slope.

Hydrothermal. Describes the processes associated with heated, or hot melted, rock rich in water.

I

Ice-Dammed Lakes. Lakes formed by glaciers or icesheets completely infilling a drainage course or flowing against higher topography, thus blocking the natural escape of meltwater.

Ignimbrites. Very hot ash can be blasted laterally across the landscape. When deposited, the ash particles fuse together to produce a hard, welded rock.

Illite. A clay mineral slightly more weathered than montmorillonite, containing hydroxyl radicals and having less iron and silica. Chemical formula $[KAl_2(OH)_2(AlSi_3(O,OH)_{10})]$.

Interfluves. The hill slopes between streams that flow in the same direction.

Intertropical Convergence. Air in the tropics rises because of heating beneath the seasonal position of the sun. Winds flow from the north and south, and converge on this zone of heating, effectively producing a barrier to the exchange of air between hemispheres.

Intra-Plate. Locations lying near the center of a plate or, at least, away from tectonic activity associated with the margins.

Inverted Barometer Effect. The height of sea level inversely relates to the pressure of the atmosphere above, such that a decrease of 1 hectopascal (hPa) in air pressure results in a rise in sea level locally of 1 cm.

Island Arc. A series of islands located on the continental plate-side of a deep ocean trench, under which gas-rich crust is being subducted to produce andesitic magma.

Isobars. Lines plotted on a map joining points of equal pressure, usually at a fixed spacing of 4 hectopascals (hPa).

Isostatic. Describing local changes in sea level induced by regional tectonic variation.

Isothermal. Having equal degrees of heat.

J

Jet Stream. A fast-moving flow of air at the top of the troposphere, usually linked to large-scale uplift or subsidence of air. The two most common jet streams are the polar jet stream, attached to uplift along the polar front, and the subtropical jet centered over the poleward and trailing side of Hadley cells.

Jökulhlaups. *See* Glacier Burst.

Jurassic. The geological period 180–135 million years BP, dominated by the dinosaurs.

K

Kaolinite. A common, clay mineral, which is resistant to further weathering, containing only metals of aluminum and silica. Chemical formula [$Al_2O_3.2SiO_2.2H_2O$].

Karst. The process in which solution is a major mechanism of erosion, or landscape developed by that process. Karst most commonly occurs in limestone, but in some instances can involve sandstone and volcanic rocks.

Katabatic Winds. Winds formed by the flow under gravity down topographic lines of denser air cooled by long-wave emission under clear skies and calm conditions. The resultant wind can reach speeds in excess of 100 km hr^{-1} and is prevalent in winter, mountainous terrain or adjacent to icecaps.

Kelvin Wave. A water wave trapped along the boundary of a coast, channel, embayment, or two bodies of water differing in some physical characteristic. The wave moves slowly parallel to the boundary and has a height that decreases rapidly away from the boundary. The movement of water across the Pacific Ocean during an ENSO event occurs as a 500–1000 km long, 10–50 cm high Kelvin wave, trapped within 5° of the equator and moving eastward at a speed of 1–3 m s^{-1}.

L

La Niña. The opposite of an ENSO event, during which waters in the west Pacific are warmer than normal; trade winds or Walker circulation is stronger; and consequently rainfalls heavier in South-East Asia.

Lahar. A mudslide induced by volcanic eruption either at the time of the eruption – by the mixing of hot gases, melted ice or water, and ash – or years later by the failure of volcanic ash deposits in the presence of heavy rain.

Landslide. The general term given to movement of material downslope in a mass.

Lapilli. Volcanic fragments about 2–60 mm in diameter.

Latent Heat. (1) of evaporation: each kilogram of water converted to water vapor at 26°C requires 2425 kilojoules of heat energy. Because this heat energy is not felt, it is termed 'latent'. (2) Of condensation: if this air is cooled and moisture condenses, the latent heat energy is released back to the atmosphere, warming the air.

Lava. Molten rock. Different terms are used to describe the nature of the lava, mainly as determined by viscosity.

Levee. When a river overflows its banks, there is an immediate decrease in velocity. This results in deposition of suspended mud, and eventually the build-up of an embankment that can contain the river above the elevation of its adjacent floodplain.

Liquefaction. If the weight of a sediment deposit or flow is not supported by the touching of individual grains against each other, but by the pressure of water or air entrapped in voids between particles, then such sediment will liquefy – that is, behave as a fluid. When liquefaction occurs, the sediment loses its ability to support objects and is able to travel at high velocities.

Lithology. Refers to the physical composition of any rock.

Little Ice Age. The period between 1650 and 1850 AD, when sunspot activity was minimal, global temperatures were 1°C cooler, and temperate mountain glaciers advanced.

Loess. Dust picked up from barren ground in front of icecaps by dense, cold winds blowing off the icecap and deposited hundreds of kilometers downwind.

Longitudinal *Or* Meridional. Any movement that occurs in a north–south direction.

Longshore Drift. The movement of water and sediment within surf zones due to waves approaching shore at an angle.

Lunar Cycle. *See* M_n Lunar Tide.

M

Magma. Liquid rock derived from the mantle.

Magnetic Reversal. The Earth has a magnetic north–south polarity centered within 1000 km of its present axis of rotation. The intensity of this magnetic field can vary and, at intervals of hundreds of thousands of years, weaken and reverse polarity very rapidly.

Magnitude–Frequency. The intensity or magnitude of all geophysical events such as floods occur at discrete time intervals in such a way that the lower the magnitude of the event, the more frequent its occurrence. The magnitude of an event can be plotted against how often that event occurs, to produce a magnitude–frequency diagram.

Mantle. That part of the Earth from the crust to depths of 3000 km, which consists of molten, dense rock made up of silicates of magnesium, iron, calcium, and aluminum.

Marigrams. Records of tsunami wave height on tide gauges.

Mass Extinctions. The dying-out or disappearance of large numbers of plant and animal species in short periods of time, supposedly because of a catastrophic event that affected the climate of the Earth in the prehistoric past.

Maunder Minimum. Refers to the lack of sunspots between 1650 and 1700 AD at the height of the Little Ice Age.

Maximum Probable Rainfalls. The amount of rain that can theoretically fall within a given time interval over a catchment under the most favorable conditions of humidity, convection, and condensation.

Meandering. The process whereby a flow of air or water tends to become unstable and travel in a winding path.

Mediterranean Climate. A seasonally dry climate having rainfall in winter months. Its average temperature for at least four months exceeds 10°C, with no monthly average falling below –3°C. Rainfall of the driest summer month is less than 40 mm and less than one-third that of the wettest winter month.

Meridional Air Flow. Movement of air in a north–south direction.

Microseisms. Very small earthquakes that can be detected only by sensitive instruments.

M_n Lunar Tide. A periodicity of 18.6 years that exists in the orbit of the moon about the Earth because of the orientation of the moon's orbit to the equatorial plane of the sun.

Mobile Polar High. Pools of cold, shallow air that move out from particular locations around the poles into the mid-latitudes. They tend to lose momentum such that successive highs stack up or agglutinate in a single location over the oceans. When their pressure is averaged over time, these locations appear as areas of constant high pressure, which have been labelled Hadley cells.

Mogi Doughnut. In some seismic areas, the epicenter of a large earthquake is ringed in the prior decade by moderate earthquakes forming a doughnut-like pattern. Named after its discoverer.

Mohr–Coulomb Equation. An equation in which the shear strength of a soil is related to soil cohesion, a force applied to the soil at right angles to the slope, and the angle at which particles move relative to each other within the soil.

Monsoon. Defines any region characterized by a distinct 180° change in wind direction between summer and winter, resulting in seasonal alternation of copious rain and aridity.

Montmorillonite. A group of clay minerals with an aluminum layer sandwiched between two silica layers bonded with magnesium or calcium cations. Such soils expand easily because they can absorb water into the layered structure when wetted. Chemical formula $[(Mg,Ca)O.Al_2O_3 5SiO_2.nH_2O]$.

Mulching. Any process that causes soils not to compact, but to become more friable towards the surface.

N

Normal Stress. Any force applied at right angles to a soil surface (*see* Figure 12.3).

Normalize. A statistical process whereby the magnitude of values in a time series are made independent of their unit of measurement. This is usually done by subtracting the mean of the data set from each value, and dividing the result by the standard deviation. In most climate time series, such as the Southern Oscillation index, the resulting value is multiplied by 10.

North Atlantic Oscillation. The tendency for air pressure to oscillate in intensity between the Icelandic Low and the Azores High. The oscillation reflects inter-annual variability in the strength of Rossby waves in the northern hemisphere. *See also* Southern Oscillation.

North Pacific Oscillation. A statistical measure of the strength and position of the Aleutian low-pressure system, mainly in winter. The oscillation is linked to short-term climate change in North America. *See also* Southern Oscillation.

Nuée Ardente. A cloud of hot ash that, blown upwards during an eruption, collapses under the weight of gravity and begins to flow downslope at great velocity suspended by the gases expelled in the eruption (see Figure 11.3). Also called pyroclastic flow.

O

Orographic. Any aspect of physical geography dealing with mountains.

P

Parent Cloud. A large cloud characterized by turbulent airflow, and subdividing into separate areas of convection that can give rise to tornadoes, intense rainfall, and hail.

Partial Pressure. The contribution made by a gas to the total pressure of the atmosphere.

Permafrost. Permanently frozen ground.

Permeability. The degree to which water can flow through the voids in sediment deposits or micro-fractures in rock.

Persistence. The tendency for an abnormal climatic situation to continue over time.

Photosynthesis. The process whereby plants convert carbon dioxide and water in the presence of sunlight, and aided by chlorophyll, into carbohydrates or biomass.

Piezoelectric. The ability of a mineral to generate an electric current when pressure or stress is applied to it.

Piezomagnetism. The generation of magnetism in iron-rich magma by increases in pressure or stress.

Plastic. A property of any solid that permanently deforms without breaking up or rupturing when subject to a load.

Plate. The Earth's crust is subdivided into six or seven large segments or plates that move independently of each other at rates of 4–10 cm per year over the underlying mantle. These plates can collide with (converge), or separate from (diverge) each other.

Pleistocene. The geological age in the Quaternary beginning approximately 2–3 million years BP. It was characterized by periods of worldwide glaciation.

Pliocene. The last period in the Tertiary, 12–2 million years BP. The Pliocene terminated when large-scale glaciation began to dominate the Earth.

Pluviometers. Instruments used to measure the rate at which rain falls over very short time spans.

Polar Front. Dense air, cooled by long wave emission of heat in the polar regions and flowing towards the equator, differs substantially in temperature and pressure from the air it displaces. This difference occurs over a short distance termed the 'polar front'.

Pore Water Pressure. The pressure that builds up because of gravity in water filling the spaces or voids between sediment particles.

Power. The rate at which energy is expended.

Precursors. Variables that give an indication of the occurrence of some future event.

Prescribed Burning. The deliberate and controlled burning of vegetation growing close to or on the ground to minimize the fuel supply for future bush or forest fires. Also called (in Australia especially) 'burning-off'.

Probability of Exceedence. The probability or likelihood of an event – with a certain magnitude or size – occurring, expressed in the range of 0–100 per cent.

Pumice. Volcanic ash filled with gas particles and hence able to float on water.

Pyroclastic. Referring to the fine, hot ash or sediment produced from the fragmentation of magma in the atmosphere by the force of a volcanic eruption, and the gases contained within.

Pyroclastic Ejecta. Pyroclastic material blown out into the atmosphere. *See also* Tephra.

Pyroclastic Flow. *See* Nuée Ardente.

Q

Quaternary. The most recent geological period, beginning approximately 2–3 million years BP and consisting of the Pleistocene and the Holocene.

Quick Clays. Clays that have been deposited with a high degree of sodium cations that tend to glue particles together, giving the deposit a high degree of stability. However, sodium cations are easily leached by groundwater, with the result that the clay deposit becomes progressively unstable over time.

R

Recurrence Interval. The average time between repeat occurrences of an event of given magnitude.

Reg Desert. The residual, stony desert formed after clay and sand-sized particles are blown away.

Regolith. The layer of weathered or unweathered, loose, incoherent rock material of whatever origin forming the surface of the Earth.

Regression Coefficient. A term used in statistics to define the relationship between two variables. The coefficient can vary from –1.0, a perfect inverse relationship, to 1.0, a perfect direct relationship.

Reservoir. (1) A structure deliberately constructed to impound water. (2) A pocket of resistance that exists in the natural environment in which an organism or pathogen can exist with impunity and can spread outwards under favorable conditions.

Resonance. A process involving the development of air or water waves in which the wavelength is equal to, or some harmonic of, the physical dimensions of a basin or embayment.

Rift Valley. A zone of near-vertical fracturing of the Earth's surface that causes one segment to drop substantially below another.

Rossby Waves. Any wave motion that develops with a long wavelength of hundreds or thousands of kilometers in a fluid moving parallel to the Earth's surface and controlled by Coriolis force. Generally restricted to the upper westerlies attached to the polar jet stream at mid-latitudes.

Run-Up. When any ocean wave reaches the shoreline, its momentum tends to carry a mass of water landward. The distance landward of the shoreline is the run-up distance, while the elevation above sea level is the run-up elevation.

Rhyolite. A fine-grained, extruded, volcanic rock with a wavy, banded texture. Mineralogically it is similar to granite, consisting of quartz, feldspar, and mica.

S

Sahel. The seven African countries – Gambia, Senegal, Mauritania, Mali, Burkina Faso, Niger, and Chad – at the southern edge of the Sahara Desert.

Saltation. The tendency for sediment particles to move from the bed into air- or water-flow and back to the bed in zigzag paths, where the distance beneath upward movement is much shorter than the distance beneath downward movement.

Scoriae. Fused volcanic slag or ash that results from a pyroclastic flow and contains a high degree of gas.

Seamounts. Volcanoes that have formed in the ocean, planed off at the top because of close proximity to the ocean's surface, and have then been lowered below the sea as the Earth's crust moves away from the center of volcanic activity.

Seiching. Excitation of a regular oscillation of water waves within a basin or embayment – caused by an earthquake, the passage of an atmospheric pressure wave, or the occurrence of strong winds.

Seismic. Related to sudden and usually large movement of the Earth's crust.

Seismic Gap. That part of an active fault zone that has not experienced moderate or major earthquake activity for at least three years.

Seismic Risk. The probability of earthquakes of given magnitude occurring in a region.

Set-Back Line. A line established shoreward of a coastline, incorporating 50–100 years of postulated shoreline retreat, and in front of which no development should proceed.

Set-Up. The enhanced elevation of sea level at a coastline due mainly to the process of wave-breaking across a surf zone, but also aided by wind, shelf waves, Coriolis force, upwelling, and current impinging along a coast.

Shallow Focus. Earthquakes occurring within 60 km of the surface.

Shear Sorting. Any process that causes coarse particles to be transported more efficiently in a liquid or air flow that is thin, full of sediment, and traveling at high velocity.

Shear Strength. The internal resistance of a body of material to any stress applied with a horizontal component.

Shear Stress. The force applied to a body of material, which tends to move it parallel to the contact with another solid or fluid.

Shearing. A stress caused by two adjacent moving objects tending to slide past each other parallel to the plane of contact.

Slump. The tendency for some landslides to undergo rotation, with the greatest failure of material at the downslope end.

Solar Flares. Large ejections of ionized hydrogen from the sun's surface accompanied by a pulse of electromagnetic radiation.

Solifluction. The slow movement of water-saturated material downslope, usually in environments where freezing and thawing are the dominant processes affecting the surface.

Southern Oscillation. The usual movement of air in the tropical Pacific is strong easterly trades reinforced by low pressure over Indonesia–Australia and high pressure over the south-east Pacific. Every two to seven years, this low and high pattern weakens or even reverses, causing the trades to fail or become westerlies. It is this switching from one pattern to another that is the Southern Oscillation. *See also* North Atlantic Oscillation *and* North Pacific Oscillation.

Spot Fires. Small fires ignited by flying embers carried on updrafts generated by heated air from the main core of a forest or bushfire.

Steric. Refers to the change in volume of seawater due to heating or cooling.

Storm Surge. The elevation of sea or lake levels above predicted or normal levels resulting mainly from a drop in air pressure, or from the physical piling up of water by wind associated with a low-pressure disturbance or storm.

Strain. The deformation or movement caused by applying a force to material on a slope.

Stratosphere. That part of the atmosphere lying 12–60 km above the Earth, and characterized by stability and an increase in temperature with altitude.

Stratospheric Fountains. Little air is exchanged between the troposphere and stratosphere. However, intense convection – generated by thunderstorms in the tropics at the west side of the Pacific – breaches this barrier and, like a fountain, injects air originating from the ground into the stratosphere.

Strip Farming. An agricultural practice used to conserve soil moisture whereby crops are alternated in strips with fallow land.

Subcritical Flow. Tranquil river flow in which velocities relative to the channel depth are low enough that sediment transport is not excessive.

Subduction Zones. The zone down which an oceanic plate passes beneath a continental plate to be consumed into the mantle.

Subsidence. The downward failure of the Earth's surface brought about mainly by removal of material from below.

Sunspot. An area 20 000 km in diameter on the sun's surface featuring strong magnetic disturbance so strong that convective movement of heat to the surface is curtailed. Hence, the spot is darker because it is cooler.

Supercell Thunderstorm. A particularly large, convective thunderstorm cell with a diameter of 50–100 km. Such a storm has a greater chance of producing hail, tornadoes, strong wind, or copious rainfall.

Supercritical Flow. River flow in which velocities relative to the channel depth are so high that the flow begins to shoot or jet. Such flow is very erosive.

Surcharging. The escape of stormwater through access-hole covers that are designed to lift off under excess pressures caused by flooding. This prevents stormwater pipes from bursting underground, which would necessitate costly repairs.

Suspension Load. That portion of sediment being carried in air or water flow by turbulence.

T

Taiga. Swampy, coniferous, evergreen forest of sub-Arctic Siberia.

Tectonic. The process characterizing the deformation of the Earth's crust because of earthquakes.

Tectonic Creep. Deformation of the Earth's crust that occurs in such small increments that it is virtually undetectable, except over time.

Teleconnection. The linkage of a measured climatic time series to another one at some distance across the globe. For instance, rainfall in Florida can be related, through the Southern Oscillation, to sea-surface temperature in northern Australia.

Teleseismic Tsunami. A tsunami generated by an earthquake from a distant source. The Chilean earthquake of 1960 generated a teleseismic tsunami that impacted on most coasts surrounding the Pacific Ocean.

Telluric Currents. Electric currents moving through the Earth.

Tephra. Volcanic ash that is disaggregated and blown by the force of an eruption, usually vertically, into the atmosphere.

Tertiary. The geological period from 65 to 2–3 million years BP.

Thermal Expansion. Water is unusual in that it has its greatest density at 4°C. Thus, it expands readily when heated. Ice, however, has its greatest volume around the freezing point. It shrinks and cracks around –10°C. If water can enter the cracks and refreeze, such ice will expand and buckle as it warms.

Thermocline. The abrupt change in water temperature with depth within a large body of water.

Thixotropy. The property of becoming fluid when stirred or agitated, especially by compressive shock waves. *See also* Liquefaction.

Threshold. The point at which the application of a force triggers a dramatic response that did not exist beforehand. For example, increasing stress applied to rock may produce no observable effect until a threshold is reached and the rock fractures.

Tidal Inlet. The opening that forms across a barrier island in which there is a free and substantial exchange of saline water between the ocean and a landward lagoon during each tide cycle.

Tidal Waves. A misnomer applied to tsunami and storm surges because these phenomena sometimes produce a slow drop in sea level followed by a rapid rise, analogous to tides.

Tilt Steps. Tilting of the Earth's crust that occurs in a series of discrete steps rather than smoothly.

Topographic Control. Any airflow whose path is deflected or altered by terrain that is higher than the surrounding landscape. For example, a moving mass of air may be blocked by high mountains and go around the obstruction.

Tornado. An extremely violent vortex or whirlwind of air measuring only hundreds of meters in diameter and with wind speeds up to 1300 km hr^{-1}.

Trade Winds. Winds on the equatorial side of mobile polar cells, blowing from east to west within 20–30° of the equator. Their steadiness for three to four months of the year greatly benefited trading ships in the days of sail.

Trench. An area of oceanic crust that has been dragged downward by the movement of oceanic crust beneath a continental plate.

Tropical Cyclone. A large-scale vortex of rising air hundreds of kilometers in diameter that forms over the tropical oceans. It is characterized by copious rain and a central area of calm surrounded by rotating winds blowing at speeds in excess of 200–250 km hr^{-1}.

Tropopause. The band separating the troposphere – where temperature decreases with altitude – from the stratosphere – where temperature increases with altitude. It forms a barrier to the ready exchange of air between these two zones.

Troposphere. The lower part of the atmosphere, 10–12 km in altitude, characterized by cloud formation and a drop in temperature with elevation.

Tsunami. A Japanese word for 'harbor' (tsu) 'wave' (nami). It is used to define a water wave generated by a sudden change in the seabed resulting from an earthquake, volcanic eruption, or landslide.

Tsunami Earthquake. Some earthquakes that are normally too small to produce any sea waves locally generate devastating and deadly tsunami.

Turbulence. Flow – also called turbulent flow – in which individual particles or molecules travel unpredictable paths apparently unaffected by gravity.

U

Ultrasound. Sound with a frequency above 20 000 vibrations per second and generally not perceivable by the human ear.

Uniformitarianism. A geological explanation developed in the eighteenth century to counteract the Catastrophists. It involves two concepts: the first implies that geological processes follow natural laws applicable to science (methodological uniformitarianism); the second implies that the type and rate of processes operative today have remained constant throughout geological time (inductive reasoning).

V

Van Der Waals' Bonds. The weak attraction exerted by all molecules to each other, caused by the electrostatic attraction of the nuclei of one molecule for the electrons of another.

Vertical Aggradation. The build-up of a floodplain in a vertical direction because of the deposition of sediment.

VHF. Abbreviation for 'very high frequency'. That part of the electromagnetic spectrum with a wavelength between 30 and 300 megahertz, characteristic of radio and television transmission.

Victim–Helper Relationship. In disasters where survivors are at the mercy of outside support for survival for their daily requirements, a strong dependence can develop between a victim and those who are providing the aid. This dependency may not be easily terminated when the necessity for relief has ended.

Viscosity. The resistance of a fluid to the motion of its own molecules because they tend to stick together.

Voids. The spaces or gaps that exist between particles in a sedimentary deposit, filled with air or water.

Vortex. A spinning or swirling mass of air or water that can reach considerable velocity.

W

Walker Circulation. The normal movement of air across the tropical Pacific is from the east, between a stationary high-pressure cell off the coast of South America, and a low-pressure zone over Indonesia–India. In the 1920s, Gilbert Walker discovered that the strength of this air movement determined the intensity of the Indian monsoon.

Wave Diffraction. The tendency for wave energy to spread out along the crest of a wave. Hence, when a wave passes through a narrow harbor entrance, the wave will spread out to affect all the coastline inside a harbor.

Wave Refraction. The tendency for the velocity at which a wave travels to reduce over shallow bathymetry. Hence, a wave crest tends to wrap around headlands and to spread out into bays.

Wave-rider Buoy. An instrument, consisting of sensitive accelerometers encased inside a buoy and floating on the ocean surface, used for measuring the height and energy of waves.

Wind-Chill Factor. The factor by which the temperature of a surface is reduced because of the ability of wind to remove heat from that surface more efficiently by convection (see Figure 3.23).

Wind Shear. The tendency for winds to move at different velocities with elevation because of strong temperature differences. The wind, as a result, may accelerate in speed over short altitudes or even change direction – a particularly hazardous situation for airplanes landing or taking off.

Y

Yield Limit. The limit beyond which a soil can no longer resist a force without permanently deforming.

Yield Strength. The maximum resistance of a solid to a force before it is permanently deformed.

Z

Zonal. Restricted to an area bounded by parallels of latitude on the Earth's surface.

Index